Marine Ecology and Fisheries

Marine Ecology and Fisheries

Edited by **Roger Creed**

Callisto Reference

New York

Published by Callisto Reference,
106 Park Avenue, Suite 200,
New York, NY 10016, USA
www.callistoreference.com

Marine Ecology and Fisheries
Edited by Roger Creed

International Standard Book Number: 978-1-63239-700-3 (Hardback)

Printed in China.

Contents

Preface

Over the recent decade, advancements and applications have progressed exponentially. This has led to the increased interest in this field and projects are being conducted to enhance knowledge. The main objective of this book is to present some of the critical challenges and provide insights into possible solutions. This book will answer the varied questions that arise in the field and also provide an increased scope for furthering studies.

This book unravels the recent studies in the field of marine ecology. It aims to give a thorough view of the different areas of this field. Marine Ecology is a part of the vast aquatic ecosystem present on Earth. It consists of salt marshes, estuaries, lagoons, water animals, etc. Fisheries science is a complex study of the life and habitat of fishes. Thus, is co-related with marine ecology. This book attempts to understand the multiple branches that fall under this discipline and how such concepts can be used to flourish the aquatic ecosystem. It is designed to provide readers with the latest advances made in this subject. This book aims to shed light on some of the unexplored aspects of marine ecology and fisheries, and present the recent researches in this discipline. Students, researchers, experts and all associated with this field will benefit alike from this book.

I hope that this book, with its visionary approach, will be a valuable addition and will promote interest among readers. Each of the authors has provided their extraordinary competence in their specific fields by providing different perspectives as they come from diverse nations and regions. I thank them for their contributions.

Editor

1

Variations of structural protein sequences among geographical isolates of white spot syndrome virus

Toms C. Joseph · L. Anbu Rajan · Roswin James · K. V. Lalitha · P. K. Surendran

Abstract White spot syndrome virus (WSSV) is the causative agent of a disease that causes severe mortalities in cultured shrimp worldwide. The sequence of five structural genes of an Indian isolate of WSSV was compared with sequences from other WSSV isolates deposited in GenBank. Among the structural genes analyzed, the sequences of vp28 and vp19 had maximum divergence with nucleotide changes at ten different positions from twenty eight submissions for vp28 and six changes from seventeen submissions for vp19. The vp68, vp26, and vp281 genes were found to be highly conserved between isolates.

Keywords WSSV · Structural protein · vp28 · vp26 · vp19 · vp68 · vp281 · vp466

Introduction

White spot syndrome virus, a major pathogen of cultured shrimp first appeared in 1990s in Taiwan and China (Zhang et al. 1994; Wongteerasupaya et al. 1995; Nadala et al. 1998). Since then WSSV has spread rapidly to shrimp farming areas around the world causing heavy economic losses. White spot syndrome virus not only infects shrimp but also has a much wider host range which includes other invertebrate aquatic organisms such as crab, crayfish, and fresh water prawn, *Macrobrachium* sp. (Ramasamy 1995; Lo et al. 1996; Flegel 1997; Wang et al. 1998; Mohan et al. 1998).

WSSV is an enveloped ovoid virus with rod-shaped nucleocapsid with flat ends (Wang et al. 1995). WSSV is a major shrimp pathogen that is highly virulent in penaeid shrimp (Lo et al. 1996; Chen et al. 1997; Flegel 1997) and can result in of up to 90–100 % mortality within 3 to 7 days (Zhan et al. 1998). The virus belongs to the genus *Whispovirus* (Van Hulten et al. 2000; Tsai et al. 2000) under the family Nimiviridae. The complete genome sequence of WSSV has been determined for three different isolates with GenBank Accession numbers AF369029, AF440570, and AF332093 (Van Hulten et al. 2001, Yang et al. 2001, Lo and Kou 2001) for

T. C. Joseph · L. Anbu Rajan · R. James · K. V. Lalitha
Microbiology, Fermentation and Biotechnology Division, Central Institute of Fisheries Technology, Cochin 682 029, Kerala, India

L. Anbu Rajan (✉)
Andaman and Nicobar Centre for Ocean Science and Technology, National Institute of Ocean Technology, Dollygunj P.O., Port Blair 744 103, Andaman and Nicobar Islands, India
e-mail: anburajanl@yahoo.co.in

P. K. Surendran
Poothuvallil, Palluruthy, Cochin, India

viruses isolated from Thailand, Taiwan, and China, respectively. The genome of WSSV is approximately 300 kb in length with 180 putative open reading frames. Most of the open reading frames of WSSV encode certain structural genes. Structural proteins of viruses are classified as envelope and non-envelope proteins. Envelope proteins play a vital role in virus entry assembly and release (Hsiao et al. 1999; Lin et al. 2000; Chiu and Chang 2002). Neutralization experiments of antibodies against six WSSV envelope proteins (vp28, vp26/P22, vp466, vp281, vp68, and vp292) revealed that four envelope proteins (vp28, vp466, vp68, and vp281) play major roles in the initial steps of WSSV infection in shrimp (Wu et al. 2005). There is little genetic variation in the structural genes of WSSV from different geographical locations. (Lo et al. 1999; Moon et al. 2003; Satish et al. 2004; You et al. 2004). In this study, sequences of the genes encoding for six major structural proteins of WSSV from different geographical locations were compared with a WSSV isolate from south India.

Materials and methods

Total DNA extraction

Black tiger shrimp (*Penaeus monodon*) naturally infected with WSSV were obtained from a shrimp farm in Cochin, India. The samples were immediately transferred to sterile polythene bags and transported to the laboratory in ice. On arrival in the laboratory, the total DNA was extracted from pleopods as described previously (Moon et al. 2003).

PCR amplification of structural genes

The viral structural protein genes: vp28, vp26, vp19, vp68, vp281, and vp466 were amplified using six primer pairs (Table 1) designed based on the sequences of a Taiwanese isolate of WSSV (Yang et al. 2001). The primers were designed so as to include both the upstream and down stream sequences of the genes to be amplified. The PCR reaction was carried out in a final reaction volume of 50 μl containing 0.5 μM each of forward and reverse primer, 100 ng of DNA, 200 μM dNTP (MBI Fermentas, USA), 2.5 mM $MgSO_4$, 1× *Pfu* buffer, 1.25 U of *Pfu* DNA polymerase (MBI Fermentas), and remaining autoclaved Millipore water. The amplification was performed in a PTC-150 Mini cycler (MJ Research, USA). The gene amplification conditions were as follows: primary denaturation at 94 °C for 5 min, followed by 30 cycles of 94 °C for 30 s, 55 °C for 30 s, and 72 °C for 2 min, and final extension at 72 °C for 5 min. The PCR amplified products were analyzed on 1.5 % agarose gel along with DNA molecular weight marker and documented using a gel documentation system (Alpha Imager 1220, Alpha Innotech, USA).

Cloning of PCR products in pTZ57R/T

An A-overhang was added to the PCR product for cloning into TA vector. The reaction was carried at 72 °C for 45 min in a reaction volume of 30 μl containing 26 μM dATP, 1× *Taq* buffer, 1.25 units of *Taq* DNA polymerase (MBI Fermentas), and the PCR product. The A-overhang added PCR product was purified using MinElute Gel purification kit (Qiagen, Germany). InsTA clone PCR cloning kit (MBI Fermentas) was used to clone the PCR product to pTZ57R/T vector. Ligation was set at 4 °C overnight in a 30 μl reaction volume using 5 U of T4 DNA ligase (MBI Fermentas). Competent *E. coli* JM109 cells were used as host system for transformation. The transformed cells were incubated at 37 °C overnight on LB agar plates containing 40 μl of 0.1 M IPTG and 4 μl of 20 mg/ml X-gal in order to enhance blue white selection by α-complementation. Plasmids were isolated from the overnight culture by alkaline lysis method (Sambrook et al. 1989).

The recombinant clones having the specific insert in the forward orientation was then sequenced from both directions to get a contiguous sequence in an automated sequencer (ABI prism, model 377, Version 3.0). The nucleotide sequences obtained and the deduced amino acid sequences were analyzed using bioinformatics tools. Sequences were compared with database using BLAST (http://www.ncbi.nlm.nih.gov/BLAST) blastn and blastp (Altschul et al. 1990).

Table 1 Primers used for amplification of WSSV structural genes

Sl. No.	Gene	Sequence (5′–3′)	Product size (bp)
1	vp28	**F**-5′TTCACGAGGTTGTCATCACC3′	799
		R-5′TGGTATAAATTTCCTCAATTGTTTT3′	
2	vp26	**F**-5′TGGATCCAACCAACACGTAA3′	713
		R-5′CTTGTATTTTTATTCAAACAAAACCTT3′	
3	vp19	**F**-5′GGTGTCCTGACAAAAAACCGTA3′	477
		R-5′TTGTCCCTGATGTTGTGTTTTC3′	
4	vp68	**F**-5′AACACTTCTGGGTGAAACCTA3′	323
		R-5′TCGGACAAATAAAAGAATTGGAA3′	
5	vp281	**F**-5′AGAAACCCAAGGAAGGGTTG3′	927
		R-5′TTTGTTTGCAACACCCTTTT3′	
6	vp466	**F**-5′TCAAGACCAGTACACGTAATTTGAT3′	1471
		R-5′ TGATGTCTGAGCCATTTTTATTATG3′	

Results and discussion

Six envelope protein genes of WSSV: vp28, vp26, vp19 vp68, vp281, and vp466 on PCR amplification and sequence analysis revealed genes of expected size 615, 615, 366, 207, 846, and 1401 bp, respectively. These sequences were submitted to the GenBank with the assigned accession numbers EF534254, EF534253, EU012447, EF534252, EF534251, and EF534255, respectively. The gene sequences were compared with reported nucleotide and amino acid sequences (Tables 2, 3, 4, 5, 6, 7). The nucleotide sequence of vp28 was compared with twenty eight GenBank sequences from various geographical locations (Table 2). Only one sequence reported from India (Accession No.DQ013883) had 100 % homology with the WSSV isolate reported in this study. There was an A → G nucleotide substitution at position 125 for all the other reported sequences, which translated into a substitution of aspartic acid for glycine. The nucleotide change at position 125 was present only in two WSSV isolates from India including the isolate reported in this paper indicating that this mutation is restricted to India. A Chinese isolate (Accession No.AY682926) and a Korean isolate (Accession No. AF380842) had a nucleotide substitution at position 234 and 444, respectively, from T → C both of which did not translate to change of amino acid. Two sequences from China (AF502435 and AY249434) had an additional mutation at nucleotide position 119 from G → A which translated into an amino acid substitution from arginine to histidine. A Thailand isolate (Accession No. EF194079), a Chinese isolate (Accession No. DQ979320), and an Indian isolate (Accession No. AY422228) had three nucleotide substitutions including that at position 125. The Thailand isolate had nucleotide substitution at position 113 from T → A and at 434 from T → C which translated into change of amino acid from valine to glutamic acid and leucine to proline, respectively. The Chinese isolate had T → C nucleotide substitutions at positions 306 and 536 with change of amino acid from valine at both positions to proline and alanine, respectively. The Indian isolate had substitutions from T → C at positions 483 and 485 with change of amino acid from phenylalanine to serine in the first position, while the second position was conserved. There was nucleotide substitutions at ten different positions with the isolates studied, seven of which translated to a change of amino acid.

The nucleotide sequence of the gene encoding for envelope protein vp26 was compared with 16 reported sequences and it was revealed that there was 100 % similarity with fourteen of the reported sequences (Table 3). The G → A nucleotide substitution at the nucleotide position 575 for the Chinese isolate (Accession No. AY220746) resulted in a change of amino acid from arginine to lysine, while a nucleotide change for the Vietnamese isolate (Accession No. AJ551446) at nucleotide position 345 from T → C was conserved and did not translate to change of amino acid. The vp26 gene was found to be highly conserved among isolates from different geographical locations with just two nucleotide variations from the sixteen isolates compared.

The vp19 sequence was compared with seventeen reported sequences, and it was found that there was 100 % sequence similarity with four WSSV isolates reported from India (DQ681071), Vietnam (AY160771),

Table 2 Nucleotide and amino acid substitution in vp28 gene

Accession nos.	Position of mutation	Base substitution	Change in amino acid
DQ013883	Nil	Nil	Nil
DQ902658, DQ681069, DQ098011, AY324881, AF332093[a], AF272979, AY249443, AY249442, AY249441, AY249440, AF227911, AJ551447, DQ013882, DQ013881, DQ007315, AF369029[a], AY168644, AF440570[a], AF173993, AY873785	125	A → G	D → G
AF380842	125	A → G	D → G
	444	T → C	Nil
AF502435, AY249434	125	A → G	D → G
	119	G → A	R → H
AY682926	125	A → G	D → G
	234	T → C	Nil
EF194079	125	A → G	D → G
	113	T → A	V → E
	434	T → C	L → P
DQ979320	125	A → G	D → G
	306	T → C	V → P
	536	T → C	V → A
AY422228	125	A → G	D → G
	483	T → C	F → S
	485	T → C	Nil

[a] Complete genome sequence of WSSV

Table 3 Changes in vp26 gene sequence in GenBank

Accession nos.	Position of mutation	Base substitution	Change in amino acid
AF308164, AF332093[a], AF272980, AY249439, AY249438, AY249437, AY249436, AY249435, AY422230, AF380841, AF369029[a], AF440570[a], AF173992, DQ681070	Nil	Nil	Nil
AY220746	575	G → A	R → K
AJ551446	345	T → C	Nil

[a] Complete genome sequence of WSSV

Table 4 Changes in vp19 gene sequence in GenBank

Accession nos.	Position of mutation	Base substitution	Change in amino acid
DQ681071, AY160771, AF440570[a], AJ937860	Nil	Nil	Nil
AY316119, AY873786	57	T → C	Nil
AF332093[a], AF402997	196	C → T	P → S
AY249448, AY249447, Ay249446, AY248445, AY249444, AY220744, AF369029[a]	218	T → A	V → D
AY422227	25	T → C	S → P
	44	T → C	V → A
DQ902655	25	T → C	S → P
	44	T → C	V → A
	287	C → T	T → M

[a] Complete genome sequence of WSSV

Table 5 Changes in vp68 gene sequence in GenBank

Accession nos.	Position of mutation	Base substitution	Change in amino acid
AF332093[a], AF369029[a], AF440570[a], AF4114664	Nil	Nil	Nil

Table 6 Changes in vp281 gene sequence in GenBank

Accession nos.	Position of mutation	Base substitution	Change in amino acid
AF332093[a], AF369029[a], AF440570[a], AF411634	807	A → T	Nil
DQ979321	807	A → T	Nil
	841	C → A	P → T
AY517490	807	A → T	Nil
	843	G → A	Nil

Table 7 Changes in vp466 gene sequence in GenBank

Accession nos.	Position of mutation	Base substitution	Change in amino acid
AF332093[a], AF440570[a], AF395545	Nil	Nil	Nil
AF369029[a]	1212	A → T	K → N

[a] Complete genome sequence of WSSV

Taiwan (AF440570), and Mexico (AJ937860) (Table 4). Point mutations: T → C, T → A and C → T were found at six different nucleotide positions with 13 reported sequences. To ascertain whether the variation at nucleotide level resulted in change in at amino acid level, the protein sequences of vp19 gene were also compared. The T → C point mutation at position 57 (Accession No. AY316119 and AY873786) did not result in change in amino acid while a C → T mutation at position 196 (Accession No. AF332093 and AF402997) resulted in change amino acid from proline to serine. A T → A mutation at position 218 (Accession No. AY249444 to AY249448 and AF369029) had a change in amino acid from valine to aspartic acid. Two reported sequences from India, (Accession Nos. AY422227 and DQ902655), gave a T → C mutation at nucleotide positions 25 and 44 that translated into change of amino acids; serine to proline and valine to alanine, respectively. The WSSV isolate with Accession No. DQ902655 had a C → T extra substitution at position 287 other than at nucleotide positions 25 and 44 that translated into methionine, a change in the amino acid from threonine.

The vp68 sequence was 100 % similar with the four reported sequences (Table 5). Comparison of vp281 sequence with six published sequences revealed A → T, C → A, and G → A mutations at three different nucleotide positions (Table 6). The A → T mutation at nucleotide position 807 present in all the reported sequences did not translate into change of amino acid. While the C → T mutation at position 841 (Accession No. DQ979321) translated into change in amino acid from proline to threonine. The G → A mutation at position 843 (Accession No. AY517490) did not translate into change of amino acid. Three reported nucleotide sequence of vp466 had 100 % homology with the sequence reported in the present study (Table 7). There was an A → T nucleotide substitution at position 1212 for a Thailand isolate, which translated into a substitution of lysine to arginine.

In the present study, the sequences of six envelope proteins were compared with reported sequences of WSSV isolates. A common ancestor for the virus isolate could not be designated as there was variation in sequence with one envelope protein or other. The vp26 gene of the Taiwanese, Thailand, Chinese, and Korean isolates were identical at both nucleotide and amino acid levels, while the vp28 gene was 99 % identical at nucleotide level and 100 % identical at amino acid level (Moon et al. 2003). Sequence of five WSSV structural proteins (vp15, 19, 24, 26, and 28) from 5 different geographical WSSV isolates when compared with four WSSV isolates reported showed that the sequence of each gene were identical among the 9 isolates with the exception of vp28 gene of a Korean isolate with 1 base change from that of other WSSV isolates. However, this base change did not result in change of amino acid sequence (You et al. 2004). The extent of sequence variation among the isolates suggests that newer strains of WSSV are evolving that may have variations in their virulence potential. Chang et al. 2002 attributed the point mutations in the envelope proteins to the non-infectious nature of the virus isolate.

Highly conserved DNA and amino acid sequence of the structural proteins of WSSV provide the basis for the development of sensitive nucleic acid and antibody-based detection methodology for WSSV (You et al. 2004). Many PCR and protein-based diagnostic tests target the envelop proteins. However, variations in sequence of envelop proteins may result in false negative results. Recombinant vp28 and vp19 were found to have potential for use as injected vaccine or oral vaccine (Witteveldt et al. 2004). The variations at nucleotide and amino acid in the structural genes can result in difference in the protection offered by these proteins when used as vaccines. Hence further studies need to be done to ascertain the efficacy of these proteins in providing protection against the virus.

Author contribution Research concept and the experiments were performed by TCJ, LAR and RJ. KVL and PKS analyzed the data and reviewed the manuscript. All authors approved the final manuscript.

Acknowledgments Authors are grateful to the Director, Central Institute of Fisheries Technology (CIFT), Cochin for providing the necessary facilities to carry out this research work and Indian Council for Agricultural Research, New Delhi, India for financial assistance.

Conflict of interest The authors declare that they have no conflict of interest in the publication.

References

Altschul SF, Gish W, Miller W, Myersand EW, Lipman DJ (1990) Basic local alignment search tool. J Mol Biol 215:403–410

Chang YS, Lo CF, Peng SE, Liu KF, Wang CH, Kou GH (2002) White spot syndrome virus (WSSV) PCR-positive Artemia cysts yield PCR-negative nauplii that fail to transmit WSSV when fed to shrimp post larvae. Dis Aquat Org 49:1–10

Chen XF, Chen C, Wu DH, Huai H, Chi XC (1997) A new baculovirus of cultured shrimp. Sci China Ser C 40:630–635

Chiu WL, Chang W (2002) Vaccinia virus J1R protein: a viral membrane protein that is essential for virion morphogenesis. J Virol 756:9575–9587

Flegel TW (1997) Major viral diseases of black tiger prawn (*Penaeus monodon*) in Thailand. World J Microbiol Biotechnol 13:433–442

Hsiao JC, Chung CS, Chang W (1999) Vaccinia virus envelope D8L protein binds to cell surface chondroitin sulfate and mediates the adsorption of intracellular mature virions to cell. J Virol 73:8750–8761

Huang CH, Zhang XB, Lin QS, Xu X, Hu ZH, Hew CL (2002) Proteomic analysis of shrimp white spot syndrome viral proteins and characterization of a novel envelope protein vp466. Mol Cell Proteomics 1:223–231

Lin CL, Chung CS, Heine HG, Chang W (2000) Vaccinia virus envelope protein H3L protein binds to cell surface heparin sulfate and is important for intracellular mature virion morphogenesis and virus infection in vitro and in vivo. J Virol 74:3353–3365

Lo CF, Kou GH (2001) Shrimp white spot syndrome virus, complete genome. GenBank, gi: 19481591, AF440570.1, 26-OCT-2001

Lo CF, Ho CH, Peng SE, Chen CH, Hsu HC, Chiu YL, Chang CF, Liu KF, Su MS, Wang CH, Kou GH (1996) Whitespot syndrome baculovirus (WSBV) detected in cultured and captured shrimp arthropods. Dis Aquat Org 27:215–226

Lo CF, Hsu HC, Tsai MF, Ho CH, Peng SE, Kou GH, Lightner DV (1999) Specific genomic DNA fragment analysis of different geographical clinical samples of shrimp white spot syndrome virus. Dis Aquat Org 35:175–185

Mohan CV, Shanker KM, Kulkarni S, Sudha PM (1998) Histopathology of cultured shrimp showing gross signs of yellow head syndrome and white spot syndrome during 1994 Indian epizootics. Dis Aquat Org 34:9–12

Moon CH, Do JW, Cha SJ, Yoon WJ, Kim SB, Ko MS, Park MA, Kim JW, Sohn SK, Lee JH, Park JW (2003) Highly conserved sequences of three major virion proteins of Korean isolates of white spot syndrome virus (WSSV). Dis Aquat Org 53:11–13

Nadala ECB, Tapay LM, Loh PC (1998) Characterization of a non-occluded baculovirus-like agent pathogenic to penaeid shrimp. Dis Aquat Org 33:221–229

Ramasamy P (1995) Diseases of shrimp in aquaculture systems; diagnosis and therapeutic measures. Vanitha Publications (Eds.), Madras, pp 1–99

Sambrook J, Fritsch EF, Maniatis T (1989) Molecular cloning: a laboratory manual, 2nd edn. Cold Spring Harbor Laboratory Press, Cold Spring Harbor

Satish S, Musthaq S, Sahul Hameed AS, Narayanan RB (2004) Production of recombinant structural proteins from the Indian WSSV isolate. Aquaculture 242:69–80

Tsai MF, Yu HT, Tzeng HF, Leu JH, Chou CM, Huang CJ, Wang CH, Lin JY, Kou GH, Lo CF (2000) Identification and characterization of a shrimp white spot syndrome virus (WSSV) gene that encodes a novel chimeric polypeptide of cellular-type thymidine kinase and thymidylate kinase. Virology 277:100–110

Tsai JM, Wang HC, Leu JH, Hsiao HH, Wang AHJ, Kou GH, Lo CF (2004) Genomic and proteomic analysis of thirty-nine structural proteins of shrimp white spot syndrome virus. J Virol 78:11360–11370

Van Hulten MC, Tsai MF, Schipper CA, Lo CF, Kou GH, Valk JM (2000) Anaysis of genomic segment of white spot syndrome virus of shrimp containing ribonucleotide reductase genes and repeat regions. J Gen Virol 81:307–316

Van Hulten MC, Witteveldt J, Peters S, Kloosterboer N, Tarchini R, Fiers M, Sandbrink H, Lankhorst RK, Valk JM (2001) The white spot syndrome virus DNA genome sequence. Virology 286:7–22

Wang CH, Lo CF, Leu JH (1995) Purification and genomic analysis of baculovirus associated with white spot syndrome (WSBV) of *Penaeus monodon*. Dis Aquat Org 23:239–242

Wang YC, Lo CF, Chang PS, Kou GH (1998) Experimental infection of white baculovirus in some cultured and wild decapods in Taiwan. Aquaculture 164:221–231

Witteveldt J, Cifuentes CC, Vlak JM, Van Hulten MCW (2004) Protection of *Penaeus monodon* against white spot syndrome virus by oral vaccination. J Virol 78:2057–2061

Wongteerasupaya C, Vickers JE, Sriurairatana S, Nash GL, Akarajamorn A, Boonsaeng V, Panyim S, Tassanakajon A, Withyachumnarnkul B, Flegel TW (1995) A non-occluded, systemic baculovirus that occurs in cells of ectodermal and mesodermal origin and causes high mortality in black tiger prawn *Penaeus monodon*. Dis Aquat Org 21:69–77

Wu W, Wang L, Zhang X (2005) Identification of white spot syndrome virus (WSSV) envelope proteins involved in shrimp infection. Virology 332:578–583

Yang F, He J, Lin XH, Li Q, Pan D, Zhang X, Xu X (2001) Complete genome sequence of the shrimp white spot bacilliform virus. J Virol 75:11811–11820

You Z, Nadala ECB, Yang J, Loh PC (2004) Conservation of the DNA sequences encoding the major structural viral proteins of WSSV. Dis Aquat Org 61:159–163

Zhan WB, Wang YH, Fryer JL, Yu KK, Fukuda H, Meng QX (1998) White spot syndrome virus infection of cultured shrimp in China. J Aquat Anim Health 10:405–410

Zhang L, Zhang J, Chen D, Xia OL (1994) Assembly of rod shaped virus of *Penaeus orientalis kishinoye* in vivo. J Chin Electron Microsc Soc 13:354–363

2

Optimization of enzymatic hydrolysis of tilapia (*Oreochromis Spp.*) scale gelatine

Abdul Wahab Mohammad · Abd Ghani Kumar · Roseliza Kadir Basha

Abstract Gelatine hydrolysate is a type of gelatine that undergoes a controlled hydrolysis treatment to further break apart the gelatine or collagen molecules. In this study, gelatine hydrolysate was produced from commercial tilapia scale gelatine via controlled enzymatic hydrolysis. Commercial Alcalase 2.4 L, a protease enzyme was used to breakdown the peptide chains present in the gelatine. Optimization of hydrolysis conditions (temperature, time and enzyme to substrate ratio) was conducted by utilizing response surface methodology (RSM). Results showed that a hydrolysis temperature of 57.6 °C together with a hydrolysis time of 80 min and enzyme to substrate ratio of 1.20 % (v/w) were the optimum conditions to obtain the highest degree of hydrolysis (10.91 %). The freeze-dried gelatine hydrolysate was characterized with respect to chemical composition:approximate composition, viscosity and molecular weight. The gelatine hydrolysate produced contained a high content of protein (85.26 %); thus, it may serve as a potential protein source for human needs.

Keywords Gelatine hydrolysate · Enzymatic hydrolysis · Optimization · RSM

Introduction

Tilapia (*Oreochromis spp.*) is a freshwater species that has become the third most widely cultured fish, after carp and salmon (El-Sayed 2006). The production of tilapia has increased steadily and become an important source of fish supply in Malaysia (Jamilah and Harvinder 2002). It is classified as a durable, fast growing species compared to others and highly resistance to disease (Foh et al. 2011a). Its high protein content is comparable with meat sources and plays an important role in human supplementation (Ghorbani and Mirakabad 2010). The crude protein content of an adult tilapia is approximately 16–25 %, which is considerably high for freshwater species, while the fat content is very low (0.5–3.0 %) (Foh et al. 2011b).

Many studies have reported on gelatine production from fish sources. Over the past several decades, fine qualities of fish gelatine production have been widely studied from species as diverse as tilapia (Jamilah and Harvinder 2002; Muyonga et al. 2004), salmon (Arnesen and Gildberg 2007), cod (Arnesen and Gildberg

A. W. Mohammad (✉) · A. G. Kumar
Department of Chemical and Process Engineering, Universiti Kebangsaan Malaysia UKM, Bangi, 43600 Selangor Darul Ehsan, Malaysia
e-mail: awm.ukm@gmail.com

R. K. Basha
Department of Process and Food Engineering, Universiti Putra Malaysia UPM, Serdang, 43400 Selangor Darul Ehsan, Malaysia

2006), and pollock (Zhou and Regenstein 2004). Another valuable product that can be derived from fish derivatives is hydrolysate. Hydrolysate, also known as gelatine hydrolysate or collagen hydrolysate, is a type of gelatine that undergoes a controlled hydrolysis treatment to further break apart the gelatine or collagen molecules. During the process, peptide chains are cleaved into smaller chains, which results in lower molecular weights, between 500 Da and 25 kDa. However, a wide variety of peptide chains will be generated depending on the enzyme's specificity, environmental conditions and extent of the hydrolysis process (Guerard et al. 2002). Compared to gelatine, hydrolysate does not gel anymore but still has an active surface behaviour exactly like gelatine. In practical terms, hydrolysate can be manufactured into two ways; either derived from gelatine itself or directly from an animal's pure collagen. However, pure collagen from animals is known for its high resistance; hence, special collagenases have to be used in large amounts, which will result in higher production costs. Therefore, in industry, gelatine is widely used as a raw material for production of gelatine hydrolysate.

Most of the gelatine hydrolysates investigated in previous studies were derived by enzymatic hydrolysis from yellowfin tuna (Guerard et al. 2001; Ovissipour et al. 2010), tilapia (Foh et al. 2011a; Yang et al. 2009), catfish (Aleman et al. 2011), salmon (See et al. 2011), persian sturgeon (Ovissipour et al. 2009), mackerel (Wua et al. 2003), cobia (Amiza et al. 2012), catla (Bhaskar et al. 2008), capelin (Shahidi et al. 1995) and cod (Slizyte et al. 2005a). Enzymatic treatment offers several advantages: the ease of controlling the reaction efficiently, minimal formation of by-products and the milder processing conditions required to run the hydrolysis (Alfonso Clemente 2000; Kristinsson and Rasco 2000a; Mannheim and Cheryan 1992). By employing enzymes to hydrolyze protein, one could improve the physiochemical, functional, sensory and also nutritional behaviour of the hydrolyzed proteins (Slizyte et al. 2005a). In addition, hydrolyzing protein can also improve its intestinal absorption (Kristinsson and Rasco 2000a). Research shows that extensive hydrolysis of proteins has a great potential to reduce immunological reactions, which is important for hyperallergic infants (Mahmoud 1994). Moreover, the products of hydrolysis, which are lower molecular weight peptides, can be easily absorbed in the gut, promoting better supplementation for humans and also animals (Slizyte et al. 2005b; Kristinsson and Rasco 2000a). In many countries, traditional and commercial preparations of fish protein hydrolysates are currently used as health foods, functional foods and nutraceuticals (Chalamaiah et al. 2012). Therefore, gelatine hydrolysate can be used as a supplement for clinical, infant and sports nutrition. Regarding the suitability of the enzymes in the protein hydrolyzation field, Alcalase, a microbial protease, is most preferred and is widely used. Microbial-derived enzymes are better suited as they offer a wide variety of available catalytic activities as well as greater pH and temperature stabilities (Diniz and Martin 1997). Previous research reported that Alcalase was the most efficient enzyme among the proteolytic enzymes studied in hydrolyzing protein, and was able to produce the highest degree of hydrolysis (DH) of hydrolysate (Ng and Mohd Khan 2012). The highest protein recovery and lowest lipid content were reported for hydrolysates prepared via hydrolysis with Alcalase compared with Papain and Neutrase (Adler-Nissen 1986). Moreover, it has been reported previously that hydrolysates from Alcalase had less bitterness compared to those prepared with Papain (Hoyle and Merritt 1994). Furthermore, Alcalase was also classified as the best choice for protein hydrolyzation based on enzyme cost per unit activity (Kristinsson and Rasco 2000b).

This study was carried out to investigate the effects of reaction parameters [temperature, time and enzyme to substrate (E:S) ratio] on degree of hydrolysis (DH) and to optimize the enzymatic hydrolysis condition for the highest degree of hydrolysis using response surface methodology (RSM).

Methodology

Chemicals and raw materials

The substrate used in this study was a commercial food grade gelatine from tilapia scales with a gel strength of 257 g. It was purchased from Halagel Sdn. Bhd. An endoproteinase from Bacillus licheniformis (Alcalase 2.4 L) with a declared activity of 2.4 AU/kg and density of 1.18 g/ml, was purchased directly from Novozymes. All reagents used were analytical grade.

Optimization of hydrolysis conditions by RSM

Response surface methodology (RSM) was utilised to obtain the optimum process parameters [hydrolysis temperature, hydrolysis time and enzyme to substrate (E:S) ratio] and maximum degree of hydrolysis (DH) for the enzymatic hydrolysis process. Hence, a number of experiments with various process parameters were carried out as an input in RSM.

Optimization of the hydrolysis conditions was accomplished by a central composite design (CCD) in RSM. CCD was efficiently applied to develop a second-order response model with only a small number of factors n; $2 \leq n \geq 6$ (Ng et al. 2011). The experiment was designed using three independent factors and five levels. As an input, the three different independent variables were designated as hydrolysis temperature (A, °C), hydrolysis time (B, minutes) and enzyme to substrate ratio (C, % v/w); while, for output, only DH (Y, %) was chosen as response. The ranges for each variable are tabulated as coded levels in Table 1. Once the optimization was performed, an enzymatic hydrolysis experiment was carried out applying the optimized parameters from RSM. The hydrolysis was done in triplicate and the hydrolysates were tested and analysed.

CCD consists of three types of design points; two-level factorial or fractional factorial design points, axial points and centre points. Centre points have to be repeated four to six times to obtain a precise estimation of experimental error. Table 2 shows the 19 total runs rendered for the optimization process using RSM together with the actual levels for each variable and the response values. In this study, randomized experimental runs were carried out due to minimizing the effects of unexpected variability in the observed responses.

Enzymatic hydrolysis

Gelatine solution (5 % w/v) was prepared by dissolving some amount of pure gelatine substrate into the required amount of distilled water at an appropriate temperature range (50–60 °C). The reaction mixture was stirred by a four-blade impeller at a speed of 300 rpm. Once it was homogeneous, the temperature was adjusted to a certain level while the pH of the gelatine solution was increased to pH 8 by adding a small amount of 2 N NaOH. Then, the Alcalase 2.4 L enzyme was added into the homogenized gelatine solution at the desired concentrations. The hydrolysis process was conducted at the required period before being heated rapidly and maintained at a temperature of 90 °C for at least 10 min (Aleman et al. 2011) to deactivate the enzyme. The hydrolysates obtained were lyophilized using freeze-drying equipment (Martin Christ, Alpha 1-4 LSC, Germany) at temperature of −40 °C for almost 24 h. The freeze-dried hydrolysates were then crushed into smaller particles before being labelled and sealed in vacuum bags for storage in a dry place.

Degree of hydrolysis

Measurement of the degree of hydrolysis (DH) was carried out using the pH–stat method according to Adler-Nissen (1977). In this method, DH can be monitored continuously by maintaining the alkaline pH (pH 8) during the enzymatic hydrolysis process via direct addition of 2 N NaOH. From the total base consumed during the hydrolysis process, DH can be calculated directly from the Eq. (1) (Adler-Nissen 1986).

$$\mathrm{DH} = \frac{\mathrm{B} \times \mathrm{N_B}}{\mathrm{M_P}} \times \frac{1}{\alpha} \times \frac{1}{\mathrm{h_{tot}}} \times 100\,\% \tag{1}$$

Table 1 Coded and actual levels for the process variables studied

Process parameter	Code	Coded level				
		−1.68(−α)	−1	0	+1	+1.68(α)
Temperature (°C)	X_1	44.89	50.0	57.5	65.0	70.11
Time (min)	X_2	26.36	40.0	60.0	80.0	93.64
E/S (% v/w)	X_3	0.16	0.5	1.0	1.5	1.84

Table 2 Experimental design and the results for optimization of enzymatic hydrolysis of tilapia gelatine

Standard order	Coded level of variable			Degree of hydrolysis (%), *Y*	
	X_1	X_2	X_3	Predicted	Experimental
1	−1.000	−1.000	−1.000	5.62	5.48
2	1.000	−1.000	−1.000	6.35	6.69
3	−1.000	1.000	−1.000	7.09	7.18
4	1.000	1.000	−1.000	7.90	7.92
5	−1.000	−1.000	1.000	8.34	7.94
6	1.000	−1.000	1.000	7.95	7.48
7	−1.000	1.000	1.000	11.21	10.49
8	1.000	1.000	1.000	10.89	10.65
9	−1.682	0.000	0.000	7.26	7.77
10	1.682	0.000	0.000	7.61	7.63
11	0.000	−1.682	0.000	7.20	7.43
12	0.000	1.682	0.000	10.91	11.23
13	0.000	0.000	−1.682	5.80	5.43
14	0.000	0.000	1.682	10.60	11.50
15	0.000	0.000	0.000	9.22	9.60
16	0.000	0.000	0.000	9.22	9.06
17	0.000	0.000	0.000	9.22	9.15
18	0.000	0.000	0.000	9.22	9.51
19	0.000	0.000	0.000	9.22	8.69

X_1—hydrolysis temperature (°C), X_2—hydrolysis time (min.), X_3—enzyme to substrate ratio (% v/w)

Where DH is the degree of hydrolysis, B is the base consumption (in ml), N_B is the normality of the base (N), α is the average degree of dissociation of the α-NH groups, M_P is the mass of protein (in g) and h_{tot} is the total number of peptide bonds in the protein substrate (11.1 meqv/g protein for gelatine). According to Kristinsson and Rasco (2000b), the degree of dissociation, α was calculated based on the Eq. (2).

$$\alpha = \frac{10^{pH-pK}}{1 + 10^{pH-pK}} \tag{2}$$

The pK value was directly dependent on the process temperature, which varied between experiments. Therefore, the Eq. (3) was used to calculate the correct value of pK (Steinhardt and Beychok 1964).

$$pK = 7.8 + \frac{298 - T}{298T} \times 2400 \tag{3}$$

Proximate analysis

The proximate analyses were carried out on the gelatine powder as a raw material and freeze-dried gelatine hydrolysate as a final product. Moisture and ash contents were determined according to Gelatine Manufacturer of Europe (GME) Monograph (2000). The oven method was used to determine moisture content, while the ash content was determined by charring the pre-dried sample in a crucible at 550 °C for about 17 h until white ash was formed. Finally, protein content was estimated by the differences. All analyses were carried out in triplicate and average values were taken.

Amino acid composition

Amino acid composition was determined after hydrolysis at 110 °C for 24 h in 6 N HCl. The amount amino acid composition was analysed by a high performance liquid chromatography (HPLC), equipped with Waters 410 Scanning Flourescence and AccQ Tag Column (3.9 × 150 mm). Amino acid composition was reported as g amino acid per 100 g protein.

Molecular weight

In order to characterize the hydrolysates based on their molecular weights, Tris–glycine sodium dodecyl sulphate polyacrylamide gel electrophoresis (Tris–Glycine SDS-PAGE) was performed on samples from the optimized enzymatic hydrolysis. The test was done using 17 % resolving gel and 5 % stacking gel according to Laemmli (1970) with some adjustments. The sample buffer was prepared from 0.5 M Tris HCl buffer pH 6.8, 10 % (w/v) glycerol, 0.5 % (w/v) SDS, bromophenol blue and 5 % β-mercaptoethanol. Samples were diluted until a concentration of 0.5 % was obtained. Three samples (S_1, S_2 and S_3) of the same hydrolysate and one sample of gelatine (S_0) were analysed for determination of their molecular weight. All samples prepared were denatured in 95 °C hot water for about 3 min. After electrophoresis, the gels were stained using GelCodeTM blue safe protein stain purchased from thermo scientific.

Viscosity

The viscosity of gelatine and gelatine hydrolysate was determined using modification method described by Cho et al. (2006) and Ninan et al. (2012). 5 % w/v gelatine solution was prepared by dissolving the dry gelatine powder in distilled water at a temperature of 60 °C. The viscosity of gelatine and gelatine hydrolysate was monitored continuously starting at 40 °C until it reached 60 °C using a computerized Brookfield digital viscometer (Model DV-II) together with spindle No. 1 at 60 rpm.

Statistical Analysis

All the results obtained from the experimentation were analysed using Design Expert 6.0.10 statistical software (Stat-Ease Inc., Minneapolis). Design-Expert fits linear, two-factor interaction (2FI), quadratic and cubic polynomials to the response. The program displays a measure of progress during the calculations.

Results and discussion

Optimization of enzymatic hydrolysis parameters

The experimental response, DH, was calculated based on the Eqs. (1–3), and inserted in the RSM to be analysed further. The response ranged from 5.43 to 11.50 %; hence, the ratio of the maximum to minimum is only 2.12. This indicates that no transformation is required ($\lambda = 1.0$), as the ratio was <10. In the Fit Summary section, certain data were shown based on the calculations carried out by the software to obtain the best model to predict the most accurate response.

Based on the Sequential Model Sum of Squares in Table 3, the best model was chosen based on the highest order polynomial criteria where the additional terms were significant and the model was not aliased. In this research, the software revealed that the linear and quadratic models were significant ($P < 0.05$). However, the quadratic model was chosen as it was a higher order polynomial compared to the linear model.

Lack of Fit in Table 4, the residual error was compared to the Pure Error from replicated design points. A high probability value showed insignificant lack of fit and it was desired. Therefore, the linear model definitely can be ruled out, because the Prob > F falls below 0.05. However, the quadratic model did not showed significant lack of fit, hence it was chosen. The result showed the quadratic model was fitted well with the DH data indicated that the generated or predicted model is able to be fitted with the data variations and significantly represents the actual relationships with the parameters of a reaction.

Table 3 Sequential Model Sum of Squares

Source	Sum of squares	DF	Mean square	*F* value	Prob > *F*
Mean	1,361.4459	1	1,361.4459	–	–
Linear	44.5439	3	14.8480	20.0174	<0.0001
2FI	1.6034	3	0.5345	0.6735	0.5846
Quadratic	6.5274	3	2.1758	6.5373	0.0122
Cubic	1.7279	4	0.4320	1.7039	0.2845
Residual	1.2676	5	0.2535	–	–
Total	1,417.1161	19	74.5851	–	–

Table 4 Lack of Fit

Source	Sum of squares	DF	Mean square	*F* value	Prob > *F*
Linear	10.5915	11	0.9629	7.2017	0.0357
2FI	8.9881	8	1.1235	8.4033	0.0281
Quadratic	2.4607	5	0.4921	3.6809	0.1154
Cubic	0.7328	1	0.7328	5.4809	0.0793
Pure error	0.5348	4	0.1337	–	–

The Model Summary Statistics in Table 5 provides additional statistical measures that were useful in comparing the available models in the software. From the data obtained, the quadratic model was chosen due to the low standard deviation (Std. Dev.), high R-Squared (R^2) values and a low PRESS. R-Squared is a correlation coefficient for the model while PRESS is a measure of how a particular model fits each design point. According to the R-Squared values, the closer the value to one, the better. In this studies showed that the highest R-Squared was obtained for the cubic model, followed by the quadratic model. However, the cubic model had been discarded earlier as it was aliased and could not be chosen; thus, the quadratic model was utilised.

Analysis of variance (ANOVA) was used to confirm the significant and adequacy of the quadratic model predicted, as shown in Table 6. The predicted quadratic model showed significant behaviour at a 95 % confidence level ($P < 0.05$). The significance of the predicted model was evaluated further for each parameter varied. From the result obtained, only X_2, X_3, X_1^2 and X_3^2 were significant ($P < 0.05$) while the rest (X_1, X_2^2 and X_1X_2, X_1X_3 and X_2X_3) showed non-significant behaviour. However, F value of 17.58 implied the model was significant ($P < 0.05$). In addition, there was only a 0.01 % chance that a model F-Value this large could occur due to noise. Hence, the adequacy of the quadratic model was confirmed from ANOVA as the model Prob > *F* was <0.05.

Regression analysis was performed for the tree difference variables (X_1, X_2 and X_3). From the regression analysis, the second-order polynomial model was predicted. It was consisted of 10 β-coefficients: three linear effects, three quadratic effects, three interaction effects and one constant.

$$Y = \beta_0 + \beta_1X_1 + \beta_2X_2 + \beta_3X_3 + \beta_{11}X_1^2 + \beta_{22}X_2^2 + \beta_{33}X_3^2 + \beta_{12}X_1X_2 + \beta_{13}X_1X_3 + \beta_{23}X_2X_3$$

Where Y represented the estimated dependent variable (DH) and X_1, X_2 and X_3 were hydrolysis temperature, hydrolysis time and enzyme to substrate ratio respectively, while β_0 was the constant term and β_i, β_{ii} and β_{ij} were the linear, quadratic and interaction terms ($i = 1-3$ and $j = 1-3$), respectively. Finally, the second-order regression equation obtained from RSM was shown below.

$$Y = -38.6269 + 1.3732X_1 + 0.0304X_2 + 7.9600X_3 - 0.0112X_1^2 - 0.0001X_2^2 - 1.4461X_3^2 + 0.0001X_1X_2 - 0.0748X_1X_3 + 0.0348X_2X_3$$

Table 5 Model Summary Statistics

Source	Std Dev.	R-squared	Adjusted R-squared	Predicted R-squared	Press
Linear	0.8613	0.8001	0.7602	0.6718	18.2686
2FI	0.8908	0.8289	0.7434	0.4283	31.8275
Quadratic	0.5769	0.9462	0.8924	0.6457	19.7245
Cubic	0.5035	0.9772	0.9180	−1.9177	162.4315

Table 6 Analysis of variance (ANOVA)

Source	Sum of squares	DF	Mean square	*F* value	Prob > *F*
Model	52.6747	9.0	5.8527	17.5846	0.0001
X_1	0.1451	1.0	0.1451	0.4359	0.5257
X_2	16.5846	1.0	16.5846	49.8285	<0.0001
X_3	27.8142	1.0	27.8142	83.5681	<0.0001
X_1^2	5.4508	1.0	5.4508	16.3771	0.0029
X_2^2	0.0449	1.0	0.0449	0.1348	0.7220
X_3^2	1.7840	1.0	1.7840	5.3599	0.0458
X_1X_2	0.0027	1.0	0.0027	0.0081	0.9303
X_1X_3	0.6302	1.0	0.6302	1.8934	0.2021
X_2X_3	0.9705	1.0	0.9705	2.9159	0.1219
Residual	2.9955	9.0	0.3328	–	–
Lack of fit	2.4607	5.0	0.4921	3.6809	0.1154
Pure error	0.5348	4.0	0.1337	–	–
Cor total	55.6702	18.0	–	–	–

X_1—hydrolysis temperature (°C), X_2—hydrolysis time (min.), X_3—enzyme to substrate ratio (% v/w)

Overall, the analysis of variance representing the second-order regression equation which exhibited the results fitted well to all the process parameters. Thus, the predicted model was statistically valid for enzymatic hydrolysis of tilapia skin gelatine. From the optimization through RSM, the optimal condition for enzymatic hydrolysis of tilapia skin gelatine was achieved at 57.6 °C of hydrolysis temperature (X_1), 80 min of hydrolysis time (X_2) and 1.20 % of enzyme to substrate ratio (% v/w) (X_3). These optimal parameters gave a maximum value of yield in terms of DH, which was 10.91 %. Finally, by employing the optimal parameters (X_1, X_2 and X_3), three sets of enzymatic hydrolysis experiments were conducted. The average results obtained from the three actual experimental gave a higher yield of DH (11.89 %).

A higher percentage of DH (19.9 %) was reported in a previous study for a controlled hydrolysis of catfish skin gelatine using high pressure (Aleman et al. 2011). The error in the yield for DH was calculated based on the predicted and actual value using the Eq. (4). The error calculated was acceptable with the value of 8.24 %.

$$\text{Error}(\%) = \left[\frac{A - B}{A}\right] \times 100 \quad (4)$$

where A is actual value and B is predicted value.

Effect of the hydrolysis parameter

The effects of hydrolysis temperature, time and enzyme to substrate ratio, upon the degree of hydrolysis (DH) were determined by employing response surface methodology (RSM). The hydrolysis temperature (X_1) and hydrolysis time (X_2) were set to top re-arranged ranges while the enzyme to substrate ratio (X_3) was set to a minimum value to achieve the maximum DH percentage. In order to obtain a clear view of the influences of these independent variables on the dependent variable, several three-dimensional views of response surface were plotted. The plots showed the effects of two independent variables toward DH by holding another independent variable at a constant value that was the optimal value.

All of the three-dimensional plots for the enzymatic hydrolysis of tilapia skin gelatine are presented in Figs. 1, 2 and 3. The plots are represented as a function of two factors and holding the other factor. Figure 1 shows the effect of hydrolysis time and temperature on DH by holding the enzyme to substrate ratio at an optimal value of 1.20 %. The result showed that DH increased as time and temperature increased. However, a decreasing trend was observed at temperatures above 57.6 °C. This finding can be related to the best range of temperature for the highest performance of Alcalase. Higher temperatures tend to deactivate the enzymes used, resulting in lower DH being achieved. This result was supported in a study on the hydrolysis of *Catla* visceral waste protein where the optimum temperature obtained was 55 °C (Bhaskar et al. 2008). In another study of hydrolysis on threadfin bream protein, 60 °C was determined as the optimum temperature (Normah et al. 2005). Figure 2 shows the effect of the enzyme to substrate ratio and hydrolysis temperature on DH by holding the value of the hydrolysis time constant at an optimum of 80 min. Here, it can be observed that the DH increased almost linearly with the increasing enzyme to substrate ratio. This finding showed that a higher enzyme concentration will eventually cleave more peptide bonds available in the substrate. The same results were reported in a previous hydrolysis study employing palm kernel expeller, where a higher concentration of Alcalase 2.4 L gave a higher value of DH (Ng and Mohd Khan 2012). In addition, a study on the hydrolysis of protein concentrate from sunflower wholemeal also supported the results (Ordonez et al. 2008). However, selection of enzyme concentration should considered the total concentration of peptide chains in the substrate and the degree of protein breakdown desired to avoid an excessive use of enzyme that will drive up the cost for industrial use. Furthermore, Fig. 3 shows both factors (enzyme to substrate ratio and hydrolysis time) were slightly linear relationships against the DH by holding the value of the hydrolysis temperature at an optimum of 57.6 °C.

Proximate composition

The composition according to the proximate analysis is tabulated in Table 7. The result showed that protein was the main constituent in both gelatine and hydrolysate (87.62 and 85.26 %, respectively). Followed by a fair amount of moisture content of gelatine and hydrolysate (12.30 and 12.35 %, respectively). The higher percentage of ash content in the hydrolysate (2.39 %) compared to the gelatine (0.08 %) indicated that it was contained greater amounts of salt. It was mainly due to the addition of sodium hydroxide and sulfuric acid during pH adjustment and pH control during the enzymatic treatment (Chen et al. 2012). Overall, the yield of

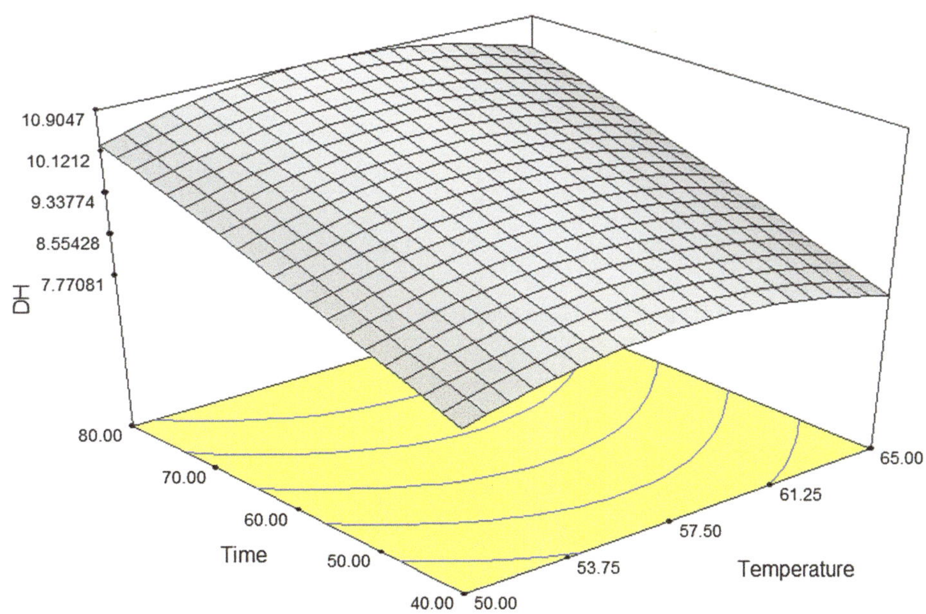

Fig. 1 Response surface graph of enzymatic hydrolysis as a function of hydrolysis time and temperature versus DH

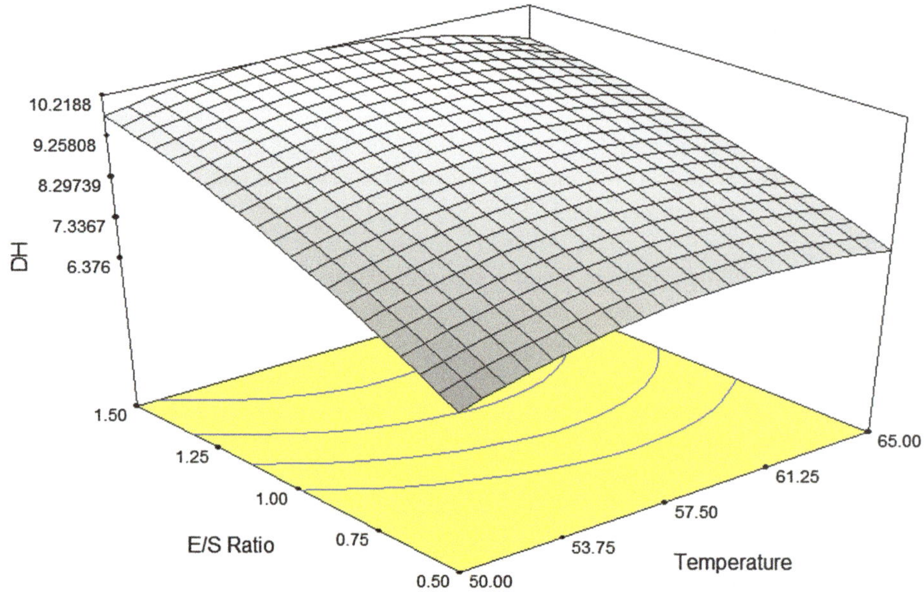

Fig. 2 Response surface graph of enzymatic hydrolysis as a function of enzyme to substrate ratio and hydrolysis temperature versus DH

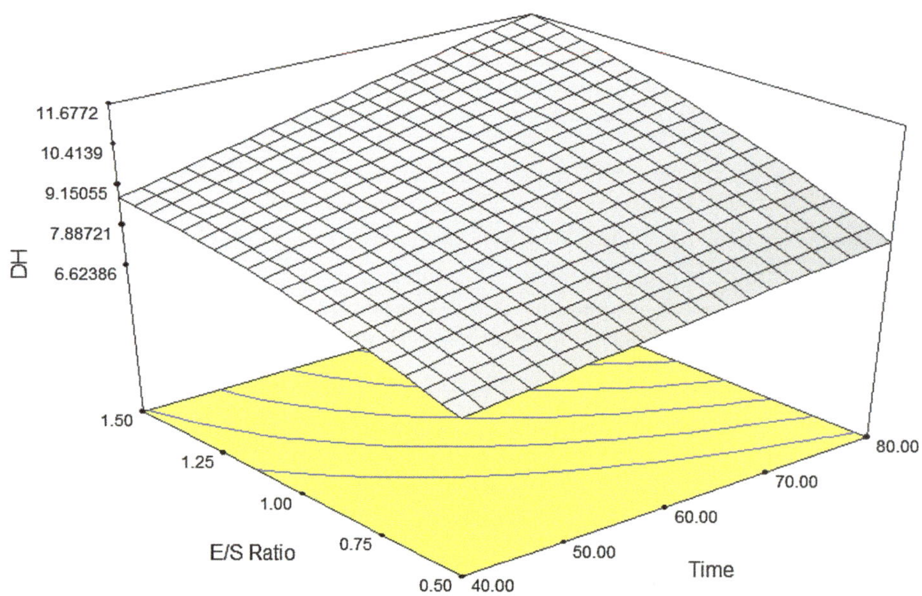

Fig. 3 Response surface graph of enzymatic hydrolysis as a function of enzyme to substrate ratio and hydrolysis time versus DH

Table 7 Proximate compositions of gelatine and gelatine hydrolysate

Analysis	Fish gelatine		Gelatine hydrolysate	
	Mean	SD (±)	Mean	SD (±)
Moisture (%)	12.30	0.024	12.35	0.039
Ash (%)	0.08	0.023	2.39	0.019
Protein (%)	87.62	0.021	85.26	0.020

Table 8 Amino acid composition of gelatine and gelatin hydrolysate

Amino acid	Gelatine (g/100 g)		Gelatine hydrolysate (g/100 g)	
	Mean	SD (±)	Mean	SD (±)
Histidine	2.6155	0.6611	3.0147	0.3194
Isoleucine	0.8193	0.0446	0.9244	0.0594
Leucine	1.8592	0.0668	1.8908	0.0149
Lysine	4.2752	0.3937	3.6450	0.0668
Methionine	0.3466	0.0074	0.2206	0.0371
Phenyl alanine	1.1765	0.0446	1.2395	0.0000
Tyrosine	0.2731	0.0149	0.1891	0.0594
Threonine	1.9748	0.1337	2.0378	0.0149
Arginine	6.7962	0.1857	4.0126	0.2080
Valine	1.3971	0.0520	1.5441	0.0074
Aspartime	4.0966	0.2080	4.2647	0.1040
Glutamin acid	7.5000	0.2971	8.0147	0.0371
Serine	2.7206	0.0817	2.1534	0.0074
Glycine	18.3403	0.3862	18.9391	0.4085
Alanine	7.4160	0.2674	8.1828	0.1560
Hydroxyproline	6.6282	0.1560	7.0693	0.0520
Proline	8.4559	0.2748	9.3277	0.0297
Total	76.6912	2.5031	76.6702	0.1708

hydrolysate was high in protein and low in inorganic minerals and it was a similar composition to that of the gelatine. This result was supported by Aleman et al. (2011), who obtained 88.11 % of protein composition and 11.52 % moisture content from catfish skin gelatine.

Amino acid composition

Table 8 shows an amino acid composition of commercial gelatine from tilapia scales and gelatine hydrolysate. After hydrolysis, the composition of total amino acid changed to be increased due to the optimum conditions during hydrolysis process. The higher amino acid contents which obtained from this study were similar to previous reports (Jongjareonrak et al. 2006; Sarabia et al. 2000 and Herpandi and Adzitey 2011). The results showed both gelatine and gelatine hydrolysate containing glycine (18.3/100 and 18.9/100 g respectively) as a main component, followed by proline, alanine and hydroxyproline.

Molecular weight

The molecular weight distribution was employed for selection of particular application based on their molecular sizes. Based on previous studies (Lim and Mohammad 2011; Gomez-Guillen et al. 2002) gelatine consisted mainly of α-chains followed by a fair number of β-chains, along with a low content of γ-chains. In this research, SDS-PAGE analysis was done on the gelatine and hydrolysate samples as mentioned previously. The electrophoretic patterns for both gelatine and hydrolysate samples are presented in Fig. 4 below. The result clearly showed that the sizes of gelatine molecules were within the range of 34–260 kDa. Whereas, the average molecular weight of three samples of the same hydrolysates was below 10 kDa. This result indicated that the hydrolysis process had successfully cleaved the peptide bonds, resulting in lower molecular weight and higher DH. In order to obtain protein hydrolysates of high nutritional value, the dietary protein in it should rich in low molecular weight species, with the amounts of free amino acids as low as possible (Vijayalakshmi et al. 1986).

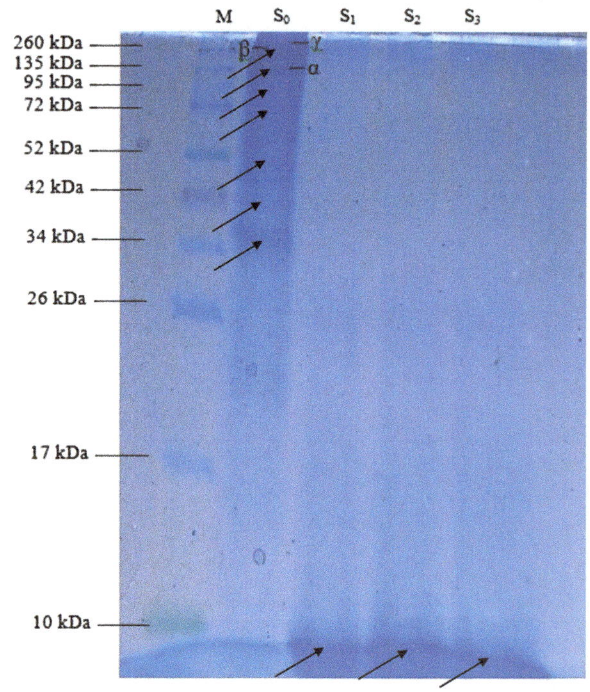

Fig. 4 Electrophoretic patterns for tilapia skin gelatine and gelatine hydrolysate (M—standard protein marker, S_0—gelatine sample, S_1, S_2 and S_3—gelatine hydrolysate samples)

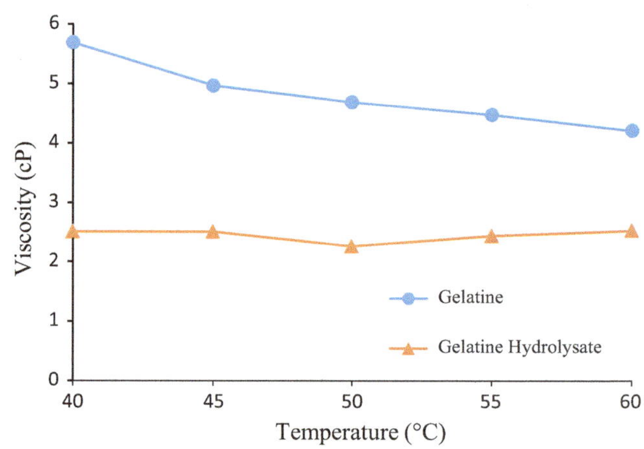

Fig. 5 Viscosity plots for gelatine and gelatine hydrolysate

Viscosity

Figure 5 shows the viscosity plots for gelatine and gelatine hydrolysate. The hydrolysate samples were obtained from triplicate runs employing the optimal parameters obtained from the optimization process. The average viscosity of the hydrolysates produced was 2.27–2.54 cP. The results showed higher viscosity for the gelatine, which was almost twice the viscosity of the hydrolysates. The viscosities showed decreasing trends as the temperature increased. In addition, the molecular weight of hydrolysates also influenced the viscosity of solution. Hydrolysates with low molecular weight tended to produce a solution of low viscosity that was easier to process even at relatively high concentration (Schrieber and Gareis 2007). Therefore, viscosity was an important measurement for the industry to monitor and control the hydrolysis process.

Conclusion

The degree of hydrolysis of gelatine hydrolysate from tilapia scales was significantly influenced by the hydrolysis conditions (hydrolysis time, hydrolysis temperature and enzyme to substrate ratio). Based on the model predicted, the optimum hydrolysis conditions were 57.6 °C for hydrolysis temperature, 80 min for hydrolysis time and 1.20 % (v/w) for the enzyme to substrate ratio. These optimum conditions gave a maximum value for yield, in terms of DH (10.91 %). The gelatine hydrolysates produced were rich in protein content and may serve as a better alternative source to fulfil human needs.

Acknowledgments The authors would like to thank the Universiti Kebangsaan Malaysia for the Grant DIP-2012-01 and Post-graduate Scholarship for AGK Abdul-Aziz and the Ministry of Higher Education, Malaysia for MyBrain Scholarship.

References

Adler-Nissen J (1977) Enzymatic hydrolysis of food proteins. Process Biochem 8:18–32
Adler-Nissen J (1986) Enzymatic hydrolysis of food proteins. Elsevier, Barking
Aleman A, Gimenez B, Gomez-Guillen MC, Montero P (2011) Enzymatic hydrolysis of fish gelatin under high pressure treatment. Int J Food Sci Technol 46:1129–1136
Amiza MA, Kong YL, Faazaz AL (2012) Effects of degree of hydrolysis on physicochemical properties of Cobia (*Rachycentron canadum*) frame hydrolysate. Int Food Res J 19(1):199–206
Arnesen JA, Gildberg A (2006) Extraction of muscle proteins and gelatine from cod head. Process Biochem 41:697–700
Arnesen JA, Gildberg A (2007) Extraction and characterisation of gelatine from Atlantic salmon (*Salmo salar*) skin. Bioresour Technol 98:53–57
Bhaskar N, Benila T, RaDHa C, Lalitha RG (2008) Optimization of enzymatic hydrolysis of visceral waste proteins of Catla (*Catla catla*) for preparing protein hydrolysate using a commercial protease. Bioresour Technol 99:335–343
Chalamaiah M, Dinesh Kumar B, Hemalatha R, Jyothirmayi T (2012) Fish protein hydrolysates: proximate composition, amino acid composition, antioxidant activities and applications: a review. Food Chem 135:3020–3038
Chen C, Yu-Jie C, Wei X (2012) Comparisons on the functional properties and antioxidant activity of spray-dried and freeze-dried egg white protein hydrolysate. Food Bioprocess Technol 5:2342–2352
Cho SH, Jahncke ML, Chin KB, Eun JB (2006) The effect of processing conditions on the properties of gelatin from skate (*Raja kenojei*) skins. Food Hydrocoll 20:810–816
Clemente Alfonso (2000) Enzymatic protein hydrolysates in human nutrition. Trends Food Sci Technol 11:254–262
Diniz FM, Martin AM (1997) Fish protein hydrolysates by enzymatic processing. Agro Food Ind Hi Tech 8:9–13
El-Sayed AM (2006) Tilapia culture. CABI publishing, Oxford, pp 1–24
Foh MBK, Amadou I, Kamara MT, Foh BM, Wenshui Xia (2011a) Effect of enzymatic hydrolysis on the nutritional and functional properties of Nile tilapia (*Oreochromis niloticus*) proteins. Am J Biochem Mol Biol 1(1):54–67
Foh MBK, Kamara MT, Amadou I, Foh BM, Wenshui Xia (2011b) Chemical and physicochemical properties of tilapia (*Oreochromis Niloticus*) fish protein hydrolysate and concentrate. Int J Biol Chem 5(1):21–36
Gelatine Manufacturer of Europe (GME) (2000) Monograph. Standardised methods for the testing of edible gelatin version 1
Ghorbani M, Mirakabad HZ (2010) Factor influencing on trout production in Khorasan Razavi Province. Trends Agric Econ 3:1–18
Gomez-Guillen MC, Turnay J, Fernandez-Diaz MD, Ulmo N, Lizarbe MA, Montero P (2002) Structural and physical properties extracted from different marine species: a comparative study. Food Hydrocoll 16:25–34
Guerard F, Dufosse L, De La Broise D, Binet A (2001) Enzymatic hydrolysis of proteins from yellowfin tuna (*Thunnus albacares*) wastes using Alcalase. J Mol Catal B Enzym 11:1051–1059
Guerard F, Guimas L, Binet A (2002) Production of tuna waste hydrolysates by a commercial neutral protease preparation. J Mol Catal B Enzym 19–20:489–498
Herpandi NH, Adzitey F (2011). Fish bone and scale as a potential source of halal gelatin. J Fish Aquat Sci, 6 (4):379–389
Hoyle NT, Merritt JH (1994) Quality of fish protein hydrolysate from Herring (*Clupea harengus*). J Food Sci 59:76–79
Jamilah B, Harvinder KG (2002) Properties of gelatins from skins of fish-black tilapia (*Oreochromis mossambicus*) and red tilapia (*Oreochromis nilotica*). Food Chem 77:81–84
Jongjareonrak A, Benjakul S, Vusessanguan W, Tanaka M (2006) Skin gelatin from bigeye snapper and brownstripe red snapper: chemical composition and effect of microbial transglutaminase on gel properties. Food Hydrocoll 20(8):1216–1222
Kristinsson HG, Rasco BA (2000a) Fish protein hydrolysates: production, biochemical and functional properties. Crit Rev Food Sci Nutr 40:43–81
Kristinsson HG, Rasco BA (2000b) Kinetics of the hydrolysis of Atlantic salmon (*Salmo salar*) muscle proteins by alkaline proteases and a visceral serine protease mixture. Process Biochem 36:131–139
Laemmli UK (1970) Cleavage of structural proteins during assembly of the head of bacteriophage T4. Nature 227:680–685
Lim YP, Mohammad AW (2011) Physicochemical properties of mammalian gelatin in relation to membrane process requirement. Food Bioprocess Technol 4:304–311
Mahmoud MI (1994) Physicochemical and functional properties of protein hydrolysates in nutritional products. Food Technol. 48:89–95
Mannheim A, Cheryan M (1992) Enzyme modified protein from corn gluten meal: preparation and functional properties. J Am Oil Chem Soc 69:1163–1169

Muyonga JH, Cole CGB, Duodu KG (2004) Extraction and physico-chemical characterisation of Nile perch (*Lates niloticus*) skin and bone gelatin. Food Hydrocoll 18:581–592

Ng KL, Mohd Khan A (2012) Enzymatic preparation of palm kernel expeller protein hydrolysate (PKEPH). Int Food Res J 19(2):721–725

Ng LY, Leo CP, Mohammad AW (2011) Optimizing the incorporation of silica nanoparticles in polysulfone/poly(vinyl alcohol) membranes with response surface methodology. J Appl Polym Sci 121(3):1804–1814

Ninan G, Joseph J, Aliyamveettil ZA (2012) A comparative study on the physical, chemical and functional properties of carp skin and mammalian gelatins. J Food Sci Technol 51:2085–2091

Normah I, Jamilah B, Saari N, Yaakob Che Man B (2005) Optimization of hydrolysis conditions for the production of threadfin bream (*Nemipterus japonicus*) hydrolysate by Alcalase®. J Muscle Foods 16:87–102

Ordonez C, Benitez C, Gonzalez JL (2008) Amino acid production from a sunflower wholemeal protein concentrate. Bioresour Technol 99:4749–4754

Ovissipour M, Abedian A, Motamedzadegan A, Rasco B, Safari R, Shahiri H (2009) The effect of enzymatic hydrolysis time and temperature on the properties of protein hydrolysates from Persian sturgeon (*Acipenser persicus*) viscera. Food Chem 115:238–242

Ovissipour M, Benjakul S, Safari R, Motamedzadegan A (2010) Fish protein hydrolysates production from yellowfin tuna (*Thunnus albacares*) head using Alcalase and Protamex. Int Aquat Res 2:87–95

Sarabia AI, Gómez-Guillén MC, Montero P (2000) The effect of added salts on the viscoelastic properties of fish skin gelatin. Food Chem 70:71–76

Schrieber R, Gareis H (2007) Gelatine Handbook Theory and Industrial Practice. Wiley, German

See SF, Hoo LL, Babji AS (2011) Optimization of enzymatic hydrolysis of Salmon (*Salmo salar*) skin by Alcalase. Int Food Res J 18(4):1359–1365

Shahidi F, Han XQ, Synowiecki J (1995) Production and characteristics of protein hydrolysates from capelin (*Mallotus villosus*). Food Chem 53:285–293

Slizyte R, Dauksas E, Falch E, Storro I, Rustad T (2005a) Characteristics of protein fractions generated from cod (*Gadus morhua*) by-products. Process Biochem 40:2021–2033

Slizyte R, Rustad T, Storro I (2005b) Enzymatic hydrolysis of cod (*Gadus morhua*) by-products Optimization of yield and properties of lipid and protein fractions. Process Biochem 40:3680–3692

Steinhardt H, Beychok S (1964) Interaction of protein with hydrogen ions and other small ions and molecules. In: Neurath H (ed) The Proteins. Academic Press, New York, vol 2. pp 139–304

Vijayalakshmi MA, Lemieux L, Amiot J (1986) High performance size exclusion liquid chromatography of small molecular weight peptides from protein hydrolysates using methanol as a mobile phase additive. J Liq Chromatogr 9:3559–3576

Wua HC, Chen HM, Shiau CY (2003) Free amino acids and peptides as related to antioxidant properties in protein hydrolysates of mackerel (*Scomber austriasicus*). Food Res Int 36:949–957

Yang J, Liang W, Chow C, Siebert KJ (2009) Process for the production of tilapia retorted skin gelatin hydrolysates withoptimized antioxidative properties. Process Biochem 44:1152–1157

Zhou P, Regenstein JM (2004) Optimization of extraction conditions for Pollock skin gelatin. J Food Sci 69:393–398

3

Evaluation of growth, survival and body composition of larval white shrimp (*Litopenaeus vannamei)* fed the combination of three types of algae

Hadi Jamali · Nasrollah Ahmadifard · Daruosh Abdollahi

Abstract The larvae of the Pacific white shrimp *Litopenaeus vannamei* are fed, from the stage of protozoea I until to protozoea III (PZ III), with microalgae. The survival rate, duration of the terms of Z_I–Z_{III} and total length of protozoea larvae (PZ) of L. *vannamei* were survived. The shrimp larvae fed with the six treatments of microalgae (*Chaetoceros muelleri, Isochrysis galbana, Tetraselmis tetrathele, C. muelleri* : *I. galbana, C. muelleri: T. tetrathele* and *I. galbana: T. tetratheie*). The biochemical composition (protein, carbohydrate, lipid and energy) of the algae and the larvae fed those algae were also measured. The largest sizes were recorded for larvae fed with the mixture *C. muelleri and I. galbana* (4.35 mm) and smaller sizes were observed on larvae fed with *I. galbana* (3.04 mm) ($P < 0.05$). The larvae fed with mixture *T. tetrathele and C. muelleri* (88.42 %) and *I. galbana and T. tetrathele* (84.50 %) had the highest survival in experimental treatments. The lowest time for development was observed in larvae fed with mixture of *T. tetrathele and C. muelleri* (92.6 h). The larvae fed with mixture *I. galbana and T. tetrathele* had the highest protein and carbohydrate levels as compared with other treatments. Also highest lipid level and gross energy were shown in larvae fed with mixture *C. muelleri and I. galbana* ($P < 0.05$). The results showed that *L. vannamei* larvae fed with *C. muelleri* and the mixed diets containing this species (*C. muelleri* + *Isochrysis* sp., *C. mueller-i* + *Tetraselmis* sp.) had the highest growth and survival among treatments.

Keywords Shrimp · Algae · Nutrition · Biochemical composition

Introduction

In production of shrimp post larvae, feeding is one of the most important aspects (Lopez-Elias et al. 2008). Cultivation of larval stages of various aquaculture species is still highly dependent on live food which is for herbivorous larvae, like molluscs and crustaceans (Boeing 2005). The variability of the nutritional value of the main live food organisms currently used in aquaculture (i.e., microalgae, rotifers and brine shrimp) is well documented, especially with respect to fatty acids (Merchie et al. 1997). Microalgae are utilized as live feed for all growth stages of bivalve molluscs (e.g., oysters, scallops, clams and mussels), early juvenile stages of abalone, crustaceans and some fish species, also for zooplankton culture (Catarina Guedes and Xavier Malcata

H. Jamali (✉) · N. Ahmadifard
Department of Fisheries, Faculty of Natural Resources, Urmia University, P.O. Box: 46414-356, Urmia, Iran
e-mail: saeed.jamali11@gmail.com

D. Abdollahi
Department of Fisheries, Gonbad Kavous University, Gonbad Kavous, Iran

2012). The most algae used in aquaculture are: *Chaetoceros*, *Thalassiosira*, *Tetraselmis*, *Isochrysis*, *Nannochloropsis*, *Pavlova* and *Skeletonema* (Brown et al. 1997; Enright et al. 1986; Thompson et al. 1993). These organisms are fed directly or indirectly through artemia, rotifers and Daphnia for feeding grown organism. It is widely accepted that microalgae play an important role in larval nutrition of shrimp; however, it is uncertain whether juveniles and adults feed on microalgae. Some reports suggest that microalgae are found in their gut because shrimp accidentally ingests them together with debris (Marinez-Cordova and Pena-Messina 2005; Kent et al. 2011). The microalgae that are usually used as feed of peneids larvae are among different genus, including *Chaetoceros* spp. and *Tetraselmis* spp. (Apt and Behrens 1999; Voltolina and Lopez-Elias 2002). This genus of algae is considered as a beneficial species, because they contain all the requirements of peneids larvae (Rodriguez et al. 1994; D'Souza and Kelly 2000). Some species of algae such as *Isochrysis* sp. (Okauchi et al. 1997), *Tetraselmis tetrathele* (Okauchi and Hirano 1986), and *T. chuii* (Tobias-Quinitio and Villegas 1982) have high nutritional value and potential growth. *Chaetoceros muelleri* has a high growth rate (Trujillo-Valle and Voltolina 1994) and it can grow in outdoor conditions with a wide range of temperature and illumination (Nelson et al. 1992). *Isochrysis galbana* is a free living marine unicellular phytoflagellate of the order Chrysomonadales. This species is rich in polyunsaturated fatty acids (Wood 1974) and has a high nutritional value for marine fish larvae (Scott and Middleton 1979) and juvenile stages of molluscs (Rhodes and Landers 1973).

Microalgae have an important role in aquaculture and also in the enrichment of zooplankton for feeding fish larvae. In addition to providing proteins (that contain essential amino acids) and energy, they carry nutrients such as vitamins, essential PUFAs, pigments and sterols—which are transferred to a high level through the food chain. For instance, PUFA-rich microalgae, such as *Pavlova* sp. and *Isochrysis* sp. to enrich the DHA, have been successfully fed to zooplankton (Nichols et al. 1989). Brown et al. (1998) described that rotifers fed with microalgae (e.g., *Isochrysis* sp. and *N. oculata*) become rapidly enriched with ascorbic acid (AsA), whereas rotifers fed on baker's yeast (which itself is deficient in AsA) contained only residual amounts of AsA.; after 16 h of starving, rotifers lost ca. 10 % of their AsA, while retaining ca. 50 % of the total AsA ingested. Similarly, the concentration of AsA in *Artemia* sp. may be increased by feeding with microalgae (Merchie et al. 1995). However, little information is available on the transfer of other vitamins from microalgae to fish larvae.

The feeding protocols of larvae in commercial shrimp hatcheries include a wide range of balanced feed and nutritional supplements, particularly in the early stages. However, live feed continues to be the principal nutritional basis for culture of larvae (Aguirre-Hinojosa et al. 1999; Voltolina and Lopez-Elias 2002; Richmond 2004). For Zoea larvae, and to a lesser degree for Mysis, phytoplankton is the main source of proteins, carbohydrates, lipids and other nutritional compounds. It has been proven that the proximal composition and growth rate of shrimp larvae are associated with the biochemical composition of microalgae used as feed (D'Souza and Loneragan 1999). The biochemical composition of microalgae varies depending on culture conditions, and is affected by factors such as light, pH, temperature and nutrients (Lopez-Elias et al. 1999). Variable culture conditions alter the biochemical composition of the microalgae and affect their quality as live feed (Lopez-Elias et al. 2003). The objective of the present study was to carry out the efficacy of micro algal species, such as *Chaetoceros* sp., *Tetraselmis* sp., *Isochrysis* sp, on shrimp growth, survival and body composition.

Materials and methods

The study was conducted in Gomishan shrimps research center (Gomishan, Golestan province, Iran). *I. galbana* was obtained from the Center of Urmia Lake Research Institute (Urmia-Iran). *C. muelleri* and *T. tetrathele* were obtained from the Gomishan shrimp production hatchery. Each algal species was grown under greenhouse conditions in 500 L opaque conical cylinders. The culture media used for the maintenance of inoculum and accomplishment of the experiments were: Guillard f/2 (Guillard 1975) for *C. muelleri* and *T. tetrathele* and Conway (Walne 1966) for *I. galbana*. Temperatures were 26 ± 1 °C, Salinity 32 ± 2.5 ppt, pH 7.2–8 and light 5,000 lux. Every day the algal cell densities were estimated using a hemocytometer in the algal culture medium and larval culture flasks so that the desired experimental algal cell densities could be maintained.

The feeding of bioassays was conducted in a completely randomized design with six treatments and four replicates per treatment for a total of 24 fiberglass tanks (each with a capacity of 10 L), at a stocking density of

100 nauplii L^{-1}, with a daily water exchange of between 20 and 50 %. A concentration of 100,000 cells mL^{-1} for monoalgal diets, and 50,000 cells mL^{-1} of each of the species for the mixed diet was used. Six treatments were fed to *Litopenaeus vannamei* larvae in the Zoea I to Zoea III stages that include three monoalgal diets (*C. muelleri, I. galbana, T. tetrathele*), and three mixture of species (*C. muelleri*:*I. galbana, I. galbana*:*T. tetrathele, T. tetrathele*:*C. muelleri*).

The protein, carbohydrate and lipid contents of the microalgae were quantified from samples collected from the conical cultivation cylinders. The dry weight was quantified gravimetrically by filtering 100–300 mL of the cultivated microalgae through a 47 mm diameter Whatman GFC glass fiber filters. For evaluation of proteins and carbohydrates, 10–30 mL and for lipids 30–50 mL of the cultivated microalgae were filtered. All samples were evaluated in quadruplicate. Microalgae cells were broken by homogenizing them for 10 min by ultrason (Branson sonifier B-12, energy of 80 W) (Molkenboer 1964). Proteins were extracted with NaOH 0.1 N (Lopez-Elias et al. 1999), according to the Lowry method (1951). The Dubois et al. method (1956) was used for carbohydrate extraction. Lipids were extracted with a mixture of methanol chloroform and water (Bligh and Dye 1959). Gross energy (GE) content was measured by combustion in a bomb calorimeter (Parr Instrument Company, Moline, IL, USA) using benzoic acid as a standard.

Larval development was evaluated by daily microscopic observation. The total length of the larvae at each developmental stage was measured under a binocular microscope from the tip of the rostrum to the end of the tail (total length = TL), and at the end of the experiment, percentage survival was quantified. The biochemical composition of the shrimp larvae was also determined at the end of each experiment by the same methods used for microalgae. To evaluate the effect of the diet treatments on proteins, carbohydrates and lipids, as well as on the growth survival and biochemical composition of the larvae, a one-way ANOVA and Tukey test for post comparison were performed (Zar 1984).

Result

The size of Zoea I was equal among experimental treatments ($P < 0.05$) with an average value of 1.05 mm. For Zoea II, the largest sizes were recorded for larvae fed with the mixture *C. muelleri*:*I. galbana* (2.32 mm) and *T. tetrathele*:*C. muelleri* (2.30 mm) ($P <$ 0/05). For Zoea III, the largest sizes were recorded for larvae fed with the mixture *C. muelleri*:*I. galbana* (4.35 mm) and smallest sizes were observed on larvae fed with *I. galbana* (3.04 mm).

The proximal composition of the algae supplied in experimental treatments was different. Protein level was highest in *T. tetrathele* (42.6 %) and *C. muelleri* (39.8 %) as compared with *I. galbana* (30. 3 %). The carbohydrate and lipid levels and GE were higher in *T. tetrathele* treatment as compared with *I. galbana* and *C. muelleri* treatment.

The larvae fed with mixture *C. muelleri*:*T. tetrathele* (88.42 %) and *I. galbana*:*T. tetrathele* (84.50 %) had the highest survival vice versa larvae fed with *I. galbana* (72.33 %) had the lowest survival. The larvae fed with *I. galbana* (109.7 h) and *T. tetrathele* (107.5 h) had the highest duration of the terms of Z_I–Z_{III} and larvae fed with mixture *T. tetrathele* : *C. muelleri* (92.6 h) had the lowest duration of the terms of Z_I–Z_{III}.

The proximal composition of Zoea III was significantly different among the diets ($P < 0.05$). The larvae fed with mixture *I. galbana*:*T. tetrathele* (45.16 %) had the highest protein level as compared with other experimental treatments. The larvae fed with mixture *I. galbana* : *T. tetrathele* (31.81 %) and *T. tetrathele* (31.42 %) had the highest carbohydrate levels. Also highest lipid level and GE were in larvae fed with mixture *C. muelleri*:*I. galbana* (24.28 %), (4.14 kcal g^{-1}) and *T. tetrathele* (24.71 %) and (4.35 kcal g^{-1}), respectively.

Discussion

In our case, the higher growth was shown in larvae fed with mixture *C. muelleri*:*I. galbana* treatment. However, highest survival and lowest time for development were obtained in larvae fed with mixture *T. tetrathele*:*C. muelleri*. Pina et al. (2006) reported that *L. vannamei* larvae fed with *C. muelleri* and the mixed diets containing this species (*C. muelleri* + *Isochrysis* sp., *C. muelleri* + *Tetraselmis* sp.) had the highest

Table 1 Larval length (mm) in Zoea I, II and III stages, survival and duration of the terms of ZI to ZIII (h) of *Litopenaeus vannamei* fed with monospecific and mixture monoalgae

Diet	Survival (%)	Duration of the terms of Z_I–Z_{III} (h)	Total length(mm)		
			Z_I	Z_{II}	Z_{III}
Chaetoceros muelleri	79.21 ± 9.5^{b}	102.4 ± 2.7^{b}	1.05 ± 0.03	2.21 ± 0.07^{b}	3.12 ± 0.11^{d}
Isochrysis galbana	72.33 ± 7.6^{c}	109.7 ± 3.1^{a}	1.06 ± 0.05	2.25 ± 0.08^{ab}	3.04 ± 0.12^{e}
Tetraselmis tetrathele	76.14 ± 10.3^{bc}	107.5 ± 2.4^{a}	1.05 ± 0.05	2.20 ± 0.10^{b}	3.08 ± 0.10^{de}
C. muelleri and *I. galbana*	80.64 ± 7.5^{b}	100.2 ± 2.8^{b}	1.05 ± 0.04	2.32 ± 0.08^{a}	4.35 ± 0.11^{a}
I. galbana and *T. tetrathele*	84.50 ± 8.2^{ab}	102.5 ± 3.2^{b}	1.04 ± 0.04	2.27 ± 0.06^{ab}	3.65 ± 0.10^{c}
T. tetrathele and *C. muelleri*	88.42 ± 7.1^{a}	92.6 ± 1.5^{c}	1.05 ± 0.05	2.30 ± 0.09^{a}	4.11 ± 0.12^{b}

Different superscript letters indicate significant differences between the treatments

survival in Z_{II} (77 %), (68 %), (67 %) and Z_{III} (66 %), (61 %), (58 %), while the highest mortality was shown in *Tetraselmis* sp. treatment. D'Souza and Loneragan (1999) reported that survival of larvae fed *C. muelleri* or the mixed diet was always higher than that of larvae fed *Tetraselmis suecica*, *Isochrysis* sp. or *Dunaliella tertiolecta* alone. Furthermore, the development of larvae fed *C. muelleri* or the mixed diet was always at least as fast as those fed *T. suecica*, and always faster than that of larvae fed the other diets. In this study, the average growth of *L. vannamei* larvae from Zoea I to III was different in size range described by Treece and Yates (1990) and Rodriguez et al. (2012). Rodriguez et al. (2012) reported that the survival was similar between the monospecific treatments (*Chaetoceros* and *Isochrysis*) and the mixed treatment (Table 1).

Several factors contribute to the nutritional value of a microalga—including its size and shape, and digestibility as related to cell wall structure and composition (as mentioned above), as well as biochemical composition (e.g., accumulation compounds, enzymes and toxins) and specific requirements of the target animal (Catarina Guedes and Xavier Malcata 2012). D'Souza and Loneragan (1999) reported that differences in rates of development and dry weight of larvae were most likely related to the biochemical composition of the diets, while differences in survival may have been related to the spawning from which the larvae came. Crocos and Coman (1997) have shown that when the diet of *Penaeus semisulcatus* broodstock is constant, the survival of the PZI stage varies with such factors as age of the broodstock and season of spawning. The nutritional condition of the broodstock also appears to influence reproductive performance: the fatty acids, 20:5($n-3$) and 22:6($n-3$), in the eggs of *P. chinensis* were correlated with fecundity and the hatch rate of eggs (Xu et al. 1994).

Some authors have attributed the poor performance of prawn larvae to the large size of the algal cells in the diet (Tobias-Quinitio and Villegas 1982; Sanchez 1986). *C. muelleri* and *Isochrysis* are small and very similar in size (3–5 μm in length), while *Tetraselmis* sp. are larger in size (5–12 μm in length). In the current study, the larvae fed both the small algae (*C. muelleri*) and the large algae (*Tetraselmis* sp.) displayed high survival and development. Furthermore, the natural diet of *L. vannamei* PZI larvae includes algae as large as 1 mm length and algae >20 μm long are often ingested (Preston et al. 1992). It is, therefore, unlikely that the differences in larval growth performance between algal diets were due to the differences in size of the algal cells alone. *Tetraselmis* is considered a good food source for several species of penaeid shrimp (Liao et al. 1993; Loya-Javellana 1989). In particular, *Tetraselmis* has been shown to rank higher than *Isochrysis* for the protozoea stages of *P. monodon*, *P. semisulcatus* and *M. japonicas* (D'Souza and Loneragan 1999), but it is clearly inadequate as food for the zoea stages of *L. vannamei* (Pina et al. 2006). This was confirmed in a later experiment by Robles-Barraza (2003), who fed a different batch of larvae with *C. muelleri* until the stages of PZ II and PZ III, and with *Tetraselmis* sp. during the successive 5 and 4 days, respectively. Most literature points out that mixtures of microalgae perform better than monospecific diets for filter feeders (Treece 1984; Brown et al. 1989; Smith et al. 1992) (Table 2).

Dietary protein is a source of essential amino acids and energy for prawn larvae. D'Souza and Kelly (2000) reported that it is unlikely that the amino acid content of the diets was responsible for the large differences in development and dry weight of the larvae. Lipids are energy source and a component of the membrane structures of prawn larvae; in the form of steroids, they are precursors to vitamins, bile acids and hormones, including ecdysone, the vital molting hormone of crustaceans (D'Souza and Kelly 2000). Carbohydrates are not a dietary requirement for crustaceans, but they can be a cheaper source of energy than protein and lipid in

Table 2 Average percentage and standard deviation (s.d.) of the energy, protein, carbohydrate and lipid composition of monoalgae

Algal species	Protein (%)	Lipid (%)	Carbohydrate (%)	Energy (Kcal g^{-1})
Chaetoceros muelleri	39.8 ± 2.4^{b}	16.3 ± 1.8^{b}	7.5 ± 0.86^{c}	2.21 ± 0.36^{b}
Isochrysis galbana	30.3 ± 3.7^{c}	17.4 ± 2.3^{b}	14.2 ± 1.2^{b}	2.18 ± 0.71^{b}
Tetraselmis tetrathele	42.6 ± 2.5^{a}	21.3 ± 1.5^{a}	17.4 ± 2.3^{a}	3.07 ± 0.24^{a}

Different superscript letters indicate significant differences between the treatments

Table 3 Average percentage and standard deviation (s.d.) of the energy, protein, carbohydrate and lipid composition of *L. vannamei* larvae fed with monospecific diets (*Chaetoceros muelleri, Isochrysis galbana* and *Tetraselmis tetratheie*) and the mixture

Algal species	Proteins (%)	Carbohydrates (%)	Lipids (%)	Energy (Kcal g^{-1})
Chaetoceros muelleri	41.82 ± 3.19^{c}	28.77 ± 1.31^{c}	15.87 ± 1.12^{e}	2.20 ± 0.17^{c}
Isochrysis galbana	42.61 ± 2.71^{bc}	30.37 ± 1.54^{b}	18.54 ± 1.74^{d}	2.73 ± 0.24^{c}
Tetraselmis tetrathele	43.22 ± 3.46^{b}	31.42 ± 2.41^{a}	24.71 ± 1.62^{a}	4.35 ± 0.57^{a}
C. muelleri and *I. galbana*	43.34 ± 2.35^{b}	30.18 ± 1.20^{b}	24.28 ± 0.85^{a}	4.14 ± 0.31^{a}
I. galbana and *T. tetrathele*	45.16 ± 2.43^{a}	31.81 ± 1.63^{a}	20.15 ± 1.25^{c}	3.75 ± 0.40^{b}
T. tetrathele and *C. muelleri*	43.54 ± 2.50^{b}	28.50 ± 2.19^{c}	21.35 ± 2.05^{b}	3.84 ± 0.38^{b}

Different superscript letters indicate significant differences between the treatments

artificial diets (D'Abramo and Conklin 1995). Several studies have indicated that, in the late-logarithmic growth phase, microalgae contain typically 30–40 % (w/w) protein, 10–20 % (w/w) lipids and 5–15 % (w/w) carbohydrates (Brown et al. 1997; Renaud et al. 1999). When cultured through the stationary phase, the proximate composition of microalgae may significantly change; nitrate limitation leads carbohydrate levels to double at the expense of protein (Brown et al. 1993; Harrison et al. 1990). Hence, a strong correlation exists between composition of microalgae and their measurable nutritional value—even though diets containing high levels of carbohydrates have been reported to produce the best growth of juvenile oysters (Enright et al. 1986) and larval scallops (Whyte et al. 1989). Conversely, high dietary protein provides maximum growth for juvenile mussels (Kreeger and Langdon 1993) and oysters (Knuckey et al. 2002).

In this study, *T. tetrathele* had the highest percentage protein and lipid. Similar values were found by Lopez-Elias et al. (1999) for the same two species: percentage of protein was 39.3 % for Chaetoceros and 30.4 % for Isochrysis; carbohydrates and lipids' recorded values were 9.1 and 12.8 % for Isochrysis and 5.4 and 8.3 % for *Chaetoceros*. Pina et al. (2006) reported that *C. muelleri* had the highest percentage lipid content (25.81 %), arachidonic (5.27 Ag mg^{-1}) and eicosapentaenoic acids (9.03 Ag mg^{-1}), while *Isochrysis* sp. had the highest percentage protein (60.93 %) and docosahexaenoic acids (0.014 Ag mg^{-1}), also *Tetraselmis* sp. was richer in carbohydrates (24.75 %). Rodriguez et al. (2012) reported that although *Isochrysis* sp. had the lowest ratio of proteins, its carbohydrate and lipid ratio was high, and it also had the highest percentage of total polyunsaturated fatty acids. The larvae fed with mixture *I. galbana*:*T. tetratheie* had the highest protein and carbohydrate levels as compared with other experimental treatments (Table 3).

D'Souza and Loneragan (1999) reported that *C. muelleri* had the highest lipid level when compared with other treatments, while mixture of *C. muelleri* + *T. suecica* had the highest Polysaccharide level when compared with other treatments. Rodriguez et al. (2012) reported that protein and lipid levels of Zoea III were not significantly different among the diets (*Chaetoceros*, *Isochrysis* and mixture), in spite of the lipid level being significantly higher in Diet III (*Chaetoceros* + *Isochrysis*). The mixed diet was more complete with regard to the major constituents and fatty acid profile, with survival and growth comparable to monospecific diets (Pina et al. 2006). A mixture of *C. muelleri* and *T. suecica* might also provide a diet encouraging high survival and fast development in penaeid prawn (Penaeus spp.) larvae because of the presence of 18:2($n-6$), 18:3($n-3$) and 22:6($n-3$) in the *T. suecica* cells (D'Souza and Loneragan 1999). Based on literature and review, the algae C. *muelleri* show considerable amount of 20:4n − 6 and the EPA (20:5$n-3$). Puello Cruz et al. (2009) stated that, monospecific diets may cause nutritional deficiencies, because of the inadequate content of one or more essential nutrients. To reduce this risk, several authors have suggested the use of mixed diets, because their combined nutrient contents are more likely to meet the nutritional requirements of the target species (Vidhya et al. 2014). The results of this research indicate that mixtures of microalgae perform better than monospecific diets for *L. vannamei*. In future studies, we can analyze the profile of fatty acids in algae and also examine the effects of *C. muelleri*, *I. galbana* and *T. tetratheie* diets supplemented with microencapsulated PUFAs poly-unsaturated fatty acids, HUFAs highly unsaturated fatty acids on growth, survival and fatty acid composition of penaeid larvae.

Acknowledgments The authors wish to thank people in Urmia Lake research center (Urmia, Iran) and the Gomishan shrimp production hatchery (Golestan-Iran) for their invaluable assistance.

References

Aguirre-Hinojosa E, Lopez-Torres M, Garza-Aguirre MC (1999) Cultivos larvarios de camarones peneidos. pp 67–104. In: Martinez-Cordova LR (ed) Cultivo de camarones peneidos. Edit. AGT, Mexico, 283 p

Apt KE, Behrens PW (1999) Commercial developments in microalgal biotechnology. J Phycol 35:215–226

Bligh EG, Dye WJ (1959) A rapid method of total lipid extraction and purification. Can J Biochem Physiol 37:911–917

Boeing P (2005) Larval feed alternatives. Marine Aquafauna Inc. Internet: www.Aquafauna.com/larvalfeedalt.htm

Brown MR, Jeffrey SW, Garland CD (1989) Nutritional aspects of microalgae used in mariculture. Rep CSIRO Mar Lab 205:1–44

Brown MR, Dunstan GA, Jeffrey SW, Volkman JK, Barrett SM, Leroi JM (1993) The influence of irradiance on the biochemical composition of the *prymnesiophyte Isochrysis* sp. (clone T-ISO). J Phycol 29:601–612

Brown MR, Jeffrey SW, Volkman JK, Dunstan GA (1997) Nutritional properties of microalgae for mariculture. Aquaculture 151:315–331

Brown MR, Skabo S, Wilkinson B (1998) The enrichment and retention of ascorbic acid in rotifers fed with microalgal diets. Aquac Nutr 4:151–156

Catarina Guedes A, Xavier Malcata F (2012) Nutritional value and uses of microalgae in aquaculture. Dr. Muchlisin Z (ed) ISBN: 978-953-307-974-5, InTech, pp 59–79

Crocos PJ, Coman GJ (1997) Temporal and age variability in the reproductive performance of *Penaeus semisulcatus* brood stock optimizing brood stock selection. Aquaculture 155:55–67

D'Abramo LR, Conklin DE (1995) New developments in the understanding of the nutrition of penaeid and caridean species of shrimp. In: Browdy CL, Hopkins JS (eds) Swimming through troubled water. In: Proceedings of the special session on shrimp farming. Aquaculture 1995, February 1–4, 1995, San Diego, CA. World Aquaculture Society, Baton Rouge, LA, USA, pp 95–107

D'Souza FML, Kelly GJ (2000) Effects of a diet of a nitrogenlimited algae (*Tetraselmis suecica*) on growth, survival and biochemical composition of tiger prawn (*Penaeus semisulcatus*) larvae. Aquaculture 181:311–329

D'Souza FM, Loneragan NR (1999) Effects of monospecific and mixed-algae diets on survival, development and fatty acid composition of penaeid prawn (*Penaeus* sp.) larvae. Mar Biol 133:621–633

Dubois M, Gilles KA, Hamilton JK, Rebers PA, Smith F (1956) Colorimetric method for determination of sugars and related substances. Anal Chem 28:350–356

Enright CT, Newkirk GF, Craigie JS, Castell JD (1986) Growth of juvenile *Ostrea edulis* L. fed *Chaetoceros gracilis* Schütt of varied chemical composition. J Exp Mar Biol Ecol 96:15–26

Guillard RRL (1975) Culture of phytoplankton for feeding marine invertebrates. In: Smith WL, Chanley MH (eds) Culture of marine invertebrate animals. Plenum Press, New York, pp 26–60

Harrison PJ, Thompson PA, Calderwood GS (1990) Effects of nutrient and light limitation on the biochemical composition of phytoplankton. J Appl Phycol 2:45–56

Kent M, Browdy CL, Leffler JW (2011) Consumption and digestion of suspended microbes by juvenile Pacific white shrimp *Litopenaeus vannamei*. Aquaculture. doi:10.1016/j.aquaculture.2011.06.048

Knuckey RM, Brown MR, Barrett SM, Hallegraeff GM (2002) Isolation of new nanoplanktonic diatom strains and their evaluation as diets for juvenile Pacific oysters (*Crassostrea gigas*). Aquaculture 211:253–274

Kreeger DA, Langdon CJ (1993) Effect of dietary protein content on growth of juvenile mussels, *Mytilus trossulus* (Gould 1850). Biol Bull 185:123–139

Liao IC, Su HM, Lin JH (1993) Larval foods for penaeid prawns. In: Mc Vey JP (ed) Handbook of mariculture, crustacean aquaculture, vol 1. CRC Press, Boca Raton, pp 43–69

Lopez-Elias JA, Encinas A, Valenzuela C, Valdes J, Hoyos F (1999) Produccion anual de *Isochrysis* sp. y *Chaetoceros muelleri* Lemmerman en un centro acuicola en Bahia Kino, Sonora. Oceanide 14(1):59–65

Lopez-Elias JA, Voltolina D, Cordero-Esquivel B, Nieves-Soto M (2003) Produccion comercial de larvas de camaron y microalgas en cuatro estados de la Republica. Mex Biotecnia 5(1):42–51

Lopez-Elias JA, Enriquez-Ocana F, Pablos-Mitre MN, Huerta-Aldaz N, Leal S, Miranda-Baeza A, Nieves-Soto M, Vasquez-salgado I (2008) Growth and biomass production of *Chaetoceros muelleri* in mass outdoor cultures: effect of the hour of the inoculation, size of the inoculum and culture medium. Rev Invest Mar 29(2):171–177

Lowry OH, Rosebrough J, Far AL, Randall J (1951) Protein measurement with the Folin phenol reagent. J Biol Chem 193:265–275

Loya-Javellana G (1989) Ingestion saturation and growth responses of *Penaeus monodon* larvae to food density. Aquaculture 81:329–336

Marinez-Cordova LR, Pena-Messina E (2005) Biotic communities and feeding habits of *Litopenaeus vannamei* (Boone 1931) and *Litopenaeus stylirostris* (Stimpson 1974) in monoculture and polyculture semi-intensive ponds. Aquac Res 36:1075–1084

Merchie G, Lavens P, Dhert P, Dehasque M, Nelis H, de Leenheer A, Sorgeloos P (1995) Variation of ascorbic acid content in different live food organisms. Aquaculture 134:325–337

Merchie G, Lavens P, Sorgeloos P (1997) Optimization of dietary vitamin C in fish and crustacean larvae: a review. Aquaculture 155:165–181

Molkenboer JA (1964) De Sonifier, een universeel ultrasonoor apporaat. Chem Weekbl 25(60):333–341

Nelson JR, Guarda S, Sowell LE, Hefferman PB (1992) Evaluation of microalgal clones of mass culture in a subtropical green house bivalve hatchery: growth rate and biochemical composition at 30 °C. Aquaculture 106:357–377

Nichols PD, Holdsworth DG, Volkman JK, Daintith M, Allanson S (1989) High incorporation of essential fatty acids by the rotifer *Brachionus plicatilis* fed on the prymnesiophyte alga *Pavlova lutheri*. Aust J Mar Fresh Res 40:645–655

Okauchi M, Hirano Y (1986) Nutritional value of *Tetraselmis tetrathele* for larvae of *Penaeus japonicus*. Bull Natl Res Ins Aquac 9:29–33

Okauchi M, Kawamuru K, Mizukami Y (1997) Nutritive value of Tahiti Isochrysis. *Isochrysis* sp. for larval greasy back shrimp *Metapenaeus emsis*. Bull Natl Res Ins Aquat 26:1–11

Pina P, Voltolina D, Nieves M, Robles M (2006) Survival, development and growth of the Pacific White Shrimp *Litopenaeus vannamei* protozoea larvae, fed with mono algal and mixed diets. Aquaculture 253:523–530

Preston NP, Burford MA, Coman FE, Rothlisberg PC (1992) Natural diet of larval *Penaeus merguiensis* (Decapoda:Penaeidae) survival. Mar Biol 113:181–191

Puello-Cruz AC, Mezo-Villalobos S, Gonzalez-Rodriguez B, Voltolina D (2009) Culture of the calanoid copepod *Pseudodiaptomus euryhalinus* (Johnson 1939) with different microalgal diets. Aquaculture 290:317–319

Renaud SM, Thinh LV, Parry DL (1999) The gross composition and fatty acid composition of 18 species of tropical Australia microalgae for possible use in mariculture. Aquaculture 170:147–159

Rhodes EW, Landers WC (1973) Growth of oyster larva *Grassorted virginica* of various sizes in different concentrations of the Chrysophyte *Isochrysis galbana*. Proc Natl Shellfish Assoc 63:53–59

Richmond A (2004) Handbook of microalgal culture: biotechnology and applied phycology. Blackwell, USA 566

Robles-Barraza M (2003) Sobrevivencia, crecimiento y desarrollo de larvas zoea de *Litopenaeus vannamei* Boone alimentadas con dietas monoalgales y mixtas. BS thesis. Universidad Autonoma de Sinaloa, Facultad de Ciencias del Mar. Mazatlan, Mexico, p 34

Rodriguez A, Le Vay L, Mourente G, Jones DA (1994) Biochemical composition and digestive enzyme activity in larvae and postlarvae of *Penaeus japonicus* during herbivorous and carnivorous feeding. Mar Biol 118:45–51

Rodriguez EO, Lopez-Elias JA, Aguirre-Hinojosa E, Garza-Aguirre MDC, Constantino-Franco F, Miranda-Baeza A, Nieves-Soto M (2012) Evaluation of the nutritional quality of *Chaetoceros muelleri* schutt (Chaetocerotales: Chaetocerotaceae) and *Isochrysis* sp. (Isochrysidales: Isochrysidaceae) grown outdoors for the larval development of *Litopenaeus vannamei* (Boone, 1931) (Decapoda: Penaeidae). Arch Biol Sci Belgrade 64(3):963–970

Sanchez R (1986) Rearing of mysis stage of *Penaeus vannamei* fed cultured algae of three species. Aquaculture 58:139–144

Scott AP, Middleton C (1979) Unicellular algae as a food for turbot (*Scophthalmus maximus* L.) larvae—the importance of dietary long chain polyunsaturated fatty acids. Aquaculture 18:227–240

Smith LL, Biedenbach JM, Lawrence AL (1992) Penaeid larviculture: Galveston method. In: Fast AE, Lester LJ (eds) Marine shrimp culture, principles and practices. Elsevier Science, Amsterdam, pp 171–191

Thompson PA, Guo M, Harrison PJ (1993) The influence of irradiance on the biochemical composition of three phytoplankton species and their nutritional value for larvae of the Pacific oyster (*Crassostrea gigas*). Mar Biol 117:259–268

Tobias-Quinitio E, Villegas CT (1982) Growth, survival and macronutrient composition of *Penaeus monodon* Fabricius larvae fed with *Chaetoceros calcitrans* and *Tetraselmis chuii*. Aquaculture 29:253–260

Treece GD (1984) Larval rearing technology. In: Chamberlain GM, Haby MG, Miget RJ (eds) Texas shrimp farming manual: an update on current technology. Texas Agricultural Extension Service, Texas A and M University, Corpus Christi, TX, USA, pp 43–64

Treece G, Yates M (1990) Laboratory manual for the culture of penaeid shrimp larvae. Marine Advisory Service Sea Grant Collage Program Texas A and M University, 95 pp

Trujillo-Valle L, Voltolina D (1994) Cultivo de microalgas para la aquacultura: Serie Cientifica, U.A.B.C. 2(1): 73–85

Vidhya K, Uthayakumar V, Muthukumar S, Munirasu S, Ramasubramanian V (2014) The effects of mixed algal diets on population growth, egg productivity and nutritional profiles in cyclopoid copepods (*Thermocyclops hyalinus* and *Mesocyclops aspericornis*). J Basic Appl Zool 67:58–65

Voltolina D, Lopez-Elias JA (2002) Cultivos de apoyo para la acuacultura: Tendencias e innovaciones, pp 23–41. In: Martinez-Cordova LR (ed) Camaronicultura. Avances y Tendencias, AGT Editor, S.A., México. 167 pp

Walne PR (1966) Experiments in the large scale culture of the larvae of *Ostrea edulis*. Fish Invest Lond 25:1–53

Whyte JNC, Bourne N, Hodgson CA (1989) Influence of algal diets on biochemical composition and energy reserves in *Patinopecten yessoensis*. Jay. larvae. Aquaculture 78:333–347

Wood BJB (1974) Fatty acids and saponifiable lipids. In: Stewart WDP (ed) Algal physiology and biochemistry. Blackwell Scientific Publications, Hoboken, pp 236–265

Xu XL, Ji WJ, Castell JD, O'Dor RK (1994) Essential fatty acid requirement of the Chinese prawn *Penaeus chinensis*. Aquaculture 127:29–40

Zar J (1984) Biostatistical analysis. Prentice-Hall Inc, USA 718

4

Efficacy of phytoremediation technology in decontaminating the toxic effluent released during recovery of metals from polymetallic sea nodules

Huma Vaseem · T. K. Banerjee

Abstract Efficacy of phytoremediation using two macrophytes *Azolla pinnata* and *Lemna minor* in decontaminating the toxic effluent released during recovery of metals from polymetallic sea nodules was analysed by applying fish bioassay. The economically important fish, *L. rohita*, was exposed to both, the *Azolla*-phytoremediated effluent (APE) and *Lemna*-phytoremediated effluent (LPE) for assessment of metal bioaccumulation (Fe, Mn, Zn, Cu, Pb, Cr and Cd) and alterations in biochemical (proteins, lipids, glycogen, cholesterol, AST (aspartate amino transferase), ALT (alanine amino transferase) and ALP (alkaline phosphatase) composition of various tissues. Accumulation of metals (e.g. Mn, Zn, Cu and Fe) decreased in most of the tissues exposed to both the phytoremediated effluents perhaps due to decontamination of metals by the two macrophytes. The significantly recovered concentrations of different biomolecules included glycogen, lipids, cholesterol and proteins. The activities of three marker enzymes (AST, ALT and ALP) in phytoremediated effluent-exposed fish also decreased due to lowering of the toxicity of the decontaminated effluents achieved by phytoremediation. The improvement in different biomolecules and reduction in metal concentration in the fish tissues were better in APE exposed fish. However, their concentrations in both the phytoremediated effluent-exposed fish failed to reach the levels of control fish. This study points towards the efficacy of phytoremediation in detoxification of metal-contaminated effluents often released following industrial activities.

Keywords Phytoremediation · Polymetallic sea nodules · Metal bioaccumulation · Biomolecules

Introduction

Various domestic, industrial and mining toxic effluents are detoxified by several chemical, physical and mechanical processes for decontamination prior to their release into the aquatic ecosystem. Such types of detoxifying processes are not enough to contain pollution. Often due to cost factor there is a tendency to evade the mandatory treatment of the effluent for detoxification by the industrial houses. Hence biologists intervened and tried to evolve cheap and effective methods of phytoremediation for decontamination of the toxic effluents polluted with heavy metals (Mishra et al. 2008; Rai and Tripathi 2009; Vaseem and Banerjee 2012; Bharti and Banerjee 2012).

H. Vaseem · T. K. Banerjee (✉)
Department of Zoology, Banaras Hindu University (BHU), Varanasi 221005, India
e-mail: humavaseembhu@gmail.com

H. Vaseem
e-mail: beauty786waseem@gmail.com

Due to acute scarcity of certain economically useful metals like Cu, Cr, Co, Mn and Ni, etc., metallurgist at National Metallurgical Laboratory, Jamshedpur, India, are trying to develop indigenous technique to extract metals from polymetallic sea nodules which are abundantly found at sea bed. For the metal extraction purpose the nodules are processed through many metallurgical methods including reduction, ammonia leaching, solvent extraction, electro winning and smelting (Kumar et al. 1990; Jana et al. 1999; Agarwal and Goodrich 2008; Biswas et al. 2009). Such metallurgical processes also generate large amount of toxic effluent (PMN effluent). The toxicity of this waste water is due to contamination of toxic metals which cause extensive damage to the aquatic fauna (Vaseem and Banerjee 2013a, b). To lower the toxicity load of metals in this effluent, Vaseem and Banerjee (2012) successfully demonstrated decontamination of the effluent by phytoremediation technology.

There are numbers of reports related to application of phytoremediation in decontaminating the effluent released from various industries (Rai, 2008, 2010; Rai and Tripathi 2009). However, there are no bioassay data using fish to analyse the improvement in effluent quality following phytoremediation. Hence to bridge this gap the decontaminated effluents were subjected to fish bioassay analysis using a major carp (*Labeo rohita)* as bioindicator. Fish have widely been used as able bioindicator of variously contaminated waters hence selected (Gupta et al. 2009; Uysal et al. 2009; Maceda-Veiga et al. 2012). Six organs systems (e.g. muscles, liver, gills, kidney, brain and skin) of the fish exposed to the *Azolla*-phytoremediated effluent (APE) and *Lemna*-phytoremediated effluent (LPE) were examined for metal (Mn, Cu, Zn, Fe, Pb, Cr and Ni) bioaccumulation and alterations in biochemical (proteins, lipids, glycogen, DNA, RNA, cholesterol, AST (aspartate amino transferase), ALT (alanine amino transferase) and ALP (alkaline phosphatase) composition.

Materials and methods

Analysis of PMN effluent and its phytoremediation

Physicochemical parameters and concentrations of different metals (Fe, Mn, Zn, Cu, Pb, Cr and Cd) in PMN effluent were examined following the standard methods for evaluation of water and waste water (APHA-AWWA-WPCF, 1998) and atomic absorption spectrophotometer (AAS), respectively (Perkin-Elmer Model 2380, Inc., Norwalk, CT, USA). The details of the physicochemical characteristics of raw PMN effluent have been given by Vaseem and Banerjee (2012) and their finding shows that this effluent has been extremely toxic.

When phytoremediated separately with two different macrophytes (*Azolla pinnata* and *Lemna minor*) for 7 days, the effluents were partially but significantly detoxified due to lowered concentration of different metal species (Vaseem and Banerjee 2012).

Experimental fish

Healthy specimens of the Indian major carp *L. rohita* were collected from the hatchery situated in Banaras Hindu University, Varanasi, India. The fish were acclimated to the laboratory conditions for 1 month in plastic tanks equipped with a continuous supply of well-aerated and dechlorinated water (room temperature: 24 ± 2 °C) and under natural photoperiod. During this period, fish were fed ad libitum with commercial fish pellets. The water was renewed after every 24 h with routine cleaning of the tanks. The examined water quality parameters were: dissolved oxygen (7.0–7.5 mg/l), pH (7.1–7.4), conductivity (125–130 μS/cm), alkalinity (35–43 mg/l as $CaCO_3$) and total hardness (39–50 mg/l as $CaCO_3$).

Experimental design

Acclimated fish were divided into four groups, each of 20 fish (weight of 28–30 g and length of 11–12 cm): one group exposed to 50 L of tap water; the second group to 50 L of the raw PMN effluent, third to 50 L of APE and the fourth to 50 L of LPE in a 80-L capacity plastic tubs for maximum period of 20 days (beyond which fish failed to survive) with regular renewal of water after every 5 days of interval (semi-static bioassay). Fish were regularly fed during experiment. Eight fish from each group were cold anesthetised and killed after 20 days of exposure. Three replicates of each experimental set were prepared. Entire brain, liver, kidney, gills, small fragments of muscle and skin were dissected out and subjected for metal bioaccumulation (Mn, Cu, Zn, Fe, Pb, Cr and Ni) and biochemical (total proteins, total lipids, glycogen, cholesterol, AST, ALT and ALP) analyses.

Metal analyses

For analyses of metal accumulation six tissues (muscle, gills, liver, kidneys, brain and skin) of control, raw effluent-exposed and APE and LPE-exposed fish were collected in petri-dish and dried in an oven at 120 °C till there was no weight loss. Subsequently, the tissue samples were transferred to digestion flasks containing an acid mixture (nitric acid and perchloric acid (4:1 v/v). The digestion flasks were further heated on a hot plate at 120 °C till the tissues were dissolved. Double distilled water was added to the digestion samples to make their volume to 25 ml. Concentration of metals in the samples were analysed using atomic absorption spectrophotometer (AAS, Perkin Elmer Model 2380, Inc., Norwalk, CT, USA) (detection limits of AAS for metals in mg/kg have been given in Table 1). Concentration of Ni was found to be below the detectable level in all the tissues of APE and LPE exposed fish hence it was not shown in the Table 2.

Biochemical analysis

Total proteins Following Lowry's (Lowry et al. 1951) method total protein level (mg/g) was estimated using bovine serum albumin as standard.

Total lipids Following the method of Folch et al. (1957) total lipids from the tissue samples were extracted in chloroform methanol mixture (2:1)

Transaminase activities For measurement of the AST and ALT activities (μmole pyruvate formed/mg/h) method of Reitman and Franckel (1957) was used by using pyruvate as a standard.

Alkaline phosphatase activity The method of Bergmeyer (1956) was applied to determine the ALP activity (μmole PNP formed/mg/h) where sodium *p*-nitrophenylphosphate was used as substrate.

Cholesterol Zlatkis et al. (1953) method was used for estimation of cholesterol concentration (mg/g).

Glucogen Glycogen content (mg/g) was analysed following the method of Caroll et al. (1956) using anthrone reagent. Glucose was used as a standard.

Statistical analysis

For statistical analyses one-way analysis of variance (ANOVA) ($p < 0.05$) was performed followed by Duncan's multiple range test (DMRT). Values in the tables and figures are given in mean ± SD. In tables and figures alphabets denote the result of DMRT. Different alphabets show significant difference ($p < 0.05$) in the different values of control fish, raw effluent exposed fish, APE exposed fish and LPE exposed fish.

Results

Metal accumulation in the fish exposed to phytoremediated effluents

PMN effluent was found to be highly toxic due to its contaminated condition (BOD: 182 ± 3 mg/l, pH: 5.2, sodium: 130.26 ± 1.96 mg/l, potassium: 3.42 ± 0.07 mg/l, sulphate: 2300 ± 4.515 mg/l, carbonate: 296 ± 2.2 mg/l, Mn 4.9 mg/l, Cu 1.432 mg/l, Zn 0.816 mg/l, Fe 0.762 mg/l, Pb 0.655 mg/l, Cr 0.07 mg/l, Cd 0.018 mg/l) (Vaseem and Banerjee (2012). Following exposure of the fish to this toxic effluent the concentration of all the studied metals increased greatly in all the six tissues. Decontamination of effluent with both

Table 1 Detection limits for metals in tissues

Metals	Zn	Mn	Cu	Fe	Pb	Cr	Ni
Detection limit (mg/kg)	0.1	0.1	0.01	0.1	0.01	0.02	0.01

Table 2 Metal concentrations (mg/kg) in different tissues of the fish exposed to raw effluent, APE and LPE as well as control (wild) fish

FAO[m] (1983)	Mn (nl[n])	Cu (10)	Zn (50)	Fe (5.6)	Pb (nl)	Cr (1)
Skin						
Cont[o]	0.52 ± 0.22[a]	7.53 ± 3.25[a]	35.75 ± 0.77[a]	21.73 ± 4.38[a]	bdl	bdl
REE[p]	16.54 ± 0.27[b]	37.99 ± 0.62[c]	50.35 ± 0.82[c]	135.00 ± 2.20[c]	bdl	7.99 ± 0.13
APEE[q]	1.23 ± 0.62[a]	9.19 ± 0.41[a]	37.01 ± 1.85[a]	45.51 ± 1.25[b]	bdl	bdl
LPEE[r]	1.45 ± 0.81[a]	13.14 ± 0.69[b]	41.62 ± 1.24[b]	44.74 ± 1.1[b]	bdl	bdl
Muscles						
Cont	0.56 ± 0.01[a]	1.8 ± 0.03[a]	5.65 ± 0.09[a]	38.45 ± 0.83[a]	bdl	bdl
REE	10.35 ± 0.22[b]	11.34 ± 0.51[c]	81.14 ± 3.69[c]	61.08 ± 2.78[c]	2.1 ± 0.03	11.83 ± 0.25[a]
APEE	0.77 ± 0.20[a]	2.97 ± 1.30[ab]	6.09 ± 0.56[a]	47.29 ± 0.31[b]	bdl	bdl
LPEE	0.82 ± 0.46[a]	3.8 ± 1.07[b]	20.12 ± 1.46[b]	47.5 ± 0.59[b]	bdl	0.64 ± 0.38[b]
Gills						
Cont	0.61 ± 0.20[a]	3.86 ± 0.17[a]	5.69 ± 2.52[a]	14.87 ± 2.79[a]	bdl	bdl
REE	48.67 ± 0.79[b]	60.55 ± 0.99[b]	98.03 ± 1.60[c]	129.36 ± 2.11[c]	0.8 ± 0.02	12.70 ± 0.20[a]
APEE	0.86 ± 0.08[a]	4.58 ± 1.09[a]	7.28 ± 0.69[b]	22.66 ± 1.16[b]	bdl	bdl
LPEE	0.92 ± 0.17[a]	5.14 ± 1.25[a]	17.36 ± 1.5[b]	20.77 ± 1.02[b]	bdl	1.54 ± 0.516[b]
Liver						
Cont	0.33 ± 0.23[a]	28.19 ± 1.06[a]	38.79 ± 0.63[a]	131.05 ± 5.96[a]	bdl	bdl
REE	91.92 ± 4.18[b]	287.36 ± 6.21[b]	175.24 ± 3.79[b]	264.16 ± 5.71[c]	2.91 ± 0.15[a]	52.77 ± 5.37[b]
APEE	0.59 ± 0.17[a]	28.9 ± 1.85[a]	39.34 ± 0.99[a]	143.82 ± 1.47[b]	bdl	bdl
LPEE	0.45 ± 0.13[a]	35.12 ± 1.69[a]	54.66 ± 2.85[c]	142.3 ± 1.15[b]	0.01 ± 0.0[b]	6.56 ± 1.16[a]
Kidney						
Cont	2.28 ± 0.04[a]	14.95 ± 0.76[a]	22.69 ± 0.49[a]	36.23 ± 0.59[a]	bdl	bdl
REE	74.56 ± 1.22[b]	44.36 ± 2.26[c]	137.92 ± 7.03[c]	254.93 ± 8.27[c]	1.1 ± 0.02	40.32 ± 10.13[b]
APEE	3.42 ± 1.26[a]	15.91 ± 1.25[a]	23.91 ± 1.61[a]	36.43 ± 1.3[b]	bdl	bdl
LPEE	2.30 ± 0.65[a]	21.26 ± 1.76[b]	34.29 ± 2.33[b]	36.563 ± 1.98[b]	bdl	1.83 ± 2.261[a]
Brain						
Cont	0.92 ± 0.02[a]	3.50 ± 0.16[a]	4.38 ± 0.22[a]	11.24 ± 1.89[a]	bdl	bdl
REE	20.87 ± 0.45[b]	19.33 ± 0.31[c]	43.53 ± 0.71[c]	71.66 ± 1.17[b]	bdl	19.66 ± 0.42
APEE	0.85 ± 0.06[a]	4.19 ± 0.42[ab]	5.72 ± 1.24[a]	13.10 ± 2.12[a]	bdl	bdl
LPEE	0.87 ± 0.14[a]	5.52 ± 1.64[b]	13.97 ± 2.21[b]	11.97 ± 1.38[a]	bdl	Bdl

Values are given in mean ± SD

Different alphabets (a, b and c) show significant difference ($p < 0.05$) among different means

[m] FAO (1983): Food and Agricultural Organization

[n] No limit has been suggested by FAO (1983)

[o] Control wild fish

[p] Raw effluent-exposed fish

[q] *Azolla* phytoremediated effluent-exposed fish

[r] *Lemna* phytoremediated effluent-exposed fish

the macrophytes caused significant depletion in concentration of these metals. The concentration (mg/l) of metals in the *Azolla*-phytoremediated effluent was: Mn 0.19 ± 0.001, Cu 0.039 ± 0.000, Zn 0.019 ± 0.000, Fe 0.228 ± 0.08, Pb 0.026 ± 0.000, Cr 0.005 ± 0.000. Similarly *Lemna*-phytoremediated effluent had: Mn 0.301 ± 0.003, Cu 0.204 ± 0.001, Zn 0.306 ± 0.002, Fe 0.2 ± 0.001, Pb 0.1 ± 0.009 and Cr 0.026 ± 0.00. The percentage of decontamination of the metals reported after phytoremediation was in the order: 96 % of Mn, 97 % of Cu, 70 % of Fe, 96 % of Pb, 93 % of Cr and 78 % of Cd by *Azolla* and 94 % of Mn, 86 % of Cu, 62 % of Zn, 74 % of Fe, 84 % of Pb, 63 % of Cr and 78 % of Cd by *Lemna* phytoremediation (Vaseem and Banerjee 2012).

In the fish exposed to both of these phytoremediated effluents, Mn concentration in all the tissues decreased significantly (Table 2). Similarly the concentration of Cu, Zn and Fe also decreased significantly in most of the tissues of exposed fish to both the phytoremediated effluents (Table 2). Cr was observed in all the raw effluent-exposed tissues and also in muscles, gills, liver and kidney of fish exposed to LPE (Table 2).

Biochemical investigation in the fish exposed to phytoremediated effluents

Following exposure to raw effluent the concentration of glycogen in the different tissues of fish decreased significantly (in muscle from 24 ± 1.09 to bdl, in liver from 142 ± 3.07 to 6 ± 0.13, in gills from 11.1 ± 0.18 to bdl, in kidney from 1 ± 0.01 to bdl) (Table 3). However, its concentration increased in the brain from 26.25 ± 1.30 to 46 ± 2.09. In the fish exposed to APE better recovery of glycogen was observed (muscles: 9.32 ± 0.58, liver: 24.69 ± 0.65 and gills: 4.98 ± 0.10) (Table 3). Similarly in LPE-exposed fish improvement in the concentration of glycogen was observed in these tissues (5.98 ± 0.37 in muscles, 15.01 ± 0.40 in liver, 1.62 ± 0.03 in gill (Table 3). In the kidney and skin the amount of the glycogen in the fish exposed to all the three types of effluents were below the detection limits. However, in the brain of both the phytoremediated effluent-exposed fish the elevated amount of glycogen continued (40.21 ± 2.23 in APE and 33.23 ± 1.85 in LPE (Table 3).

In comparison to wild control fish the loss of lipids (mg/g) following raw effluent exposure was highly significant in all the tissues (in muscles from 7.13 ± 0.32 to 0.2 ± 0.01, in liver from 8.99 ± 0.19 to

Table 3 Glycogen, lipid and cholesterol concentration (mg/g) in different tissues of the fish exposed to raw effluent, APE and LPE as well as control wild fish ($n = 3$)

	Control wild fish	REE fish[p]	APEE fish[q]	LPEE fish[r]
Glycogen				
Muscle	24 ± 1.09^{a}	Bdl	9.32 ± 0.58^{b}	5.98 ± 0.37^{c}
Liver	142 ± 3.07^{a}	6 ± 0.13^{d}	24.69 ± 0.65^{b}	15.01 ± 0.40^{c}
Gill	11.1 ± 0.18^{a}	Bdl	4.98 ± 0.10^{b}	1.62 ± 0.032^{c}
Brain	26.25 ± 1.30^{d}	46 ± 2.09^{a}	40.21 ± 2.24^{b}	33.23 ± 1.9^{c}
Kidney	1 ± 0.02	Bdl	Bdl	Bdl
Skin	Bdl	Bdl	Bdl	Bdl
Lipid				
Muscle	7.13 ± 0.32^{a}	0.2 ± 0.009^{d}	1.01 ± 0.05^{b}	0.60 ± 0.04^{c}
Liver	8.99 ± 0.19^{a}	1.42 ± 0.03^{d}	4.63 ± 0.10^{b}	1.23 ± 0.03^{c}
Gill	6.99 ± 0.11^{a}	Bdl	2.69 ± 0.04^{b}	2.38 ± 0.70^{b}
Brain	10 ± 0.45^{a}	2 ± 0.09^{d}	$4.86\ 0.22^{b}$	1.32 ± 0.07^{c}
Kidney	7.91 ± 0.13^{a}	0.98 ± 0.016^{d}	$5.11\ 0.08^{b}$	1.99 ± 0.040^{a}
Skin	0.32 ± 0.007	Bdl	Bdl	Bdl
Cholesterol				
Muscle	3.91 ± 0.18^{a}	1.71 ± 0.08^{b}	1.67 ± 0.03^{b}	1.79 ± 0.04^{b}
Liver	6.33 ± 0.14^{a}	2.4 ± 0.05^{d}	4.42 ± 0.12^{b}	3.16 ± 0.08^{c}
Gill	4.8 ± 0.08^{a}	1.23 ± 0.02^{c}	2.91 ± 0.06^{b}	2.62 ± 0.28^{b}
Brain	7.8 ± 0.35^{a}	Bdl	1.68 ± 0.09^{b}	Bdl
Kidney	4.1 ± 0.03^{a}	0.82 ± 0.01^{d}	2.02 ± 0.04^{b}	1.81 ± 0.04^{c}
Skin	Bdl	Bdl	Bdl	Bdl

Values are given in mean ± SD

Different alphabets (a, b, c and d) show significant difference ($p < 0.05$) among different means

Bdl below detectable level

[p] Raw effluent exposed

[q] *Azolla*-phytoremediated effluent exposed

[r] *Lemna*-phytoremediated effluent exposed

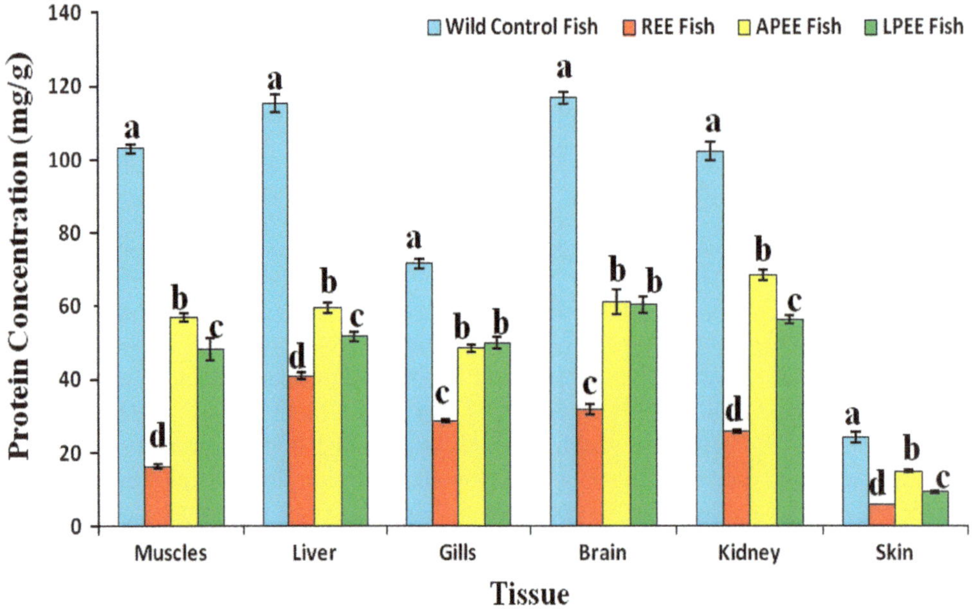

Fig. 1 Protein concentration in different tissues of the fish exposed to raw effluent, APE and LPE as well as control wild fish. *REE* raw effluent exposed, *APEE Azolla*-phytoremediated effluent exposed, *LPEE Lemna*-phytoremediated effluent exposed Different alphabets (*a*, *b* and *c*) show significant difference ($p < 0.05$) among different means

1.42 ± 0.03, in gills from 6.99 ± 0.11 to bdl, in brain from 10 ± 0.45 to 2 ± 0.09, in kidney from 7.91 ± 0.13 to 0.98 ± 0.02 and in skin from 0.32 ± 0.01 to bdl). There was marked improvement of this biomolecule in all tissues of the fish exposed to APE and the concentration of the lipids was 1.01 ± 0.05 in muscles, 4.63 ± 0.10 in liver, 2.69 ± 0.04 in gills, 4.86 ± 0.22 in brain, 5.11 ± 0.08 in kidney (Table 3). The recovery of lipids in the tissues of fish exposed to LPE was relatively less and their concentrations were 0.603 ± 0.04 in muscles, 1.23 ± 0.03 in liver, 2.38 ± 0.71 in gills, 1.32 ± 0.07 in brain, 1.99 ± 0.04 in kidney (Table 3). In the skin, however, there was complete loss of lipids in the fish exposed to all the three effluent types.

Following exposure to raw effluent the amount of cholesterol (mg/g) in different tissues of fish decreased substantially (from 3.91 ± 0.18 to 1.71 ± 0.08 in muscles, from 6.33 ± 0.14 to 2.4 ± 0.05 in liver, from 4.8 ± 0.08 to 1.23 ± 0.02 in gills, from 7.8 ± 0.35 to bdl in brain, from 4.1 ± 0.03 to 0.821 ± 0.01 in kidney) (Table 3). In the skin, however, the cholesterol level was beyond detectable limits in control as well as in all three effluent-exposed fish groups. The recovery of cholesterol in the muscular tissue in both the phytoremediated effluent-exposed fish was statistically insignificant in relation to raw effluent exposed fish. In liver, gills, brain and kidney significant recovery was noticed (4.42 ± 0.15 in liver, 2.91 ± 0.06 in gills, 1.68 ± 0.09 in brain, 2.02 ± 0.04 in kidney of APE exposed fish and 3.16 ± 0.08 in liver, 2.62 ± 0.278 in gills, bdl in brain, 1.81 ± 0.04 in kidney of LPE exposed fish (Table 3).

When exposed to raw effluent the amount of proteins present in the fish depleted extensively in all the tissues in comparison to the control ones (Fig. 1) (in muscles from 103.07 ± 1.19 to 16.28 ± 0.61, in liver from 115.35 ± 2.45 to 40.99 ± 0.89, in gills from 71.56 ± 1.14 to 28.61 ± 0.47, in brain from 116.97 ± 1.63 to 31.88 ± 1.45, in kidney from 102.36 ± 2.48 to 25.81 ± 0.42 and in skin from 24.13 ± 1.34 to 6 ± 0.00). Due to substantial decontamination of APE, its toxic activity also was lowered as manifested in the recovery of concentration of proteins in various tissues (in muscles 57.02 ± 1.14, in liver 59.48 ± 1.57, in gills 48.56 ± 0.97, in brain 61.23 ± 3.41, in kidney 68.49 ± 1.37, in skin 15 ± .40) from those of raw effluent-exposed fish (Fig. 1). Similarly concentration of proteins in the LPE-exposed fish also improved (48.34 ± 3.02 in muscles, 51.64 ± 1.37 in liver, 49.99 ± 1.71 in gills, and 60.6 ± 2.23 in brain, 56.42 ± 1.19 in kidney and 9.29 ± 0.25 in skin (Fig. 1).

PMN effluent exposure caused significant decrease in the AST activity (μmole pyruvate formed/mg/h) in all the tissues excepting skin (7.87 ± 0.13 to 1.02 ± 0.014 in muscle, 18.62 ± 0.95 to 3.43 ± 0.15 in liver, 5 ± 0.08 to 1.16 ± 0.01 in gills, 8.99 ± 0.07 to 2.08 ± 0.02 in brain, 15.91 ± 0.39 to 1.37 ± 0.02 in kidney

Table 4 AST, ALT and ALP activities in different tissues of the fish exposed to raw effluent, APE and LPE as well as control wild fish ($n = 3$)

	Control wild fish	REE fish[p]	APEE fish[q]	LPEE fish[r]
AST				
Muscle	7.87 ± 0.13[a]	1.02 ± 0.01[b]	4.56 ± 0.28[c]	4.22 ± 0.53[c]
Liver	18.62 ± 0.95[a]	3.43 ± 0.16[d]	10.24 ± 0.27[b]	7.16 ± 0.19[c]
Gill	5 ± 0.08[a]	1.37 ± 0.02[c]	2.1 ± 0.04[b]	2.03 ± 0.17[b]
Brain	8.99 ± 0.07[a]	2.08 ± 0.02[d]	6.18 ± 0.34[b]	2.22 ± 0.12[c]
Kidney	15.91 ± 0.34[a]	1.16 ± 0.02[d]	10.98 ± 0.22[b]	6.72 ± 0.13[c]
ALT				
Muscle	11.54 ± 0.17[a]	2.53 ± 0.04[b]	5.63 ± 0.35[c]	5.24 ± 0.23[c]
Liver	21.98 ± 0.47[a]	3.81 ± 0.08[d]	16.28 ± 0.43[b]	10.93 ± 0.29[c]
Gill	11.72 ± 0.20[a]	0.82 ± 0.01[d]	8.67 ± 0.17[b]	3.34 ± 0.07[c]
Brain	12.69 ± 0.11[a]	7.91 ± 0.06[c]	9.89 ± 0.55[b]	10.28 ± 0.57[b]
Kidney	4.81 ± 0.10[a]	0.13 ± 0.002[d]	1.68 ± 0.03[b]	0.61 ± 0.01[c]
ALP				
Muscle	0.53 ± 0[a]	0.20 ± 0.003[c]	0.39 ± 0.02[b]	0.36 ± 0.02[b]
Liver	1.29 ± 0.02[a]	0.56 ± 0.01[d]	0.96 ± 0.02[b]	0.73 ± 0.02[c]
Gill	0.71 ± 0.01[a]	0.26 ± 0.004[c]	0.53 ± 0.01[b]	0.53 ± 0.05[b]
Brain	0.97 ± 0.01[a]	0.24 ± 0.01[d]	0.76 ± 0.04[b]	0.54 ± 0.03[c]
Kidney	0.89 ± 0.01[a]	0.60 ± 0.009[c]	0.73 ± 0.01[b]	0.71 ± 0.01[b]

Values are given in mean ± SD

Different alphabets (a, b, c and d) show significant difference ($p < 0.05$) among different means

[p] Raw effluent exposed

[q] *Azolla*-phytoremediated effluent exposed

[r] *Lemna*-phytoremediated effluent exposed

(Table 4). Significant recovery in the enzyme activity in all the five tissues was noticed following exposure to both the decontaminated effluents. Following phytoremediation activity of the enzyme improved in liver (10.24 ± 0.27 in APE and 7.16 ± 0.19 in LPE), kidneys (10.98 ± 0.22 in APE and 6.72 ± 0.13 in LPE) and brain (6.18 ± 0.34 in APE and 2.22 ± 0.12 in LPE), in the muscles (4.56 ± 0.28 in APE and 4.22 ± 0.54 in LPE) and gills (2.1 ± 0.04 in APE and 2.03 ± 0.18 in LPE) (Table 4). In the skin, however, activity of this enzyme was neither demonstrated in the control nor in the phytoremediated effluent-exposed fish tissues. However, the activity of the enzyme in none of the tissues of the phytoremediated effluent-exposed fish reached the level of untreated control fish. The activity of the enzyme in the LPE showed similar pattern of fluctuation in different tissues even though its activity was comparatively less than APE exposed fish.

Like AST the activity (μmole pyruvate formed/mg/h) of the ALT also decreased in the raw effluent treated fish tissues (11.54 ± 0.173 to 2.53 ± 0.03 in muscles, 21.98 ± 0.47 to 3.81 ± 0.08 in liver, 11.72 ± 0.20 to 0.823 ± 0.01 in gills, 12.69 ± 0.11 to 7.91 ± 0.06 in brain and 4.81 ± 020 to 0.132 ± 0.002 in kidney) (Table 4). In both the phytoremediated exposed fish the enzyme activity recovered significantly in the muscle (5.63 ± 0.35 in APE and 5.23 ± 0.23 in LPE), liver (16.28 ± 0.43 in APE and 10.93 ± 0.29 in LPE), gills (8.67 ± 0.17 in APE and 3.338 ± 0.07 in LPE) and kidneys (1.68 ± 0.03 in APE and 0.612 ± 0.01 in LPE) (Table 4). Recovery in the ALT activity in the brain (9.89 ± 0.55 in APE and 10.28 ± 0.57 in LPE) of the phytoremediated effluent-exposed fish was not so remarkable. Excepting brain the efficacy of *Azolla* decontamination was better in rest of the four tissues as illustrated by improved recovery in this enzyme activity in these tissues.

The decrease in the ALP activity (μmole PNP formed/mg/h) in the different tissues was also extensive following exposure of the wild control fish to raw PMN effluent (0.53 ± 0.00 to 0.196 ± 0.003 in muscles, 1.294 ± 0.02 to 0.564 ± 0.01 in liver, 0.712 ± 0.01 to 0.263 ± 0.004 in gills, 0.970 ± 0.01 to 0.239 ± 0.01 in brain and 0.892 ± 0.01 to 0.599 ± 0.009 in kidney) (Table 4). In this case, also the recovery in the enzyme activity was more following APE exposure (0.389 ± 0.02, 0.961 ± 0.02, 0.532 ± 0.01, 0.765 ± 0.04 and 0.728 ± 0.01 in muscles, liver, gills, brain and kidney, respectively) than LPE exposure (0.364 ± 0.02 in muscles, 0.729 ± 0.01 in liver, 0.53 ± 0.04 in gills, 0.539 ± 0.03 and 0.713 ± 0.01 in kidney) (Table 4). However, the difference was more pronounced in the liver and brain. In the skin the enzyme activity could not be identified in any of the control as well as experimental tissues.

Discussion

Exposure of fish to the raw PMN effluent caused accumulation of significant amount of different metals (Fe, Zn, Cu, Cr, Mn and Pb) in various tissues. Bioaccumulation of metals in the different tissues of the fish exposed to various waste water has also been reported by Vinodhini and Narayanan (2008), Dube et al. (2005) and Zyadah and Abdel-Bakey (2000). Recently, Vaseem and Banerjee (2012) successfully decontaminated this PMN effluent by phytoremediation technique. Decontamination of metals by phytoremediation is extensively been used (Rai 2008, 2010). To validate and quantify the degree of detoxification by phytoremediation, both the phytoremediated effluents (LPE and APE) were subjected to bioassay analyses using fish tissues as bioindicator.

Concentration of several metals in most of the tissues of both the phytoremediated effluent-exposed fish reached quite nearer to the levels of the control (wild) fish. Accumulation of Mn, Cu, Zn and Fe in all the tissues of both the phytoremediated effluent-exposed fish decreased significantly and their concentration reached nearer to the levels of the untreated control fish (Table 2). Significantly lowered levels of all these metals in both of the phytoremediated effluents exposed fish confirm the effectiveness of phytoremediation in decontamination of Mn, Fe, Cu and Zn by *Azolla pinnata* and *Lemna minor*. In APE-exposed fish accumulation of Cr was not detected in any tissues of the fish. It might be due to almost complete decontamination of this metal from the effluent by phytoremediation. But in case of LPE-exposed fish, significant accumulation of Cr was noticed in most of the tissues (Table 2). This indicates incomplete detoxification of Cr from the effluent by phytoremediation. Analysis of the Table 2 also indicates that the level of metal accumulation in the tissues of LPE exposed fish was higher than those of APE exposed ones. This also points towards the better efficacy of phytoremediation by *Azolla pinnata*. Accumulation of many metals in the tissues of phytoremediated effluent-exposed fish failed to reach the levels of untreated control (Table 2); however, the accumulation levels reached to the safe limits suggested by FAO.

Although the metal load decreased in the phytoremediated effluents, the toxic impact of the decontaminated effluents in the fish tissues continued. This was manifested by the decreased concentration of various biomolecules of the tissues in the fish exposed to the phytoremediated effluents and the reasons for continued toxicity of phytoremediated effluents might also be due to: (1) presence of the non-metallic toxicants and (2) bioconcentration of the metals by the fish tissues due to prolonged exposure even though under greatly lowered metallic stress. Hence fish raised in the phytoremediated effluent still may not be safe for consumption

Significant decrease in the concentration of different biomolecules was noticed in the raw PMN effluent-exposed fish. Exposure of fish to phytoremediated effluent caused significant improvement in the various biomolecules. Improvement in the concentration of protein, glycogen, lipids, DNA and RNA in the fish *Heteropneustes fossilis* exposed to phytoremediated coal mine effluent has also been reported by Bharti and Banerjee (2013). In the present study the concentration of the glycogen in the phytoremediated effluent-exposed fish tissues increased greatly from that of raw effluent exposed ones. However, the concentration of this carbohydrate moiety continued to be below the levels of untreated (wild) fish tissues. This might perhaps be due to continued utilisation of the glycogen molecules to meet additional energy requirement for combating the toxicity of the existing levels of the metals (Table 3) (Singh and Banerjee 2009 and Lin et al. 2011).

Due to decrease in severity of toxicity following phytoremediation by the macrophytes the total lipid (Table 3) and cholesterol concentrations (Table 3) increased significantly in comparison to the raw effluent exposed tissues. This might perhaps be due to lowered requirement of energy with consequent decreased utilisation of these biofuels. However, the lipid concentration did not reach the level of control ones. This might perhaps be due to incomplete detoxification of the effluent by both the macrophytes which caused continuation of additional requirement of energy of the fish even though in smaller amount.

The increase in protein level in several organ systems of both the phytoremediated effluent-exposed fish also suggests partial improvement in the water quality (Fig. 1) in which the fish were retained.

The activities of the three enzymes AST, ALT and ALP also decreased substantially in different tissues of the raw effluent exposed fish. In the phytoremediated effluent-exposed ones, the activities of these enzymes in many tissues increased (Table 4) substantially suggesting the improvement in the quality of the effluent due to bioremediation. However, due to incomplete detoxification by the macrophytes the activities of these enzymes continued to remain below the levels of untreated fish.

Conclusion

Due to significant depletion in metal concentration from the APE and LPE, the toxic stress of the effluent decreased causing significant reduction in metal accumulation and recovery of the various macromolecules in the different tissue systems of the fish. Recovery in amounts of different biomolecules and depletion in metal load were higher in APE exposed fish. However, due to incomplete phytoremediation of effluents, the metals continued to accumulate in the fish tissues. This was reflected in the quantitative alteration of various biomolecules and bioaccumulation of the toxic metals even in the tissues of APE and LPE-exposed fish.

Acknowledgments Huma Vaseem is highly thankful to University Grant Commission for providing Senior Research Fellowship. Financial support from Department of Ocean Development, Ministry of Earth Science, Government of India, is also greatly acknowledged.

References

Agarwal HP, Goodrich JD (2008) Extraction of Copper, Nickel, and Cobalt from Indian ocean polymetallic nodules. Can J Chem Eng 81:303–306

APHA/AWWA/WPCF. Standard methods for the examination of water and wastewaters (1998) Twentieth ed. American Public Health Association, American Water Work Association and Water Pollution Control Federation, Washington (DC), New York

Bharti S, Banerjee TK (2012) Phytoremediation of the coalmine effluent. Ecotoxicol Environ Saf 81:36–42

Bharti S, Banerjee TK (2013) Analysis of fish tissues under stress of coal mine effluent. Department of Zoology, Banaras Hindu University, Thesis, pp 53–73

Biswas A, Chakraborti N, Sen PK (2009) Multiobjective optimization of manganese recovery from sea nodules using genetic algorithms. Mater Manuf Process 24:22–30

Caroll WV, Longly RW, Roe JH (1956) The determination of glycogen in the liver and muscle by the use of anthrone reagent. J Biol Che 220:583–593

Dube MG, MacLatchyb DL, Kiefferb JD, Glozierc NE, Culpd JM, Casha KJ (2005) Effects of metal mining effluent on Atlantic salmon (*Salmo salar*) and slimy sculpin (*Cottus cognatus*): using artificial streams to assess existing effects and predict future consequences. Sci Total Environ 343:135–154

FAO (1983) Compilation of legal limits for hazardous substances in fish and fishery products. Food Agric Organ Fish Circ No 464

Folch JL, Sloane M, Stanley GH (1957) A simple method for isolation and purification of total lipids from animal tissues. J Biol Chem 226:497–5075

Gupta A, Rai D, Pandey R, Sharma B (2009) Analysis of some heavy metals in the riverine water, sediments and fish from river Ganges at Allahabad. Environ Monit Assess 157:449–458

Jana RK, Pandey BD, Premchand (1999) Ammoniacal leaching of roast reduced deep-sea Manganese nodules. Hydrometallurgy 53:45–56

Kumar V, Pandey BD, Akerkar DD (1990) Electrowinning of nickel in the processing of polymetallic sea nodules. Hydrometallurgy 4:189–201

Lin YS, Tsai SC, Lin HC, Hsiao CD, Wu SM (2011) Changes of glycogen metabolism in the gills and hepatic tissue of Tilapia (*Oreochromis mossambicus*) during short-term cd exposure. Comp Physiol Biochem 48:517–527

Lowry OH, Rosenbrough NG, Farr AL, Randall RG (1951) Protein measurements with folin phenol reagent. J Biol Chem 193:265–275

Maceda-Veiga A, Monroy M, De Sostoa A (2012) Metal bioaccumulation in the Mediterranean barbell (*Barbus meridionalis*) in a Mediterranean river receiving effluents from urban and industrial wastewater treatment plants. Ecotoxicol Environ Saf 76:93–101

Mishra VK, Upadhyaya AR, Pandey SK, Tripathi BD (2008) Heavy metal induced due to coal mining effluent on surrounding aquatic system and its management through naturally occurring aquatic macrophytes. Bioresour Technol 99:930–936

Rai PK (2008) Phytoremediation of Hg and Cd from industrial effluents using an aquatic free floating macrophyte *Azolla pinnata*. Int J Phytorem 10:430–439

Rai PK (2010) Phytoremediation of heavy metals in a tropical impoundment of industrial region. Environ Monit Assess 165:529–537

Rai PK, Tripathi BD (2009) Comparative assessment of *Azolla pinnata* and *Vallisneria spiralis* in Hg removal from G.B. Pant Sagar of Singrauli industrial region. India Environ Monit Assess 148:75–84

Singh AK, Banerjee TK (2009) A study on carbohydrate moieties of gills and air-breathing organs of the walking catfish *Clarias batrachus* (Linn.) following exposure to arsenic. Toxicol Environ Chem 91:43–52

Uysal K, Kose E, Bulbul M, Donmez M, Koyun M, Erdogan Y, Omeroglu A, Ozmal F (2009) The comparison of heavy metal accumulation ratios of some fish species in Enne Dame Lake (Kutahya/Turkey). Environ Monit Assess 157:355–362

Vaseem H, Banerjee TK (2012) Phytoremediation of the toxic effluent generated during recovery of precious metals from polymetallic sea nodules. Int J Phytorem 14:457–466

Vaseem H, Banerjee TK (2013a) Effect of effluent released during recovery of precious metals from polymetallic sea nodules on biochemical changes in freshwater fish *Labeo rohita*. Clean Soil Air Water 41:1–6

Vaseem H, Banerjee TK (2013b) Metal accumulation in fish *Labeo rohita* exposed to effluent generated during recovery of precious metals from polymetallic sea nodules. Int J Environ Sci Technol. doi:10.1007/s13762-013-0381-2

Vinodhini R, Narayanan M (2008) Bioaccumulation of heavy metals in organs of fresh water fish *Cyprinus carpio* (Common carp). Int J Environ Sci Tech 5:179–182

Zlatkis A, Zak B, Boyle AJ (1953) A new method for direct determination of serum cholesterol. J Lab Clin Med 41:481–492

Zyadah MA, Abdel-Bakey TC (2000) Toxicity and bioaccumulation of copper, zinc and cadmium in some aquatic organism. Bull Environ Contam Toxicol 64:740–747

5

Open pond cultures of indigenous algae grown on non-arable land in an arid desert using wastewater

Dominik Winckelmann · Franziska Bleeke · Bimiya Thomas · Clemens Elle · Gerd Klöck

Abstract The algae production on non-arable land in the al-Wusta region of the Sultanate of Oman was evaluated. Pre-cleaned production water (pPW) from the oil production was enriched with nutrients and used as growth medium. The indigenous isolate *Scenedesmus* sp. MKB was grown in open ponds under semi-continuous conditions. A productivity of 16.1 g/m^2/day in 2013 and 15.4 g/m^2/day in 2014 was reached during four different experiments during March and April of 2013 and 2014. Thereby, no influences of light intensity (maximum daily values in between 2400 and 3050 μmol photons/m^2/s) or temperature (maximum daily temperatures were between 29.9 and 41.9 °C) on productivity were observed (correlation coefficient below ±0.5). Weed algae, mainly Cyanobacteria and diatoms were detected microscopically in all cultures by the first to seventh day of growth. The increase in weed algae had a negative influence on the maximum photosynthetic efficiency (*Y*). *Y* decreased after a weed algae concentration of ca. 2 % (cell/cell) was reached.

Keywords *Scenedesmus* · Open pond · Arid environment · Produced water · Weed algae · PAM

Introduction

The possible use of non-arable land (Wagener 1983) would increase the overall land area available for biomass production. A high annual temperature regime, low cloud coverage, and a maximum light exposure, which are typical for semi-arid regions, together with an already existing infrastructure are said to be favorable factors to increase the probability of economical viable microalgal production processes (Grobbelaar 2009). Algae capable of growing in brackish and non-potable water can be used as production strain, and such a choice would result in a reduced water footprint and therefore might result in an increase in potable or agriculturally usable water (Guieysse et al. 2013). However, the use of brackish water might lead to an increase in salt concentration due to evaporation which may cause possible salt stress (Winckelmann et al. 2014).

One of the major issues with the growth of microalgae is the possibility of a sudden and inevitable culture collapse (Hamilton et al. 2014), which can be caused by different means beginning with grazers (Becker 1994; Tillmann 2004), contaminations and pathogens (Gachon et al. 2010), and climatological factors (Grobbelaar

D. Winckelmann · F. Bleeke
Jacobs-University Bremen, School of Engineering and Science, Campus Ring 1, 28759 Bremen, Germany

D. Winckelmann (✉) · F. Bleeke · B. Thomas · C. Elle · G. Klöck
University of Applied Sciences Bremen, Am Neustadtswall 30, 28199 Bremen, Germany
e-mail: d.winckelmann@jacobs-university.de

2010). Another reason for the loss of revenue are weed algae overgrowing the algae to be produced. Weed algae are indigenous algae more suitable for the conditions at the production site (Becker 1994). Depending on the final use of the algae biomass, even small amounts of foreign algae can render an algae culture unsuitable for further processing (Gilroy et al. 2000; Gantar and Svirčev 2008).

The use of indigenous species as production strains is said to reduce the risk of culture loss. Their origin at the same climate as the production site will reduce the risk of culture crash due to climatological and other abiotic factors, and it is assumed that they will be able to outcompete or at least suppress other indigenous algae (Mutanda et al. 2011; Odlare et al. 2011; Rawat et al. 2013).

The aim of this study was to test if indigenous algae grown in industrial waste water can be grown on non-arable land in an arid region.

The crude oil production in the al-Wusta region of the Sultanate Oman results in 800,000 m^3 production water per day (Breuer and Al-Asmi 2010). A fraction of approximately 100,000 m^3 is treated in a reed bed water treatment plant in Nimr. Uses for the resulting hydrocarbon-free brackish water (pPW) to increase revenue are sought for and the location of Nimr being a semi-arid region led to the consideration to establish an algae production facility.

In February 2014, an open pond culture of *Cyanobacterium aponinum* grown in Nimr was invaded by green algae. The algae used for this study, *Scenedesmus* Sp. (SC) strain MKB, was isolated from the mixed culture.

To evaluate the stability of the production process, SC was grown in 2013 and 2014. The productivity and the overall efficiency of photon capture (Grobbelaar 2009) were chosen as an indicator for the efficiency of the production process. The introduction of weed algae might lead to unwanted competition for nutrients and sunlight. It was monitored microscopically and using a multi-channel pulse-amplitude-modulation chlorophyll fluorometer [PAM; (Kolbowski and Schreiber 1995)].

Methods

Analytical methods

Temperature (°C), wind velocity (m/S), and light intensity (W/m^2) were measured by an adjacent weather station (data kindly provided by Bauer Nimr LLc). Light (µmol photons/m^2/s) was measured using a Mini Quantum Sensor MQS-B (Heinz Walz GmbH, Germany). A conversion factor from W/m^2 to µmol photons/m^2/s of 2.9 was calculated. The mass transfer coefficient for oxygen [$k_La(O_2)$] of the open ponds was determined using the dynamic method (Doran 1995). The $k_La(O_2)$ can be used to quantify the oxygen transfer from the liquid phase to the surrounding air. It is also used to calculate the $k_La(CO_2)$ using the equations from Talbot et al. (1990). Flow velocity inside an open pond was measured using a current meter (C2 with a 30 mm blade (gradient 0.1) and counter Z30 OTT Hydromet GmbH, Germany).

The biomass was measured by means of optical density at 660 nm (Lovibond MaxiDirect photometer, Tintometer GmbH, Germany). Culture composition was determined microscopically (using a Zeiss Axiostar Plus light microscope, Carl Zeiss GmbH, Germany) and a Thoma Neu chamber with a depth of 0.1 mm (Paul Marienfeld GmbH & Co. KG, Germany). For the determination of bio dry weight (dw), part of a culture was washed and filtered. The filter was dried at 105 °C until the weight was constant. Increase of salinity was measured by means of conductivity (Water quality meter U-52, Horiba Ltd., Japan). The photosynthetic yield (*Y*) and all photosynthetic parameters were measured using a phyto-PAM with phyto-EDF (Heinz Walz GmbH, Germany) as described previously (Winckelmann et al. 2014). Rapid light curves (Ralph and Gademann 2005), the illumination of a sample with increasing light intensities, were measured to estimate the photosynthetic performance. Thereby, light intensities from 0 to 197 µmol photons/m^2/s were used in nine steps lasting 1 min each.

Isolation and characterization

Scenedesmus Sp. MKB was isolated from *Cyanobacterium aponinum* cultures and characterized as described earlier (Winckelmann et al. 2014). It was identified as described earlier (Winckelmann et al. 2014) using primer pair ITS4 and ITS5 (Connell 2000) (Results: GenBank KM873329) and EukA and EukB without GC-clamp (Medlin et al. 1988) (results: GenBank KM873328).

Biomass growth for inoculation

The initial inoculum was cultured in Wuxal-media (Winckelmann et al. 2014) until a biomass of above 0.5 g dry weight/L was reached. The cultures were transferred to 1.5 L bottle reactors which were situated outside; the place selected was in full sunlight until midday and covered in shade afterward. The cultures were aerated with a rate of 30 L/h. For the first 4 days of outdoor cultivation, unsterilized bottled water with 0.05 % (v/v) Wuxal (Wuxal® Universaldünger liquid plant fertilizer, Wilhelm Haug GmbH & Co. KG, Germany) was used as growth media. The cultures were divided each time so the biomass reached a dry weight of above 0.5 g dry weight/L and diluted with unfiltered (2013) or filtered (2014) pPW. The experiment performed in April 2014 was inoculated with sedimented culture from the experiment conducted in March 2013.

Production site and composition of produced water

Algae ponds were constructed on site of the Bauer Nimr water treatment plant(Abed et al. 2014). The growth media used was made up of pPW (440–550 mg/L SO_4, N–NH_3 0.3–0.6 mg/L, N–NO_3 0.2–1.0 mg/L, P–PO_4 <0.2 mg/L, salinity 7–12 ppt; analysis kindly provided by Bauer Nimr LLc; values measured before the addition of fertilizer) and 0.5 % (v/v) Wuxal.

Pond construction

Open ponds (4 m long, 80 cm wide, and 38 cm deep) were constructed on a leveled area. Building bricks (39 cm long, 19 cm wide, and 19 cm deep) were used for the outer walls and 2 mm thick high density polyethylene (HDPE) sheets were used as liner. A centrifugal pump (2 in. diameter; Pedrollo, HFm 5AM) was used for mixing.

Biomass production in open ponds

Growth experiments were conducted with a volume of 600 L. The growth medium used was pPW, which was left standing for at least 2 days in a one cubic meter container, only the top 70 % was used as feed water and 0.05 % (v/v) Wuxal was added directly into the pond. The ponds were filled to a depth of 20 cm and inoculated with 18 L inoculum (biomass content of at least 1 g dry weight/L). The cultures were mixed 24 h per day if not stated otherwise.

Harvesting

Part of the culture was exchanged on a regular basis. Therefore, the culture was removed until the depth was 10 cm. Fresh pPW containing 0.05 % v/v fertilizer was added until a depth of 20 cm was reached, and on days without removal the evaporated water was replaced with fresh pPW.

Results

Environmental conditions

The light intensity during all experiments reached medium values over the course of the sunlight hours of a day, which were in between 975 and 1740 μmol photons/m^2/s and maximum values in between 2400 and 3050 μmol photons/m^2/s. The duration of sunlight was between 12 and 14 h per day. The maximum temperature was between 29.9 and 41.9 °C (Fig. 1). The monthly mean temperatures did not differ significantly between 2013 and 2014. The monthly mean and maximum light intensities were 770 and 3030 in March 2013 and 810 and 3010 in April 2013. These values were higher than the recorded values of 2014 which were 730

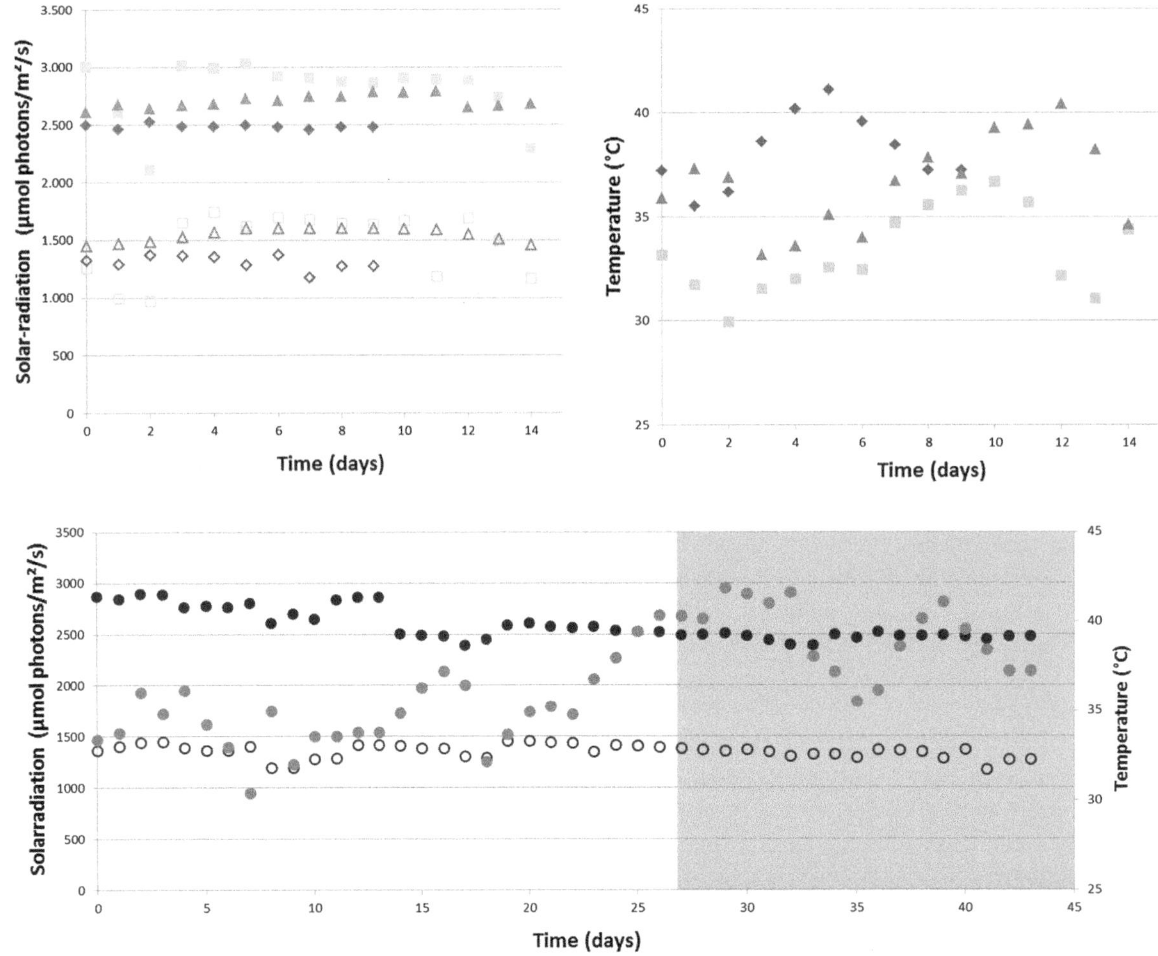

Fig. 1 Progression of light intensity and temperature. *Top row* shows three experiments (*square* March 2013; *triangle* April 2013; *diamond* April 2014) with durations of 14 days or less; *left* light intensity (*solid symbols* maximum intensity, *open symbols* medium intensity); *right* maximum temperature. *Lower row* shows the maximum light intensity (*black solid circle*), medium light intensity (*black open circle*), and maximum temperature (*gray-closed circle*) of a growth experiment lasting 43 days (March–April 2014). The area shaded in *gray* indicates a decreased water circulation

and 2890 in March and 710 and 2430 in April (all measurements in µmol photons/m^2/s). During both periods, days with dust or cloud coverage were observed (9 days in 2013 and 5 days in 2014). Heavy winds and dust led to the termination of the experiment in March 2013.

Open pond characteristics

To be able to assess CO_2 supply, oxygen degassing, and biomass sedimentation, the open pond characteristics were determined. The ponds showed a $k_La(O_2)$ value of 0.04/s (SD ±0.01), a $k_La(CO_2)$ of 0.033/s, and current velocities 5 cm above the ground were between 0.2 and 2.3 m/s (Fig. 2).

Culture composition

The cultures were examined microscopically to observe the influence of weed algae or grazers on the culture's stability. The culture's health was examined using the chlorophyll fluorescence as the parameter. We found that all cultures were mixed cultures and bacteria were found. During all experiments, three different groups of photosynthetic active organisms were found (Fig. 3) in different abundance (Fig. 4). In single cases during pump outage, the appearance of small dark brown flakes, with small air bubbles attached to it, on the surface

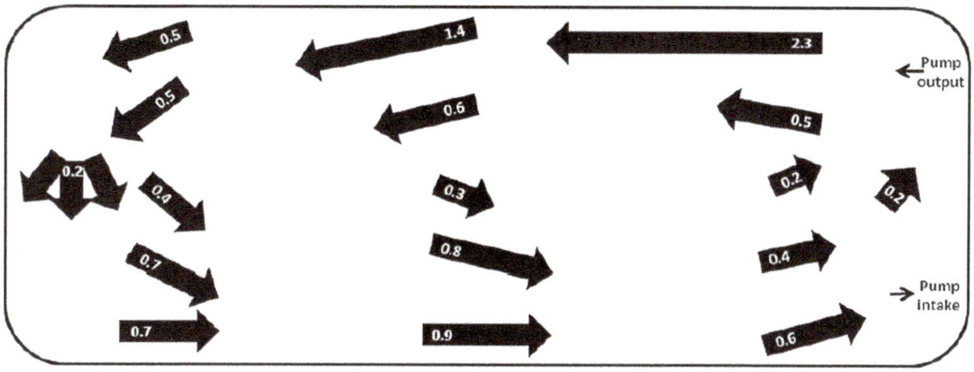

Fig. 2 Flow velocities and directions in *open* pond. *Unit* m/s. Measurement depth was 5 cm above the *bottom*. *Pond length* 4 m, *pond width* 0.8 m, *water level* 20 cm. Figure is not up to *scale*

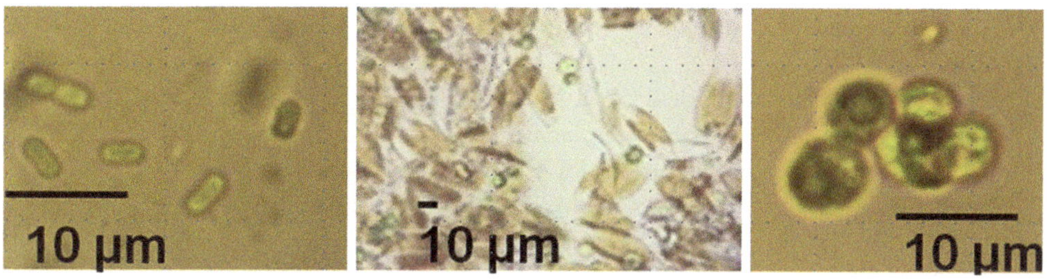

Fig. 3 Photosynthetically active organisms found in all growth experiments; from *left* to *right*: single-celled cyanobacteria, most likely *Cyanobacterium aponinum*; Diatoms; *Scenedesmus* spec

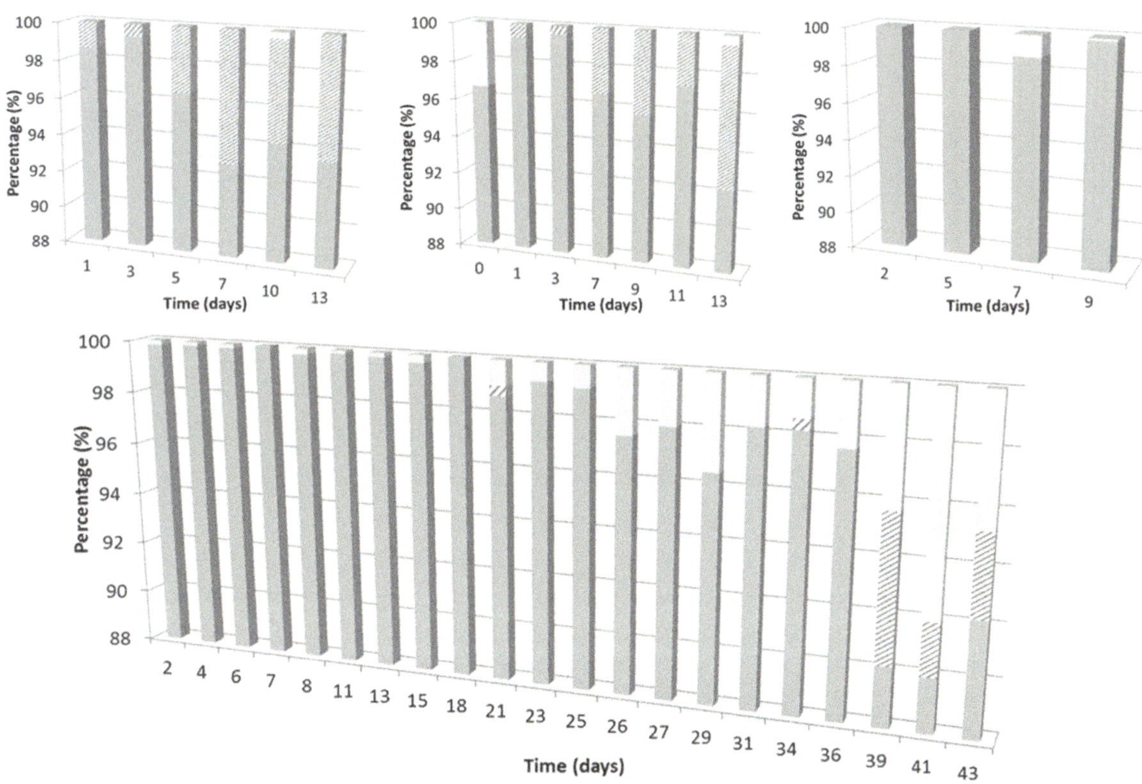

Fig. 4 Distribution of different photosynthetic active groups over time. First *row left* experiment March 2013; *middle* experiment April 2013; *right* experiment April 2014; *lower row* March–April 2014. *Graphs* show the microscopically determined share of *Scenedesmus* (*gray*, *bottom*), Diatoms (*stripes*, *middle*), and cyanobacteria (*white*, *top*)

of the culture was observed. The flakes and swabs from biofilms formed in pond regions with insufficient water movement were microscopically observed and showed a mixture of organisms and inorganic particles dominated by diatoms (Fig. 5). The amount of biofilm formed was not evaluated.

The maximum photosynthetic yield of all cultures increased above 0.7 during the first 2 days of cultivation, and it decreased right after a harvest was performed but recovered within hours. The non-photochemical quenching (NPQ) increased in the beginning above two. The overall photosynthetic characteristics in the morning, before dilution, did not differ significantly in between experiments (Table 1).

The experiment conducted in March 2013 showed signs of contamination from the early beginning. It was contaminated with diatoms and the contamination concentration breached the two percent threshold on day 5 and on the same day decreased *Y* below 0.7 and the maximum NPQ reached only 1.5 (Fig. 6). After the restart of the experiment in April 2014, the culture showed a high contamination with cyanobacteria on the day of inoculation; however, over the course of the cultivation nearly no cyanobacteria were detected until day 13. Although diatoms were detected on day 1 and after day 5, *Y* decreased below 0.7 and the maximum NPQ did only reach a maximum of around 1.5 after day 7.

The experiment conducted in April 2014 showed no sign of contamination in the beginning but a small amount of cyanobacteria was observed on day 7 and decreased until day 9. No change in *Y* or NPQ was observed.

The experiments conducted in March and April 2014 were contaminated with cyanobacteria and diatoms from the early beginning. While the diatoms were nearly undetectable using a microscope, the concentration of cyanobacteria breached the two percent threshold on day 26. The pump failed and had to be replaced with a smaller one, which only had 20 percent of performance rate on day 27 (part in Figs. 1, 3, 6, 7 lower row, which is underlayed gray). On day 36, the *Y* dropped permanently under 0.7 and the diatom concentration became detectable on day 39. The maximum NPQ was between 1.5 and 2 for most part of this experiment but dropped to values below 1 after day 36.

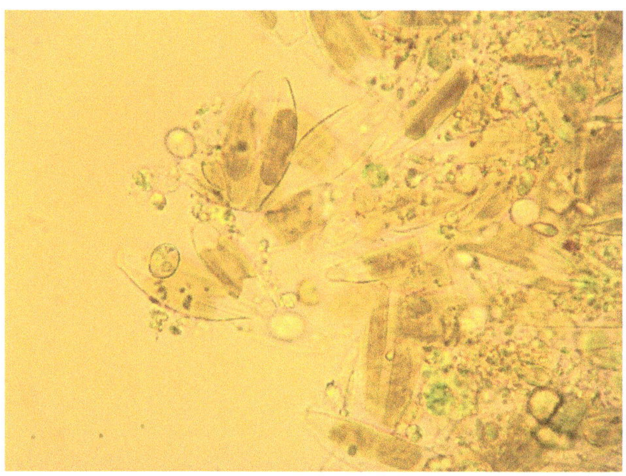

Fig. 5 Exemplary picture of community found in biofilm developed in regions with insufficient culture movement

Table 1 Overall photosynthetic performance as determined using photosynthesis-irradiance curves

	March 2013	April 2013	April 2014	March–April 2013—till day 27	March–April 2013—after day 27
Alpha (SD)	0.27 (0.02)	0.26 (0.02)	0.32 (0.07)	0.30 (0.01)	0.28 (0.03)
I_K (SD)	14 (3)	12 (5)	16 (2)	14 (2.6)	17 (5)
P_{max} (SD)	51 (12)	47 (18)	52 (13)	45 (10)	61 (19)

All data were measured in the morning between 9 and 10 a.m. before dilution of the culture. Data are not normalized against biomass

Alpha maximum light use coefficient for photosystem II, I_K transition point between light limited and light saturated photosynthesis, P_{max} maximum photosynthetic rates

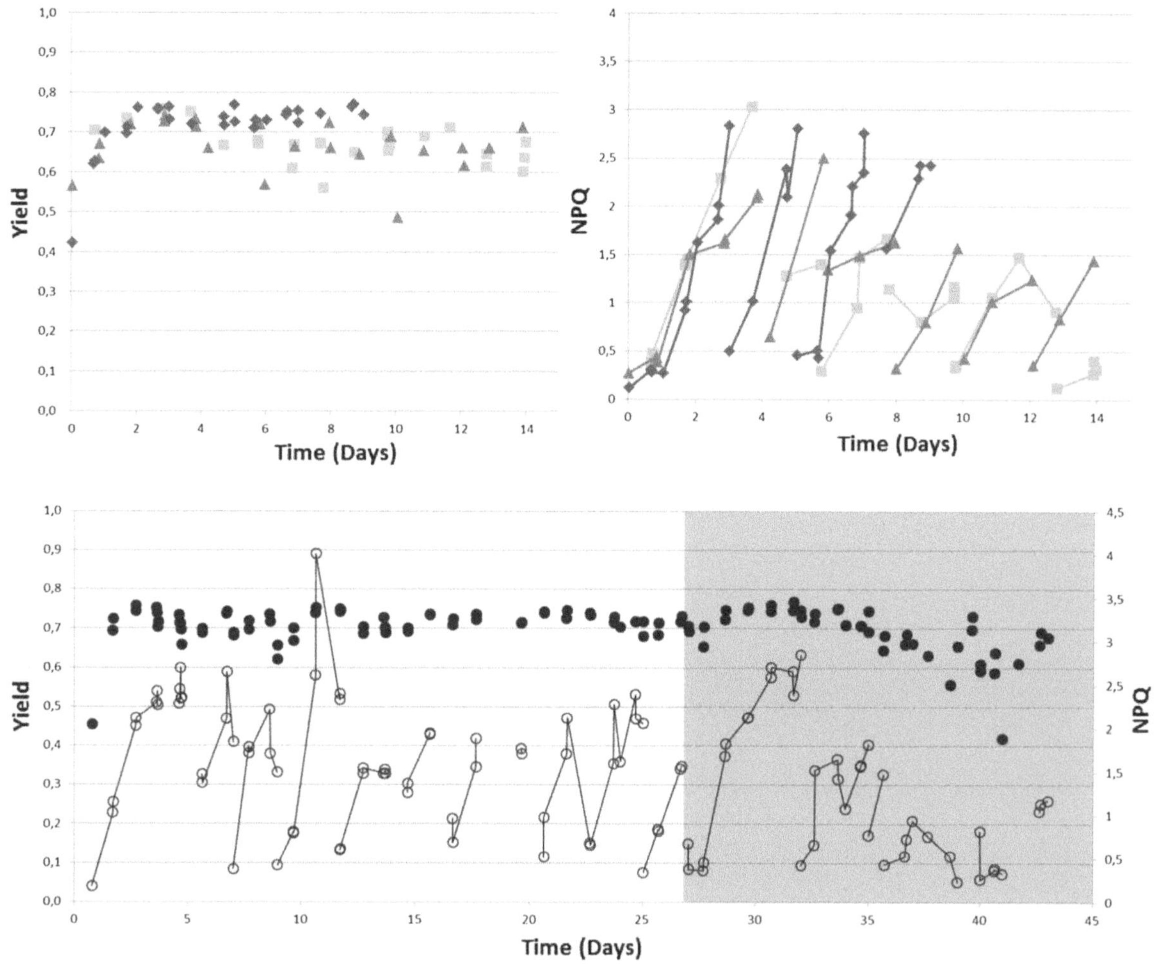

Fig. 6 Development of maximum quantum yield and non-photochemical quenching over time; samples were *dark* adapted for 5 min. Light of a wavelength of 645 nm was used for excitation. *Unlinked data points* in NPQ indicate dilution of the culture. *Top row* shows three experiments (*square* March 2013; *triangle* April 2013; *diamond* April 2014) with durations of 14 days or less; *Left* maximum quantum yield; *right* NPQ. *Lower row* shows the maximum quantum yield (*black solid circle*) and NPQ (*black open circle*) of a growth experiment lasting 43 days (March–April 2014). The area shaded in *gray* indicates a decreased water circulation

Productivity and media parameter

The overall biomass productivity and the pH and conductivity of the culture media were measured to determine the influence of a high pH and the salinity to be reached on the productivity.

All growth experiments were harvested every other day (Fig. 7) and showed a mean productivity of 16.1 g/m^2/day (SD ±1.7) in 2013 and 15.4 g/m^2/day (SD ±4.7) in 2014. The pH showed values of between 8.3 and 10.0 in the morning (Fig. 7) rising between 0.6 and 0.8 points over the day and decreasing overnight. The conductivity alternated during all experiments in various degrees between 15 and 25 mS/cm depending on the conductivity of the feed water. The dilution of the culture led to a drop in pH, and after addition of fertilizer the slope of pH-increase declined. The maximum measured pH-values of all cultures were in the afternoon after a full day of growth between 10.1 and 10.8.

During the different experiments, the circulation of the cultures stopped irregularly due to the maintenance of the generator or failure of mechanical parts (March 2013: 21 h in 14 days; April 2013: 3 h in 14 days; April 2014: 31 h in 9 days; March–April 2014: 82 h during the first 27 days and 31 h during the 16 days with reduced pump performance). Single power outages lasted as long as 3–12 h. During this time, no decrease in productivity was observed but the pH of cultures increased faster and if stopped longer than 6 h from the morning on it plateaued at values around 10.3 (SD ±0.1).

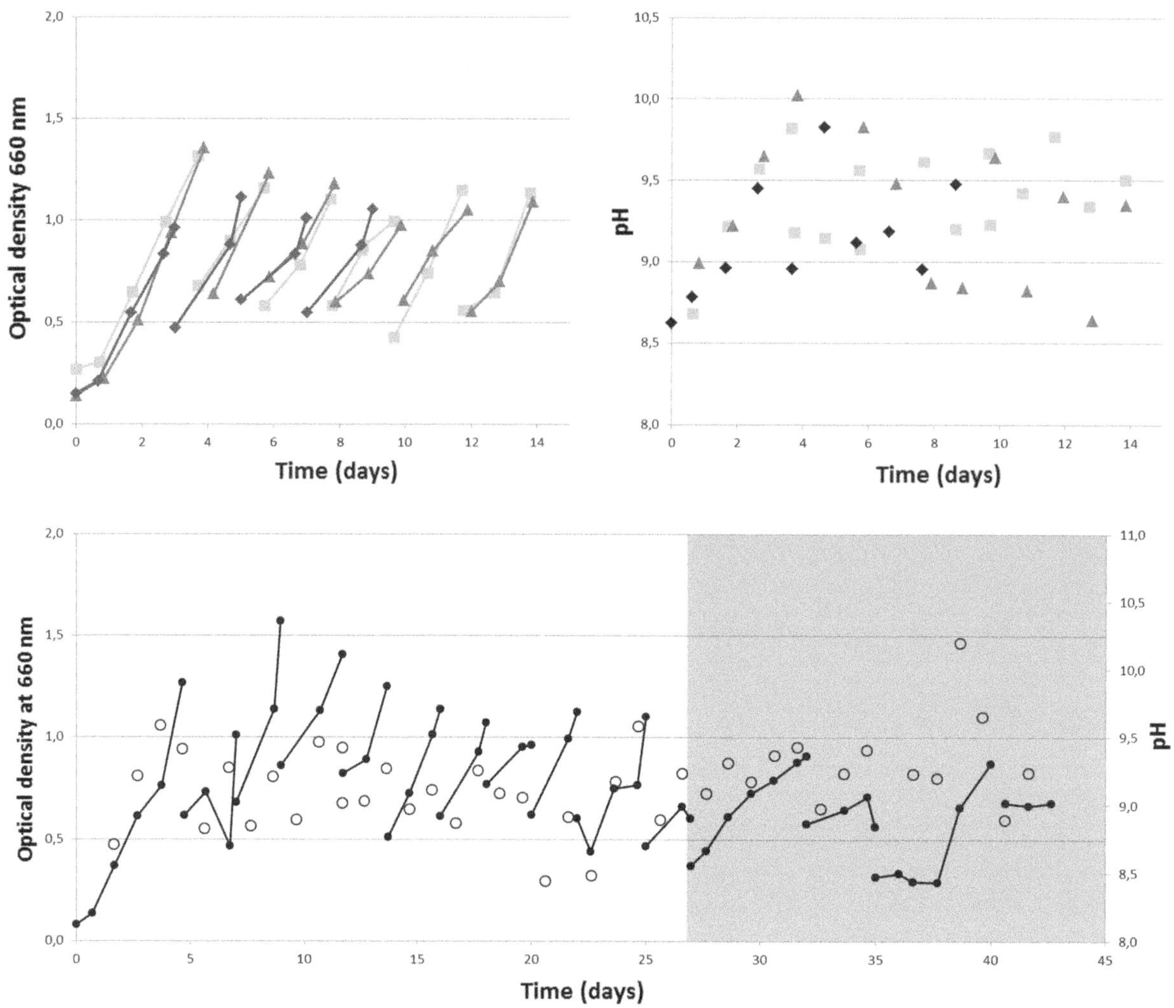

Fig. 7 Growth behavior of *Scenedesmus* Spec. MKB in 600L pre-cleaned PW while diluted by ca. 50 % every other day. Fertilizer was added after each harvest. Biomass increase is shown as determined by optical density measurement, *unlinked data points* indicate dilution. The pH shows only the mean values of 3 min measured in the morning between 8 and 10 a.m. *Top row* shows three experiments (*square* March 2013; *triangle* April 2013; *diamond* April 2014) with durations of 14 days or less; *left* biomass; *Right* pH. *Lower row* shows the biomass (*black solid circle*) and pH (*black open circle*) of a growth experiment lasting 43 days (March–April 2014). The area shaded in *gray* indicates a decreased water circulation

The productivity of the experiment conducted in March and April 2014 dropped to 6.3 g/m^2/day (SD $\pm$2.0) when the pump was replaced by a smaller one on day 27.

Discussion

Culture composition

The biomass monitored in 2013 was contaminated by weed algae, mainly diatoms, from the early beginning. This is accounted to the lack of a suitable filtration system for treatment of the pPW. The change in biomass composition from the end of the experiment in March 2013 and the beginning culture composition of the experiment performed in April 2013 showed that a significant amount of diatoms was removed from the culture via sedimentation. In 2014, diatoms were not detected (April 2014) or the earliest after 3 weeks of growth (March–April 2014), while single-celled cyanobacteria were detected in small amount during most parts of the experiments. The possible use of photosynthetic characteristics for the detection of weed algae was examined in more detail. The maximum values of NPQ decreased when the culture matured and the maximum NPQ of the experiment in April 2014 stayed around 2.5; the NPQ of all other experiments declined. An axenic culture of *Cyanobacterium aponinum* grown under the same circumstances had a NPQ of 0.5 which increased with invading green algae and diatoms (Paper submitted for publication). During times when the NPQ did not increase as high as before, the photosynthetic yield decreased as well. The photosynthetic yield (*Y*) was proven to be influenced by numerous different factors like nutrients (Lippemeier et al. 2001), light, culture density (Schreiber and Klughammer 2013), and salt stress (Lu and Vonshak 2002) and culture composition (Kolbowski and Schreiber 1995). Since measurements were taken daily at the same time, we do not consider the difference of daylight intensity between measurement days to be a considerable influence. The culture density and the salt content of the experiment conducted April 2014 alternated in between OD 0.5 and 1.3 and 15–25 mS/cm, respectively, but no significant change in *Y* was monitored, it was therefore concluded that cell density and salinity seem to have only a small influence at best. Nutrients were added after every dilution and it is assumed that they were available in abundance. *Y* is influenced by toxic substances as well (Schmitt-Jansen and Altenburger 2008; Peña-Vázquez et al. 2010), and algae cultures are suspected to accumulate physiological products inside the medium, which might inhibit growth (Gustavs et al. 2010). It was concluded that, since the change in Y during the trial conducted in March–April 2014 occurred after 34 days, the build-up of inhibiting physiological byproducts will be neglected. Cyanobacteria and diatoms showed during laboratory studies (Winckelmann et al. 2014) a maximum *Y* of 0.45 and 0.6, respectively. The change in *Y* occurred within 2 days of the microscopically detected weed algae, it is thought the lower capability of cyanobacteria and diatoms to use light energy for fluorescence might lead to a quenching of the overall *Y*.

The use of the phyto-PAM for the detection of weed algae and overall stress in an open pond as a stand-alone method is not advised. While it would be possible to establish an online sensor, it would be influenced by a range of different factors and the growth system to be monitored has to be well understood. A well-trained operator might be capable to distinguish the influence of nutrient deprivation from the afternoon photoinhibition(Lu and Vonshak 1999), but the sensitivity is insufficient to detect weed algae in an early stage. While it was possible to detect cyanobacteria concentration of 5 × 10^3 cells/mL cells/mL in 2.5 × 10^6 cells/mL with a microscope, the phyto-PAM started to detect cyanobacteria at 5 × 10^4 cells/mL. Other techniques for the detection of weed algae and even non-photosynthetic organisms like grazers can be applied (Havlik et al. 2013a). Flow cytometry was found to detect weed algae with a concentration as low as 0.01 % (Fulbright et al. 2014), and even capable to monitor algae; which change shape during their growth cycle (Havlik et al. 2013b). Real-Time quantitative PCR (qPCR) could be used to monitor production systems as well; qPCR systems were able to detect known possible contaminants at levels of 0.00001 % (Fulbright et al. 2014).

Grazers were only detected in matured cultures living in the biofilm; the abundance was too low to be quantified. Only bacteria grazing was monitored. It is assumed that the high pH and temperature did minimize the growth of grazers (Bartley et al. 2013).

Productivity

The overall production of all experiments, excluding the lag phase and the time with reduced pump performance, was 15.7 g/m^2/day (SD ±3.6). The average light intensity in Oman was reported to be around 1800 μmol photons/m^2/s (Al-Rawahy et al. 2003). Values of around 1500 μmol photons/m^2/s were measured during this study. Using the calculations proposed by Grobbelaar (2009) and assuming 40 % of algal biomass to be carbon, the photosynthetic efficiency was 2.4 % and decreased to 1 % during prolonged periods of reduced mixing.

There are no data on *Scenedesmus* productivity comparable to the set-up used during this study. The productivity of a non-axenic *Scenedesmus obliquus* dominated culture was reported to range from 1.7 g/m^2/day to 16.92 g/m^2/day in Germany (Grobbelaar et al. 1990) and a *Scendesmus* sp. culture was reported to have reached 28–82 g/m^2/day (Mendoza et al. 2013). While the first study was conducted over a range of 16 months, the second was only conducted for 6 days, both used CO_2 or flue gas as additional carbon sources. Weissman et al. (1989) conducted experiments in New Mexico, under similar temperature regimes and adding CO_2 productivities of 33 g/m^2/day, reaching the photosynthetic efficiency of 6.7 %.

The overall photosynthetic characteristics as measured in the morning (Table 1) did not change significantly with increasing sun light, biomass density, or biomass composition, so it is assumed that the sun light is not the limiting factor during the growth experiments. However, the culture's Y usually declined over the course of the day (data not shown) indicating light inhibition or damage. An alteration of the pond design might lead to a decrease in light inhibition and thereby increasing the productivity(Vonshak et al. 2013), also the ability to cope with excess light of high intensity varies over the experiment. After the inoculation and dilution, the NPQ was below 0.5 but increased up to 3 before the next dilution. The drop in NPQ can be explained by the dilution and the decreased biomass concentration reducing the cultures self-shading (Küster et al. 2004).

It is hypothesized that the high pH plays a major role in the limitation of growth. The availability of carbon as substrate for photosynthesis is crucial. The maximum concentration of carbon dioxide in liquid with a salt content of 1 % w/v is 1.2 g/L at 30 °C (Duan and Sun 2003). With the measured $k_La(O_2)$, not only would be all oxygen produced during photosynthesis be transferred out of the media, but the high $k_La(CO_2)$ would be sufficient to supply a constant stream of carbon. On the supposition of a gradient favoring the transition of CO_2 from air to liquid and using the calculations provided by Talbot et al. (1990), a maximum of 143 g of carbon dioxide could be transferred per hour into a 600 liter culture in the ponds described above.

The uptake of HCO_3^- and CO_2 can occur simultaneously in some green algae via different pathways, leading to OH^- formation and alkalization of the medium, even if $CO_2^{(aq.)}$ is present. See Badger and Price for a review regarding the CO_2 concentrating mechanism (Badger and Price 1992). Six different green algae have shown to stop photosynthesis at a pH above 10 (Shiraiwa et al. 1993) at which most of the dissolved inorganic carbon has the form of CO_3^{2-} (Markou et al. 2014). pKa of CO_3^{2-} is 10.3 at 25 °C but decreases with increasing temperature (Rabinowitch 1945). CO_3^{2-} is unavailable for most microalgae (Markou et al. 2014). While it is known that *Scenedesmus* and some marine algae are capable of thriving at pH of 8–9, it is unknown how the excess CO_3^{2-} influences the growth (Shiraiwa et al. 1993).

Scenedesmus obliquus showed an increase of lipid content after addition of 5 g/L NaCl (Gorain et al. 2013) but were also negatively influenced by further increased NaCl level (Affenzeller et al. 2009; Salama et al. 2013). The NaCl level reported as being sufficient for repressing growth had a conductivity of 11.1 mS/cm (Salama et al. 2013). No work about salt stress and green algae isolated from brackish water was found. The fluctuating salinity during this experiment was always higher than 15 m/S/cm, but no significant correlations between conductivity and productivity were found.

Economic feasibility and suggested improvements

If algae were to be produced under similar conditions as experienced during this study and the production facility would have a pond surface of 80 % and could operate on 300 days per year, it would produce 28 tons (dw) biomass per year. It was shown that *Scenedesmus* sp. biomass, grown phototrophically outdoors in Arizonas climate, consisted of 42 % (dw) carbohydrates [7 % (dw) were starch], 15 % (dw) lipids, and 27 % (dw) proteins, which was altered after 14 days of nutrient depletion to 45 % (dw) carbohydrates (20 % (dw) were starch), 30 % (dw) lipids, and 10 % (dw) proteins (Laurens et al. 2014). Under different conditions, lipid accumulations as high as 53 % (dw) were reported (Xin et al. 2010). The lipid composition of Scenedesmus sp. has been reported to be suitable for biodiesel production (Gouveia and Oliveira 2009; Makarevi et al. 2011; Jena et al. 2012). If *Scenedesmus* sp. grown in pPW would have the same composition as shown after growth in modified BG-11 media(Laurens et al. 2014), it would be possible to harvest 9500 L of lipids (assuming a density of 800 g/L) leading to 7600 L of biodiesel per hectare per year (Chisti 2007). This would be only a fraction of the 58,700 L/ha/year predicted, but still higher than the oil yield reported for palm oil (Chisti 2007). There are multiple possibilities to increase productivity and to make the production of green algae on site more sustainable. The biomass residues not sold can be digested for biogas, which can be used as CO_2 and energy source, while the residues can be used instead of fertilizer (Chisti 2013). A change in the mixing system will change the productivity; the use of a centrifugal pump as opposed to a paddle wheel will lead to a well-mixed culture and a high turbulence which is stated to prevent algal sedimentation, nutritional gradients, and to expose the algae to variations in the quantity and quality of light (Grobbelaar 2010). But it is also less energy efficient. The floating of algae flocks at times of pump failure showed that there was a biofilm formation. The biofilm mainly consisted of diatoms and dust particles and was stronger in areas with less turbulence. It is strongly suggested that the biofilm formation is influenced by the amount of dust, the turbulence of the culture, and the roughness of the pond liner. A change in pond design or mixing might lead to less biofilm formation and thereby reduce the risk of product contamination. It was shown that wave formation inside of a pond led to an increased productivity and a better nitrogen uptake (Chiaramonti et al. 2013). If wave formation would be obtained by intermittent reduction of the pumps delivery rate, a decrease in energy consumption might be achieved.

A final statement of economic feasibility is not possible at this stage. The energy consumption for all steps involved in upstream and downstream processing has to be included into the calculation as well as possible nutrient recycling, nutrient costs, the usability of algae residues on site, and the work hours to be invested (Collet et al. 2013).

Conclusion

It was shown that it is possible to grow indigenous algae in open ponds in an arid region using pre-cleaned waste water as growth medium. The concerns regarding salt stress due to evaporation were unwarranted. The green algae isolated on-site were suited to the conditions of high temperature and increasing salt content, and dominated the culture, while no signs of decreasing health were detectable.

High turbulence and flow regime are necessary to ensure high production rates. The design of the ponds and the means used to establish the flow regime have to be changed during scale-up to increase or at least stabilize productivity, while reducing the applied energy and thereby make economic viability possible.

Acknowledgments The authors would like to gratefully thank Rayan Al-Kabani and Heiko Patzelt from Mazoon Environmental and Technological Services LLC, Thomas Headley, Stéphane Prigent and the whole treatment plant team of Bauer Nimr LLC for their help, time, and support, and Thomas Parry for the critical reading of the manuscript. The project underlying this report was funded by the Federal Ministry of Education and Research of Germany under grant number FHprofUnt 17003X11. The responsibility for the content of this publication lies with the author.

References

Abed RMM, Al-Kharusi S, Prigent S, Headley T (2014) Diversity, distribution and hydrocarbon biodegradation capabilities of microbial communities in oil-contaminated cyanobacterial mats from a constructed wetland. PLoS One 9(12):e114570. doi:10.1371/journal.pone.0114570

Affenzeller MJ, Darehshouri A, Andosch A, Lütz C, Lütz-Meindl U (2009) Salt stress-induced cell death in the unicellular green alga Micrasterias denticulata. J Exp Bot 60(3):939–954. doi:10.1093/jxb/ern348

Al-Rawahy SH, Al-Dhafri KS, Al-Bahlany SS (2003) Germination, growth and drought resistance if native and alien plant species of the genus *Prosopis* in the sultanate of Oman. Asian J Plant Sci 2(14):1020–1023. doi:10.3923/ajps.2003.1020.1023

Badger MR, Price GD (1992) The CO2 concentratiing mechanism in cyanobacteria and microalgae. Physiol Plant 84:606–615

Bartley ML, Boeing WJ, Dungan BN, Holguin FO, Schaub T (2013) pH effects on growth and lipid accumulation of the biofuel microalgae Nannochloropsis salina and invading organisms. J Appl Phycol 26(3):1431–1437. doi:10.1007/s10811-013-0177-2

Becker EW (1994) Large-scale cultivation. In: Biotechnology and microbiology: microalgae. First Edition.Cambridge University Press, New York, pp 63–171

Breuer R, Al-Asmi SR (2010) Nimr water treatment project—up scaling a reed bed trail to industrial scale produced water treatment. Soc Petrol Eng SPE Int Conf Health Saf Environ Oil Gas Explor Prod 1:286–296

Chiaramonti D, Prussi M, Casini D, Tredici MR, Rodolfi L, Bassi N, Zittelli GC, Bondioli P (2013) Review of energy balance in raceway ponds for microalgae cultivation: Re-thinking a traditional system is possible. Appl Energy 102:101–111. doi:10.1016/j.apenergy.2012.07.040 **(Elsevier Ltd)**

Chisti Y (2007) Biodiesel from microalgae. Biotechnol Adv 25(3):294–306. doi:10.1016/j.biotechadv.2007.02.001 **(Elsevier Inc.)**

Chisti Y (2013) Constraints to commercialization of algal fuels. J Biotechnol 167(3):201–214. doi:10.1016/j.jbiotec.2013.07.020 **(Elsevier B.V.)**

Collet P, Spinelli D, Lardon L, Hélias A, Steyer J, Bernard O (2013) Life-cycle 7 assessment of microalgal-based biofuels. In: Pandey A, Lee D-J, Chisti Y, Soccol CR (eds): Biofuels from Algae. pp 287–312, (Elsevier B.B. Amsterdam). doi: 10.1016/B978-0-444-59558-4.00013-9

Connell LB (2000) Nuclear ITS region of the alga Heterosigma akashiwo (Chromophyta: Raphidophyceae) is identical in isolates from Atlantic and Pacific basins. Mar Biol 136:953–960. doi:10.1007/s002270000314

Doran PM (ed) (1995) Bioprocess engineering principles, pp 129–163. doi:10.1016/B978-012220855-3/50007-9 **(Elsevier B.V.)**

Duan Z, Sun R (2003) An improved model calculating CO_2 solubility in pure water and aqueous NaCl solutions from 273 to 533 K and from 0 to 2000 bar. Chem Geol 193:257–271. doi:10.1016/S0009-2541(02)00263-2

Fulbright SP, Dean MK, Wardle G, Lammers PJ, Chisholm S (2014) Molecular diagnostics for monitoring contaminants in algal cultivation. Algal Res. doi:10.1016/j.algal.2013.11.008 **(Elsevier B.V.)**

Gachon CMM, Sime-Ngando T, Strittmatter M, Chambouvet A, Kim GH (2010) Algal diseases: spotlight on a black box. Trends Plant Sci 15(11):633–640. doi:10.1016/j.tplants.2010.08.005

Gantar M, Svirčev Z (2008) Microalgae and *Cyanobacteria*: food for Thought. J Phycol 44(2):260–268. doi:10.1111/j.1529-8817.2008.00469.x

Gilroy DJ, Kauffman KW, Hall RA, Huang X, Chu FS (2000) Assessing potential health risks from microcystin toxins in blue-green algae dietary supplements. Environ Health Perspect 108(51):435–439

Gorain PC, Bagchi SK, Mallick N (2013) Effects of calcium, magnesium and sodium chloride in enhancing lipid accumulation in two green microalgae. Environ Tech 34:37–41. doi:10.1080/09593330.2013.812668

Gouveia L, Oliveira AC (2009) Microalgae as a raw material for biofuels production. J Ind Microbiol Biotechnol 36(2):269–274. doi:10.1007/s10295-008-0495-6

Grobbelaar JU (2009) Upper limits of photosynthetic productivity and problems of scaling. J Appl Phycol 21(5):519–522. doi:10.1007/s10811-008-9372-y

Grobbelaar JU (2010) Microalgal biomass production: challenges and realities. Photosynth Res 106(1–2):135–144. doi:10.1007/s11120-010-9573-5

Grobbelaar JU, Soeder CJ, Stengel E (1990) Modeling algal productivity in large outdoor cultures and waste treatment systems. Biomass 21(4):297–314. doi:10.1016/0144-4565(90)90079-Y

Guieysse B, Béchet Q, Shilton A (2013) Variability and uncertainty in water demand and water footprint assessments of fresh algae cultivation based on case studies from five climatic regions. Biores Technol 128:317–323. doi:10.1016/j.biortech.2012.10.096

Gustavs L, Eggert A, Michalik D, Karsten U (2010) Physiological and biochemical responses of green microalgae from different habitats to osmotic and matric stress. Protoplasma 243(1–4):3–14. doi:10.1007/s00709-009-0060-9

Hamilton CE, Rossmeissl N, Ruddick C (2014) Exploring the utilization of complex algal communities to address algal pond crash and increase annual biomass production for algal biofuels. In: p 21

Havlik I, Lindner P, Scheper T, Reardon KF (2013a) On-line monitoring of large cultivations of microalgae and cyanobacteria. Trends Biotechnol 31(7):406–414. doi:10.1016/j.tibtech.2013.04.005

Havlik I, Reardon KF, Ünal M, Lindner P, Prediger A, Babitzky A, Beutel S, Scheper T (2013b) Monitoring of microalgal cultivations with on-line, flow-through microscopy. Algal Res 2(3):253–257. doi:10.1016/j.algal.2013.04.001 **(Elsevier B.V.)**

Jena J, Nayak M, Panda HS, Pradhan N, Sarika C, Panda PK, Rao VSK, Prasad BBN, Sukla LB (2012) Microalgae of Odisha coast as a potential source for biodiesel production. World Environ 2(1):12–17. doi:10.5923/j.env.20120201.03

Kolbowski J, Schreiber U (1995) Computer-controlled phyto- plankton analyzer based on a 4-wavelength PAM Chl fluoro meter. In: Mathis, P. (ed.): Photosynthesis: From Light to Biosphere. Kluwer Academic Publishers, Dordrecht

Küster A, Schaible R, Schubert H (2004) Light acclimation of photosynthesis in three charophyte species. Aquat Bot 79(2):111–124. doi:10.1016/j.aquabot.2004.01.010

Laurens LML, Van Wychen S, McAllister JP, Arrowsmith S, Dempster TA, McGowen J, Pienkos PT (2014) Strain, biochemistry, and cultivation-dependent measurement variability of algal biomass composition. Anal Biochem 452:86–95. doi:10.1016/j.ab.2014.02.009

Lippemeier S, Hintze R, Vanselow K, Hartig P, Colijn F (2001) In-line recording of PAM fluorescence of phytoplankton cultures as a new tool for studying effects of fluctuating nutrient supply on photosynthesis. Eur J Phycol 36(1):89–100. doi:10.1080/09670260110001735238

Lu C, Vonshak A (1999) Photoinhibition in outdoor Spirulina platensis cultures assessed by polyphasic chlorophyll fluorescence transients. J Appl Phycol 11:355–359

Lu C, Vonshak A (2002) Effects of salinity stress on photosystem II function in cyanobacterial Spirulina platensis cells. Physiol Plant 114(3):405–413. doi:10.1034/j.1399-3054.2002.1140310.x

Makarevi V, Andrulevi V, Skorupskait V (2011) Cultivation of Microalgae Chlorella sp. and Scenedesmus sp. as a potentional biofuel feedstock. Environ Res Eng Manag 57(3):21–27

Markou G, Vandamme D, Muylaert K (2014) Microalgal and cyanobacterial cultivation: the supply of nutrients. Water Res 65:186–202. doi:10.1016/j.watres.2014.07.025 **(Elsevier Ltd)**

Medlin L, Elwood HJ, Stickl S, Sogin ML (1988) The characterization of enzymatically amplified eukaryotic 16S-like rRNA-coding regions. Gene 71(2):491–499. doi:10.1016/0378-1119(88)90066-2 **PO BOX 211, 1000 AE AMSTERDAM, NETHERLANDS (ELSEVIER SCIENCE BV)**

Mendoza JL, Granados MR, de Godos I, Acién FG, Molina E, Heaven S, Banks CJ (2013) Oxygen transfer and evolution in microalgal culture in open raceways. Biores Technol 137:188–195. doi:10.1016/j.biortech.2013.03.127 **(Elsevier Ltd)**

Mutanda T, Ramesh D, Karthikeyan S, Kumari S, Anandraj A, Bux F (2011) Bioprospecting for hyper-lipid producing microalgal strains for sustainable biofuel production. Biores Technol 102(1):57–70. doi:10.1016/j.biortech.2010.06.077 **(Elsevier Ltd)**

Odlare M, Nehrenheim E, Ribé V, Thorin E, Gavare M, Grube M (2011) Cultivation of algae with indigenous species—Potentials for regional biofuel production. Appl Energy 88(10):3280–3285. doi:10.1016/j.apenergy.2011.01.006

Peña-Vázquez E, Pérez-Conde C, Costas E, Moreno-Bondi MC (2010) Development of a microalgal PAM test method for Cu(II) in waters: comparison of using spectrofluorometry. Ecotoxicol 19(6):1059–1065. doi:10.1007/s10646-010-0487-y

Rabinowitch EI (1945) Photosynthesis and related processes. Interscience Publishers inc, New York. doi:bhl.title.5714

Ralph P, Gademann R (2005) Rapid light curves: a powerful tool to assess photosynthetic activity. Aquat Bot 82(3):222–237. doi:10.1016/j.aquabot.2005.02.006

Rawat I, Ranjith Kumar R, Mutanda T, Bux F (2013) Biodiesel from microalgae: a critical evaluation from laboratory to large scale production. Appl Energy 103:444–467. doi:10.1016/j.apenergy.2012.10.004

Salama E-S, Kim H-C, RaI Abou-Shanab, Ji M-K, Oh Y-K, Kim S-H, Jeon B-H (2013) Biomass, lipid content, and fatty acid composition of freshwater Chlamydomonas mexicana and *Scenedesmus* obliquus grown under salt stress. Bioprocess Biosyst Eng 36(6):827–833. doi:10.1007/s00449-013-0919-1

Schmitt-Jansen M, Altenburger R (2008) Community-level microalgal toxicity assessment by multiwavelength-excitation PAM fluorometry. Aquat Toxicol 86(1):49–58. doi:10.1016/j.aquatox.2007.10.001

Schreiber U, Klughammer C (2013) Wavelength-dependent photodamage to Chlorella investigated with a new type of multi-color PAM chlorophyll fluorometer. Photosynth Res 114(3):165–177. doi:10.1007/s11120-013-9801-x

Shiraiwa Y, Goyal A, Tolbert NE (1993) Alkalization of the medium by unicellular green algae during uptake of dissolved inorganic carbon. Plant Cell Physiol 34(5):649–657

Talbot P, Lencki MPGRW, Noiie J De (1990) Absorption of CO_2 in algal mass culture systems: a different characterization approach. Biotech Bioeng 37(9):834–842. doi:10.1002/bit.260370907

Tillmann U (2004) Interactions between planktonic microalgae and protozoan grazers. J Eukaryot Microbiol 51(2):156–168

Vonshak A, Laorawat S, Bunnag B, Tanticharoen M (2013) The effect of light availability on the photosynthetic activity and productivity of outdoor cultures of Arthrospira platensis (Spirulina). J Appl Phycol. doi:10.1007/s10811-013-0133-1

Wagener K (1983) Mass cultures of marine algae for energy farming in coastal deserts. –. Int J Biometeorol 27(3):227–233. doi:10.1007/BF02184238

Weissman JC, Tillet DM, Goebel RP (1989) Design and operation of an outdoor microalgae test facility final subcontract report design and operation of an facility, vol 1. Solar Energy Research Institute, Golden Colorado, pp 1–68

Winckelmann D, Bleeke F, Bergmann P, Klöck G (2014) Growth of cyanobacterium aponinum influenced by increasing salt concentrations and temperature. Biotech. doi:10.1007/s13205-014-0224-y

Xin L, Hu H, Ke G, Sun Y (2010) Effects of different nitrogen and phosphorus concentrations on the growth, nutrient uptake, and lipid accumulation of a freshwater microalga Scenedesmus sp. Biores Technol 101(14):5494–5500. doi:10.1016/j.biortech.2010.02.016

6

Chromosome preparation in fish: effects of fish species and larval age

Ali Karami · Peyman Eghtesadi Araghi · Mohd Arif Syed · Scott P. Wilson

Abstract To date, several protocols have been developed to achieve clear and identifiable metaphase chromosome spreads from larvae of a single fish species. However, the efficiency of these protocols in more than one fish species has barely been compared within a single study. This work investigated the dependency of chromosome preparation parameters including colchicine concentration (0.01, 0.025, 0.05 %) and exposure duration (3, 5 h), hypotonic solution (distilled water, 0.075 M KCl solution), and Giemsa stain solution concentration (6, 8, 10, 11, 12, and 14 %) and incubation period (15, 30, 45, and 60 min) to two species of fish, the African catfish (*Clarias gariepinus*) and the zebrafish (*Danio rerio*) at different larval ages (0, 2, and 4 days post-hatch, dph). Results indicated that larval age, colchicine concentration and/or incubation time, and/or the type of hypotonic solution varied with fish species while staining the chromosomes with 11 % Giemsa solution for 45 min can be maintained regardless of the species or larval age. Interestingly, employing the selected values from diploid *C. gariepinus* experiment to prepare metaphase chromosomes from larvae of their triploid siblings proved to be efficient.

Keywords Metaphase chromosome spread · Colchicine · Hypotonic solution · Giemsa stain · Zebrafish · African catfish

Introduction

Cytogenetic methods, such as taxonomic works, and karyological analysis are essential in providing basic information on fish breeding programs such as inter-specific hybridization (Crego-Prieto et al. 2013), chromosome manipulation techniques (Christopher et al. 2010; Gilna et al. 2014; Thresher et al. 2014), and genetic

---

A. Karami (✉)
Laboratory of Aquatic Toxicology, Faculty of Medicine and Health Sciences, Universiti Putra Malaysia, 43400 Serdang, Selangor, Malaysia
e-mail: alikaramiv@gmail.com

P. Eghtesadi Araghi
Iranian National Institute for Oceanography and Atmospheric Science (INIOAS), Tehran, Iran

M. A. Syed
Department of Biochemistry, Faculty of Biotechnology, University Putra Malaysia, 43400 Serdang, Selangor, Malaysia

S. P. Wilson
School of Medical and Applied Sciences, CQ University, Rockhampton, Australia

improvement of commercial fish stocks (Gui and Zhu 2012). To date, a variety of karyotyping techniques such as tissue cultures (Lomax et al. 2000), squashing techniques (Armstrong and Jones 2003), and cell suspensions of the tissues undergoing mitosis (Fan and Fox 1990; Henegariu et al. 2001) have been developed to visualize chromosomes of fish at different developmental stages.

Nevertheless, methods targeting embryonic and larval stages of fish have faced difficulties in achieving a reliable number of identifiable and wide-spread metaphase chromosomes probably due to the variations in mitotic cell division rates among different fish species (Shao et al. 2010). In conventional chromosome preparation protocols, a spindle poison (e.g., colchicine) is employed to arrest the cells at their metaphase stage (Kligerman and Bloom 1977). To achieve clear and identifiable metaphase chromosome spreads, it is essential to select the proper concentration and the correct incubation period of the poison (Rieder and Palazzo 1992). While insufficient concentration and/or duration of exposure to the spindle poisons fails to arrest the cells at the metaphase stage, very high concentrations and/or overly long exposure duration may lead to chromosomal condensation (Rieder and Palazzo 1992; Wood et al. 2001). Following mitotic spindle inhibition, cells or larvae must be incubated in a hypotonic solution to swell the nuclei and scatter the chromosomes on the slides (Moore and Best 2001). Choosing an improper hypotonic solution and incubation period would either make chromosomes knotted and overlapped or result in chromosomal loss (Baksi and Means 1988).

The hypotonic solution step is further followed up by fixing the samples using Carnoy's fixative solution, and preparing a cell suspension. Furthermore, during slide preparation, factors such as fixative evaporation rate and preheating of slides affect the quality and quantity of mitotic chromosomal spread (Moore and Best 2001). Therefore, the approach to slide preparation is another important step as employing improper spreading techniques would cause the chromosomes to be washed out during the staining process.

Following slide preparation, various staining techniques such as classic staining (e.g., aceto-orcein, haematoxylin, Giemsa, Wright and Leishman stains) or banding techniques (e.g., Q-banding, G-banding, R-banding, C-banding, and high-resolution banding) (Calado et al. 2013; Moore and Best 2001; Wang et al. 2010) are used to stain chromosomes for different purposes. Concentrated staining solutions and/or over-incubation would result in dark background filling the space between chromatids, whereas diluted staining solutions and/or a limited incubation period creates indistinguishable chromosomal spreads.

Table 1 provides detailed information on colchicine treatments, hypotonic solutions, fixatives, and Giemsa staining used by previous studies with fish embryos and larvae. To the best knowledge of the researchers, very limited studies have examined the requirements to alter chromosomal procedures for different species of fish and/or larval age. To fill this gap, 0, 2, and 4 days post- hatch (dph) zebrafish (*Danio rerio*), a widely used vertebrate model, and African catfish (*Clarias gariepinus*), a widespread tropical/subtropical species, were used and the results compared. A stepwise procedure was followed to determine the key elements in obtaining the highest number of mitotic chromosome spreads pertaining to the species of fish and larval age.

Materials and methods

Chemicals

Colchicine, methanol (99 % HP grade), glacial acetic acid, and KCl were purchased from R&M chemicals (UK), and Giemsa stain from Merck. Distilled water (HPLC grade) was produced daily in the laboratory.

Fish

Immature African catfish *C. gariepinus* and wild-type *D. rerio* were purchased from local suppliers and reared for 3 months at 28 °C on a 12:12 dark/light photoperiod in a recirculation system, and fed three times daily. 0, 2, and 4 days post-hatch (dph) larvae of *C. gariepinus* and *D. rerio* were obtained according to Karami et al. (2010) and Nasiadka and Clark (2012), respectively. Triploid *C. gariepinus* were produced as described by Karami et al. (2011).

Table 1 Comparison of chromosome preparation methods using fish embryos and larvae

Fish species	Developmental stage	Colchicine	Hypotonic solution: incubation period (min)	Fixation method	Giemsa concentration (%), incubation period (min)	References
Cyprinodon variegatus and *Morone saxatilis*	Embryo and 0 dph larvae	0.05 % for 4 h	Distilled water: 90	Carnoy's fixative	10, 20	Baksi and Means (1988)
Acipenser baeri × *Acipenser gueldenstaedti*	0 dph larvae	0.025 % for 4 h	KCl (0.075 M): 60	Carnoy's fixative	5, 20	Fopp-Bayat and Woznicki (2006)
Oreochromis niloticus, *Slamo salar*, and *Salmo trutta*	0 dph larvae	0.002–0.005 % for 4–6 h	Distilled water: 8–12	4:1 methanol acetic acid	10, 10–20	Hussain and McAndrew (1994)
Oncorhynchus mykiss, Oncorhynchus musou, Oncorhynchus keta	Embryos	0.0012–0.0025 % for 6–12 h	0.8 % tri-sodium citrate dihydrate: 15–30	Carnoy's fixative	4, 30	Inokuchi et al. (1994)
Clarias gariepinus	0 dph larvae	0.05 % for 3 h	Distilled water: 30	Carnoy's fixative	14, 45	Karami et al. (2010)
Oreochromis mossambicus × *O. niloticus*	0 dph larvae	0.01 % for 4–6 h	KCl (0.075 M): 40	Carnoy's fixative	10, 20	Pradeep et al. (2011)
Cynoglossus semilaevis	Embryos and 0 dph larvae	0.02 % for 1–2 h	KCl (0.075 M): 30	Carnoy's fixative	10, 10	Shao et al. (2010)

dph days post hatch

Chromosomal preparation procedure

In this study, the method of Shao et al. (2010; Table 1), which is a modified method of Kligerman and Bloom (1977), was used as the basis for chromosomal preparation. Accordingly, all the variables have been optimized in a stepwise manner. In all stages, five larvae were used for each treatment and two slides were prepared from each larva. Larvae were killed by smashing the head with forceps and the yolk sac was removed carefully prior to starting the experiments (Thorgaard et al. 1981).

Colchicine

Larvae were incubated in 0.01, 0.025, or 0.05 % colchicine for 3 or 5 h. The procedure was followed as described by Shao et al. 2010; Table 1). Finally, on each slide, clear metaphase chromosome spreads were counted. In all stages, the metaphase spreads were photographed under 1000 X (oil immersion) using a Motic BA400 microscope.

Hypotonic solution

Using the selected values of colchicine concentration and incubation period, colchicine-treated larvae were incubated in distilled water (Karami et al. 2010) or 0.075 M KCl solution (Pradeep et al. 2011) for 30 min, followed by the procedure of Shao et al. (2010; Table 1). Finally, clear chromosome spreads were counted on each slide.

Fixation

Following the optimization of colchicine concentration and incubation period, and hypotonic solution for the different larval age in each species, larvae were fixed in chilled Carnoy's fixative solution (ethanol:glacial acetic acid 3:1 vol/vol) for a total of 40 min. The solution was renewed every 20 min. No alternative fixative solution was considered as Carnoy's solution is the most widely used fixative found in the previous studies (Fopp-Bayat and Woznicki 2006; Karami et al. 2010; Pradeep et al. 2011, Table 1).

Slide preparation method

Each *C. gariepinus* and *D. rerio* larva was minced gently in 40 and 20 μl of 50 % acetic acid, respectively. Afterwards, 10 μl of the cell suspension was pipetted onto a ethanol-cleaned slide. Slide were held 3–5 cm above the flame while the droplet was continuously trundled on the slide in a dilatory spiral movement making a 0.75- to 1-cm diameter circle (Fig. 1).

Fig. 1 Spiral movement of dropped suspension on the slide during flame drying

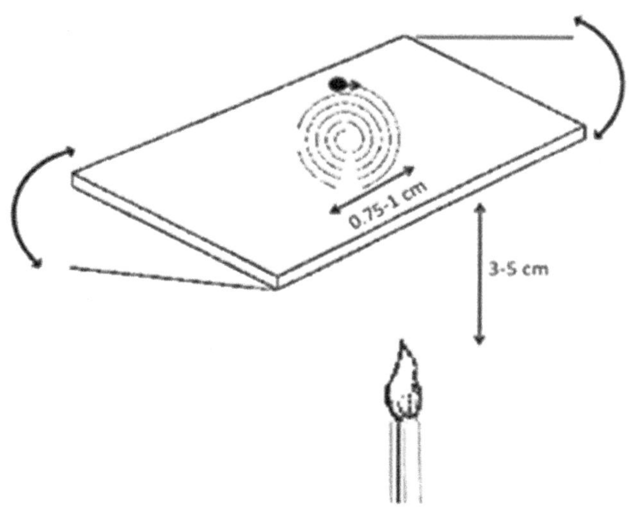

Giemsa stain

Slides were incubated in freshly prepared 6, 8, 10, 11, 12, or 14 % Giemsa solution for 15, 30, 45, or 60 min. The optimum concentration and duration of Giemsa stain were chosen based on the visibility of chromosomes, background clarity, and distinction between chromatids within a single chromosome.

Due to the wide application of chromosomal preparation technique in ploidy determination, the optimized protocol for 2 dph *C. garipinus* larvae was tested on 2 dph triploid *C. garipinus*.

Statistical analyses

To improve normality and homogeneity of variance, chromosome spread counts were arcsin-square-root transformed. Prior to running ANOVA analyses, normality and homogeneity of data had been tested (Tabachnick and Fidell 2001). A three-way ANOVA was employed to evaluate the interactions and effects of colchicine concentration, exposure duration, and larval age in *C. gariepinus* and *D. rerio*. A two-way ANOVA was applied to elucidate the interactions and effects of hypotonic treatments and larval age. The ANOVA tests were followed by post hoc Duncan's multiple range tests when significant differences ($P < 0.05$) were detected. All statistical comparisons were performed through IBM SPSS Statistics Version 21.

Ethics statement

Animal ethics approval was obtained from the Animal Ethics Committee, UPM (reference no: UPM/IACUC/AUP- R076/2013).

Results

Throughout the study, 4 dph larvae failed to produce countable mitotic chromosome spreads in both *C. gariepinus* and *D. rerio*; therefore, they were excluded from the statistical analyses. Colchicine concentration and duration, hypotonic solution, cell suspension density, Giemsa stain concentration and duration were optimized according to the following results:

Colchicine

African catfish

Except for one interaction (larval age × colchicine incubation period), the interactions among colchicine concentration, incubation period, and larval age on the number of chromosome spreads were non-significant ($P > 0.05$). The effect of larval age on the number of metaphase chromosome spreads differed with colchicine incubation period (Three-way ANOVA, $F_{1,24} = 18.28$, $P < 0.01$). In 2 dph larvae, 3 h of incubation in colchicine provided significantly higher number of chromosome spreads; however, the incubation period did not significantly alter the number of chromosome spreads in 0 dph larvae. Figure 2a depicts changes in the number of metaphase chromosome spreads within different colchicine concentrations. Figure 2b shows the number of metaphase chromosome spreads of 0 and 2 dph larvae following 3- and 5-h incubation in colchicine solution.

Zebrafish

The interactions between larval age × colchicine concentration (Three-way ANOVA, $F_{1,24} = 3.37$) and larval age × colchicine incubation period (Three-way ANOVA, $F_{1,24} = 14.37$) proved to be significant ($P < 0.05$). Figure 2c, d displays changes in the number of metaphase chromosome spreads within different colchicine concentrations and colchicine incubation time, respectively. The results from a three-way ANOVA failed to show significant interaction between the effects of colchicine concentration, exposure duration, and larval age on the number of chromosome spreads.

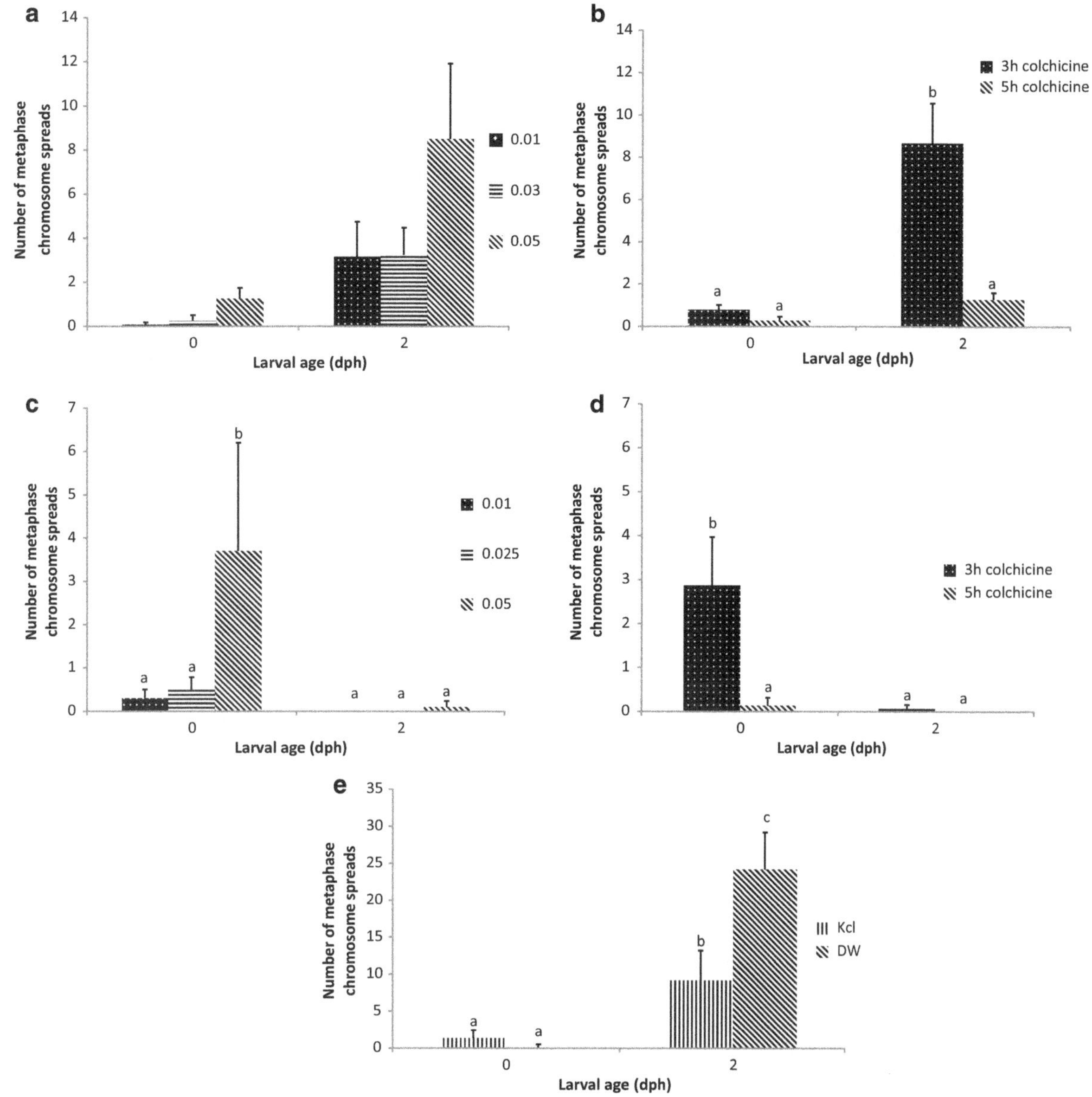

Fig. 2 Effects of different treatments on the number of clear and identifiable metaphase chromosome spreads **a** larval age × colchicine concentration interaction in *C. gariepinus*, **b** larval age × colchicine incubation period interaction in *C. gariepinus*, **c** larval age × colchicine concentration interaction in *D. rerio*, **d** larval age × colchicine incubation period interaction in *D. rerio*, **e** hypotonic treatments in *C. gariepinus*. The data shown here are Mean + SE. *Bars* noted by *different letters* are significantly different from each other ($P < 0.05$, Duncan's multiple range test)

Hypotonic solution

African catfish

The interaction between hypotonic treatment and larval age on the number of metaphase chromosome spreads was significant (Two-way ANOVA, $F = 8.65$, $P < 0.05$). Duncan's multiple range tests proved that for 0 dph *C. gariepinus*, there was no significant difference in using distilled water or KCl solution (Fig. 2e). However, using distilled water would significantly increase the number of metaphase chromosome spreads in 2 dph larvae (Fig. 2e).

Zebrafish

The result of this experiment showed the types of hypotonic solution do not have significant interactions with larval age to augment the number of metaphase chromosome spreads in zebrafish.

Giemsa staining

Regardless of the duration of incubation, using a Giemsa stain at concentrations lower than 11 % produced non-visible chromosomes in all the treatments. In contrast, using 12 % and particularly 14 % Giemsa caused sedimentation, dark background and filled the space between chromatids. In general, in both species desirable results were achieved when the spreads were stained with 11 % Giemsa for 45 min making chromosome spreads suitable for karyotyping (Fig. 3a–c).

Discussion

Most chromosomal preparation protocols for fish have been designed for a single species. To the best knowledge of the researchers, to date, no study has evaluated the modification of these protocols according to the fish species and larval age. For the first time, this study proved the necessity of modifying colchicine concentration and incubation period, and the type of hypotonic solution according to the species of fish and the larval age.

Obtaining a desirable number of mitotic chromosome spread is directly related to the number of cells undergoing mitotic division. In turn, mitotic rate is a factor influenced by species and environmental parameters (Shelton et al. 1997). By comparing metaphase chromosome numbers and mitotic rates in different life stages of *Cynoglossus semilaevis*, Shao et al. (2010) found a higher mitotic rate in the larvae compared to the juveniles and adults. In a similar way, differences in mitotic division rates may explain why in this study different aged *C. gariepinus* and *D. rerio* larvae responded differently to the colchicine concentration and duration, and to the hypotonic solutions. In another study, Wakahara (1972) found a higher mitotic rate in the ventral tail-fin epidermis of the larval African clawed frog (*Xenopus laevis*) during the night time compared to the day time, thus highlighting the importance of environmental factors in getting a reliable number of chromosome spreads.

Chourrout and Happe (1986) pinpointed the necessity of using less than 170-day-old larvae of *Salmo gairdneri* to achieve a high number of metaphase chromosomes. In the current study, the highest number of metaphase chromosome spread was recorded in 0 dph *D. rerio* and 2 dph *C. gariepinus*. Therefore, different aged larvae may be used when different fish species are involved.

Colchicine is the most common microtubular poison used (Table 1). It has been reported that before nuclear envelope breakdown (NEB) in the metaphase cells, sufficient concentration of colchicine inhibits spindle microtubules and disperses the metaphase chromosomes in the cytoplasm (Caperta et al. 2006), whereas in *D. rerio* both colchicine concentration and incubation period were significantly influenced by larval age, and in *C. gariepinus* only colchicine incubation period was dependent on the larval age. The sensitivity of microtubule polymers to physical and chemical parameters has been shown before (Tilney and Porter 1967; Weisenberg 1972). Therefore, age- and species-dependent cellular parameters may influence the sensitivity of cells towards depolymerizing effects of colchicine. According to the results of this study, to achieve clear and identifiable metaphase chromosome spreads, colchicine concentration and/or incubation period should be altered according to the fish species and/or larval age.

Using an appropriate hypotonic solution is the other critical element that has been emphasized in the current study. Potassium chloride (KCl 0.075 M) is one of the most commonly used hypotonic solutions in chromosomal preparation protocols (Table 1). Analogously, the efficacy of distilled water as a hypotonic treatment has been shown in some other protocols (Table 1). In *C. gariepinus*, the number of clear metaphase chromosome spreads was significantly increased when distilled water was used compared to KCl. Using KCl caused extensive cell burst and chromosomal loss. In contrast, in *D. rerio* the changes in the hypotonic solution did not significantly influence the number of metaphase chromosome spreads. This study showed that

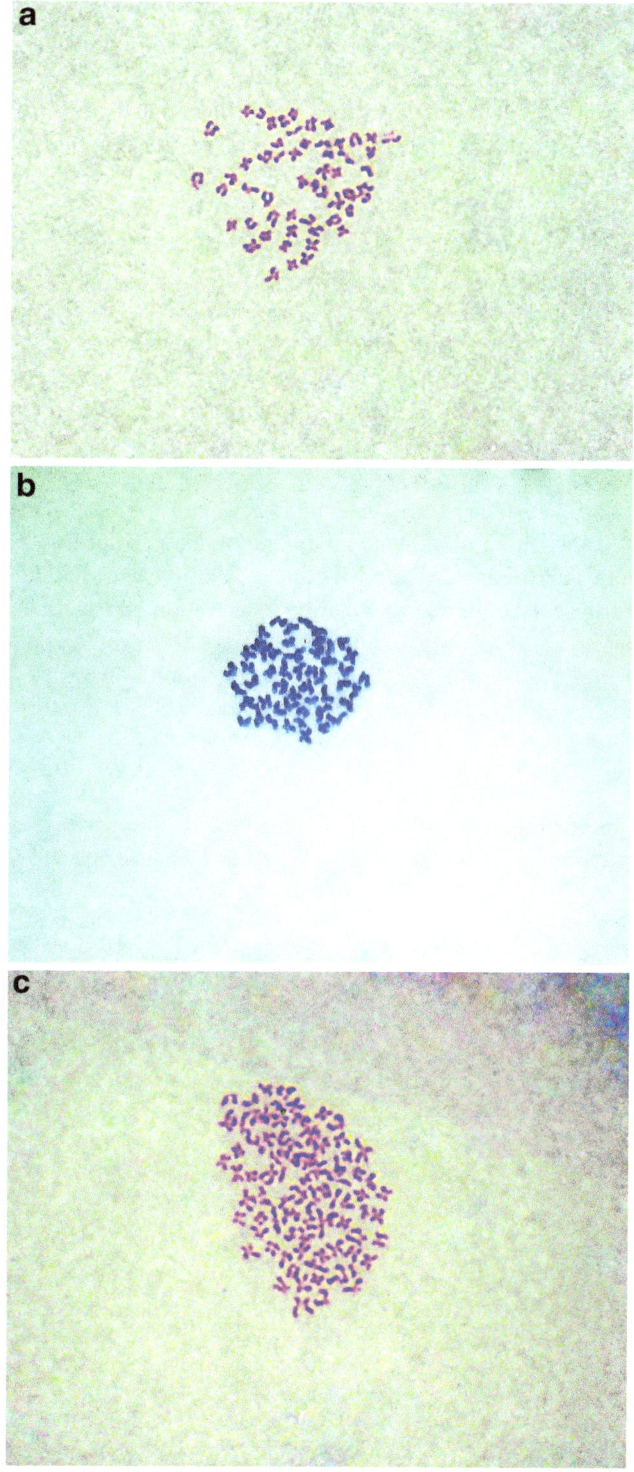

Fig. 3 Metaphase chromosome spreads of **a** diploid *C. gariepinus* ($2n = 56$); **b** diploid *D. rerio* ($2n = 50$); and **c** triplioid *C. gariepinus* ($n = 84$)

to procure a desirable number of clear chromosome spreads, the type of hypotonic solution should be modified according to the species of fish and/or larval age.

Besides the aforementioned elements, this study attempted to modify other important factors of chromosomal preparation protocols. Our preliminary studies proved that the larvae should be killed prior to the incubation in colchicine as incubating live larvae in colchicine solution failed to produce chromosome spread. Furthermore, to obtain clear chromosome spread, the yolk sac should be removed prior to the incubation in colchicine (Hussain and McAndrew 1994; Pradeep et al. 2011) because high lipophilicity of the yolk may restrict the penetration of colchicine or hypotonic solution to the cells (Baksi and Means 1988).

The quality of slides is affected by the proper staining solution concentration and the duration of incubation. This study showed that using 11 % of Giemsa solution for 45 min could lead to better quality of stained chromosomes in both fish species and in the polyploid fish. Most of the previous methods employed concentrations similar to this study but with a lower incubation period (e.g., Hussain and McAndrew 1994; Pradeep et al. 2011; Shao et al. 2010). Some other studies used lower concentrations such as 4–5 % (Fopp-Bayat and Woznicki 2006; Inokuchi et al. 1994; Kligerman and Bloom 1977) or higher concentrations like 14 % (Karami et al. 2010). The results of this study showed that concentrations less than 11 % and incubation period below 45 min would lead to unclear chromosomes. In contrast, higher concentrations encouraged sedimentation; therefore, dark background was achieved and the spaces between chromatids were stained. Concentration and incubation period were neither affected by larval age nor fish species; therefore, incubating the slides at 11 % of Giemsa for 45 min could be extended to other fish species. Despite having the chromosome preparation parameters optimized for the two species, several specimens of each species did not produce well-spread metaphase chromosomes, which are likely due to the interference of physiological factors. Further studies are required to understand the role of these elements in cytogenetic studies.

This study proved that depending on the fish species and the age of larvae, colchicine concentration and/or exposure duration, and/or hypotonic solution must be altered to achieve clear metaphase chromosome spread. Furthermore, the Giemsa stain concentration and incubation period proved to be independent of species and larval age. Probably significant differences on the amount of yolk content and its rate of absorption, and the differences on the rate of cell division among larvae of different fish species are responsible for differences among chromosome preparation protocols. These factors are either absent or less effective when juvenile or adult fish are of concern.

Author's contributions AK and PEA formulated the hypothesis and designed the study. AK conducted the experiments. AK and PEA analyzed the data. AK wrote the manuscript with assistance from SPW and MAS. All authors read and approved the final manuscript.

Acknowledgments This paper was supported by E-Science grant (No: 01-02-05-SF0017), Ministry of Science and Technology and Innovation (MOSTI), Government of Malaysia. Special thanks go to Mr. Hamzah Adam and Mr. Ramli Suhaimi for their assistances during the experiments.

Conflict of interest The authors declare that they have no conflict of interests.

References

Armstrong SJ, Jones GH (2003) Meiotic cytology and chromosome behaviour in wild-type *Arabidopsis thaliana*. J Exp Bot 54:1–10

Baksi S, Means J (1988) Preparation of chromosomes from early stages of fish for cytogenetic analysis. J Fish Biol 32:321–325

Calado LL, Bertollo LAC, Costa GWWFd, Molina WF (2013) Cytogenetic studies of Atlantic mojarras (*Perciformes–Gerreidae*): chromosomal mapping of 5S and 18S ribosomal genes using double FISH. Aquac Res 44:829–835

Caperta A, Delgado M, Ressurreição F, Meister A, Jones R, Viegas W, Houben A (2006) Colchicine-induced polyploidization depends on tubulin polymerization in c-metaphase cells. Protoplasma 227:147–153

Chourrout D, Happe A (1986) Improved methods of direct chromosome preparation in rainbow trout, *Salmo gairdneri*. Aquaculture 52:255–261

Christopher JSG, Murugesan AG, Sukumaran N (2010) Induction of meiotic gynogenesis in the stinging catfish *Heteropneustes fossilis* (Bloch) and evidence for female homogamety. Aquac Res 42:129–138

Crego-Prieto V, Danancher D, Campo D, Perez J, Garcia-Vazquez E, Roca A (2013) Interspecific introgression and changes in population structure in a flatfish species complex after the *Prestige* accident. Mar Poll Bull 74:42–49

Fan Z, Fox D (1990) A new method for fish chromosome preparation. J Fish Biol 37:553–561

Fopp-Bayat D, Woznicki P (2006) Verification of ploidy level in sturgeon larvae. Aquac Res 37:1671–1675

Gilna B, Kuzma J, Otts SS (2014) Governance of genetic biocontrol technologies for invasive fish. Biol Invasion 16:1299–1312

Gui J, Zhu Z (2012) Molecular basis and genetic improvement of economically important traits in aquaculture animals. Chin Sci Bull 57:1751–1760

Henegariu O, Heerema NA, Lowe Wright L, Bray-Ward P, Ward DC, Vance GH (2001) Improvements in cytogenetic slide preparation: controlled chromosome spreading, chemical aging and gradual denaturing. Cytometry 43:101–109

Hussain M, McAndrew B (1994) An improved technique for chromosome karyotyping from embryonic and soft tissues of tilapia and salmonids. Asian Fish Sci 7:187–190

Inokuchi T, Abe S, Yamaha E, Yamazaki F, Yoshida MC (1994) BrdU replication banding studies on the chromosomes in early embryos of salmonid fishes. Hereditas 121:255–265

Karami A, Christianus A, Ishak Z, Courtenay S, Syed MA, Noor A, Noorshinah H (2010) Effect of triploidization on juvenile African catfish (*Clarias gariepinus*). Aquac Int 18:851–858

Karami A, Christianus A, Zokaeifar H, Saad KZ, Imraan FTJ, Shakibazadeh S, Negarestan H, Courtenay SC (2011) Ovaprim treatment promotes oocyte development and milt fertilization rate in diploid and triploid African catfish (*Clarias gariepinus*). Aquac Int 19:1025–1034

Kligerman AD, Bloom SE (1977) Rapid chromosome preparations from solid tissues of fishes. J Fish Res Board Can 34:266–269

Lomax B, Tang S, Separovic E, Phillips D, Hillard E, Thomson T, Kalousek DK (2000) Comparative genomic hybridization in combination with flow cytometry improves results of cytogenetic analysis of spontaneous abortions. Am J Hum Genet 66:1516–1521

Moore CM, Best RG (2001) Chromosome preparation and banding. eLS. doi:10.1038/npg.els.0001444

Nasiadka A, Clark MD (2012) Zebrafish breeding in the laboratory environment. ILAR J 53:161–168

Pradeep PJ, Srijaya TC, Zain RBM, Papini A, Chatterji AK (2011) A simple technique for chromosome preparation from embryonic tissues of teleosts for ploidy verification. Caryologia 64:235–241

Rieder CL, Palazzo RE (1992) Colcemid and the mitotic cycle. J Cell Sci 102:387–392

Shao C-W, Wu P-F, Wang X-L, Tian Y-S, Chen S-L (2010) Comparison of chromosome preparation methods for the different developmental stages of the half-smooth tongue sole, *Cynoglossus semilaevis*. Micron 41:47–50

Shelton WL, Mims SD, Clark JA, Hiott AE, Wang C (1997) A temperature-dependent index of mitotic interval ($\tau 0$) for chromosome manipulation in paddlefish and shovelnose sturgeon. Prog Fish Cult 59:229–234

Tabachnick BG, Fidell LS (2001) Using multivariate statistics

Thorgaard GH, Jazwin ME, Stier AR (1981) Polyploidy induced by heat shock in rainbow trout. Trans Am Fish Soc 110:546–550

Thresher RE, Hayes K, Bax NJ, Teem J, Benfey TJ, Gould F (2014) Genetic control of invasive fish: technological options and its role in integrated pest management. Biol Invasion 16:1201–1216

Tilney LG, Porter KR (1967) Studies on the microtubules in Heliozoa II. The effect of low temperature on these structures in the formation and maintenance of the axopodia. J Cell Biol 34:327–343

Wakahara M (1972) Daily variation in mitotic rate in tail-fin epidermis of larval *Xenopus laevis* and its modification by pineal organ-subcommissural organ system and photoperiods. Neuroendocrinol 9:267–277

Wang S, Su Y, Ding S, Cai Y, Wang J (2010) Cytogenetic analysis of orange-spotted grouper, *Epinephelus coioides*, using chromosome banding and fluorescence in situ hybridization. Hydrobiol 638:1–10

Weisenberg RC (1972) Microtubule formation in vitro in solutions containing low calcium concentrations. Science 177:1104–1105

Wood KW, Cornwell WD, Jackson JR (2001) Past and future of the mitotic spindle as an oncology target. Curr Opin Pharmacol 1:370–377

7

Demonstration of virulent genes within *Listeria* and *Klebsiella* isolates contaminating the export quality frozen shrimps

Rashed Noor · Md. Faqrul Hasan · Md. Sakil Munna · M. Majibur Rahman

Abstract In the previously studied export quality shrimp samples, presence of a range of pathogenic bacteria including *Listeria* spp. and *Klebsiella* spp. were detected. The presence of *eae* gene in *Escherichia coli*, *aero* specific gene in *Aeromonas* spp., and *sodB* gene in *Vibrio* spp. were observed which might confer the associated virulence. Present study further attempted to detect the existence of virulent genes in *Listeria* spp. and *Klebsiella* isolates from the same shrimp samples through the gene specific polymerase chain reaction (PCR). Visualization of bands of *iap* (457 bp) and *gyr*B2 (711 bp) genes after PCR amplification referred the presence of virulence strains of *Listeria* spp. and *Klebsiella* spp., respectively in the export quality shrimp samples. Together with the previous findings of virulence genes of several pathogenic bacteria, the outcomes of current study further conferred the possible detrimental impact of the export quality frozen shrimps as and when consumed.

Keywords Consumer health safety · Export quality frozen shrimps · *Listeria* spp. · *Klebsiella* spp. · *lap* · *gyr*B2

Findings

Although the export quality frozen shrimps cover the prime economic importance in Bangladesh, they are not unlikely to be microbiologically contaminated resulting in various food-borne disease outbreaks (Rahman et al. 2012; Noor et al. 2013; Hassan et al. 2013; Ahmed et al. 2013; World Health Organization 2008, 2013). Inappropriate conditions of transport, handling and processing may impart such problem associated with quality of shrimps (Noor et al. 2013; Hassan et al. 2013; Ahmed et al. 2013; Sultana et al. 2014; Ghalis et al. 2010; Rahman et al. 2012). Besides, the subsistence of virulent genes could be of clinical importance

R. Noor (✉) · Md. Sakil Munna
Department of Microbiology, Stamford University Bangladesh, 51 Siddeswari Road, Dhaka 1217, Bangladesh
e-mail: noor.rashed@yahoo.com

Md. Sakil Munna
e-mail: sakilmunna@gmail.com

Md. Faqrul Hasan · M. Majibur Rahman
Department of Microbiology, University of Dhaka, Dhaka 1000, Bangladesh
e-mail: faqrulhasan10225@gmail.com

M. Majibur Rahman
e-mail: majibrahman@gmail.com

contributing to the pathogenesis of the contaminating bacteria (Hotopp 2011; Amin and Salem 2012; Yuan et al. 2013; Postollec et al. 2011; Nakamura et al. 2013).

Listeria monocytogenes is known to be an opportunistic foodborne pathogenic bacterium generally triggering listeriosis, and also may lead to the onset of meningitis, sepsis, etc. (Painter and Slutsker 2007; Scallan et al. 2011). Among several food borne complications like diarrhoea, cholera, salmonellosis, etc., commencement of listeriosis also emerged as a major food borne health threat round the globe (Ross 2014). *L. monocytogenes* have been vastly reported to contaminate raw and undercooked meats, raw vegetables and fruits, unpasteurized milk, soft cheeses, etc. (Senjuti et al. 2014; Tahera et al. 2014; Marjan et al. 2014; Sarker et al. 2013). The *iap* gene of *L. monocytogenes* has been reported to be essential for maintaining the cell viability (Krawczyk-Balska et al. 2005).

Besides *Listeria* spp., bacteria belonging to the genus *Klebsiella* are well known for the onset of common food borne diseases (Abdaslam et al. 2014). Seafoods have been found earlier to be contaminated by *Klebsiella* spp. at the processing stages, possibly due to the poor handling and hygiene maintenance (Hamdan et al. 2008). DNA gyrase in *Klebsiella* spp., encoded by *gyrB2* gene, is a prokaryotic type II topoisomerase which is required for DNA replication as well as to maintain the cell viability (Nitiss 2009; Gore et al. 2006).

The molecular techniques have been proven to be promising in the disease diagnosis and estimating the prophylaxis in recent years for the purpose of detection of pathogens as well as the associated virulent genes (Liu et al. 2008). The polymerase chain reaction (PCR) has been reported to be the most useful and cost effective tool (Noor et al. 2014; Liu et al. 2008; Yu et al. 2007). In Bangladesh, the molecular aspect of pathogenic diagnosis of the export quality shrimps still remains scarce (Noor et al. 2014). In our recent study, the gene specific PCR method revealed the presence of *eae* gene in *E. coli*, *aero* specific gene in *Aeromonas*, *sodB* gene in *Vibrio* spp. (Noor et al. 2014). Besides, as stated earlier, some other genes from *Listeria* spp. and *Klebsiella* spp. could also be important for imparting the virulence of the respective bacteria as isolated in our previous study (Noor et al. 2014; Nawaz et al. 2012). Based on this rationale, current investigation further intended to detect the presence of virulent genes in the pathogenic isolates of *Listeria* spp. and *Klebsiella* spp.

Methods

As has been stated in our recent study (Noor et al. 2014), a total of 30 export quality frozen shrimp samples were collected and was processed for conducting the associated microbiological load estimation including *Listeria* spp. and *Klebsiella* spp. (Noor et al. 2014). To meet the objective of the current study, the genomic DNA from each of the two bacterial isolates was extracted through the modified boiling method and used for specific gene amplification (Noor et al. 2014). The primers used for *iap* gene (457 bp) in *Listeria* isolates were 5′-ATGTCATGGAATAA-3′ and 3′-GCTTTTCCAAGGTGTTTTT-5′; and for the amplification of *gyr*B2 gene (711 bp) within *Klebsiella* isolates, 5′-TCCGGCGGTCTGCACGGCGT-3′ and 3′-TTGTCCGGG TTGTACTCGTC-5′ (Liu et al. 2008; Yu et al. 2007). Amplification of *lap* gene through PCR was conducted employing the initial denaturation at 94 °C for 1 min followed by further denaturation at 92 °C for 1 min, primer annealing at 60 °C for 1 min and extension at 72 °C for 1 min, and a final extension at 72 °C for 10 min. Thirty (30) PCR cycles were executed in preface experiment within the exponential phase of amplification. Subsequently, amplification of *gyr*B2 was done with an initial denaturation at 94 °C for 4 min followed by the denaturation step at 94 °C for 1 min, primer annealing at 62 °C for 1–2 min and extension at 72 °C for 90 s followed by a terminating extension step at 72 °C for 10 min. PCR products were resolved through 1.2 % gel electrophoresis, stained with ethidium bromide, and visualized by UV trans-illuminator (Gel Doc, Bio-Rad Laboratories, In., USA).

Results and discussion

Although fish has been known to serve as the major nutrient supplements for human and other animals, the associated microbial spoilage of fish during the harvesting, handling, transportation, storage, and unhygienic maintenance is too common to affect the global food safety as well as the economy (Noor et al. 2013;

Acharjee et al. 2014). Considering the popularity for fish consumption together with the business perspectives, it is worth to control the microbiological quality of the fish to ensure the public health safety.

The bacterial bio-burden of *Listeria* spp. and *Klebsiella* spp. in the export quality frozen shrimp samples have already been stated in our recent study (Noor et al. 2014). In the present study, the gene specific PCR study revealed the presence of *Listeria* specific gene *iap* (encoding the invasion associated protein p60) in the tested *Listeria* isolates (as shown for two samples in Fig. 1). The isolates of *Klebsiella* spp. exhibited the presence of *gyr*B2 gene encoding DNA gyrase βII subunit (Fig. 2).

Our earlier microbiological and molecular studies on the export quality shrimps demonstrated a huge microbial contamination, possibly due to the lack of hygienic maintenance during the finished product processing and inappropriate storage (Rahman et al. 2012; Hassan et al. 2013; Ahmed et al. 2013; Sultana et al. 2014; Noor et al. 2014). The tested bacterial isolates were found almost to be drug-resistant (Noor et al. 2014). Also, our investigation on the propagation of the virulent genes (presence of *eae*, *sodB* and *aero* specific genes) within the export quality shrimps demonstrated the molecular basis of shrimp related fatality upon consumption (Noor et al. 2014). According to present study, export quality shrimps have been found to harbor the additional virulent genes in *Listeria* spp. and *Klebsiella* spp. which further added the new molecular insight into the current knowledge on the shrimp quality.

Considering the food safety as well as the public health significance, molecular study of the virulence traits of fish borne pathogens besides the pathogenic load is of significance in light of the knowledge on the harmful gene propagation among the fish. Besides bacterial burden, the successful detection of the virulent genes in fish may further forecast on the possible remedies with accuracy. Together with our recent findings (Noor et al. 2014), the results of the current study further revealed the existence of the virulent genes in the export quality

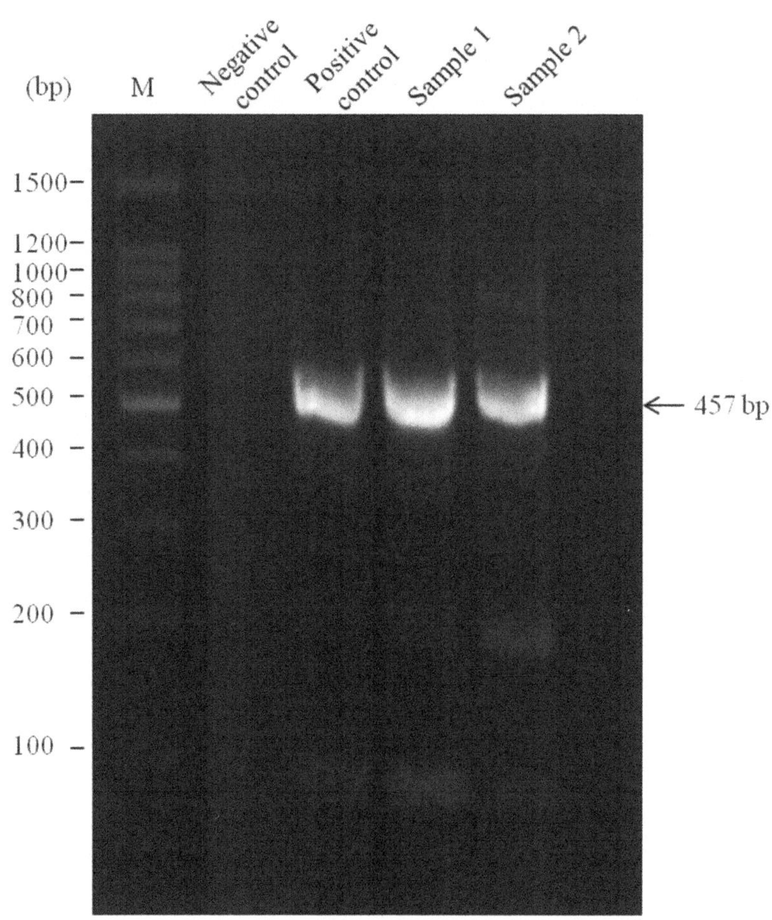

Fig. 1 Detection of the existence of *iap* gene in *Listeria* isolates. The PCR products of *iap* gene were resolved through 1.2 % agarose gel electrophoresis. As the positive control, the amplified DNA of *Listeria monocytogenes* (ATCC 19115) was used. *M* Marker

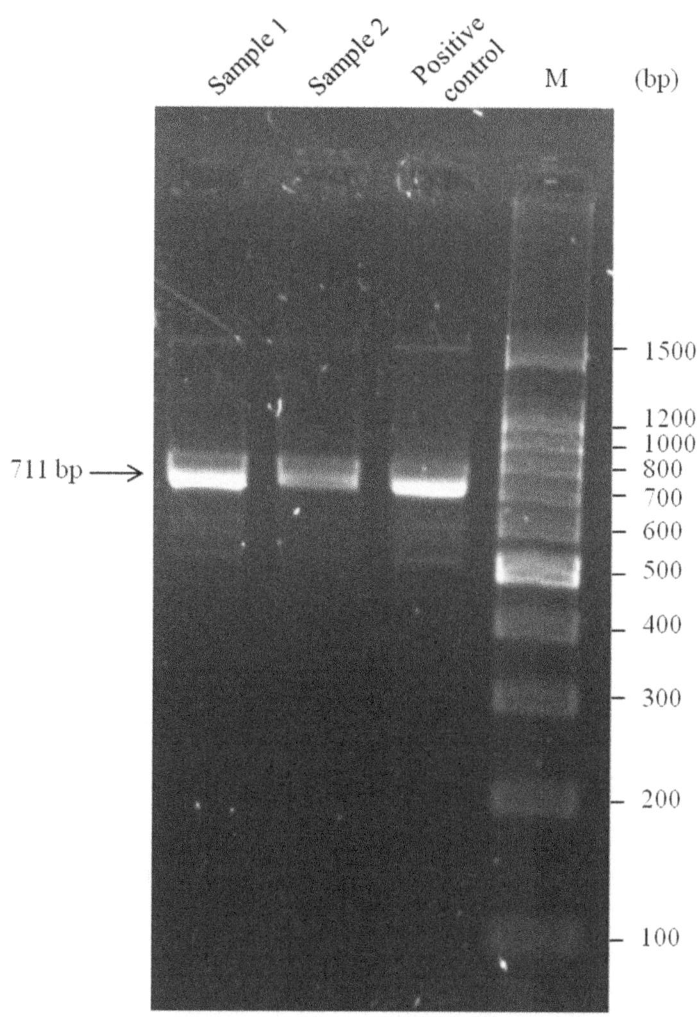

Fig. 2 Detection of *gyr*B2 gene within the *Klebsiella* isolates. The PCR products of *gyr*B2 gene were resolved through 1.2 % agarose gel electrophoresis. As the positive control, *Klebsiella pneumoniae* (ATCC 700603) was used. *M* Marker

shrimps contaminated with *Listeria* and *Klebsiella* isolates, thereby establishing the necessity of the routine molecular diagnosis for the sake of precise regulation of the microbiological quality of fish.

Author's contribution This work was carried out in collaboration between all authors. Author RN managed literature searches during preparation of the manuscript draft, analyzed the data, revised and approved the final manuscript. Author MFH conducted the experiments. Author MSM initially drafted the manuscript. Author MMR designed the study and supervised the related experiments.

Acknowledgments Authors are thankful to Saurab Kishore Munshi, Farjana Rahman and Mrityunjoy Acharjee for their thoughtful advices and technical assistance during the experiments. The work was financed by the Ministry of Education, People's Republic of Bangladesh. Author RN is the Chairman and Associate Professor of the Department of Microbiology, Stamford University Bangladesh. Author MFH is a researcher in the Department of Microbiology, University of Dhaka. Author MSM is an MS student of the Department of Microbiology, Stamford University Bangladesh. Author MMR is the Professor of the Department of Microbiology, University of Dhaka.

Conflict of interest The authors have declared that they have no competing interests.

References

Abdaslam SA, Hassan MA, Kaheel HH, Abobaker TM, Alnourain TA, Hamdan HA, Shankar SG, Thambirajah JJ (2014) Isolation of *Escherichia coli* O157 and other food borne pathogens from meat products and their susceptibility to different antimicrobial agents. Curr Microbiol Biotechnol 2(3):391–397

Acharjee M, Ahmed E, Munshi SK, Noor R (2014) Validation of γ-irradiation in controlling microorganisms in fish. Nutr Food Sci 44(3):258–266

Ahmed T, Baidya S, Sharma BC, Malek M, Das KK, Acharjee M, Munshi SK, Noor R (2013) Identification of drug-resistant bacteria among export quality shrimp samples in Bangladesh. Asian J Microbiol Biotechnol Env Sci 15:31–36

Amin RA, Salem AM (2012) Specific detection of pathogenic *Vibrio species* in shellfish by using multiplex polymerase chain reaction. Global Vet 8:525–531

Ghalis AE, Dave D, Budge S, Brooks MS (2010) Fish spoilage mechanisms and preservation techniques: review. Am J Appl Sci 7:859–877

Gore J, Bryant Z, Stone MD, Nollmann M, Cozzarelli NR, Bustamante C (2006) Mechanochemical analysis of DNA gyrase using rotor bead tracking. Nature 5(439):100–110

Hamdan RR, Musa N, Nadirah Musa N, Wei LS, Sarman A (2008) Isolation and enumeration of coliform bacteria and *Salmonella* spp. from short necked clam orbicularia orbiculata at East Coast. Malaysia. Int J Food Safety 10:58–64

Hassan MR, Acharjee M, Das E, Das KK, Ahmed T, Akond MA, Fatema KK, Noor R (2013) Microbiological study of sea fish samples collected from local markets in Dhaka city. Int Food Res J 20:1491–1495

Hotopp JCD (2011) Horizontal gene transfer between bacteria and animals. Trends Genet 27:157–163

Krawczyk-Balska A, Markiewicz Z, Bielecki J (2005) Listeria monocytogenes protein p60 affects hemolytic activity and uptake of bacteria by macrophages. Folia Microbiol 50(6):515–518

Liu Y, Liu C, Zheng W, Zhang X, Yu J, Gao Q (2008) PCR detection of *Klebsiella pneumoniae* in infant formula based on 16S-23S internal transcribed spacer. Intl J Food Microbiol 125(3):230–235

Marjan S, Das KK, Munshi SK, Noor R (2014) Drug-resistant bacterial pathogens in milk and some milk products. Nutr Food Sci 44(3):241–248

Nakamura Y, Takano T, Yasuike M, Sakai T, Matsuyama T, Sano M (2013) Comparative genomics reveals that a fish pathogenic bacterium *Edwardsiella tarda* has acquired the locus of enterocyte effacement (LEE) through horizontal gene transfer. BMC Genom 14:642

Nawaz M, Khan SA, Tran Q, Sung K, Khan AA, Adamu I, Steele RS (2012) Isolation and characterization of multidrug-resistant *Klebsiella* spp. isolated from shrimp imported from Thailand. Int J Food Microbiol 155(3):179–184

Nitiss JL (2009) DNA topoisomerase II and its growing repertoire of biological functions. Nature 16:1–6

Noor R, Acharjee M, Ahmed T, Das KK, Paul L, Munshi SK, Urmi NJ, Rahman F, Alam MJ (2013) Microbiological analysis of major sea fish collected from local markets in Dhaka city, Bangladesh. J Microbiol Biotechnol Food Sci 2:2420–2430

Noor R, Faqrul MH, Rahman MM (2014) Molecular characterization of the virulent microorganisms along with their drug-resistance traits associated with the export quality frozen shrimps in Bangladesh. Springer Plus 3:469

Painter J, Slutsker L (2007) Listeriosis in humans. In: Ryser ET, Marth EH (eds) Listeria, listeriosis and food safety, 3rd edn. Boca Raton, Florida, Taylor and Francis Group, pp 85–110

Postollec F, Falentin H, Pavan S, Combrisson J, Sohier D (2011) Recent advances in quantitative PCR (qPCR) applications in food microbiology. Food Microbiol 28:848–861

Rahman MM, Rahman F, Afroze F, Yesmin F, Fatema KK, Das KK, Noor R (2012) Prevalence of pathogenic bacteria in shrimp samples collected from hatchery, local markets and the shrimp processing plant for export quality frozen shrimps. Bangladesh J Microbiol 29:7–10

Ross T (2014) Risk analysis: risk assessment: microbiological hazards. Encycl Food Saf 1:80–92

Sarker N, Islam S, Hasan M, Kabir F, Uddin MA, Noor R (2013) Use of multiplex PCR assay for detection of diarrheagenic *Escherichia coli* in street vended food items. Am J Life Sci 1(6):267–272

Scallan E, Hoekstra RM, Angulo FJ, Tauxe RV, Widdowson MA, Roy SL, Jones JL, Griffin PM (2011) Foodborne illness acquired in the United States–major pathogens. Emerg Infect Dis 17(1):7–15

Senjuti JD, Feroz F, Tahera J, Das KK, Noor R (2014) Assessment of microbiological contamination and the in vitro demonstration of the anti-bacterial traits of the commonly available local fruit blend within Dhaka Metropolis. J Pharmacogn Phytochem 3(1):73–77

Sultana S, Tarafder GH, Siddiqui TA, Shaha BC, Walliullah M, Ahmed T, Munna MS, Noor AF, Popy SK, Das KK, Acharjee M, Urmi NJ, Rahman T, Noor R (2014) Microbiological quality analysis of shrimps collected from local market around Dhaka city. Int Food Res J 21:33–38

Tahera J, Feroz F, Senjuti JD, Das KK, Noor R (2014) Demonstration of anti-bacterial activity of commonly available fruit extracts in Dhaka, Bangladesh. Am J Microbiol Res 2(2):68–73

World Health Organization (2008) Foodborne disease outbreaks: guidelines for investigation and control. WHO Press, Geneva

World Health Organization (2013) Research priorities for the environment, agriculture and infectious diseases of poverty. WHO Press, Geneva

Yu VL, Hansen DS, Ko WC, Sagnimeni A, Klugman KP (2007) Virulence characteristics of *Klebsiella* and clinical manifestations of *K. pneumonia* bloodstream infections. Emerg Infect Dis 13:986–993

Yuan JB, Zhang XJ, Liu CZ, Wei JK, Li FH, Xiang JH (2013) Horizontally transferred genes in the genome of pacific white shrimp, *Litopenaeus vannamei*. BMC Evol Biol 13:165

8

Mesoglein differences between two jellyfish populations of the genus Aurelia

Anastasiya V. Kotova · Oleg A. Kovtun · Olga I. Podgornaya · Leonid S. Adonin

Abstract The medusa, *Aurelia aurita* (Scyphozoa, Cnidaria), is counted as a cosmopolitan species with a worldwide distribution in most seas, from the poles to the tropics. Cnidarian is thought to possess two tissue layers: endoderm (gastroderm) and ectoderm, which are separated by huge mesoglea in medusa. The main medusas' morphology is similar in different populations. Previously the new protein "mesoglein" was determined as one of the main components of mesoglea. Deduced amino acid sequence of mesoglein contains Zona Pellucida (ZP) domain. The comparison of mesoglein and its gene from three habitats White Sea (WsA), Black Sea (BsA), Japonic Sea (JsA) has been done in the current work. The set of the mesoglea protein bands after SDS-PAGE is similar in all samples. Nevertheless, JsA mesogleins' M_r is 53\55 kDa, while WsA and BsA mesogleins have M_r of 47 kDa. Antibodies raised against WsA mesoglein recognize only mesoglein with M_r of 47 kDa, but not 53\55 kDa, both on immunoblot and immunocytochemistry. Mesogleal cells and "elastic" fibrils are stained intensively in the mesoglea both from WsA and BsA but not from JsA. The possibility of gene diversity was checked by PCR with primers specific for WsA mesoglein gene. PCR products of expected length were obtained on cDNA from polyA-RNA template from mesogleal cells of WsA and BsA medusa but not on cDNA of JsA medusa. Our results evidence that there are two different species in genus Aurelia: Aurelia aurita inhabits White and Black Seas while Aurelia sp.1 inhabits Japonic Sea. Such a suggestion is in agreement with the one previously done on the base of comparison by other molecular biology methods.

Keywords *Aurelia aurita* · Jellyfish · Mesoglein · Evolution

Introduction

Cnidaria is one of the most ancient animal groups which arose ~550 millions year ago during the Cambrian explosion. Cnidaria are aquatic organisms. Some of the Cnidaria representatives inhabit fresh or less salted

---

A. V. Kotova · O. I. Podgornaya · L. S. Adonin (✉)
Institute of Cytology RAS, St. Petersburg, Russia
e-mail: leo.adonin@gmail.com

O. A. Kovtun
Hydrobiological Station of Odessa National I. I. Mechnikov University, Odessa, Ukraine

O. I. Podgornaya · L. S. Adonin
Department of Cytology and Histology, Faculty of Biology and Soil Sciences, St-Petersburg State University, St.Petersburg, Russia

water, but the vast majority is the marine animals, which populated ocean waters from littoral up abyssal deeps both in plankton and in benthos (Dogel 1981). Cnidarians were for a long time grouped with *Ctenophores* in the phylum Coelenterata, but increasing awareness of their differences caused them to be placed in separate phyla (Daly et al. 2007). The defining Cnidarian feature is their possession of stinging cells, or cnidocytes, which is used for feeding (Kayal et al. 2013). Cnidarian is thought to possess two tissue layers: endoderm (gastroderm) and ectoderm, which are separated by mesoglea in medusa (Chapman 1974; Dogel 1981). The cnidarian radiation is relatively modest, giving rise to two major body plans (polyp and medusa) and some ten thousand extant species. The main difference between polyp and medusa is the degree of the mesoglea development. *Aurelia aurita* mesoglea contains a number of mesogleal cells (Mc). This feature is not unique but rather rare (Chapman 1966). Mesoglea, with its mesogleal cells, acquires a similarity with the connective tissues, derivate of mesoderm, of more advanced animals (Hyman 1940; Chapman 1974; Arai 1997).

Genus *Aurelia* belongs to the family Ulmaridae, order Semaeostomeae, class Scyphozoa (Kramp 1961). *Aurelia* is the best studied of all scyphozoans, yet its taxonomy and evolution are still under discussion (Dawson and Jacobs 2001). The cosmopolitan moon jelly *Aurelia* is characterized by high degrees of morphological and ecological plasticity, and subsequently by an unclear taxonomic status. The latter has been revised repeatedly over the last century, dividing the genus *Aurelia* into as many as 12 or as little as two species (Schroth et al. 2002).

The degree of Aurelia morphological plasticity never influences the main medusa feature—huge mesoglea. Previously the new protein "mesoglein" was determined as one of the main components of mesoglea. Mesoglein mRNA cloned sequence is 1421 bp long. The deduced amino acid (aa) sequence of 416 aa contains Zona Pellucida (ZP) domain and Delta/Serrate/Lag-2 domain. According to reverse transcription PCR, mesoglein is expressed in the mature medusa exclusively in the mesogleal cells (Matveev et al. 2007). The presence of the ZP domain in mesoglein allows to consider it as a member of the big family of ZP domain-containing proteins, which includes more than a thousand known members. The domain is thought to mediate the formation of extracellular protein fibers or meshworks (Jovine et al. 2005), which is in full agreement with mesoglein's position in the mesoglea (Shaposhnikova et al. 2005). The name of the family originates from Zona Pellucida, which is built up of ZP domain proteins and play crucial role in fertilization (Bork and Sander 1992).

Autoradiographic analysis and immunostaining indicate that Mc are typical protein-synthesizing cells (Napara et al. 1996). Immuno staining demonstrates the presence of the mesoglein in Mc granules and in the mesoglea "elastic" fibers; apical part of the ectodermal cell is also stained. So, it is confirmed by several techniques that Mc synthesize and excrete the main mesoglea protein which is included in the "elastic" fibers (Shaposhnikova et al. 2005).

Comparison mesogleins of medusas from White Sea (WsA), Black Sea (BsA) and Japonic Sea (JsA) has been the aim of the current work. Based on morphological data the bulk of original species is summarized as a single species, *A. aurita,* which is deemed to be an ecological generalist and occupy worldwide habitats of all but north polar oceans. (Schroth et al. 2002). According to this assumption we did not expect to find dramatic differences in mesoglea composition but it happens not to be so.

Materials and methods

Animals

A. aurita medusas were collected in the vicinity of the Hydrobiological Station of Odessa I.I.Mechnikov national university, Black Sea (BsA); Marine Biologcal Station "Vostok" of AV Zhirmunsky Institute of Marine Biology, Japan Sea (JsA); the White Sea Biological Station of the Zoological Institute RAS ''Kartesh'' (Chupa Inlet, Kandalaksha Bay in the White Sea (WsA) (http://www.zin.ru/kartesh/default_en.asp) during the summers of 2009–2013. The methods of mesoglea and mesogleal cells (Mc) isolation have been published (Shaposhnikova et al. 2005).

SDS electrophoresis and immunoblot

Mesoglea polypeptides were separated by SDS-PAGE (Laemmli 1970). Acrylamide concentration was from 7 to 10 % in gels. Gels were stained with 0.1 % Coomassie Brilliant Blue R-250.

Polypeptides separated by SDS-PAGE were transferred to polyvinylidene difluoride (PVDF, Sigma, USA) membrane or nitrocellulose membrane (Sigma) in a Trans-blot (Towbin et al. 1979). The membranes were blocked with 5 % skimmed milk at PBS-Tw for 1 h or overnight. The membrane was then immersed in PBS-Tw with Gpαmes-a (1:5000 dilution) for 1 h, then washed with PBS-Tw three times. The secondary antibody commercial stock (GAGp-HRP, Sigma, USA) was diluted 1/10,000 (v:v) in TBS-Tw. After incubation for 1 h at room temperature with shaking, the membranes were washed twice with PBS-Tw (10 min each). Finally, the sites of enzyme binding were developed with DAP–H_2O_2. All operations were carried out at room temperature with shaking. Guinea pig pre-immune serum instead of Gpαmes-a was used as the control and it did not produce any signal.

AU electrophoresis and antibodies production

Mesoglein is a charged protein so acid-urea polyacrylamide gel electrophoresis (AU-PAGE; Podgornaya and Shaposhikova 1998; Waterborg 2002) was used to separate the protein for immunization. Acrylamide concentration in gels was 7 %. Gels were stained with 0.1 % Coomassie Brilliant Blue R-250.

New polyclonal serum with antibodies against mesoglein of Aurelia aurita (WsA) was obtained. In this paper, it is named Gpαmes-a (Guinea pig anti-mesoglein from AU-PAGE). Mesoglein was cut out from the slab gel, homogenized with 0.9 % NaCl and mixed with Freund adjuvant (1:1); this mixture was injected hypodermically into male Guinea pig. Boosting with the same mixture was performed intramuscularly in 3 weeks. A week later about 8 ml of blood was taken from the Guinea pig's heart. The antiserums were mixed with glycerol (1:1) and 0.02 % NaN_3, aliquoted and stored at −20 °C. The samples of pre-immune serums were obtained from the same Guinea pig a week before antigen injection for control staining. The optimal working dilution of Gpαmes-a is 1:30,000 for immunoblots and 1:10,000/5000 for immunohistochemistry.

Immunohistochemistry

Small pieces (about 0.5 cm^3) of medusa's body including ectodermal layer and mesoglea were fixed in 4 % paraformaldehyde in filtered seawater for 24 h and rinsed sequentially in 30, 50 and 70 % ethanol (for 1 h each). The fixed samples were embedded in paraffin blocks to be cut into 3–5-μm-thick sections. The sections were preincubated with 5 % skimmed milk in TBS-Tw for 1 h, washed 4 times for 10 min with TBS-Tw, incubated with Gpαmes-a at 1:5000 dilution (for 1 h), washed and incubated with HRP-conjugated goat-anti-Guinea pig Ig (GAGp-HRP, Sigma), and visualized with microscope (Leica DM6000). For controls, some sections were incubated with non-immune GP serum followed by the same subsequent steps.

RNA isolation and cDNA synthesis

About 10 ml mesoglea jellyfish from each population (BsA, JsA, WsA) was cut into pieces and treated with 1 mg/ml collagenase (ICN, cat N 150705) solution in sea water at 37 °C for 30 min. The suspension of mesogleal cells (Mc) was centrifuged at 800*g* for 20 min. Supernatant was discarded and RNA was isolated from pelleted cells by Trizol (Invitrogen) solution according to the manufacturer's instructions. RNA samples were treated with RNAse-free DNAse (Roche) to avoid DNA contamination. DNAse was removed by phenol:chloroform (2:1) extraction. About 1 μg of the RNA was reverse transcribed using oligo-dT primer (5′ T(12-20)V) by MMLV RT kit (Evrogen) according to the manufacturer's recommendations. The RT reaction product was diluted fivefold and stored at −20 °C. For controls, PCR with primers GAPDH of Aurelia aurita (Aa_GAPDH_F: 5′caaattgccttgcacccttagcc; Aa_GAPDH_R: 5′gttctggaatgcaccacgtcca) were used.

PCR amplification

All PCRs were amplified in 25 μl reactions composed of 0.2 μM mesoglein (WsA, GenBank: ABK88269.1; Matveev et al. 2007)-specific primers (Table 1), 2.5 μl of 10 × PCR buffer, 3 mM $MgCl_2$, 0.2 mM dNTPs, and 0.1 μl of Taq polymerase (Evrogen) on Eppendorf Mastercycler Personal 5332 Thermocycler. All PCRs began with a 5-min denaturation step at 94–95 °C, followed by 25–30 thermal cycles including a 94 °C for 20 s denaturation step, 57 °C for 20 s primer annealing and elongation step—72 °C for 1.5 min. All PCRs terminated with a 3-min extension step at 72 °C and refrigeration at 4 °C and immediately used for agarose electrophoresis.

Results

The total set of the main mesoglea proteins is similar in medusas from all habitats on SDS-PAGE. Medusas from White Sea (WsA) initial mesogleins' apparent molecular mass (M_r) is 47 kDa (Figs. 1a, 2; Matveev et al. 2007); same M_r revealed for mesoglein of medusa from the Black Sea (BsA). Mesoglein of medusa from the Japonic Sea (JsA) differs: M_r is 53/55 kDa (Fig. 1a). At the next step we checked by immunoblot whether JsA mesoglein has the same antigenic determinants despite the M_r difference. Antibodies Gpαmes-a against WsA mesoglein were used (Adonin et al. 2009). It happens that both BsA and WsA mesogleins do react with the antibodies, but there is no staining with any proteins in JsA mesoglea (Fig. 1b).

Same antibodies were used to reveal mesogleins' position on paraffin sections. Gpames is the newly obtained serum and it is visible that it stained Mc, "elastic" fibers inside mesoglea and apical parts of the ectodermal cells (Figs. 2, 1) according to the position already known (Shaposhnikova et al. 2005). Same picture of mesoglea staining was obtained for BsA, though the staining is completely absent on JsA sections (Fig. 2b).

The absence of the antibodies cross reactivity makes us suppose that the degree of the gene diversity could be quite high. To check it, we produce the set of primer pairs based on the known WsA mesoglein sequence. The primers are situated in mesogleins' different parts and expected to produce the product of counted length in PCR (Fig. 3). The primer sequences are given in Table 1 and primers 2,3,4 are directed to the functional mesogleins' part—its' ZP domain, primers 1 and 5 directed against flanked regions. Primers for WsA GAPDH gene were used as the control. cDNA from total polyA-RNA from Mc was used as the template for PCR.

The PCR results are shown in the Table 2. It happens that PCR products of the length expected appear only on cDNA templates of WsA and BsA but not on JsA. So, the difference between mesoglein is obvious not only at protein level but at the level of RNA as well. The divergence of mesoglein genes in Aurelia from different populations is high enough not to give products with primers based on mesoglein initial sequence.

Table 1 Mesoglein (WsA) specific primers

No. pairs	Primer name	5′-3′ primer sequence
1	bDSL1_F	ACAGAAGAGGCCCCTGGTGCA
	bDSL1_R	TGCAAGCAAGAAAACGATGGAGACA
2	ZP1_F	CCCGCATCAGCTTGAAGGACGA
	ZP1_R	ACTGGAATGCGCCAAATCCATCT
3	ZP2_F	GGATTCAGCACCACAGAATTGTGGA
	ZP2_R	GCGTTCCATGGTAGGTGTTGCAT
4	ZP3_F	GCAACACCTACCATGGAACGCA
	ZP3_R	TCACCGCCAGCGTAGCTTGAA
5	aZP_F	TGGTCGCTGGAGTCTGCAGTC
	aZP_R	TGTGGCTCAGCAATGACTTCATTCA

Fig. 1 Mesogleas' protein composition on SDS-PAGE (a) and immunoblot with Gpαmes-a against WsA mesoglein (b). M_r in kDa of the marker proteins is given at the left. Mesoglea of medusa from JsA—*1*, WsA—*2*, BsA—*3*. Mesoglein staining is visible on *2* and *3* but not *1* on b

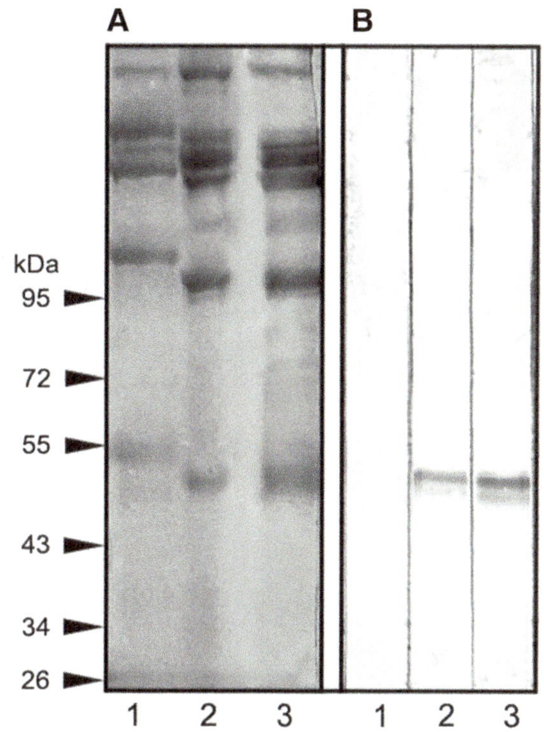

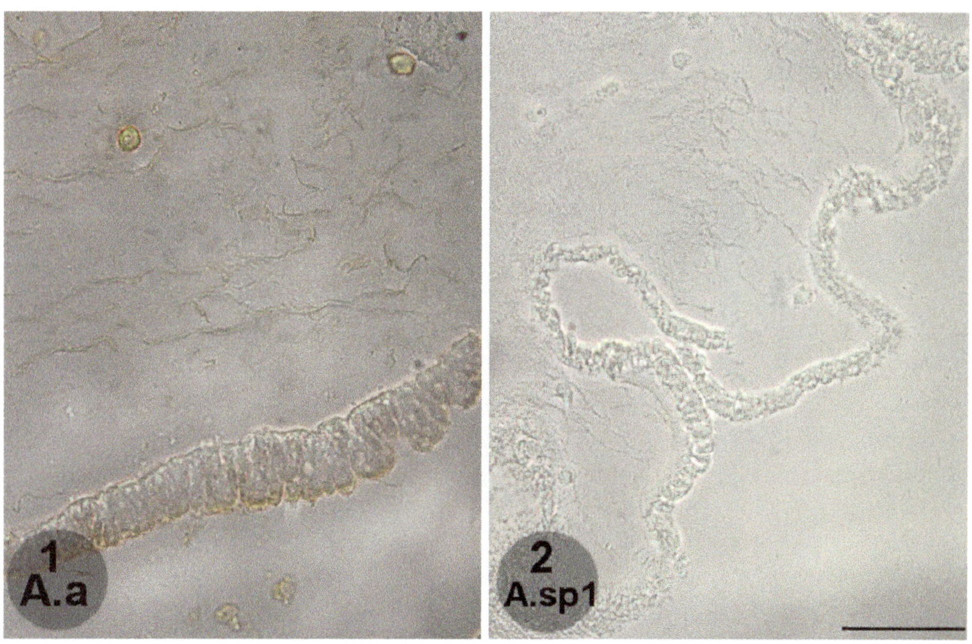

Fig. 2 Immunocytochemistry staining with Gpαmes-a antibodies on paraffin sections of WsA medusa (*1*) and JsA (*2*). *Bar* 50 μm

Discussion

The jellyfish *Aurelia aurita* (Linnaeus 1758) is usually considered to be a cosmopolitan species. Ecological differentiation in populations may force diversification and speciation (Schluter 1996) and it has to be determined whether phenotypic variation correlates to genetic and reproductive isolation. In marine environments invertebrate taxa in general show weak genetic structuring due to large population sizes, high larval

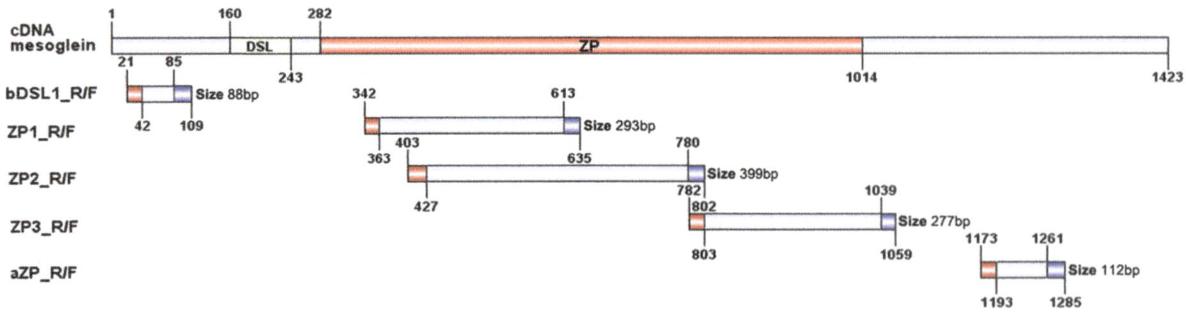

Fig. 3 Scheme of the mesoglein cDNA with the domain indicated and position of the pairs of primers. Primers' names are given according to the Table 1, first and last nucleotide position of each primer is indicated, *lines* between primers indicate the PCR product of the length counted

Table 2 PCR products obtained on Mc cDNA from medusas of three populations

Template	Primer pairs					
	1	2	3	4	5	GAPDH
WsA	+	+	+	+	+	+
BsA	+	+	+	+	+	+
JsA	–	–	–	–	–	+

Ns of primer pairs are given according to Table 1. (+) product of expected length, (–) no product

dispersal capacities, the lack of physical barriers, and sometimes interspecific hybridization (Palumbi 1992; Gosling 1994). (Schroth et al. 2002). Designating a species status can be a problematic task and the underlying concepts are subject of ongoing discussions (e.g., Templeton 1998; Avise 2000; Schroth et al. 2002).

At the beginning of the twentieth century there were three Aurelia species—*A. aurita, A. labiata, A. solida* (Mayer 1910). In the middle of the twentieth century seven species were separated: *A. aurita, A. coerulea, A. colpota, A. labiata, A. limbata, A. solida, Aurelia spp.* (Kramp 1961). Next decade brings up new classification—only two species were counted as established: *A. aurita* and *A. limbata* (Russell 1970). The rest of the species were combined in *A. aurita.* At the end of the century again one of the species from "Synopsis of the medusae of the world" (Kramp 1961), namely *A. labiata,* came back (Wrobel and Mills 1998).

The end of the century and, moreover, twenty-first century brings up molecular biology methods. In allozyme electrophoresis, significant differences at 12 of 14 polymorphic loci strongly suggest that *Aurelia* sp. from Foster City and Tokyo Bay belong to one species, while *Aurelia* sp. from Monterey Bay and Vancouver Island belong to a second species. The identities and taxonomic affinities of the two *Aurelia* defined in this study, and their relationships with the Linnaean *A. aurita* described from the North Atlantic, will require genetic and morphological study of the currently recognized species *A. aurita* and *A. limbata* (Brandt 1838) from several zoogeographical provinces. Nevertheless, on the protein level two groups, probably species, was distinguished—North American and Asian (Greenberg et al. 1996).

The sequencing approach again makes the story more complicated. Molecular phylogeny of Aurelia was built up on the base of nuclear ribosomal DNA (nrDNA), mitochondrial 16S and mitochondrial cytochrome *c* oxidase subunit I (COI) (Dawson 2003). *A. limbata* and *A. labiata* were defined as distinct species, but *A. aurita* was divided into 11 cryptic species A.sp.1–A.sp.11 (Fig. 4) (Dawson and Jacobs 2001).

The name *A. aurita* was saved to the initial population described by Linnaeus at the European North coast (Dawson 2003; Dawson et al. 2005). Comparison of mitochondrial and nuclear DNA data (16S and ITS-1/5.8S rDNA) strongly suggests historic speciation events and the reconstruction of at least seven different species within *Aurelia.* This type of analysis reveals the Aurelia similarity according to climatic zones defined by minimum (winter) surface water temperatures: boreal 0–8 °C (West-Atlantic, East-Atlantic/Northern Sea/ Baltic Sea, and Black Sea) and temperate–mediterranean 8–19 °C (Mediterranean locations, Australia, Japan, and American Pacific) (Schroth et al. 2002).

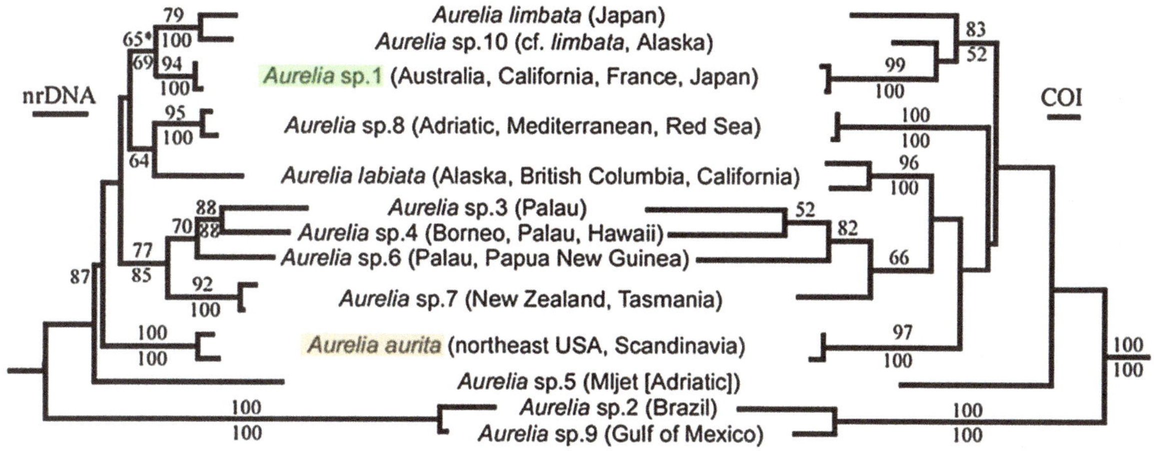

Fig. 4 Molecular phylogeny of *Aurelia* with maximum likelihood nrDNA and COI gene trees (Dawson et al. 2005, changed). *Aurelia aurita* (WsA) is in *orange rectangle*, *Aurelia sp.*1 (JsA) is in *green frame*

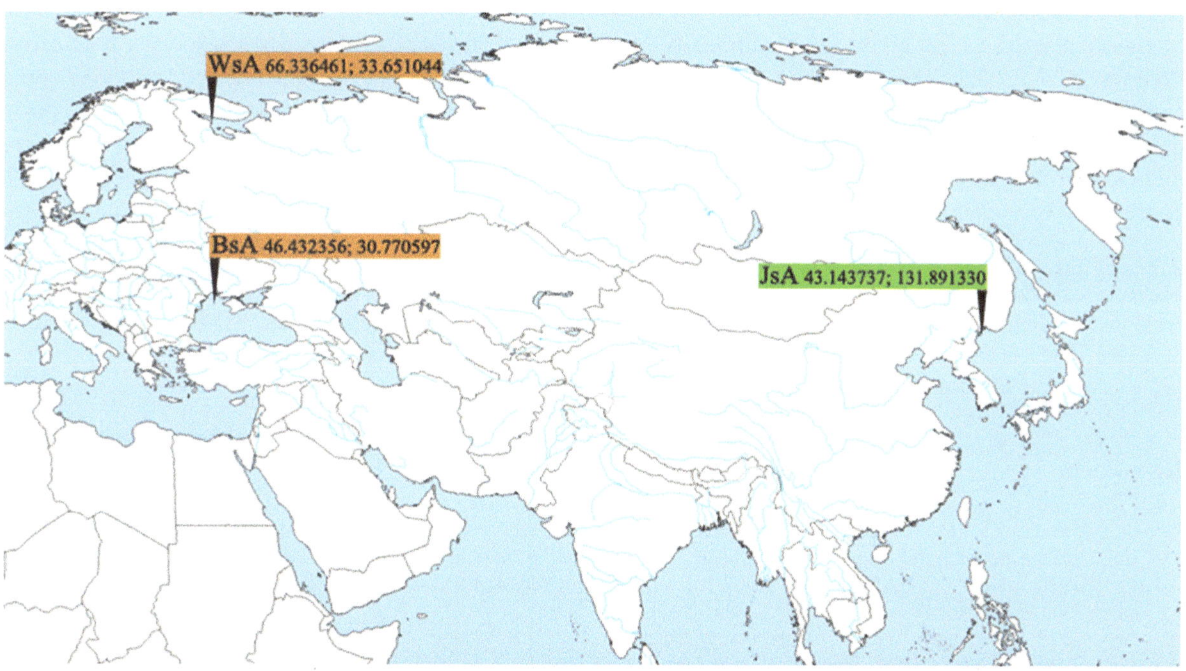

Fig. 5 Schematic representation of Aurelia distribution according to difference in mesoglein. *Orange flags*—WsA and BsA, *Aurelia aurita*; *green flag*—JsA, *Aurelia sp.1*. Species names are according to Fig. 4 (Dawson et al. 2005)

All species of genus Aurelia possess prominent mesoglea, i.e., extracellular matrix (ECM). At the morphological level two types of fibers, which are embedded in a jelly-like ground substance, have been described in cnidarians mesoglea: collagen-like and the so-called "elastic" or vertical fibers (Chapman 1959; Elder 1973; Bouillon and Coppois 1977; Weber and Schmid 1985). Medusa locomotion is attained by muscular contractions of the bell margin when water is ejected from the oral surface and the animal is propelled in the aboral direction, while the recovery is affected by elastic recoil of the mesoglea (Naumov 1961; Tardent 1978). Mesoglein is one of the main mesoglea ECM proteins and the essential component of the elastic fibers (Shaposhnikova et al. 2005; Matveev et al. 2007). It was reasonable to suppose that it is a conservative protein, especially due to its ZP domain. ZP proteins are quite an ancient purchase in the evolution: they are found even in Trichoplax genome (Matveev et al. 2012) not to mention their key role in fertilization (Jovine et al.

2005; Adonin et al. 2012). Surprisingly, we found differences at the level of M_r, (Fig. 1a), in immunoreactivity (Figs. 1b, 2) and in mesoglein gene cDNA (Table 2). The difference in sequence affects not only flanking sequences but ZP domain itself. The mesoglein gene from JsA is under cloning now to clarify the degree of diversity. Mesoglein comparison produced clear difference of two Aurelia populations: WsA and BsA belong to the boreal area, while JsA separated from its (Fig. 5).

Though one protein analysis is not enough to distinguish the species, at least two of them could be expected due to our data. Most of the contemporary animal groups appear for the first time in the fossil record some 545 million years ago (MYR) on the geological time scale in a relatively short period of time known as the Cambrian explosion (Benton and Harper 2009). Cnidaria could be even older for they could not leave the fossil record. Cnidaria do not disappear in the Permian extinction ($\sim$300 MYR) but continue their evolution. Aurelia's complex life cycle consists of several stages including alternating generations of sexual adult-stage medusa and asexual-stage polyp (scyphistoma) sitting at the sea bottom. Medusas also prefer coastal waters (Naumov 1961; Tardent 1978). So, the history of the Earth coasts is important for medusa evolution. At the board of the Jurassic and Cretaceous in Mesozoic about 150 MYR ago, the gradual rifting of the supercontinent Pangaea into separate landmasses that would eventually move into their current positions occurs. The Mesozoic was a time of significant tectonic, climate and evolutionary activity, which leads to the separation of the two oceans—Atlantic and Pacific. These two groups of genus Aurelia are clearly visible both in the current research and previously published (Greenberg et al. 1996; Schroth et al. 2002). BsA also belongs to the Atlantic population.

Genus Aurelia as the taxon has traditionally experienced difficulties in demarcating species boundaries (Schroth et al. 2002). The compound action of hybridization and temperature-dependent adaptations could contribute to the formation of the high species richness in *Aurelia,* but our data and the data previously published rather evidenced for the historically formed several groups, two of which (Atlantic and Pacific) are included in the present study. The traditional methods of molecular phylogeny analysis are necessary to confirm the existence of such groups and, probably, the precise species will be distinguished at the end.

Acknowledgments This work was supported by the Russian Foundation for Basic Research (grants 02-04-49420-a, 05-04-49578-a, 09-04-01145-a, 13-04-01739-a) and grant from presidium RAS (MCB).

References

Adonin LS, Podgornaya OI, Matveev IV, Shaposhnikova TG (2009) Plate in the zone of oocyte and germinal epithelium contact in Scyphomedusa *Aurelia aurita* binds antibodies to ZP-domain protein mesoglein. Cell Tissue Biol 3(3):283–288

Adonin LS, Shaposhnikova TG, Podgornaya OI (2012) Aurelia aurita (Cnidaria) Oocytes contact plate structure and development. PLoS One 7(11):e46542

Arai MN (1997) A functional biology of scyphozoa. Chapman and Hall, London, pp 28–206

Avise JC (2000) Phylogeography. The history and formation of species. Harvard University Press, Cambridge, p 464

Benton MJ, Harper DAT (2009) Introduction to paleobiology and the fossil record. Wiley-Blackwell, Hoboken, p 608

Bork P, Sander (1992) A large domain common to sperm receptors (Zp2 and Zp3) and TGF-beta type III receptor. FEBS Lett 300:237–240

Bouillon J, Coppois G (1977) Etude comparative de la mesoglée des cnidaires. Cah Biol Mar 18(1–3):339

Chapman G (1959) The mesoglea of Pelagia noctiluca. Q J Microsc Sci 100:599–610

Chapman G (1966) The structure and function of the mesoglea. In: The cnidaria and their evolution. London: Acad. Press. pp 147–168

Chapman DM (1974) Cnidarian histology. In: Coelenterate biology. Reviews and new perspectives. New York. Acad. Press. pp 1–93

Daly M, Brugler MR, Cartwright P, Collins AG, Dawson MN, Fautin DG, France SC, McFadden C, Opresko DM, Rodrigues E, Romano SL, Stake JL (2007) The phylum Cnidaria: a review of phylogenetic patterns and diversity 300 years after Linnaeus. Zootaxa 1668:127–182

Dawson MN (2003) Macro-morphological variation among cryptic species of the moon jellyfish, Aurelia (Cnidaria:Scyphozoa). Mar Biol 143(2):369–379

Dawson MN, Jacobs DK (2001) Molecular evidence for cryptic species of *Aurelia aurita* (Cnidaria, Scyphozoa). Biol Bull 200:92–96

Dawson MN, Gupta AS, England MH (2005) Coupled biophysical global ocean model and molecular genetic analyses identify multiple introductions of cryptogenic species. Proc Natl Acad Sci USA 102(34):11968–11973

Dogel VA (1981) Zoology of invertebrates. M., Vysshaya Shkola [in Russian]
Elder HY (1973) Distribution and functions of elastic fibers in the invertebrates. Biol Bull 144:43–63
Gosling EM (1994) Speciation and species concepts in the marine environment. In: Genetics and evolution of aquatic organisms. London, Chapman & Hall. p 539
Greenberg N, Garthwaite RL, Potts DC (1996) Allozyme and morphological evidence for a newly introduced species of Aurelia in San Francisco Bay, California. Mar Biol 125(2):401–410
Hyman LH (1940) The invertebrates, Vol. 1. Protozoa through ctenophora. McGraw Hill, New York, p 726
Jovine L, Costel CD, Litscher ES, Wassarman PM (2005) Zona pellucida domain proteins. Annu Rev Biochem 74:83–114
Kayal E, Roure B, Philippe H, Collins AG, Lavrov DV (2013) Cnidarian phylogenetic relationships as revealed by mitogenomics. BMC Evol Biol 13(1):5
Kramp PL (1961) Synopsis of the medusae of the world. J Mar Biol Soc UK 40:1–469
Laemmli UK (1970) Cleavage of structural proteins during the assembly of the head of bacteriophage T4. Nature 227(5259):680–685
Linnaeus C (1758) Systema naturae (Ed. 10), vol. 1. Salvii, Holmiae. 824 p
Matveev IV, Shaposhnikova TG, Podgornaya OI (2007) A novel *Aurelia aurita* protein mesoglein contains DSL and ZP domains. Gene 399:20–25
Matveev IV, Adonin LS, Shaposhnikova TG, Podgornaya OI (2012) Aurelia aurita–cnidarian with a prominent medusoid stage. J Exp Zool Part B Mol Dev Evol 318(1):1–12
Mayer AG (1910) Medusae of the World: the scyphomedusae. Carnegie institution of Washington, Washington, p 735
Napara TO, Oskol'sky AA, Shaposhnikova TG, Chaga OY (1996) Mesogleal cells of *Aurelia aurita* (Cnidaria:Scyphozoa) III. An autoradiographic analysis of the synthesis of extracellular matrix of mesoglea. Tsitologiya 38(4/5):465–474
Naumov DV (1961) Scyphomedusae of the seas of the USSR. Opred Faune SSSR 75:1–98 [in Russian]
Palumbi SR (1992) Marine speciation on a small planet. Trends Ecol Evol 7:114–117
Podgornaya OI, Shaposhikova TG (1998) Antibodies with the cell-type specificity to the morula cells from haemolymph of ascidian Styela rustica. Cell Struct Funct 23:349–355
Russell FS (1970) The Medusae of the British Isles volume II:pelagic scyphozoa, with a supplement to the first volume of Hydromedusae. Cambrige University Press, New York 284
Schluter D (1996) Ecological causes of adaptive radiation. Am Nat 148:40–64
Schroth W, Jarms G, Streit B, Schierwater B (2002) Speciation and phylogeography in the cosmopolitan marine moon jelly *Aurelia* sp. Biomed Cent Evolut Biol 2:1–10
Shaposhnikova T, Matveev I, Napara T, Podgornaya O (2005) Mesogleal cells of the jellyfish *Aurelia aurita* are involved in the formation of mesogleal fibres. Cell Biol Int 29:952–958
Tardent P (1978) Coelenterata, Cnidaria. In: Seidel T (ed) Morphogenesis der Tiere. Gustav Fischer Verlag, New York, pp 71–391
Templeton AR (1998) The role of molecular genetics in speciation studies. Molecular approaches to ecology and evolution. Birkhauser Verlag, Berlin, pp 131–156
Towbin H, Staehelin T, Gordon J (1979) Electrophoretic transfer of proteins from polyacrylamide gels to nitrocellulose sheets: procedure and some applications. Proc Natl Acad Sci 76(9):4350–4354
Waterborg JH (2002) The protein protocols handbook. Humana Press, New Jersey, pp 7–9
Weber C, Schmid V (1985) The fibrous system in the extracellular matrix of hydromedusae. Tissue Cell 17:811–822
Wrobel D, Mills C (1998) Pacific coast pelagic invertebrates:a guide to the common gelatinous animals. Sea Challengers, Monterey 108p

9

Optimization of freshwater microalgal biomass harvest using polymeric flocculants

Franziska Bleeke · Malgorzata Milas · Dominik Winckelmann · Gerd Klöck

Abstract Although microalgae show a great potential in the biotechnology sector, high production costs have limited industrial applications. Biomass harvest is one of the major bottlenecks in microalgae cultivation due to high energy inputs which are needed to separate the cells from the surrounding media. Chemical flocculation is considered to be a reliable resource to improve cost-effectiveness in the downstreaming processing. Flocculation efficiency is dependent on several factors such as the polymer type and charge as well as on the microalgae species. In the present study, 15 polyelectrolytes were tested for their potential to harvest algal biomass. Cationic, anionic and nonionic flocculants were tested in different amounts at varying incubation times to determine the adequate conditions needed. By testing the three chlorophytes, *Chlorella* sp., *Scenedesmus acuminatus* and *Chlamydomonas reinhardtii*, the influence of different sizes, morphologies and motilities of the flocculation efficiency was verified. Furthermore, the biocompatibility of an efficient flocculant was tested in a recycling experiment over a period of 8 weeks.

Keywords Microalgae · Biomass harvest · Chemical flocculation · Polyelectrolytes · Microalgal biotechnology · Polymeric flocculants

Introduction

Microalgae are microscopically small photosynthetic protists with rising significance in the biotechnology sector. According to Posten and Walter (2012), microalgae have a five times higher biomass productivity per hectare compared to terrestrial crops (Posten and Walter 2012). Furthermore, they can be grown on non-arable land and produce biomass which can serve as food or feedstock or as potential substrate for biofuel production (Chisti 2007; Mata et al. 2010).

Although microalgae cultivation displays great potential, high production costs have limited industrial application. A major factor is the cell harvest, which contributes to 20–30 % of the biomass production costs (Grima et al. 2003; Mata et al. 2010; Uduman et al. 2010). Large volumes have to be processed, since the

F. Bleeke (✉)
Jacobs University, 28759 Bremen, Germany
e-mail: franziskableeke@hotmail.de

F. Bleeke · D. Winckelmann · G. Klöck
Institute of Environmental Biology and Biotechnology, University of Applied Sciences Bremen, 28199 Bremen, Germany

M. Milas
Gdansk University of Technology, 80-233 Gdansk, Poland

concentration of cells is usually low at 0.5–2.5 g l^{-1} (Grima et al. 2003). Commonly used separation processes are a combination of filtration, flotation or flocculation, followed by a final centrifugation step. To separate the cells from the surrounding media, high-energy inputs are needed, often exceeding the energy content of the harvested biomass (Grima et al. 2003; Uduman et al. 2011; Wijffels and Barbosa 2010).

Increasing the efficiency at low-energy demands within the harvesting process is a major challenge in microalgal biotechnology. In flocculation, the cells coagulate and larger particles are produced with a higher settling velocity due to the higher density of the flocs. Flocculation can be achieved in several ways. Numerous studies have been published on the use of electrocoagulation (Lee et al. 2013; Uduman et al. 2011; Vandamme et al. 2011), pH-induced flocculation (Vandamme et al. 2012; Wu et al. 2012; Zheng et al. 2012) or bioflocculation using bacteria (Lee et al. 2009), filamentous fungi (Zhou et al. 2013) or another flocculating algae to floc a non-flocculating algae (Salim et al. 2011) for biomass harvest.

In chemical flocculation, a chemical coagulant is added to the algal suspension (Xu et al. 2010). Organic and inorganic flocculants like polyelectrolytes and metal salts were investigated. Aluminum, ferric and zinc salts were tested (Papazi et al. 2010) and compared with polyelectrolytes (Gerde et al. 2014; Granados et al. 2012; Papazi et al. 2010; Sirin et al. 2012). Comparative studies showed that higher flocculation efficiencies (FE) were achieved with polyelectrolytes than with metal salts (Gerde et al. 2014; Granados et al. 2012). Besides the lower FE, another disadvantage of the use of metal salts is the high concentration of metals in the algal biomass after harvest. Metal residues may impede the use of certain applications such as animal feed (Grima et al. 2003). Polymers are commonly used in wastewater purification processes and preferred for algal harvest, because lower amounts are needed and non-toxic and biodegradable substances are available (Granados et al. 2012). Polymeric flocculants are commercially available with cationic, anionic and nonionic charges in different charge densities, whereas cationic flocculants are considered to be the most effective for algal harvest (Granados et al. 2012; Tenney et al. 1969). The effect of different flocculants on several algae species has also been described. The results showed high variability with respect to the algae species and the nature of the flocculant (Gerde et al. 2014; Granados et al. 2012; Harith et al. 2009; Papazi et al. 2010; Rashid et al. 2013; Sirin et al. 2012).

Chlorella sp., *Scenedesmus acuminatus* and *Chlamydomonas reinhardtii* are commonly used laboratory strains and model organisms for algal research (Harris 2001; Mandal and Mallick 2009; Safi et al. 2014; Sanchez et al. 2008). By testing these three chlorophytes the influence of different sizes, morphologies and motilities of the flocculation efficiency were analyzed.

In the present study, 15 polymeric flocculants were compared to test the following hypotheses. Are cationic flocculants more effective compared to other polymeric flocculants (H1)? Do closely related species show similar flocculation behavior (H2)? Is a reuse of the media after cell separation by flocculation applicable (H3)?

Materials and methods

Microalgae and culture conditions

The experiments were carried out using the freshwater chlorophytes *Scenedesmus acuminatus*, *Chlorella* sp. and *Chlamydomonas reinhardtii* from the culture collection of the University of Applied Sciences Bremen. The cells were grown at 22 °C ± 1 in a culture volume of 2 l in Wuxal liquid medium (WM)(Winckelmann et al. 2014) without pH control. Bottles were equipped with aeration hoses providing continuous bubbling with compressed air at a flow rate of approximately 2.3 l min^{-1}. Every 3 days, 600 ml of the culture was used for the experiments and fresh medium was added to the remaining culture. Illumination was offered 24 h per day by fluorescent lamps (OSRAM L 30 W, warm white) placed in front and behind the algae cultures (light intensity, 50 μmol photons m^{-2} s^{-1}).

Flocculants

Stock solutions of 15 polymeric flocculants with cationic, anionic and nonionic characteristics were prepared according to the manufacturer's recommendation, dissolved in water and stored at 4 °C for a maximum of 2 weeks.

Flocculation experiments

Before each experiment, the pH of the sample from the algae culture was carefully adjusted to 7.0 by adding 1 N HCl. The cell number was determined using a Thoma counting chamber (Paul Marienfeld GmbH & Co. KG, Lauda-Königshofen, Germany) and subsequently adjusted to 1×10^7 cells ml^{-1} by dilution with tap water. Flocculation experiments were evaluated using jar tests (Granados et al. 2012; Hudson and Wagner 1981; Vandamme et al. 2010). 100 ml Erlenmeyer flasks were filled with 100 ml of algal suspension (Fig. 2). The flasks were stirred at 150 rpm. One type and concentration was used in each flask. Once the flocculant was added, the culture was mixed for 1 min at 250 rpm, to allow flocculant distribution. After 1 min, the mixing speed was reduced to 50 rpm for 2 min, to support floc formation, followed by a settling phase without agitation. Samples were taken directly after the flocculant distribution and after 2, 10, 30 and 60 min from 3 cm under the culture surface. Optical density was measured at 750 nm (Genesys 20, Thermo Scientific, Walthman, USA) and the flocculation efficiency (FE) was determined as follows:

$$\text{Flocculation efficiency} = \left(\frac{OD_{T0} - OD_{T1}}{OD_{T0}}\right) \times 100,$$

where OD_{T0} is the initial optical density before starting the flocculation and OD_{T1} is the optical density of the sample at a certain point of time during the process. Samples showing OD values above 0.5 were diluted with tap water for the measurements to assure linearity.

Media reuse after flocculation

A recycling experiment was conducted to investigate the potential negative impact of flocculant residues on algal growth when the media are recycled. Flocculant PK55H was chosen for the media reuse experiment. The experiment was conducted in triplicate using the microalgae *Chlamydomonas reinhardtii* under the same cultivation setup as described in "Microalgae and culture conditions". Over a period of 8 weeks, the cultures were harvested weekly by flocculation (Fig. 1a) or centrifugation (Fig. 1b), whereas the centrifugation served as positive control displaying the growth without the impact of the flocculant. The volume of the recycled media was dependent on the optical density of the culture. After cell separation, the culture supernatant was enriched with concentrated WM and led back into the culture vessel. The final optical density in the culture after media recycling was 1.5 at 750 nm.

Cell growth was monitored by optical density measurements at 750 nm, cell count determinations and dry biomass determination. For dry biomass determination, 10 ml of the algal culture was filtered through a previously balanced Whatman GF/C glass fiber filter (Whatman, Maidstone, UK). The filters were dried in an oven at 80 °C for 12 h and weighed afterward. The difference in weight was calculated and multiplied by 100 resulting in the dry biomass expressed in g l^{-1}.

Result and discussion

The flocculation process consists of different stages as displayed in Fig. 2. This shows the different phases during a flocculation process using *S.acuminatus* with the flocculant PK55H. After the addition of the polymer (a), the mixing speed was increased for 1 min to allow flocculant distribution (b). The flocculant adsorbs to the cell surface and the suspension is destabilized (c) and floc formation can be observed (d, e). The flocs grow

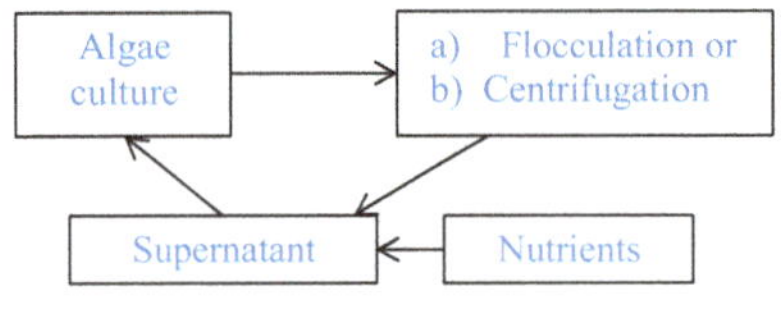

Fig. 1 Media reuse experiment. Over a period of 8 weeks, the cultures were harvested weekly by **a** flocculation or **b** centrifugation. After cell separation, the culture supernatant was enriched with concentrated WM and led back into the culture vessel

Fig. 2 Flocculation of *Scenedesmus acuminatus* using flocculant PK55H. **a** Addition of flocculant, **b** flocculant distribution, **c** destabilization of the suspension, **d**, **e** floc formation, **f–i** floc growth and **j** settling of the flocs

due to successive collisions and adsorption of microflocs (f–i). In a last phase, the agitation is stopped and the flocs are allowed to settle (j). According to the manufacturer, PK55H is a strong cationic polymer producing large dense flocs (Fig. 5i, c).

The flocculant concentration in the suspension significantly affects the FE. If a destabilization of the algal culture is not visible after flocculant addition, the amount of flocculant can be increased.

Flocculation experiments were conducted determining the FE at different dosages for all 15 flocculants investigated (Table 1). Furthermore, the optimal incubation time was identified by monitoring the FE at different time points. In each experiment, the flocculant was added and the optical density was measured after 0, 2, 10, 30 and 60 min. The settling of the microalgae was monitored as a result of flocculation. Figure 3a shows the flocculation of *Chlorella* sp., *Scenedesmus acuminatus* and *Chlamydomonas reinhardtii* with the cationic flocculant KW100. Increasing flocculant concentration causes the FE to rise. Figure 3a shows the results for *Chlorella* sp. at concentrations between 5 and 40 mg l^{-1} of flocculant in the algae culture. Whereas 5 mg l^{-1} did not even reach a FE of 50 %, concentrations above 25 mg l^{-1} resulted in a recovery rate of 90 %. The incubation time, however, differed significantly. Increasing the amount of flocculant added reduces the incubation time needed to reach 90 % FE. 60 min was needed at 25 mg l^{-1}, 10 min at 30 mg l^{-1} and only 2 min at 35 mg l^{-1} to reach 90 %. Higher amounts were needed for the flocculation of *S.acuminatus* and *C.reinhardtii* (Fig. 3b, c). An FE of 90 % was reached after 10 and 60 min, respectively, only after adding 60 mg l^{-1}. Lower doses of 15 mg l^{-1} for *Chlorella* and 30 mg l^{-1} for *Scenedesmus* and *Chlamydomonas* resulted in insufficient algae recovery. Different behaviors in flocculation performance between the algae species were also observed in experiments displayed in Fig. 4.

Cationic, anionic and nonionic flocculants were compared with regard to the flocculation efficiency of the three algal species. First, the minimum concentration needed to achieve the highest FE was determined.

Figure 4 shows the highest FE achieved within the experiments, together with the corresponding amount of flocculant applied. The concentration is given in mg l^{-1} and is indicated on top of each bar. Whereas Fig. 4a, b shows the results of cationic flocculants, Fig. 4c shows the FE for anionic and nonionic flocculants.

Table 1 The 15 tested flocculants. Commercial name, polymer type, net charge and supplier are listed

Commercial name	Polymer type	Net charge	Stock solution [g/l]	Supplier	Price [€/kg]
Emfloc KC 750	Starch	Cationic	10	Emsland-Stärke GmbH D-49824 Emlichheim	1.25
Magnafloc LT 22S DWI	Polyacrylamide	Cationic	5	BASF Corporation Florhalm Park, USA	NA
POLY SEPAR® CFL 25	Tannin, quaternary ammonia compound	Cationic	10	Separ Chemie GmbH D-22926 Ahrensburg	2.20
POLY SEPAR® KW 100	Quaternary ammonia compound, free of polyacrylamide	Cationic	10		2.40
POLY SEPAR® KW 45	Polyacrylamide	Cationic	10		2.40
POLY SEPAR® PK 55 H	Polyacrylamide	Cationic	2		3.20
POLY SEPAR® SK 72	Starch	Cationic	10		3.00
POLY SEPAR® KW 745 H	Polyacrylamide	Cationic	10		2.40
CFL 217	Poly DADMAC	Cationic	10		2.20
CFL 229	Poly DADMAC	Cationic	10		2.20
POLY SEPAR® AN 10 TW	Polyacrylamide	Anionic	10		3.50
Magnafloc LT 27	Polyacrylamide	Anionic	5	BASF Corporation	NA
Magnafloc LT 25	Polyacrylamide	Anionic	5	Florhalm Park, USA	NA
Magnafloc LT 20	Polyacrylamide	Nonionic	5		NA
POLY SEPAR® AN20	Polyacrylamide	Nonionic	10	Separ Chemie GmbH D-22926 Ahrensburg	2.80

The prices stated are price indications subjected to daily fluctuations and dependent on the ordered quantities
NA price not available

PK55H reached the best FE at the lowest flocculant concentration. According to the manufacturers' information, PK55H is a strongly cationic polymer of high molecular weight. Even at low PK55H concentrations of 1.5 mg l^{-1} for *Chlorella* and 2- and 4 mg l^{-1} for *Scenedesmus* and *Chlamydomonas*, high FE of <95 % was attained (Fig. 4b). This flocculant is suitable for all the three cell types, resulting in large and dense flocs (Fig. 5).

The cationic charge of a flocculant is dependent on the amount of cationic charged monomers bound to the polymer chain and can vary between 0 and 100 % (Gerde et al. 2014). According to the manufacturers' information, KW 100 and KW 45 are both polymers with similar polymer structure, merely differing in the amount of cationic groups attached to the polymer chain. The polymer chain of KW100 is completely substituted with cationic groups resulting in FE <90 %. In the case of KW45, only 35 % of the available sites are covered with cationic groups showing lower FE between 40 and 80 %. This result confirms the assumption that the degree of cationic ionization affects harvesting efficiencies. A higher cationic charge of a polymer results in better FE.

CFL217, CFL 229 and CFL 25 belong to the group of polyDADMACS (Diallyldimethyl ammonium chloride). According to the manufacturer's information the molecular weight of these polymers is low with a high cationic charge. The results showed FE of 80–100 % with the best results for *Scenedesmus*. In the case of CFL25, the results for the three tested microalgae differ (Fig. 4). *Chlorella* cells were coagulated with an FE of <90 % resulting in very fine and small flocs (Fig. 5). *Chlamydomonas* cells, however, only achieved FE of 20 %. The specific surface area of the cells is another important factor to consider. The smaller size of *Chlorella* cells (2–10 μm) give them a smaller surface area compared to *Scenedesmus* and *Chlamydomonas* cells (10–30 μm). Polymers of low molecular weight can entirely absorb onto the cell surface forming regions with cationic nature. These regions can bind negatively charged regions of other algal cells. Larger cells might not be bound effectively resulting in lower FE.

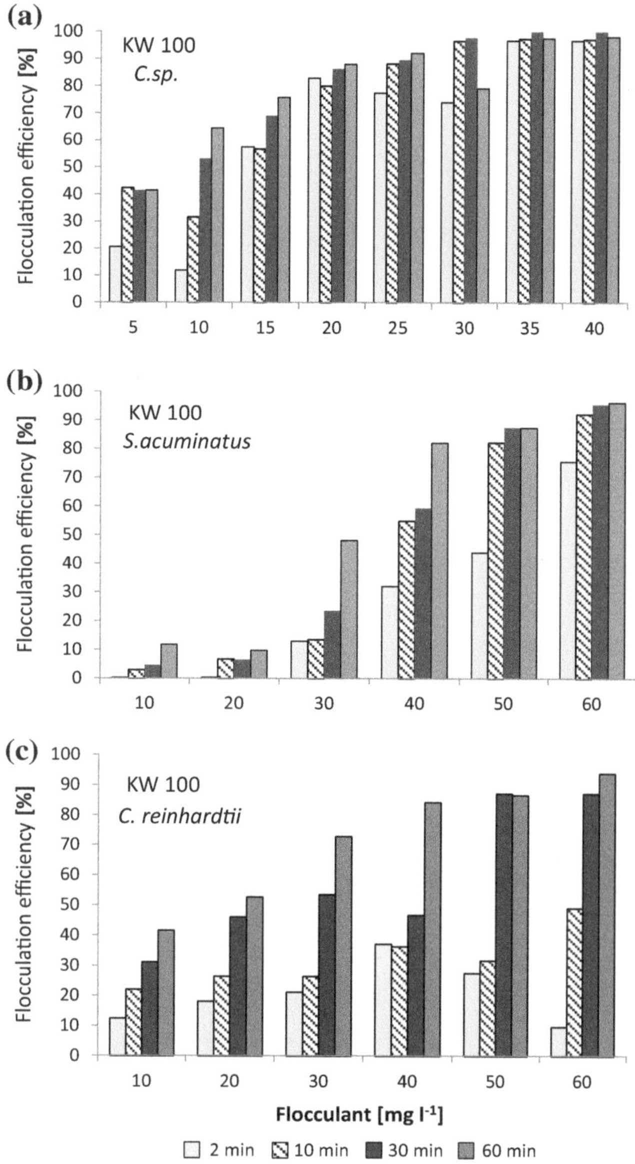

Fig. 3 Example of a flocculation experiment using the cationic flocculant KW100. Flocculation efficiencies for harvesting *Chlorella* sp. (**a**), *Scenedesmus acuminatus* (**b**) and *Chlamydomonas reinhardtii* (**c**) cultures are shown with different doses of flocculant (mg l^{-1}). Optical density measurements were conducted after 2, 10, 30 and 60 min

Several microalgae have been subjected to flocculation experiments such as *Chlorella*, *Chlamydomonas*, *Scenedesmus*, *Schizochytrium*, *Muriellopsis*, and *Phaeodactylum* resulting in different FE per milligram of flocculant (Gerde et al. 2014; Granados et al. 2012; Harith et al. 2009; Papazi et al. 2010; Rashid et al. 2013; Sirin et al. 2012). *Chlorella* sp., *Scenedesmus acuminatus* and *Chlamydomonas reinhardtii* are commonly used laboratory strains and model organisms for algal research (Harris 2001; Mandal and Mallick 2009; Safi et al. 2014; Sanchez et al. 2008). By testing these three chlorophytes, the influence of different sizes, morphologies and motilities of the flocculation efficiency should be verified. Considering the varying flocculant concentrations needed to floc the different algal species, the hypothesis that related species (Chlorophyceae) show similar flocculation behavior cannot be positively proved.

Figure 4c shows the results gained with nonionic (AN 20, LT20) and anionic flocculants (LT25, LT27 and AN10TW). No flocculation was observed for these flocculants, whichever dose was used.

At neutral pH, the cell surface of microalgae is usually negatively charged due to functional groups on the algal surface (Golueke and Oswald 1970). Therefore, generally positively charged flocculants are

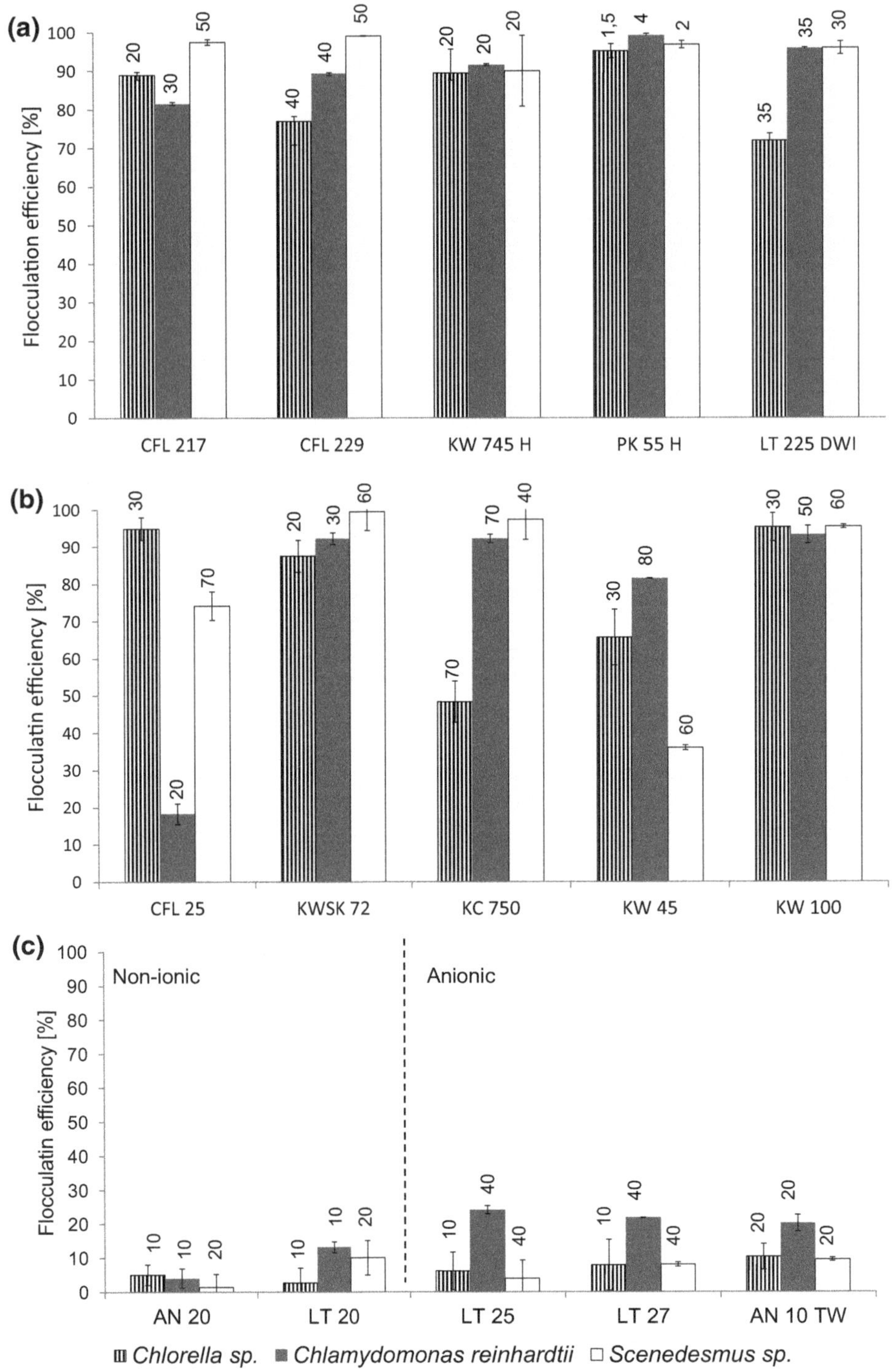

Fig. 4 Flocculation efficiencies of 15 polyelectrolytes with *Chlorella* sp., *Chlamydomonas reinhardtii* and *Scenedesmus acuminatus*. Cationic (**a**, **b**), anionic and nonionic (**c**) flocculants were tested and the most efficient concentration (mg l^{-1}) determined and displayed on *top of each bar*. The incubation time was 30 min in all experiments

recommended to neutralize the negatively charged cell surface of the microalgae (Granados et al. 2012; Tenney et al. 1969). Flocculation with polymers of the same charge can also be achieved by other mechanisms than charge neutralization, such as inter-particle bridging with long polymers. Tenney and co-workers reported the attachment of anionic polymers to the algal surface, but no floc formation in the solution was observed. Strongly anionic polymers can be effective in basic media. Harith et al. described an effective use of

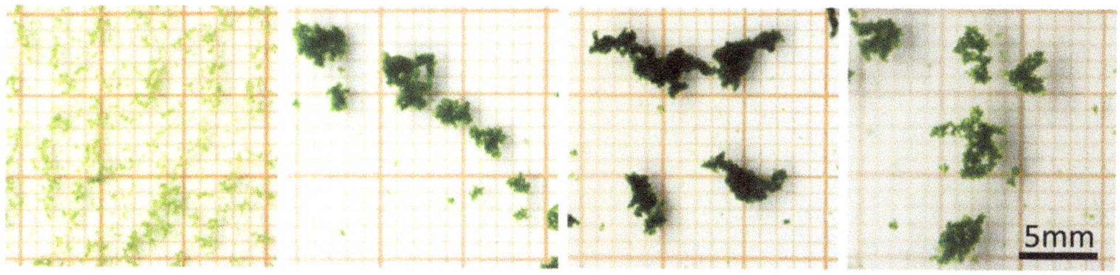

Fig. 5 Flocs produced by the flocculants CFL25, KW100, PK55H and KC750 (from *left to right*). The picture shows the result of a flocculation using *Scenedesmus acuminatus*

anionic flocculants only after algal surface charge neutralization by pH adjustments. The application of flocculant LT25 and LT 27 at a pH of 10.2 with the microalgae *Chaetoceros calcitrans* resulted in flocculation efficiencies above 90 % (Harith et al. 2009). The increase of pH before flocculation might improve FE, but negatively affect media recycling and thereby cost-effectiveness.

Based on the results, the use of anionic and nonionic flocculants cannot be recommended for harvesting of *Chlorella*, *Scenedesmus* and *Chlamydomonas* cells. The results gained positively prove the hypothesis that cationic flocculants are more effective than anionic or nonionic polymers in respects of FE.

For industrial application, the recycling of the growth media after cell harvest is an important factor in cost-effectiveness. Residues and color changes after flocculation experiments were reported, indicating that medium recycling after flocculation was not feasible (Sirin et al. 2012). It was necessary to verify if remaining flocculant particles do not negatively affect cell growth if the media are led back into the culture vessel after cell separation. A recycling experiment was conducted over a period of 8 weeks using *Chlamydomonas* cultures. The biomass was harvested weekly using the flocculant PK55H or centrifugation and the supernatant was returned into the bioreactor. Figure 6 shows the results of the growth of the microalgae by optical density (OD) measurements. After each harvesting, the optical density (750 nm) was adjusted to 1.5 to improve the light conditions for the algae. No differences between the flocculated cultures and the cultures harvested by centrifugation were visible by comparing OD values.

Although the OD values between flocculated and centrifuged cultures remained similar, the biomass concentration changed. After 10 days, the formation of aggregates was visible under the microscope and the ratio between OD and BTM changed (Table 2; Fig. 6). Although the OD values remained the same, due to the aggregates, more cells were present in the cultures, resulting in higher dry biomass values.

Table 2 lists the dry biomass values measured every week before harvesting. The data from the control and the recycling experiment differed significantly. After 8 weeks of cultivation, the dry biomass of the recycling

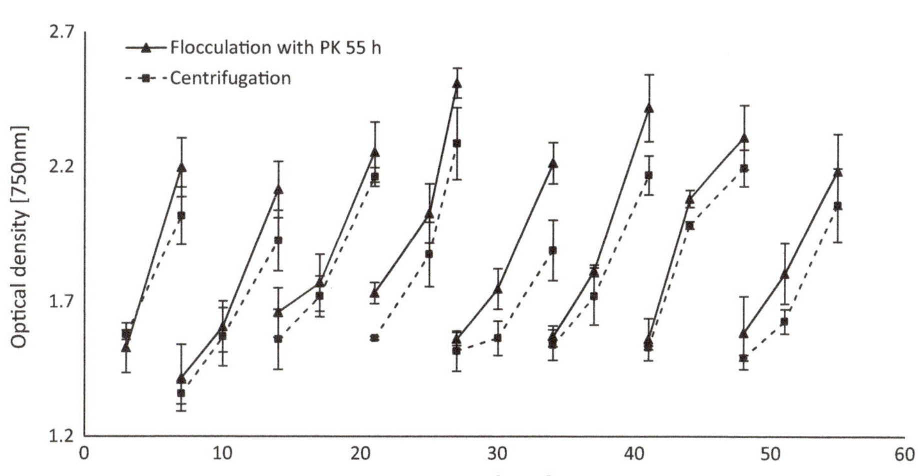

Fig. 6 Media reuse after flocculation. Over a period of 8 weeks the culture was harvested weekly by flocculation or centrifugation and the supernatant enriched with nutrients was led back into the bioreactor. The graph displays the optical density measurements registered three times per week

Table 2 Dry biomass concentration in g l^{-1}

Day	Flocculation MV ± SD	Centrifugation MV ± SD
0	1.52 ± 0	1.52 ± 0
3	1.49 ± 0.16	1.47 ± 0.04
10	2.18 ± 0.22	1.73 ± 0.01
17	2.14 ± 0.03	1.60 ± 0.18
25	2.36 ± 0.17	1.50 ± 0.11
30	2.09 ± 0.04	1.34 ± 0.10
37	2.42 ± 0.12	1.59 ± 0.04
44	2.53 ± 0.08	1.68 ± 0.22
51	2.75 ± 0.18	1.72 ± 0.14

During the media reuse growth experiment, the dry biomass was determined every week before harvesting

experiment was 1 g l^{-1} higher than that of the control cultures, although the OD values remained similar. The growth rates, however, indicate that neither the added flocculant nor a possible accumulation of algal products negatively affected cell growth.

Over a period of 8 weeks, the culture media were successfully recycled after flocculation. The hypothesis that the reuse of the media is applicable was positively proved.

Conclusion

The flocculation efficiency of microalgae for cell harvest was significantly influenced by the net charge of the tested polymer. Cationic flocculants achieved the highest flocculation efficiencies, whereas anionic and nonionic flocculants resulted in no or insufficient flocculation. The amount of flocculant needed varied among the chemicals from 1.5 to 70 mg l^{-1}, but also between the algae species (e.g., CFL25 with 30 mg l^{-1} for *Chlorella* and 70 mg l^{-1} for *Scenedesmus*). Because of high flocculation efficiencies above 95 % at low concentrations of 1.5 mg l^{-1}, the cationic polyelectrolyte PK55H was selected for recycling experiments. No limitation in culture growth and biomass production was detected, although an aggregation of the cells was visible. The results showed that chemical flocculation is a simple, efficient and inexpensive method for harvesting different types of microalgae.

Acknowledgments We thank Separ Chemie GmbH and Emslandstärke GmbH for supplying the tested flocculants. Furthermore, we would like to thank Angela Rosien for critical reading of our manuscript. This research was financially supported by the project ModuLES funded by the German aerospace association (DLR)(Grant number: 50WB1265).

References

Chisti Y (2007) Biodiesel from microalgae. Biotechnol Adv 25:294–306

Gerde JA, Yao LX, Lio JY, Wen ZY, Wang T (2014) Microalgae flocculation: impact of flocculant type, algae species and cell concentration. Algal Res Biomass Biofuels Bioprod 3:30–35

Golueke CG, Oswald WJ (1970) Surface properties and ion exchange in algae removal. J Water Pollut Control Fed 42:R304

Granados MR, Acien FG, Gomez C, Fernandez-Sevilla JM, Grima EM (2012) Evaluation of flocculants for the recovery of freshwater microalgae. Bioresour Technol 118:102–110

Grima EM, Belarbi EH, Fernandez FGA, Medina AR, Chisti Y (2003) Recovery of microalgal biomass and metabolites: process options and economics. Biotechnol Adv 20:491–515

Harith ZT, Yusoff FM, Mohamed MS, Din MSM, Ariff AB (2009) Effect of different flocculants on the flocculation performance of microalgae, Chaetoceros calcitrans, cells. Afr J Biotechnol 8:5971–5978

Harris EH (2001) Chlamydomonas as a model organism. Ann Rev Plant Physiol Plant Mol Biol 52:363–406
Hudson HE, Wagner EG (1981) Conduct and uses of jar tests. J Am Water Works Assoc 73:218–223
Lee AK, Lewis DM, Ashman PJ (2009) Microbial flocculation, a potentially low-cost harvesting technique for marine microalgae for the production of biodiesel. J Appl Phycol 21:559–567
Lee AK, Lewis DM, Ashman PJ (2013) Harvesting of marine microalgae by electroflocculation: the energetics, plant design, and economics. Appl Energy 108:45–53
Mandal S, Mallick N (2009) Microalga *Scenedesmus obliquus* as a potential source for biodiesel production. Appl Microbiol Biotechnol 84:281–291
Mata TM, Martins AA, Caetano NS (2010) Microalgae for biodiesel production and other applications: a review. Renew Sustain Energy Rev 14:217–232
Papazi A, Makridis P, Divanach P (2010) Harvesting *Chlorella minutissima* using cell coagulants. J Appl Phycol 22:349–355
Posten C, Walter C (2012) Microalgal biotechnology: potential and production. de Gruyter, Berlin
Rashid N, Rehman SU, Han JI (2013) Rapid harvesting of freshwater microalgae using chitosan. Process Biochem 48:1107–1110
Safi C, Zebib B, Merah O, Pontalier PY, Vaca-Garcia C (2014) Morphology, composition, production, processing and applications of Chlorella vulgaris: a review. Renew Sustain Energy Rev 35:265–278
Salim S, Bosma R, Vermue MH, Wijffels RH (2011) Harvesting of microalgae by bio-flocculation. J Appl Phycol 23:849–855
Sanchez JF, Fernandez JM, Acien FG, Rueda A, Perez-Parra J, Molina E (2008) Influence of culture conditions on the productivity and lutein content of the new strain Scenedesmus almeriensis. Process Biochem 43:398–405
Sirin S, Trobajo R, Ibanez C, Salvado J (2012) Harvesting the microalgae *Phaeodactylum tricornutum* with polyaluminum chloride, aluminium sulphate, chitosan and alkalinity-induced flocculation. J Appl Phycol 24:1067–1080
Tenney MW, Echelber WF, Schuessl RG, Pavoni JL (1969) Algal flocculation with synthetic organic polyelectrolytes. Appl Microbiol 18:965
Uduman N, Qi Y, Danquah MK, Forde GM, Hoadley A (2010) Dewatering of microalgal cultures: a major bottleneck to algae-based fuels. J Renew Sustain Energy 2:15
Uduman N, Bourniquel V, Danquah MK, Hoadley AFA (2011) A parametric study of electrocoagulation as a recovery process of marine microalgae for biodiesel production. Chem Eng J 174:249–257
Vandamme D, Foubert I, Meesschaert B, Muylaert K (2010) Flocculation of microalgae using cationic starch. J Appl Phycol 22:525–530
Vandamme D, Pontes SCV, Goiris K, Foubert I, Pinoy LJJ, Muylaert K (2011) Evaluation of electro-coagulation-flocculation for harvesting marine and freshwater microalgae. Biotechnol Bioeng 108:2320–2329
Vandamme D, Foubert I, Fraeye I, Meesschaert B, Muylaert K (2012) Flocculation of Chlorella vulgaris induced by high pH: role of magnesium and calcium and practical implications. Bioresour Technol 105:114–119
Wijffels RH, Barbosa MJ (2010) An outlook on microalgal biofuels. Science 329:796–799
Winckelmann D, Bleeke F, Bergmann P, Klöck G (2014) Growth of *Cyanobacterium aponinum* influenced by increasing salt concentrations and temperature. 3 Biotech 5:253–260
Wu ZC, Zhu Y, Huang WY, Zhang CW, Li T, Zhang YM, Li AF (2012) Evaluation of flocculation induced by pH increase for harvesting microalgae and reuse of flocculated medium. Bioresour Technol 110:496–502
Xu L, Wang F, Li HZ, Hu ZM, Guo C, Liu CZ (2010) Development of an efficient electroflocculation technology integrated with dispersed-air flotation for harvesting microalgae. J Chem Technol Biotechnol 85:1504–1507
Zheng HL, Gao Z, Yin JL, Tang XH, Ji XJ, Huang H (2012) Harvesting of microalgae by flocculation with poly (gamma-glutamic acid). Bioresour Technol 112:212–220
Zhou WG, Min M, Hu B, Ma XC, Liu YH, Wang Q, Shi J, Chen P, Ruan R (2013) Filamentous fungi assisted bio-flocculation: a novel alternative technique for harvesting heterotrophic and autotrophic microalgal cells. Sep Purif Technol 107:158–165

10

Using stable isotopes to investigate movement of fish in rice paddy fields

Jaeyong Lee · Jae-Ok Kim · Jeffrey S. Owen · Bomchul Kim

Abstract We evaluated movement of fish, especially *Misgurnus* spp. (loach), in paddy fields and irrigation ponds by conducting an inventory of $\delta^{13}C$ and $\delta^{15}N$ of fish, potential food sources, and soil organic matter in two irrigation pond–paddy field systems in Korea. The pond–paddy systems differ with respect to the presence or the absence of a ridge between the paddy field and the irrigation pond and also whether or not livestock are present in their watersheds. The ridge prevents the free movement of fish between paddy field and irrigation pond habitats in one of the pond–paddy watersheds, but not in the other watershed. We found differences in loach $\delta^{13}C$ and $\delta^{15}N$ inhabiting the paddy fields compared to those in loach $\delta^{13}C$ and $\delta^{15}N$ in the irrigation ponds. In irrigation ponds, loach $\delta^{13}C$ was lower in September (average −27.9 and −27.7 ‰) compared to July (average −26.2 and −26.3 ‰) in the watershed with a ridge (station 1) and without a ridge (station 6), respectively. Loach $\delta^{13}C$ in irrigation ponds in September was similar to loach $\delta^{13}C$ in the paddy field in July at both sampling sites, indicating loach might have moved into irrigation pond from paddy field. Loach $\delta^{15}N$ in the watershed with livestock was significantly higher (average 18.2 ‰) in the irrigation ponds than loach $\delta^{15}N$ in the watershed with no livestock present (average 11.3 ‰), probably reflecting higher anthropogenic nitrogen inputs from livestock. Differences in $\delta^{13}C$ and $\delta^{15}N$ in the loach reflected changes in habitat utilization of loach between paddy fields and irrigation ponds. Aquatic insect $\delta^{13}C$ differed from loach $\delta^{13}C$ but were more similar to carp $\delta^{13}C$. The stable isotope inventory approach used in this study could be used to augment or replace a more traditional field based mark–recapture approach.

Keywords Fish habitat · Irrigation pond · Misgurnus spp. · Paddy field · Stable isotopes

J. Lee
Environmental Research Institute, Nature and Human Co. Ltd, Wonju 220-844, Korea

J.-O. Kim
Rural Research Institute Korea Rural Community Corporation, Ansan 426-908, Korea

J. S. Owen (✉)
Department of Environmental Science, Hankuk University of Foreign Studies, Yongin 449-791, Korea
e-mail: owenjef2002@gmail.com

B. Kim
Department of Environmental Science, Kangwon National University, Chuncheon 200-701, Korea

Introduction

The inclusion of paddy fields as wetlands in the list of protected Ramsar Convention sites has led to a focus on restoration and maintenance of biodiversity in these agricultural landscapes, including rice paddy fields in Korea (Yamazaki et al. 2004; Kim et al. 2006a, b). Throughout east-Asia, paddy fields are recognized as important landscape features related to biodiversity of wetlands. Freshwater fish populations in paddy fields are a food resource, help control insect pests, and contribute to soil fertility in paddy fields (Vromant and Chau 2005; Aditya et al. 2010; Chen et al. 2013; Kim et al. 2011).

Typically, rice paddy fields in Korea are tilled between April and June and are irrigated prior to planting. Paddy fields are supplied with water from spring to early summer, irrigated intermittently after a short drainage period in summer, and drained in autumn. These paddy fields, together with the irrigation ponds, provide habitats for aquatic insects, amphibians, and fish (Hata 2002; Fujimoto et al. 2008). In particular, *Misgurnus* spp. (loach) is an important commercial and recreational fishery in Korea which is widely distributed in streams and ponds draining agricultural lands. Loach is often among the dominant fish species because loach can thrive in habitats with partial desiccation, low predation pressure, and low dissolved oxygen concentrations (Harding et al. 2007).

One typical field-based method of estimating population size and identifying movement of fish is the mark–recapture method (Pine et al. 2003). However, mark–recapture methods require tagging individuals and can be labor intensive because several repeated sampling periods may be needed. Other approaches to investigate the movement of fish in aquatic ecosystems utilize carbon and nitrogen stable isotopes ($\delta^{13}C$ and $\delta^{15}N$). Stable isotope measurements have been used to study fish migration between freshwater and estuarine habitats (Cunjak et al. 2005). Stable isotopes are especially useful as tracers of animal movement between habitats because $\delta^{13}C$ can trace energy and carbon sources and $\delta^{15}N$ can identify feeding relationships and trophic position (Vinagre et al. 2008). If a stable isotope approach can be shown to be as effective as mark–recapture methods at identifying movement and understanding fish life-history, then incorporating less field-intensive stable isotope approaches in future studies on fish movement in rice paddy field ecosystems would be desirable.

The aim of this study was to use patterns in carbon and nitrogen stable isotope composition of loach (*Misgurnus* spp.) to determine whether loach move from paddy fields to irrigation ponds during the autumn drainage period. These type of studies are needed in Korea and the region to improve our understanding of the importance of small irrigation ponds in supporting biodiversity in agricultural landscapes. This study can also provide important fundamental data on patterns in carbon and nitrogen stable isotope composition in rice paddy field ecosystems in the east-Asia region.

Materials and methods

Study sites

This study was conducted in two paddy fields with irrigation ponds located in Chungcheongnam Province, Korea (Fig. 1). Samples were collected during two seasons (July and September) during 2011. At both study sites, irrigation ponds are surrounded by rice paddy fields, but only the irrigation pond at station 1 is perennial and is characterized by a drainage channel through the adjacent paddy fields. At station 6, the water in the irrigation pond is supplied from an upper paddy field. The pond surface areas are about 14.1 and 11.4 m^2 at stations 1 and 6, respectively. The average water depths are approximately 163 cm at station 1 and 30 cm at station 6. Both paddy fields are fertilized with organic compost during the rice-production season. Grazing by livestock occurs within the station 1 watershed but not in the station 6 watershed.

The dominant fish species in the irrigation ponds are loach (*Misgurnus* spp.), but carp (*Carassius auratus*) can only be found at station 1. There is a ridge between the paddy field and irrigation pond at station 1, but at station 6, the ridge is lacking. At station 6, loach can freely move between habitats (paddy field and irrigation pond; Fig. 1).

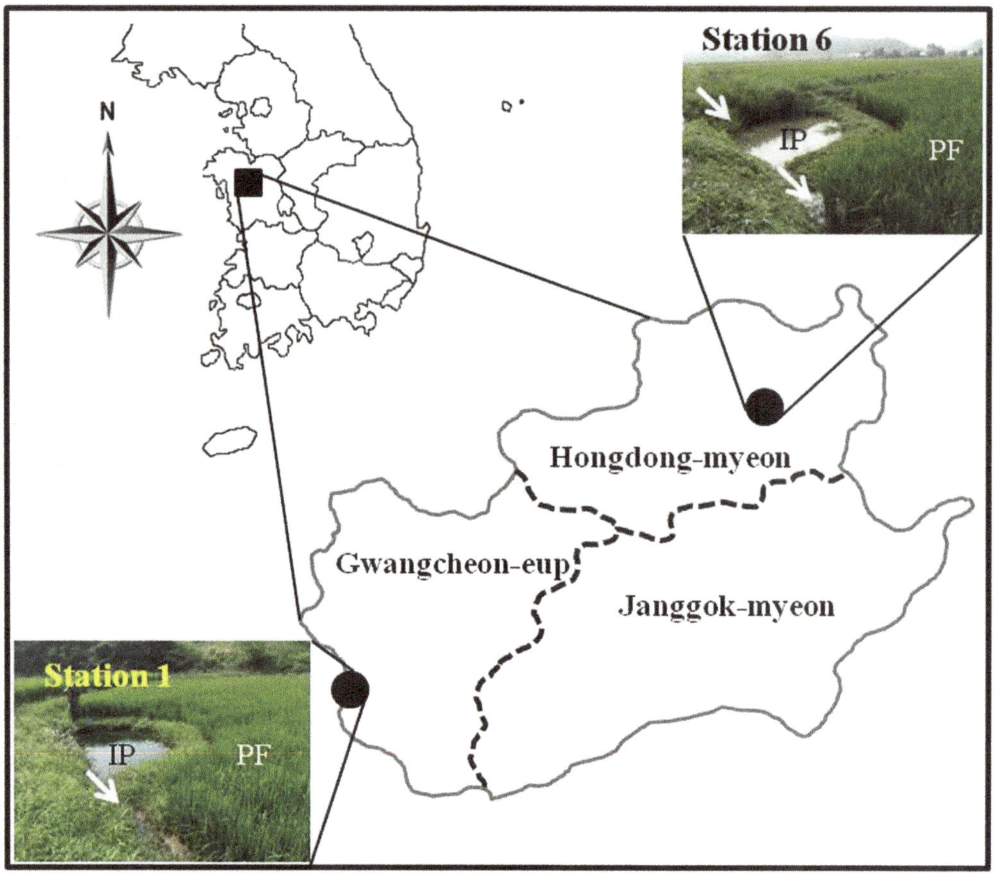

Fig. 1 Location of study sites (station 1, 6). *Arrow* represents direction of water flow. *IP* and *PF* refer to irrigation pond and paddy field

Methods

Fish were sampled using a kick net (width, 1.8 m; mesh size 4 mm) and fish traps (length 25 cm; ring length, 1.5 m; mesh size 5 mm) in the irrigation pond and paddy field. Carp were found only in the perennial irrigation pond at station 1. Macroinvertebrates in the rice paddy fields were collected using a D-frame aquatic insect net (40-cm width; mesh size 1 mm). Water samples were analyzed for concentrations of total nitrogen and total phosphorus measured according to American Public Health Association (APHA) methods (APHA 1998). Water samples for Chl-*a* determination were filtered (GF/F) and processed using 90 % acetone, and Chl-*a* concentration was measured according to standard methods (APHA 1998).

Soil organic matter was collected with a hand trowel (depth 5 cm) from three representative sites adjacent to each irrigation pond. In the laboratory, soil organic matter samples and samples for aquatic insects such as chironomids and *Tubifex* spp. were sieved using standard sieves (mesh size 250 μm).

Samples for carbon stable isotope analysis were acid fumed in a dessicator (1 N HCl for 24 h) to remove inorganic carbon, rinsed in deionized water, and finally dried at 45 °C. Samples for nitrogen isotopic measurement were not acid treated to avoid any treatment effects (Carabel et al. 2006). Samples were loaded into tin capsules, and C and N isotopes were measured using an IsoPrime isotope ratio mass spectrometer (Manchester, UK). The natural abundances of ^{13}C and ^{15}N are expressed as the per mil (‰) difference from the international standard as $\delta^{13}C$ and $\delta^{15}N = (R_{sample}/R_{standard} - 1) \times 10^3$ where $R = {}^{13}C/{}^{12}C$ or ${}^{15}N/{}^{14}N$ and Vienna PeeDee Belemnite (VPDB) and atmospheric N_2 as standards for $\delta^{13}C$ and $\delta^{15}N$, respectively. The analytic variations for replicate measurements were typically better than ±0.3 ‰ for both $\delta^{13}C$ and $\delta^{15}N$. Stable isotope compositions were compared using Student's *t* tests to test for a significant difference between the two sites using a significance level of 0.05 (Zar 1999).

Table 1 Physical and chemical characteristics of irrigation ponds

	Station 1	Station 6
Location	36°31′16.6″N	36°32′21.2″N
	126°38′24.3″E	126°42′02.9″E
Surface area (m^2)	14.1	11.4
Water depth (cm)	163	30
Water temperature (°C)	10.5–17.9 (15.0 ± 2.1)	16.2–28.4 (22.4 ± 4.0)
pH	6.3–8.0 (6.7 ± 0.5)	6.8–9.2 (7.6 ± 0.7)
Conductivity (μS cm^{-1})	484–829 (601 ± 122)	133–570 (384 ± 114)
DO (mg L^{-1})	2.7–19.3 (9.0 ± 5.5)	0.5–14.3 (6.4 ± 4.1)
TP (mg L^{-1})	0.023–0.200 (0.100 ± 0.05)	0.091–0.346 (0.225 ± 0.11)
TN (mg L^{-1})	1.09–27.55 (21.1 ± 10.66)	1.61–4.29 (2.51 ± 0.86)
Chl-*a* (μg L^{-1})	1.0–43.3 (15.0 ± 16.0)	4.2–310.0 (82.8 ± 126.5)

WT, DO, TN, TP, and Chl-*a* indicate water temperature, dissolved oxygen, total nitrogen, total phosphorus, and chlorophyll *a* concentrations, respectively ($n = 12$; range and mean ± SD shown)

Results

Characteristics of irrigation ponds

We found some differences in water quality parameters between the irrigation ponds (Table 1). At station 1, TN concentrations were generally higher (average 21.1 mg L^{-1}) compared to station 6 (average 2.5 mg L^{-1}; Table 1). TP in the irrigation pond at station 1 ranged from 0.023 to 0.200 mg L^{-1} (average 0.100 mg L^{-1}); at station 6, TP ranged from 0.091 to 0.346 mg L^{-1} (average 0.225 mg L^{-1}).

Relationship between isotopic composition and total length of fishes

At both stations 1 and 6, loach was present in the paddy field in July; in September, we did not find loach in the paddy fields as the paddy fields were drained. The lengths of loach at station 1 ranged from 62 to 137 mm (average 101.6 ± 27.8 mm, $n = 9$) and from 60 to 113 mm (average 81.2 ± 23.7 mm, $n = 5$), in the irrigation pond and paddy field, respectively (Fig. 2). At station 6, loach varied in length from 42 to 133 mm (average 79.8 ± 30.9 mm, $n = 16$) and from 78 to 113 mm (average 97.5 ± 16.1 mm, $n = 4$) in the irrigation pond and paddy field, respectively (Fig. 2).

The average lengths of loach in the irrigation ponds in September (station 1, 88.2 ± 32.1 mm; station 6, 63.9 ± 19.9 mm) were smaller than the average length in July (station 1, 118.3 ± 3.0 mm; station 6, 106.3 ± 28.3 mm) at both sampling sites (Fig. 2).

In July, loach $\delta^{13}C$ was higher compared to loach $\delta^{13}C$ in September in the irrigation ponds (Fig. 2). Loach $\delta^{13}C$ in September were more similar to the $\delta^{13}C$ of loach in the paddy field in July at both stations. In July, at station 1, loach $\delta^{13}C$ in the irrigation pond ranged from −26.9 to −25.7 ‰ (average −26.2 ± 0.5 ‰). At station 1, loach $\delta^{13}C$ in the paddy field ranged from −28.3 to −26.5 ‰ (average −27.3 ± 0.7 ‰). At station 6, loach $\delta^{13}C$ in the irrigation pond and paddy field in July ranged from −27.4 to −24.5 ‰ (average −26.3 ± 1.1 ‰), and −28.1 to −26.8 ‰ (average −27.4 ± 0.5 ‰), respectively. In September, loach $\delta^{13}C$ in the irrigation ponds in station 1 and station 6 ranged from −29.5 to −27.1 ‰ (average −27.9 ± 0.9 ‰), and −28.4 to −26.6 ‰ (average −27.7 ± 0.5 ‰), respectively.

In July, irrigation pond loach $\delta^{15}N$ was significantly higher than paddy field loach $\delta^{15}N$ in station 1, but not at station 6 (station 1, $P = 0.006$; station 6, $P = 0.078$). However, in September, irrigation pond loach $\delta^{15}N$ was similar to loach $\delta^{15}N$ in the paddy field in July in both stations (Fig. 3; station 1, $P = 0.786$; station 6, $P = 0.476$). In July, irrigation pond loach $\delta^{15}N$ and paddy field loach $\delta^{15}N$ at station 1 ranged from 18.9 to 21.0 ‰ (average 20.2 ± 0.9 ‰) and from 15.4 to 18.3 ‰ (average 17.0 ± 1.5 ‰), respectively. At station 6 in July, irrigation pond loach $\delta^{15}N$ and paddy field loach $\delta^{15}N$ ranged from 9.7 to 18.5 ‰ (average 13.9 ± 4.0 ‰) and 9.7–10.4 ‰ (average 10.2 ± 0.4 ‰), respectively. In September, irrigation pond loach

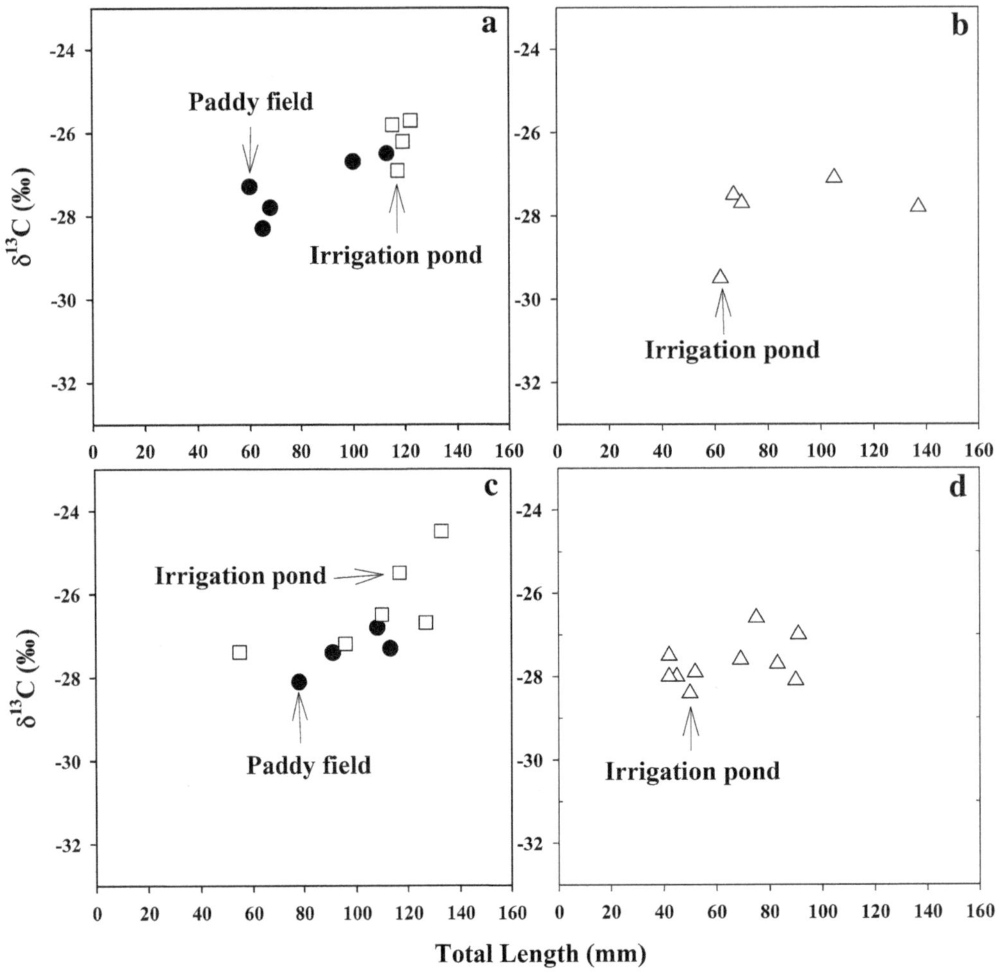

Fig. 2 Relationship between length and $\delta^{13}C$ for *Misgurnus* spp. (loach). **a** Station 1, July 2011; **b** station 1, September 2011; **c** station 6, July 2011; **d** station 6, September 2011

$\delta^{15}N$ at station 1 and station 6 varied from 14.0 to 18.7 ‰ (average 16.7 ± 2.0 ‰) and from 8.2 to 14.6 ‰ (average 9.8 ± 1.9 ‰), respectively.

Carp were only found in the irrigation pond at station 1. The length of carp varied from 26 to 79 mm (average 48.1 ± 22.0 mm, $n = 7$) in July. In September, the length of carp ranged from 21 to 79 mm (average 41.8 ± 17.6 mm, $n = 15$) (Fig. 4).

At station 1, carp $\delta^{13}C$ and $\delta^{15}N$ did not differ between July and September ($P = 0.481$ and $P = 0.795$), respectively. Overall, in the irrigation pond at station 1, carp $\delta^{13}C$ in July and September ranged from −32.0 to −27.9 ‰ (average −30.3 ± 1.7 ‰) and from −33.8 to −28.6 ‰ (average −29.9 ± 1.3 ‰), respectively (Fig. 4). In the irrigation pond at station 1, carp $\delta^{15}N$ in July and September ranged from 15.0 to 17.9 ‰ (average 16.6 ± 0.9 ‰) and from 15.0 to 17.9 ‰ (average 16.6 ± 0.9 ‰), respectively (Fig. 4).

$\delta^{13}C$ and $\delta^{15}N$ in food web components

We examined $\delta^{13}C$ and $\delta^{15}N$ of potential food sources for fish at stations 1 and 6 (Fig. 5). The bivariate plots show that the potential food sources for loach and carp were Tubifex worms (*Tubifex* spp.) and chironomids, respectively. Aquatic insect $\delta^{13}C$ was not clustered near the loach, but at station 1, carp $\delta^{13}C$ was similar to aquatic insect $\delta^{13}C$ (Fig. 5). Carp $\delta^{15}N$ was higher (average 16.3 ‰) than aquatic insect $\delta^{15}N$ (average 11.0 ‰).

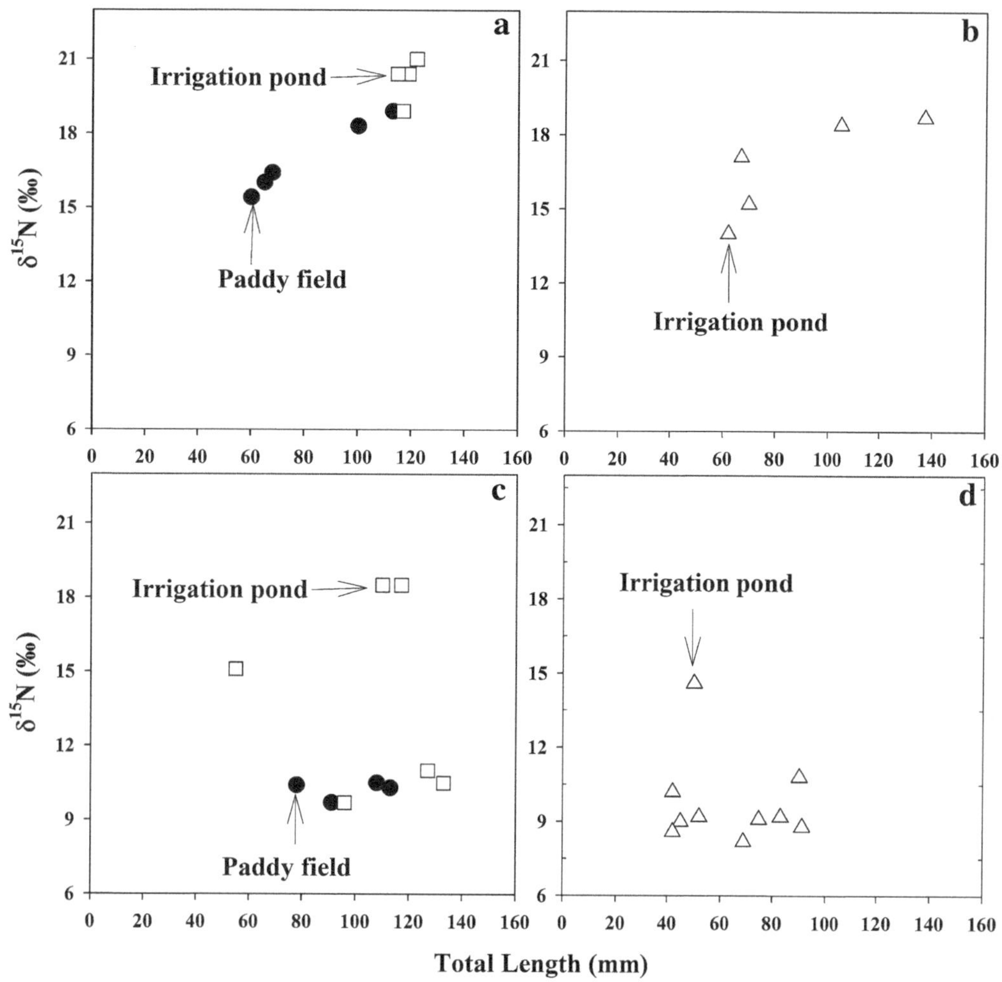

Fig. 3 Relationship between length and $\delta^{15}N$ for *Misgurnus* spp. (loach). **a** Station 1, July 2011; **b** station 1, September 2011; **c** station 6, July 2011; **d** station 6, September 2011

Discussion

Fish movement and hydrologic connectivity

Loach can readily move between different habitats (irrigation pond, connecting channel) according to hydrologic conditions and when irrigation water is drained from paddy fields. The difference in the hydrologic connection (channel width and height) between the two stations is important because there is no ridge between the irrigation pond and paddy field at station 6, allowing fish to move between habitats. However, at station 1, there is a ridge between the irrigation pond and paddy field. Thus, fish movement is influenced by the connectivity in addition to the drainage of standing water from paddy fields.

Other studies have used stable isotopes to identify patterns in fish migration between different habitats (Grey 2001; Maruyama et al. 2001; Kim et al. 2011). In our study, the difference in loach $\delta^{13}C$ between the irrigation pond and paddy field at station 1 (Fig. 2) shows the possibility to use detailed $\delta^{13}C$ measurements of fish and other food web components to study fish movement even at the scale of these small irrigation ponds and paddy field ecosystems. In July, the pattern in loach $\delta^{13}C$ likely reflected the difference in diet for loach living in the paddy fields and loach living in the irrigation pond. In September, regardless of size, loach $\delta^{13}C$ in the irrigation ponds were similar to loach $\delta^{13}C$ in paddy field in July (Fig. 2), indicating that loach changed their habitat from the paddy field to the irrigation pond at station 1. Loach in the paddy field at station 1 cannot move to the irrigation pond until drainage of standing water in paddy field because there is the ridge between the irrigation pond and paddy field. However, at station 6, there is an open connection between the irrigation

Fig. 4 Relationship between length and stable isotope composition for *Carassius auratus* (carp) at station 1. **a** Carp $\delta^{13}C$ and total length (mm); **b** Carp $\delta^{15}N$ and total length (mm) at station 1. *Filled circles* indicate July and *open squares* indicate September

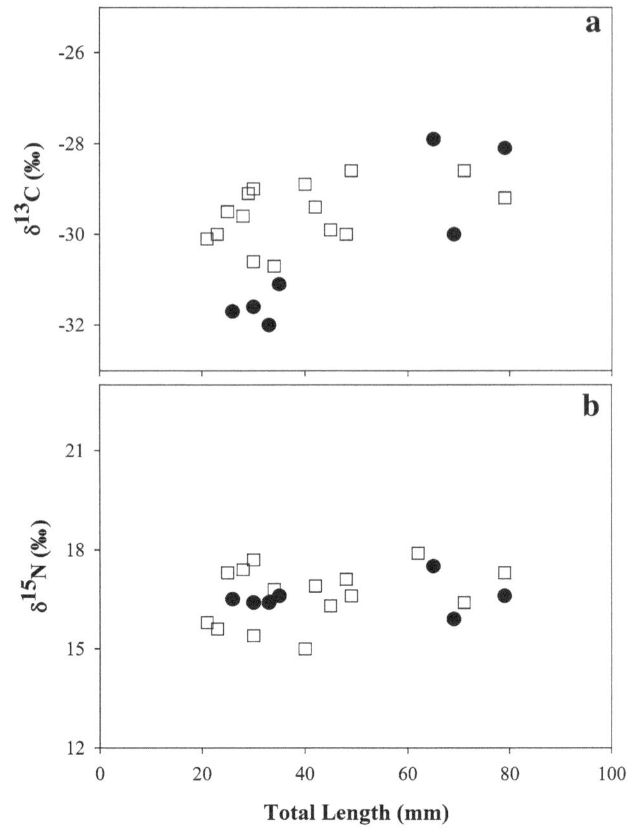

pond and paddy field. Therefore, small-size loach can easily move from the paddy field to the irrigation pond, but large loach may not be able to inhabit shallow water depths for overwintering. The spawning season of the loach extends from mid-May to August (Fujimoto et al. 2008). In these irrigation ponds, small size loach had spawned before the water was drained from the paddy fields, and a number of larger size loach might have moved into the irrigation pond or downstream through a channel (Kim et al. 2011).

We found only minor variation in loach $\delta^{15}N$ with fish length (Fig. 3), suggesting that loach used similar food sources among all size classes. Previous studies have found positive relationships between $\delta^{15}N$ and total length regardless of taxa. The isotopic composition of fish feeding in different habitats can vary with their foraging behaviors and food sources (Yoshioka et al. 1994; Maruyama et al. 2001; Cocheret de la Moriniere et al. 2003). A change in isotopic composition can result from ontogenetic shifts and will reflect trophic relationships among species if, during distinct developmental stages, different food sources are exploited.

Carp $\delta^{15}N$ did not show a difference between July and September, unlike for carp $\delta^{13}C$, indicating little variation in food source $\delta^{15}N$ for carp regardless of body size. The pattern in loach $\delta^{15}N$ seems more related to movement with drainage water to the irrigation ponds or into the channel adjacent to the paddy fields. Loach $\delta^{15}N$ was lower in September than in July, and loach $\delta^{15}N$ in the irrigation ponds in September was similar to paddy field loach $\delta^{15}N$ in July, reflecting that loach in the irrigation pond in September had likely migrated from the paddy field. In these paddy fields and irrigation ponds, fish movement is also related to habitat selection for overwintering during the dry season, emphasizing that even small irrigation ponds provide important habitat for overwintering of fish in these agricultural landscapes.

$\delta^{15}N$ and watershed characteristics

The pattern for loach, aquatic insects, and soil organic matter $\delta^{15}N$ in the two sampling sites (station 1 and station 6) might be related to the difference in land use in the watersheds (Fig. 5). The irrigation pond at station 1 is impacted by livestock in the surrounding watershed and had higher nitrogen concentrations in the irrigation pond. At station 6, there are no livestock grazing in the watershed, and nitrogen concentrations in the

Fig. 5 Bivariate plot showing $\delta^{13}C$ and $\delta^{15}N$ of food web components at station 1 and station 6. *IP* indicates the irrigation pond, and *PF* indicates the paddy field as shown in the legend. **a** station 1; **b** station 6. *Vertical* and *horizontal bars* indicate ± 1 SD. The *boxed area* indicates *Misgurnus* spp. (loach); the *circled area* indicates *Carassius auratus* (carp). Aquatic insects included *Muljarus japonicus* (giant water bug), *Notonecta triguttata* (backswimmer), *Ranatra chinensis* (water stick insect)

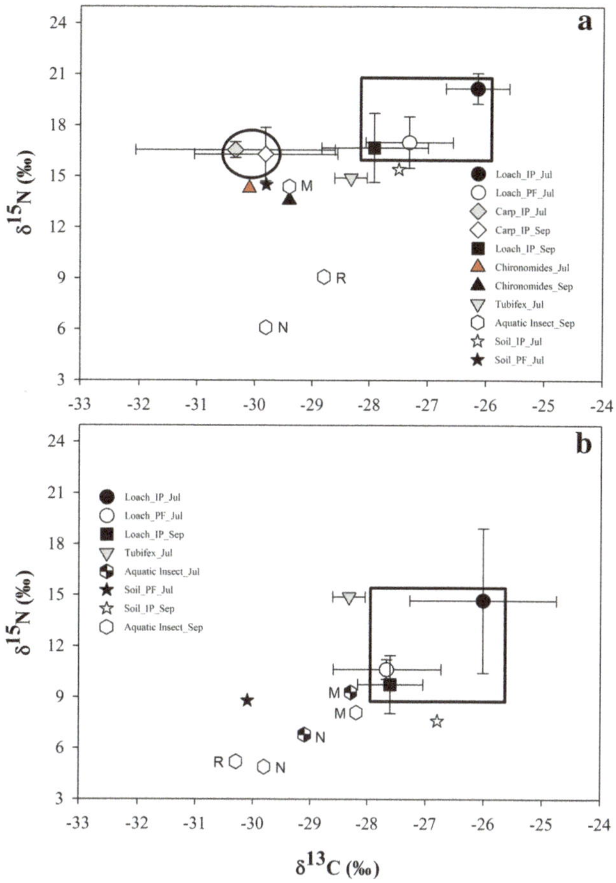

irrigation pond were generally lower. Unlike for $\delta^{13}C$, we found a significant difference (about 6 ‰) in loach $\delta^{15}N$ between the sampling stations (Fig. 5). Overall, our results show higher $\delta^{15}N$ for loach and aquatic insects at station 1 compared to station 6 (Fig. 5). The higher $\delta^{15}N$ in these aquatic organisms at station 1 might reflect higher nitrogen inputs from the watershed at station 1. With higher N inputs from livestock manure, increased nitrate, ammonium, or total N concentrations in surface water or groundwater could lead to increased $\delta^{15}N$ of food sources, and fish tissue would have higher $\delta^{15}N$ in the food web (Anderson and Cabana 2006).

Studies in other regions have shown that differences in $\delta^{15}N$ of fauna can reflect anthropogenic N loading to aquatic ecosystems. $\delta^{15}N$ of aquatic consumers can generally be used as an indicator of the magnitude of anthropogenic N loading from watersheds, but food webs in Korea have not been thoroughly studied using stable isotopic methods. In these paddy field ecosystems, the higher $\delta^{15}N$ at station 1 might reflect higher rates of one or more specific mechanisms of nitrogen loss from the system, such as denitrification, a process which results in a large isotopic fractionation (Wada et al. 1984; Diebel and Vander Zanden 2009).

In July, carp $\delta^{13}C$ differed from loach $\delta^{13}C$ in station 1. In September, loach $\delta^{13}C$ and $\delta^{15}N$ were lower at both stations (Fig. 5). The bivariate $\delta^{13}C$ versus $\delta^{15}N$ plot clearly shows a difference in $\delta^{15}N$ between loach and carp at station 1 (Fig. 5). Loach $\delta^{13}C$ was similar to Tubifex worm $\delta^{13}C$ but were less similar to chironomid $\delta^{13}C$. Using stomach content analysis, loach have been shown to consume chironomids, zooplankton, protozoa, and macroinvertebrates (Nakamura and Oda 2003; Kim et al. 2011). The majority of fish species in paddy fields can consume zooplankton and aquatic insects (Knight et al. 2003). Though our study did not identify specific food sources (zooplankton, insects, protozoa), the preliminary patterns we identified in this study show that there is a clear evidence for a difference in $\delta^{13}C$ between carp and loach in these irrigation ponds. The lack of detailed studies of patterns in $\delta^{13}C$ and $\delta^{15}N$ in these irrigation pond and paddy field systems highlights the need for future studies using stable isotopes in paddy field food webs within this important agricultural landscape.

This also suggests that immigrant individuals might be distinguishable between these sites at this small scale using more detailed $\delta^{15}N$ and $\delta^{13}C$ measurements. For the next phase of this research, a study comparing results from mark–recapture methods to quantify fish movement with the stable isotope approach would be valuable.

The presence or the absence of a ridge between paddy field and irrigation pond and the depths of irrigation ponds are important factors determining the distribution of fish populations. Even small irrigation ponds in agricultural landscapes with rice paddy fields can play an important role in providing fish habitat. Continued habitat conservation of irrigation ponds is important to sustain resident fish populations and maintain biodiversity in these agricultural landscapes and expanded use of stable isotope inventory approaches will be useful to quantify fish movement in rice paddy field ecosystems.

Conclusion

In agricultural landscapes with paddy fields and irrigation ponds, the movement of fish is affected by hydrologic and structural characteristics (water depth and characteristics of drainage channels) of the system. Differences in $\delta^{13}C$ and $\delta^{15}N$ in the two fish species reflect changes in the habitat utilization of fish between paddy fields and irrigation ponds. The stable isotope approach used here can be used to identify fish movement to irrigation ponds during water drainage from paddy fields and utilization of irrigation ponds for overwintering during the dry season.

The results of this study highlight some of the dynamic nature of habitat utilization occurring even in small irrigation ponds and paddy fields within rice production agricultural landscapes in Korea. This study has also provided some important basic data for understanding patterns in trophic relationships in small irrigation ponds and shows the need for additional studies using stable isotopes to investigate fish life history and habitat utilization in agricultural landscapes in east-Asia. It will be especially be useful to conduct studies comparing results from mark–recapture methods to study fish movement with a more complete stable isotope inventory. Detailed studies will be needed to more completely understand patterns in fish movement and behavior in irrigation ponds and paddy fields, food web structure, and trophic relationships in paddy field ecosystems in these agricultural landscapes.

Acknowledgments This study was supported by the Rural Research Institute Korea Rural Community Corporation. Jaeyong Lee was supported by the Rural Research Institute Korea Rural Community Corporation. The authors appreciate the constructive comments from two anonymous reviewers.

References

Aditya G, Pal S, Saha GK (2010) An assessment of fish species assemblages in rice fields in West Bengal, India: implications for management. J Appl Ichthyol 26:535–539

Anderson C, Cabana G (2006) Does $\delta^{15}N$ in river food webs reflect the intensity and origin of N loads from the watershed? Sci Total Environ 367:968–978. doi:10.1016/j.scitotenv.2006.01.029

APHA (American Public Health Association) (1998) Standard Methods for the Examination of Water and Wastewater, 20th edn. American Public Health Association, Washington, DC

Carabel S, Godinez-Dominguez E, Verisimo P, Fernandez L, Freire J (2006) An assessment of sample processing methods for stable isotope analyses of marine food webs. J Exp Mar Biol Ecol 336:254–261. doi:10.1016/j.jembe.2006.06.001

Chen SK, Jang CS, Chen SM, Chen KH (2013) Effect of N-fertilizer application on return flow water quality from a terraced paddy field in Northern Taiwan. Paddy Water Environ 11:123–133

Cocheret de la Moriniere E, Pollux BJA, Nagelkerken I, Hemminga MA, Huiskesa HL, Vander Velde G (2003) Ontogenetic dietary changes of coral reef fishes in the mangrove-seagrass-reef continuum: stable isotopes and gut-content analysis. Mar Ecol Prog Ser 246:279–289

Cunjak RA, Roussel JM, Gray MA, Dietrich JP, Cartwright DF, Munkittrick KR, Jardine TD (2005) Using stable isotope analysis with telemetry or mark-recapture data to identify fish movement and foraging. Oecologia 144:636–646

Diebel M, Vander Zanden MJ (2009) Nitrogen stable isotopes in streams: effects of agricultural sources and transformations. Ecol Appl 19:1127–1134

Fujimoto Y, Ouchi Y, Hakuba T, Chiba H, Iwata M (2008) Influence of modern irrigation, drainage system and water management on spawning migration of mud loach, *Misgurnus anguillicaudatus* C. Environ Biol Fish 81:185–194. doi:10.1007/s10641-007-9188-7

Grey J (2001) Ontogeny and dietary specialization in brown trout (*Salmo trutta* L.) from Loch Ness, Scotland, examined using stable isotopes of carbon and nitrogen. Ecol Fresh Fish 10:168–176

Harding JS, Norton DA, McIntosh AR (2007) Persistence of a significant population of rare Canterbury mudfish (*Neochanna burrowsius*) in a hydrologically isolated catchment. N Z J Mar Fresh Res 41:309–316. doi:10.1080/00288330709509918

Hata K (2002) Field experiment on the migration of fishes to a paddy field with a small fishway. Jpn Agric Res Q 36:219–225

Kim TC, Gim US, Kim JS, Kim DS (2006a) The multi-functionality of paddy farming in Korea. Paddy Water Environ 4:169–179. doi:10.1007/s10333-006-0046-6

Kim KG, Park MY, Choi HS (2006b) Developing a wetland-type classification system in the Republic of Korea. Landsc Ecol Eng 2:93–110. doi:10.1007/s11355-006-0012-x

Kim JO, Shin HS, Too JH, Lee SH, Jang KS, Kim BC (2011) Functional evaluation of small-scale pond at paddy field as a shelter for mudfish during midsummer drainage period. Korean J Environ Agric 30:37–42 (in Korean with English summary)

Knight RL, Walton WE, O'Meara GF, Reisen WK, Wass R (2003) Strategies for effective mosquito control in constructed treatment wetlands. Ecol Eng 21:211–232

Maruyama A, Yamada Y, Yuma M, Rusuwa B (2001) Stable nitrogen and carbon isotope ratios as migration tracers of a landlocked goby, *Rhinogobius* sp. (the orange form), in the Lake Biwa water system. Ecol Res 16:697–703. doi:10.1046/j.1440-1703.2001.00427.x

Nakamura T, Oda N (2003) Seasonal changes in fishes ascending a paddy field ditch in the Naka River system, central Japan. Jpn J Ichthyol 50:25–33

Pine WE, Pollock KH, Hightower JE, Kwak TJ, Rice JA (2003) A review of tagging methods for estimating fish population size and components of mortality. Fisheries 28:10–23

Vinagre C, Salgado J, Costa MJ, Cabral HN (2008) Nursery fidelity, food web interactions and primary sources of nutrition of the juveniles of *Solea solea* and *S. senegalensis* in the Tagus estuary (Portugal): a stable isotope approach. Estuar Coast Shelf Sci 76:255–264

Vromant N, Chau NTH (2005) Overall effect of rice biomass and fish on the aquatic ecology of experimental rice plots. Agric Eco Environ 111:153–165. doi:10.1016/j.agee.2005.06.015

Wada E, Imaizumi R, Takai Y (1984) Natural abundance of ^{15}N in soil organic matter with special reference to paddy soils in Japan: biogeochemical implications on the nitrogen cycle. Geochem J 18:109–123

Yamazaki M, Hamada Y, Kamimoto N, Momii T, Kimura M (2004) Composition and structure of aquatic organism communities in various water conditions of a paddy field. Ecol Res 19:645–653. doi:10.1111/j.1440-1703.2004.00679.x

Yoshioka T, Wada E, Hayashi H (1994) A stable isotope study on seasonal food web dynamics in a eutrophic lake. Ecology 75:835–846

Zar JH (1999) Biostatistical analysis, 4th edn. Prentice Hall Inc., Upper Saddle River

11

Cortisol levels and expression of selected stress- and apoptosis-related genes in the embryos of Atlantic cod, *Gadus morhua* following short-term exposure to air

Christopher Marlowe A. Caipang · Fernand F. Fagutao · Effrosyni Fatira · Carlo C. Lazado · Michail Pavlidis

Abstract Embryos (morula stage) of Atlantic cod, *Gadus morhua* L., were collected and subjected to air exposure for 2 min. followed by recovery at ambient conditions in the rearing container. Total immunoreactive cortisol and transcription of selected stress- and apoptosis-related genes of the embryos were determined before the application of the stressor and at 0.5, 1 and 24 h post-exposure. There was no significant difference in the total cortisol levels of the fertilized eggs before and after handling stress. There was high expression level of *hsp70* and *sod* before application of the stressor and significantly increased at 0.5 h post-exposure. The expression levels of *cat* and *gpx* were weak to moderate and were not affected by the stressor. The apoptotic genes, *mcl1* and NR-13 were highly expressed and significantly increased after exposure to air. Bcl-X1 and Bcl-X2 were moderately expressed in the control samples, but only the expression level of Bcl-X1 significantly increased following exposure to air. Cluster analysis of the different gene expression levels indicated three categories: those genes that did not show any change in the expression levels post-air exposure; those that had low expression level in the control followed by a significant increase after air exposure; and those that had high expression levels in the control followed by a further increase in expression after air exposure. These results clearly demonstrate that there are potential molecular biomarkers of the response in cod embryos as a consequence of air exposure at a time when cortisol is not fully active.

Keywords Atlantic cod · Cortisol · Early development · Gene expression · Stressor

---

C. M. A. Caipang (✉)
School of Applied Science, Temasek Polytechnic, Singapore 529757, Singapore
e-mail: cmacaipang@yahoo.com

F. F. Fagutao
Laboratory of Aquatic Animal Diseases, College of Veterinary Medicine, Gyeongsang National University, Jinju, Gyeongnam 660-701, Republic of Korea

E. Fatira · M. Pavlidis
Department of Biology, University of Crete, 71409 Heraklion, Crete, Greece

C. C. Lazado
Aquaculture Genomics Research Unit, Faculty of Biosciences and Aquaculture, University of Nordland, 8049 Bodo, Norway

Present Address:
C. C. Lazado
National Institute of Aquatic Resources, Technical University of Denmark, 9850 Hirtshals, Denmark

Introduction

The culture of Atlantic cod, *Gadus morhua,* has gained much attention in recent years. The interest in commercial aquaculture of this species is brought about by various factors including reduced capture of cod from the wild, good market value, high growth rates, anti-freeze protein production and the integration of this species for grow-out in existing culture operations for Atlantic salmon, *Salmo salar* (Goddard et al.1999; Otterlei et al. 1999; Brown et al. 2003). Artificial fertilization has been developed and routinely used to produce embryos and larvae of Atlantic cod to supply the growing demand of the seedstock for cod aquaculture (Trippel and Neilson 1992; Rideout et al. 2004). At present, there are commercial cod hatcheries that are located in the USA, UK, Canada, Iceland and Norway (Rosenlund and Halldórsson 2007).

Modern hatchery methods for Atlantic cod include supplemental feeding with cultivated plankton such as rotifers and *Artemia* (Howell 1984). The provision of supplemental feeds ensures year-round mass production of larvae. However, bulk production of cod eggs and larvae is also confronted with high mortality rates during the early stages of embryonic and larval development (Seppola et al. 2009). These high mortality rates in cod hatcheries are due to poor egg quality and infection with either bacterial or viral pathogens. The identification of the different defense mechanisms, particularly maternal-derived immune factors in cod during early embryogenesis has been explored in previous studies (Seppola et al. 2009; Lanes et al. 2012; Rise et al. 2012). In those studies, the various immune response genes that are related to innate immune defense have been identified and believed to have roles in protecting the early embryos and the larvae against those pathogens. Aside from identifying the key immune factors that will improve survival of the embryos and larvae of cod, egg quality is also one aspect that is considered when predicting survival rates in cod juveniles.

The variability in the quality of eggs and embryos is one of the most important constraints on the yield of production in any aquaculture facility (Bromage et al. 1991). As such, it is important that good quality eggs are produced to ensure good quality larvae and juveniles that will be used for grow-out culture operations. Good quality eggs exhibit low mortality at fertilization, eyeing, hatching and at first feeding (Bromage and Shepherd 1992). There are several factors that contribute to egg quality in fish, and these include: nutrition, stress, genetics, egg buoyancy, egg overripening, fertilization rates,age of the broodstock, hormonal induction of spawning, environmental factors and the like (Brooks et al. 1997; Bobe and Labbé 2010). The effects of these different factors on egg and embryo quality are species-specific (Rideout et al. 2004). In Atlantic cod, the preceding factors contribute to good quality eggs as well as other requirements for a successful production of juveniles (Rosenlund and Halldórsson 2007).

One of the factors that are likely to affect egg quality in fish is the exposure of the eggs to a variety of stressors. There have been previous studies on the effects of stress on the broodstock and spawners and the resulting progenies that are produced (Stratholt et al. 1997; Mileva et al. 2011; Li and Leatherland 2012). On the other hand, the primary response of the eggs and embryos during exposure to various stressors is somewhat difficult to assess because cortisol is not synthesized during early embryogenesis due to a lack of functional interrenal cells (Barry et al. 1995; Szisch et al. 2005; Auperin and Geslin 2008). Hence, there is no direct measure of what is happening to eggs and early embryos immediately after application of the stressor. The effects of stress are only seen at later development of the fish larvae, including decreased hatching rate, higher incidence of deformities and higher larval mortality, which were observed in milkfish, *Chanos chanos,* eggs following exposure to mechanical stress and simulated transport (Hilomen-Garcia 1998).

Exposure of Atlantic cod eggs and embryos to elevated temperature, ultraviolet and environmental contaminants resulted in high mortality of the embryos, differential expression of heat shock proteins and stress-related genes, changes in the activities of antioxidant enzymes and DNA damage (Lesser et al. 2001; Olsvik et al. 2012; Skjærven et al. 2011, 2013). These are few of the stressors that cod eggs and embryos are exposed to and there are other hatchery-related activities that could contribute to embryonic stress. Routine collection and handling of cod eggs including the pouring of fertilized eggs from one container to the other, siphoning as well as dispensing eggs in the incubation tanks make the fertilized eggs as well as the developing embryos vulnerable to being exposed to air even for a short period of time. There have been no studies conducted on the effects of these handling procedures on the fertilized eggs of cod; hence, this study aimed to find out the effects of short-term air exposure on the fertilized eggs of cod by measuring the cortisol content of the early embryos and profiling the transcription of selected oxidative stress- and apoptosis-related genes that could serve as biomarkers of the response to a stressor and immune status during the early stages of embryonic development in cod.

Materials and methods

Sampling of eggs

Fertilized Atlantic cod eggs (morula stage), approximately 36–48 h post-fertilization, were kindly provided by a commercial cod hatchery in Bodø, Norway. These eggs were obtained from the same batch of fertilized eggs and hence had more or less similar stage of development. The fertilized eggs (approximately 150–200 eggs) were placed in a 500-ml beaker (filled up to 80 % capacity) containing seawater (salinity 32–35 ppt) at 6 °C and provided with ample aeration. The eggs were immediately transported to the laboratory for the experiment and used immediately for the assays. The beakers were placed on top of a wooden plank during transport to prevent severe mechanical disturbance to the eggs. Upon reaching the laboratory, the eggs were subjected to handling stress by scooping them out with a fine-meshed net and exposing them to air for 2 min. The duration of 2 min is most likely the time that the eggs are exposed to air during normal hatchery operations when they are transferred from one tank to the other. Except for a study done by Ramsay et al. (2009) on the effects of air exposure for 3 min on the resting egg cortisol of adult zebrafish, no such data on the length of exposure to air as a potential stress factor in fertilized fish eggs have been collected. After exposure to air, the eggs were placed in individual beakers to allow recovery. Each beaker contained 50–60 eggs and four beakers with eggs were sampled at each time point. Samplings were done at 0.5, 1 and 24 h post-exposure to air. Analyses of gene expression were done over a 24-h period to ensure that the fertilized eggs are still within the morula stage. Egg samples that were not exposed to air were used as control for the experiment.

Determination of egg cortisol

Total immunoreactive cortisol from the fertilized eggs (20–30 eggs per cortisol extraction) was extracted following the procedures described by de Jesus et al. (1991). Egg cortisol was quantified by Enzyme-linked Immunosorbent Assay (ELISA) with a commercial kit (Cortisol EIA Kit, Oxford Biomedical Research, MI, USA) following the protocol supplied by the supplier. A series of dilutions of the egg cortisol samples were tested to determine the effective range for cortisol determination.

Isolation of total RNA and analyses for gene expression

Total RNA was extracted from fertilized eggs using TriZol® (Invitrogen) following standard procedures. To obtain significant amounts of RNAs, 20–30 eggs were pooled for each extraction. The samples for each time point as well as the control had four batches of fertilized eggs for the extraction. The RNA purity of all the samples was >1.85 and hence were suitable for the synthesis of the complementary DNA. First-strand complementary DNA (cDNA) was synthesized from the total RNA (1 μg) using qScript™ cDNA SuperMix (Quanta Biosciences, USA). The cDNAs were quantified and normalized at a concentration of 10 μg mL^{-1} with 1× TE buffer. The samples were stored at −20 °C until use for subsequent PCR reactions. The expression of selected genes that are related to oxidative stress response (catalase, *cat*; glutathione peroxidase, *gpx*; Hsp70, *hsp70*; and Cu,Zn-SOD, *sod*) and apoptotic activity (Bcl-X1; Bcl-X2; Mcl-1, *mcl1*; and NR-13) was determined using the published primers in previous studies (Caipang et al. 2008a, b; Feng and Rise 2010).

PCR amplification was carried out at the following conditions: an initial denaturation at 95 °C for 3 min, followed by 30 cycles of denaturation at 95 °C for 30 s, annealing at 53 °C for 30 s and elongation at 72 °C. A final elongation step at 72 °C for 5 min was done to complete the reaction. Five microliters (5 μl) of the PCR product was used for electrophoresis on a 0.8 % agarose gel and stained with SYBR Safe™ (Molecular Probes, USA). The intensity of the band on the image was visualized using a gel documentation system (Kodak Electrophoresis Documentation and Analysis System 120, Eastman Kodak Co., USA) and compared with the intensity of the band from the reference gene, β-actin using Image J software (http://rsb.info.nih.gov/nih-image/). The ratio of expression for each gene was measured and used for the Cluster analysis.

Cluster and TreeView analysis

Relative expression levels of the genes were grouped into categories based on the patterns of their expression using Cluster Program 3.0 (Clustering library version 1.27) (De Hoon et al. 2004) and visualized with TreeView program (Version 1.60) (Saldanha 2004). The genes that belong to specific clusters were grouped based on their expression patterns in the fertilized eggs before and after air exposure.

Data analysis

The data are represented as mean + SD. One-way ANOVA was used to determine significant differences in the total cortisol in the eggs. For the gene expression analyses, the assumptions for ANOVA were checked prior to carrying out the computations. If the data did not follow the Gaussian distribution, transformation was conducted prior to parametric one-way ANOVA; thereafter, the differences between the groups were checked by Tukey's Multiple Comparison Test. Kruskal–Wallis test followed by Dunn's Multiple Comparison Test was used for non-parametric data. All significance levels were set at $P < 0.05$.

Results

Cortisol content of the fertilized eggs

Fig. 1 shows the cortisol content of the fertilized eggs before exposure to air and several time points during recovery. The cortisol content of the fertilized eggs was less than 1 ng per gram of fertilized eggs and the levels were not significantly affected upon exposure of the eggs to air as a stressor.

Gene expression

The expression levels of the different stress-related genes in the fertilized eggs before and after handling stress are shown in Fig. 2. The levels of *hsp70* and *sod* prior to application of the stressor were high, whereas there was weak expression of *cat* and *gpx*. The expression level of *hsp70* significantly increased at 0.5 h post-air exposure and maintained until 24 h post-exposure. On the other hand, significant increase in the expression of *sod* was observed at 0.5 h post-air exposure and then the expression returned to its pre-exposure level. The expression level of *cat* was still weak but significantly increased at 0.5 h post-air exposure. However, the levels of expression returned to their pre-exposure levels after 1 h post-exposure to air. No significant change in the levels of expression in *gpx* was observed before and after exposing the fertilized eggs to air.

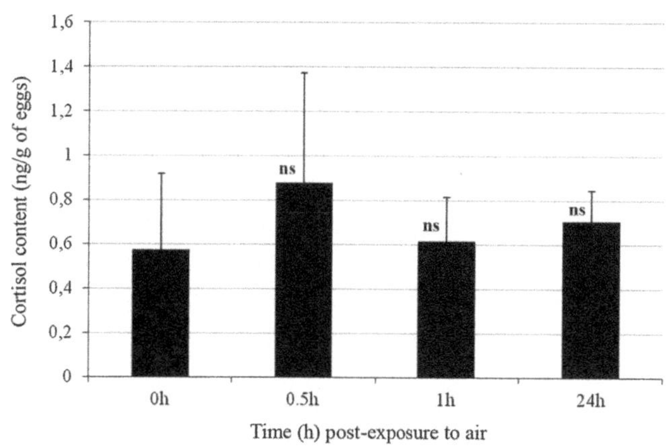

Fig. 1 Immunoreactive cortisol content of the fertilized eggs of Atlantic cod before and after exposure to air for 2 min. Each bar represents the mean ± SD of $n = 4$ egg batches. *Ns* denotes no significant difference from the initial value (time 0) at $p > 0.05$

Fig. 3 shows the expression profile of the different apoptosis-related genes in the fertilized eggs of cod as a consequence of handling stress due to air exposure. The expression levels of Bcl-X1 and Bcl-X2 in the eggs were weak before air exposure; however, the expression of Bcl-X1 significantly increased at 0.5 h post-air exposure and gradually returned to the pre-exposure levels. No significant change in expression of Bcl-X2 in the fertilized eggs was observed before and after exposure to air. There were high expression levels of both *mcl1* and NR-13 in the fertilized eggs before exposure to air. The expression levels of both genes significantly increased at 0.5 h after exposing the fertilized eggs to air and continued until 24 h post-exposure.

Cluster analysis

The expression levels of the different transcripts that were analyzed in this study are categorized into three categories (Fig. 4). Category 1 included the genes that did not show significant change in their expression levels post-exposure to air. These were the *cat*, *gpx* and Bcl-X2. Category 2 included the genes that had low expression level in the control followed by a significant increase after air exposure; and this included Bcl-X1. Category 3 is classified as the group of genes that had high expression levels in the control followed by a further increase in the expression after air exposure. The genes in this category included: *hsp70*, *sod*, *mcl1* and NR-13.

Discussion

The present study demonstrated that immunoreactive cortisol can be detected in fertilized eggs of Atlantic cod and corroborated the earlier findings of Kleppe et al. (2013). However, the egg cortisol was not responsive when the eggs are exposed to air for a short period of time as shown by no significant changes in the cortisol levels following exposure to air. In several species of fish including rainbow trout, *Oncorhynchus mykiss* (Barry et al. 1995; Auperin and Geslin 2008) and gilthead sea bream, *Sparus aurata* (Szisch et al. 2005), cortisol was detected in the fertilized eggs. It is believed that the hormones, which are present in the oocytes and during the early embryonic stages in teleosts are of maternal origin (Greenblatt et al. 1989; Hwang et al. 1992). The data that were obtained in this study together with the findings obtained in other species of fish are concrete evidence that the cortisol in the fertilized eggs of Atlantic cod is maternal in origin and is utilized during early embryonic development similar with other species of fish (de Jesus et al. 1991; Yamano et al. 1991; Hwang et al. 1992). The cortisol level of the fertilized eggs in Atlantic cod was comparable to the cortisol levels in the fertilized eggs of gilthead sea bream (Szisch et al. 2005). The fertilized eggs of rainbow trout had higher cortisol content than Atlantic cod (Barry et al. 1995; Auperin and Geslin 2008) and this could be due to differences in egg size and the amount of hormones and other bioactive substances that are transferred from the mother to the developing oocytes. The amount of cortisol in the oocytes as well as in the fertilized eggs is largely determined by the amount of the circulating cortisol in the female spawner as shown in coho salmon, *O. kisutch* (Stratholt et al. 1997); thus, if the female spawner is subjected to stressors, the level of circulating cortisol could increase and subsequently be transferred to the eggs. Hence, the initial cortisol level in the fertilized eggs is reflective of the condition of the female spawner during oocyte production. It is difficult to ascertain whether the low level of egg cortisol in cod obtained in the present study is an intrinsic factor or due to the physiological status of the female spawners because the condition of the spawners was not assessed in the study. It is important to consider this aspect of stress physiology in female spawners when assessing cortisol content during embryogenesis.

The effects of a hatchery-related stressor on the fertilized eggs of cod were determined by measuring the amount of circulating cortisol before and after application of a potential stressor. Exposure of the fertilized eggs to air for 2 min followed by placing the embryos back in the rearing container for recovery was the method of choice as a stressor for the embryos because this practice is common in hatcheries, especially when fertilized eggs are scooped out from the spawning tanks and transferred to the hatching tanks. Exposing the eggs to air during handling is inevitable and is a likely stress factor. No significant changes in the cortisol content of the embryos were observed following air exposure, indicating that there was no mechanism in the developing cod embryo that can synthesize cortisol. This is expected because at this stage in cod development, the embryos have the ability to metabolize cortisol but do not have the capacity to produce this corticosteroid

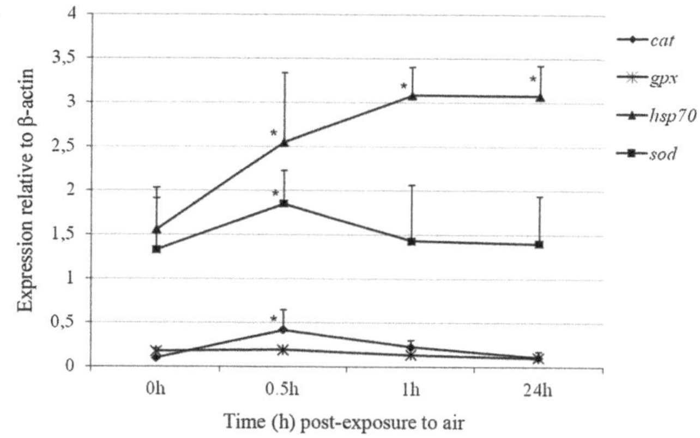

Fig. 2 Changes in the expression levels of oxidative stress-related genes in fertilized eggs of Atlantic cod before and after exposure to air for 2 min. Each value represents Mean + SD of $n = 4$ egg batches. An *asterisk* denotes significant differences from the initial value (time 0) at $p < 0.05$

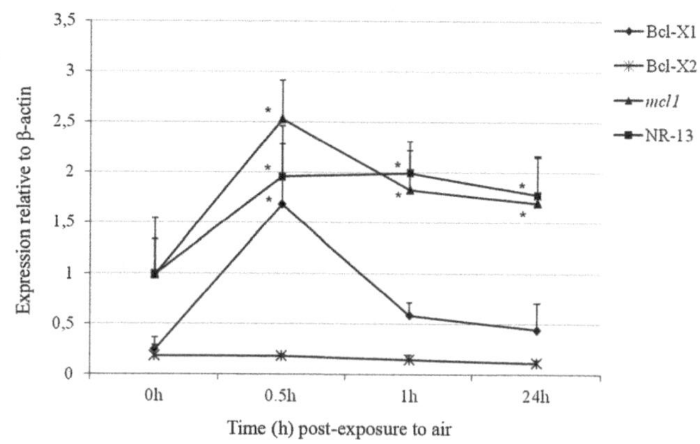

Fig. 3 Changes in the expression levels of apoptosis-related genes in fertilized eggs of Atlantic cod before and after exposure to air for 2 min. Each value represents Mean + SD of $n = 4$ egg batches. An asterisk denotes significant differences from the initial value (time 0) at $p < 0.05$

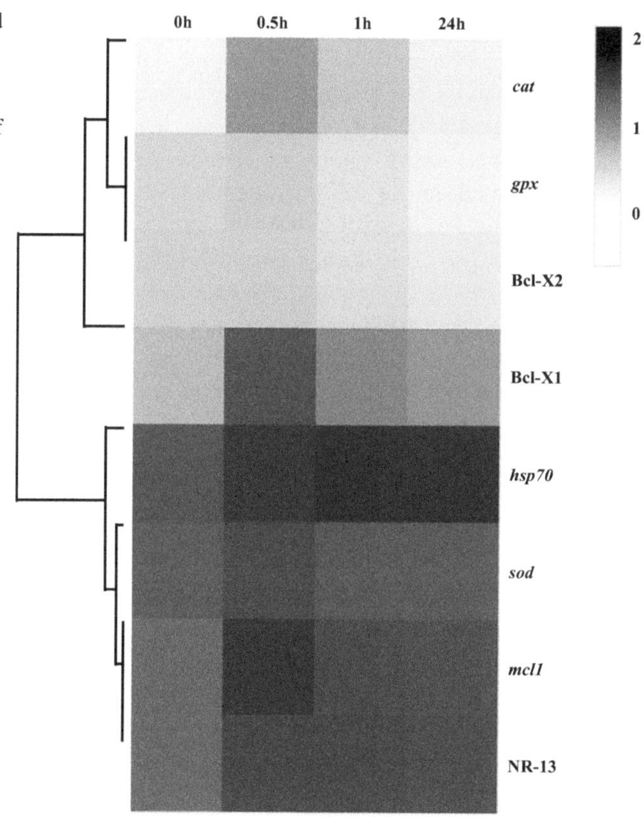

Fig. 4 Clustering of genes involved in oxidative stress and apoptosis in the fertilized eggs of Atlantic cod following exposure to air for 2 min. *Black shade* indicates highest expression levels and *light gray shade* indicates lowest expression levels. Analyses were done for the expression of the individual gene from $n = 4$ egg batches

(Stratholt et al. 1997). Studies in other fish species including the temperate flatfishes, tilapias, salmonids and sea breams also demonstrated the inability of embryos to synthesize cortisol (de Jesus et al. 1991; Hwang and Wu 1993; Barry et al. 1995; Tanaka et al. 1995; Sampath-Kumar et al. 1997; Szisch et al. 2005; Auperin and Geslin 2008). In rainbow trout, changes in the cortisol levels due to acute stress have been observed 9 days after hatching, and this species of fish is capable of cortisol synthesis at least 6 days before hatching (Auperin and Geslin 2008). This was also observed in chum salmon at 1 week before hatching (de Jesus and Hirano 1992). In this study, the cod embryos were at the morula stage; hence, the cells that are responsible for cortisol production are not likely to be fully functional. It is unlikely that the hypothalamus-pituitary-interrenal axis (HPI) that produces and regulates cortisol in fish is fully functional during the embryonic stage (McCormick and Nechaev 2002). Because cortisol is regulated by the hypothalamus and pituitary in fish (Mommsen et al. 1999), the changes in the levels of this hormone in the fertilized eggs as a consequence of exposure to various exogenous factors may not be possible.

At several time points after exposing the embryos to air, there were changes in the expression levels of genes related to oxidative stress and apoptosis. This indicates that even if there was no increase in cortisol production, the embryonic cells were already capable of synthesizing transcripts that respond to an exogenous stimulus. The changes in the expression levels of these different genes are likely a result of the response of the embryos to an external stimulus rather than a response due to a shift in the developmental stage. This is because embryonic development in cod takes a longer time as compared to warm-water fish, and the duration of sampling that was done over a 24-h period ensures that the stage of the embryonic development was more or less similar over that time period. All the genes that were used in the analyses were upregulated following exposure of the eggs to air, and it would be interesting to determine what other genes that could be down-regulated as a consequence of air exposure. Large-scale analysis using microarray or bulk transcript sequencing could be done in future studies.

The oocytes and early embryos largely depend on the barrier defense that is provided by the vitelline envelope as well as by the innate immune factors in the ooplasm (Li and Leatherland 2012). These defense systems are maternal in origin and protect the eggs from opportunistic pathogens in the aquatic environment. Moreover, the transcriptional profile of the eggs includes various mRNA transcripts of the immune response that contribute to the proteomic profile of early embryos (Aegerter and Jalabert 2004; Keyvanshokooh and Vaziri 2008; Li and Leatherland 2012). In cod, maternal-derived transcripts that are involved in immune defenses have been detected in developing embryos (Seppola et al. 2009; Lanes et al. 2012; Rise et al. 2012). Lysozyme activity was also observed even in unfertilized eggs, suggesting the presence of a functional protein (Seppola et al. 2009). In the present study, the oxidative stress-related genes, *sod* and *hsp70,* responded to the handling stress, where significant upregulation of the expression has been observed for both genes. The increased expression of *sod* is a likely consequence of the response of the embryos against oxidative stress. The by-products of oxidative stress are known as reactive oxygen species (ROS) and these include superoxide anion radical (O_2^-), hydrogen peroxide (H_2O_2) and the hydroxyl radical (HO.) (Halliwell and Gutteridge 2007). These ROS by-products are degraded or converted to non-toxic metabolites by both enzymatic and non-enzymatic antioxidant defense systems (Romero et al. 2011). Superoxide dismutase (SOD) is an anti-oxidant enzyme that prevents oxidative stress by transforming O_2^- to H_2O_2 (Paital et al. 2013). Exposure of embryos to a stressor leads to increased production of ROS as demonstrated in zebrafish embryos when exposed to an environmental contaminant (Xu et al. 2013). Although in both studies the early embryos were subjected to different stressors, the responses that were observed for *sod* expression were similar. Unlike in aquatic invertebrates where there have been some studies done on the effects of air exposure to the production of antioxidant enzymes (Almeida and Bainy 2006; Weihe et al. 2010; Romero et al. 2011), in fish very few studies have been done in such effects. Hence, the results generated on this study will provide a platform for future studies on assessing gene expression and activity of these oxidative stress-related genes when early embryos are exposed to various handling stressors.

There was also increased expression of *hsp70* when the early embryos were exposed to air for 2 min. Heat shock proteins function as molecular chaperones during embryonic development and they are involved in cell movement, proliferation, morphogenesis and apoptosis (Rupik et al. 2011). In fertilized eggs of cod, basal expression of *hsp70* was detected (Lanes et al. 2012), indicating the role of this gene during early development. Exposing fertilized eggs of cod to heat resulted in the enhanced expression of *hsp70* and *hsp90a* transcripts (Skjærven et al. 2011). While *hsp70* is constitutively expressed in fertilized eggs of cod, this gene had low expression level in any developmental stage of *Xenopus laevis* but was induced upon heat exposure in the embryos at the post-blastula stage (Lang et al. 2000). Although the specific function of *hsp70* during embryogenesis is not clearly understood, it was clearly evident that various kinds of stressors upregulated the expression of *hsp70* in fish (Deane and Woo 2003). Whether the upregulation of *hsp70* in early embryogenesis would protect the embryonic cells or ensure proper folding of proteins that are required for normal metabolism and development needs further studies.

In addition to the upregulation of key oxidative stress-related genes in the fertilized eggs of cod during exposure to air, there was also increased transcription of apoptosis-related genes, *mcl1*, Bcl-X1 and NR-13. These genes belong to the anti-apoptotic gene sub-family, Bcl-2. These anti-apoptotic genes bind and sequester the pro-apoptotic Bcl-2 proteins, thus preventing mitochondrial membrane permeabilization (MMP)- induced apoptosis (Brunelle and Letai 2009). These genes have been identified in cod and were differentially expressed when the fish were exposed to both bacteria and virus (Feng and Rise 2010). In the fertilized eggs of cod, *mcl1* and Bcl-X2 were detected and believed to be maternal-derived transcripts (Lanes et al. 2012). The roles of these anti-apoptotic genes during short-term exposure of fertilized eggs of cod to air are not clearly defined due to a limited number of studies. However, it is known that the damaged, unwanted and misplaced cells are continuously replenished through apoptosis during early development in multicellular organisms (Krumschnabel and Podrabsky 2009), but such process may only occur during the gastrula stage as demonstrated in zebrafish (Hensey and Gautier 1997; Ikegami et al. 1999). The early embryos that were used in this study were at the morula stage, which is a few stages earlier than the gastrula stage; hence it is possible that the upregulation of the anti-apoptotic genes following air exposure prevented the unwarranted programmed cell death during the morula stage. This is also to ensure that the early embryos undergo normal growth and development. Additional studies are needed to firmly establish the roles of these anti-apoptotic genes during early embryonic development of cod and in other fish species as well.

Conclusion

In conclusion, the study demonstrated that the embryos of Atlantic cod during early development contained immunoreactive cortisol, which has been derived from the female spawner. The embryonic cortisol does not show changes in the amount when the fertilized eggs are exposed to air for a short period of time, indicating this hormone is not synthesized by the cod embryos during the early stages of development. On the other hand, there are potential molecular biomarkers in the cod embryos that could offer clues on how these embryos respond to a potential stressor. Transcription analyses of two oxidative stress-related genes, *sod* and *hsp70,* demonstrated increased expression levels following exposure of the embryos to air. In addition, the apoptotic genes: *mcl1*, Bcl-X1 and NR-13 showed enhanced expression levels after exposure to air. These are anti-apoptotic genes and their upregulation in the embryos following air exposure could indicate their role in arresting premature cell death during early embryogenesis. These molecular biomarkers that are involved in stress response and apoptosis could be potentially used to assess the impacts of any drastic conditions that the developing embryos are subjected to as well as the immune status of these eggs during early development. Transcription profiles and the subsequent products of translation can be determined in the embryos in lieu of measuring cortisol at a time when this hormone does not exhibit a dynamic response during the early stages of development. Future studies should focus on establishing the relationship between gene expression and stress-induced abnormal development in the fish following exposure of the fertilized eggs to various stressors at a longer duration as well as the nutritional status and stress levels of the female spawners.

Acknowledgments A postdoctoral research fellowship at the Faculty of Biosciences and Aquaculture, University of Nordland, Norway and a travel grant awarded by the Short-term Scientific Mission of the COST scientific programme (FA0801) on "Criticial success factors for fish larval production in European Aquaculture: a multidisciplinary network" (LARVANET) awarded to the first author are greatly acknowledged.

Conflict of interest On behalf of the co-authors, I declare that there is no conflict of interest regarding the manuscript.

References

Aegerter S, Jalabert B (2004) Effects of post-ovulatory oocyte ageing and temperature on egg quality and on the occurrence of triploid fry in rainbow trout, *Oncorhynchus mykiss*. Aquaculture 231:59–71

Almeida EA, Bainy ACD (2006) Effects of aerial exposure on antioxidant defenses in the brown mussel *Perna perna*. Braz Arch Biol Technol 49:225–229

Auperin B, Geslin M (2008) Plasma cortisol response to stress in juvenile rainbow trout is influenced by their life history during early development and by egg cortisol content. Gen Comp Endocrinol 158:234–239

Barry TP, Malison JA, Held JA, Parrish JJ (1995) Ontogeny of the cortisol stress response in larval rainbow trout. Gen Comp Endocrinol 97:57–65

Bobe J, Labbé C (2010) Egg and sperm quality in fish. Gen Comp Endocrinol 165:535–548

Bromage NR, Shepherd CJ (1992) Intensive fish farming. BSP Professional Books, Oxford 404 p

Bromage NR, Jones J, Randall C, Thrush M, Springate J, Duston J, Barker G (1991) Broodstock management, fecundity, egg quality and the timing of egg production in the rainbow trout *Oncorhynchus mykiss*. Aquaculture 100:141–166

Brooks S, Tyler CR, Sumpter JP (1997) Egg quality in fish: what makes a good egg? Rev Fish Biol Fish 7:387–416

Brown JA, Minkoff G, Puvanendran V (2003) Larviculture of Atlantic cod (*Gadus morhua*): progress, protocols, and problems. Aquaculture 227:357–372

Brunelle JK, Letai A (2009) Control of mitochondrial apoptosis by the Bcl-2 family. J Cell Sci 122:437–441

Caipang CMA, Brinchmann MF, Berg I, Iversen M, Eliassen R, Kiron V (2008a) Changes in selected stress and immune-related genes in Atlantic cod, *Gadus morhua* following overcrowding. Aquac Res 39:1533–1540

Caipang CMA, Brinchmann MF, Kiron V (2008b) Short-term overcrowding of Atlantic cod, *Gadus morhua*: effects on serum-mediated antibacterial activity and transcription of glucose transport and antioxidant defense related genes. Comp Biochem Physiol A 151:560–565

Deane EE, Woo NYS (2003) Ontogeny of thyroid hormones, cortisol, hsp70 and hsp90 during silver sea bream larval development. Life Sci 72:805–818

De Hoon M, Imoto S, Nolan J, Miyano S (2004) Open source clustering software. Bioinformatics 20:1453–1454

De Jesus EG, Hirano T (1992) Changes in whole body concentration of cortisol, thyroid hormones, and sex steroids during early development of the chum salmon, *Oncorhynchus keta*. Gen Comp Endocrinol 85:55–61

De Jesus EG, Hirano T, Inui Y (1991) Changes in cortisol and thyroid hormone concentrations during early development and metamorphosis in the Japanese flounder, *Paralichthys olivaceus*. Gen Comp Endocrinol 82:369–376

Feng CY, Rise ML (2010) Characterization and expression analyses of anti-apoptotic Bcl-2-like genes NR-13, Mcl-1, Bcl-X1, and Bcl-X2 in Atlantic cod (*Gadus morhua*). Mol Immunol 47:763–784

Goddard SV, Kao MH, Fletcher GL (1999) Population differences in antifreeze production cycles of juvenile Atlantic cod (*Gadus morhua*) reflect adaptations to overwintering environment. Can J Fish Aquat Sci 56:1991–1999

Greenblatt M, Brown CL, Lee M, Dauder S, Bern HA (1989) Changes in thyroid hormone levels in eggs and larvae and in iodine uptake by eggs of coho and Chinook salmon *Oncorhynchus kisutch* and *O. tschawytscha*. Fish Physiol Biochem 6:261–278

Halliwell B, Gutteridge JMC (2007) Free radicals in biology and medicine, 4th edn. Oxford University Press, New York

Hensey C, Gautier J (1997) A developmental timer that regulates apoptosis at the onset of gastrulation. Mech Dev 69:183–195

Hilomen-Garcia GV (1998) Sensitivity of fertilized milkfish (*Chanos chanos* Forsskal) eggs to mechanical shock and simulated transport. Aquaculture 159:239–247

Howell BR (1984) The intensive rearing of juvenile cod, *Gadus morhua* L. Fløder Rapp Ser 1:657–675

Hwang PP, Wu SM (1993) Role of cortisol in hypoosmoregulation in larvae of the tilapia (*Oreochromis mossambicus*). Gen Comp Endocrinol 92:318–324

Hwang P, Wu S, Lin J, Wu L (1992) Cortisol content of eggs and larvae of teleosts. Gen Comp Immunol 86:189–196

Ikegami R, Hunter P, Yager TD (1999) Developmental activation of the capability to undergo checkpoint-induced apoptosis in the early zebrafish embryo. Dev Biol 209:409–433

Keyvanshokooh S, Vaziri B (2008) Proteome analysis of Persian sturgeon (*Acipenser persicus*) ova. Anim Reprod Sci 109:287–297

Kleppe L, Karlsen O, Edvardsen RB, Norberg B, Andersson E, Taranger GL, Wargelius A (2013) Cortisol treatment of prespawning female cod affects cytogenesis related factors in eggs and embryos. Gen Comp Endocrinol 189:84–95

Krumschnabel G, Podrabsky JE (2009) Fish as model systems for the study of vertebrate apoptosis. Apoptosis 14:1–21

Lanes CFC, Fernandes JMO, Kiron V, Babiak I (2012) Profiling key apoptotic, stress, and immune-related transcripts during embryonic and postembryonic development of Atlantic cod (*Gadus morhua* L.). Theriogenology 78:1583–1596

Lang L, Miskovic D, Lo M, Heikkila JJ (2000) Stress-induced, tissue-specific enrichment of hsp70 mRNA accumulation in *Xenopus laevis* embryos. Cell Stress Chaperones 5:36–44

Lesser MP, Farrell JH, Walker CW (2001) Oxidative stress, DNA damage and p53 expression in the larvae of Atlantic cod (*Gadus morhua*) exposed to ultraviolet (290–400 nm) radiation. J Exp Biol 204:157–164

Li M, Leatherland JF (2012) The interaction between maternal stress and the ontogeny of the innate immune system during teleost embryogenesis: implications for aquaculture practice. J Fish Biol 81:1793–1814

McCormick MI, Nechaev IV (2002) Influence of cortisol on developmental rhythms during embryogenesis in a tropical damselfish. J Exp Zool 293:456–466

Mileva VR, Gilmour KM, Balshine S (2011) Effects of maternal stress on egg characteristics in a cooperatively breeding fish. Comp Biochem Physiol A 158:22–29

Mommsen TB, Vijayan MM, Moon TW (1999) Cortisol in teleosts: dynamics, mechanism of action, and metabolic regulation. Rev Fish Biol Fish 9:211–268

Olsvik PA, Berntssen MHG, Hylland K, Eriksen DØ, Holen E (2012) Low impact of exposure to environmentally relevant doses of ^{226}Ra in Atlantic cod (*Gadus morhua*) embryonic cells. J Environ Radioact 109:84–93

Otterlei E, Nyhammer G, Folkvord A, Stefansson SO (1999) Temperature- and size dependent growth of larval and early juvenile Atlantic cod (*Gadus morhua*): a comparative study of Norwegian coastal cod and northeast Arctic cod. Can J Fish Aquat Sci 56:2099–2111

Paital B, Kumar S, Farmer R, Tripathy NK, Chainy GBN (2013) *In silico* prediction of 3D structure of superoxide dismutase of *Scylla serrata* and its binding properties with inhibitors. Interdiscip Sci Comput Life Sci 5:69–76

Ramsay JM, Feist GW, Varga ZM, Westerfield M, Kent ML, Schreck CB (2009) Whole-body cortisol response of zebrafish to acute net handling stress. Aquaculture 297:157–162

Rideout RM, Trippel EA, Litvak MK (2004) Predicting haddock embryo viability based on early cleavage patterns. Aquaculture 230:215–228

Rise ML, Hall JR, Alcock BP, Hori TS (2012) Dynamic expression profiles of virus-responsive and putative antimicrobial peptide-encoding transcripts during Atlantic cod (*Gadus morhua*) embryonic and early larval development. Gene 509:232–246

Romero MC, Tapella F, Sotelano MP, Ansaldo M, Lovrich GA (2011) Oxidative stress in the subantarctic false king crab *Paralomis granulosa* during air exposure and subsequent re-submersion. Aquaculture 319:205–210

Rosenlund G, Halldórsson O (2007) Cod juvenile production: research and commercial developments. Aquaculture 268:188–194

Rupik W, Jasik K, Bembenek J, Widłak W (2011) The expression patterns of heat shock genes and proteins and their role during vertebrate's development. Comp Biochem Physiol A Mol Integr Physiol 159:349–366

Saldanha AJ (2004) Java TreeView—extensible visualization of microarray data. Bioinform Appl Note 20:3246–3248

Sampath-Kumar R, Lee STL, Tan CH, Munro AD, Lam TJ (1997) Biosynthesis in vivo and excretion of cortisol by fish larvae. J Exp Zool 277:337–344

Seppola M, Johnsen H, Mennen S, Myrnes B, Tveiten H (2009) Maternal transfer and transcriptional onset of immune genes during ontogenesis in Atlantic cod. Dev Comp Immunol 33:1205–1211

Skjærven KH, Olsvik PA, Finn RN, Holen E, Hamre K (2011) Ontogenetic expression of maternal and zygotic genes in Atlantic cod embryos under ambient and thermally stressed conditions. Comp Biochem Physiol A 159:196–205

Skjærven KH, Penglase S, Olsvik PA, Hamre K (2013) Redox regulation in Atlantic cod (*Gadus morhua*) embryos developing under normal and heat-stressed conditions. Free Radic Biol Med 57:29–38

Stratholt ML, Donaldson EM, Liley NR (1997) Stress induced elevation of plasma cortisol in adult female coho salmon (*Oncorhynchus kisutch*), is reflected in egg cortisol content, but does not appear to affect early development. Aquaculture 158:141–153

Szisch V, Papandroulakis N, Fanouraki E, Pavlidis M (2005) Ontogeny of the thyroid hormones and cortisol in the gilthead sea bream, *Sparus aurata*. Gen Comp Endocrinol 142:186–192

Tanaka M, Tanangonan JB, Tagawa M, de Jesus EG, Nishida H, Isaka M, Kimura R, Hirano T (1995) Development of the pituitary, thyroid and interrenal glands and application of endocrinology to the improved rearing of marine fish larvae. Aquaculture 135:111–126

Trippel EA, Neilson JD (1992) Fertility and sperm quality of virgin and repeat-spawning Atlantic cod (*Gadus morhua*) and associated hatching success. Can J Fish Aquat Sci 49:2118–2127

Weihe E, Kriews M, Abele D (2010) Differences in heavy metal concentrations and in the response of the antioxidant system to hypoxia and air exposure in the Antarctic limpet, *Nacella concinna*. Mar Environ Res 69:127–135

Xu H, Shao X, Zhang Z, Zou Y, Wu X, Yang L (2013) Oxidative stress and immune related gene expression following exposure to di-n-butyl phthalate and diethyl phthalate in zebrafish embryos. Ecotoxicol Environ Saf 93:39–44

Yamano K, Tagawa M, de Jesus EG, Hirano T, Miwa S, Inui Y (1991) Changes in whole body concentration of thyroid hormones and cortisol in metamorphosing conger eel. J Comp Physiol B 161:371–375

12

Viability of sweet potato flour as carbon source for the production of biofloc in freshwater culture of tilapia, *Oreochromis* sp.

Christopher Marlowe A. Caipang · Han Xuan Choo · Zemin Bai · Huilin Huang · Clara M. Lay-yag

Abstract The use of sweet potato, *Ipomoea batatas* flour as a carbon source for biofloc production in freshwater culture of tilapia, *Oreochromis* sp., was evaluated in terms of its effects on water quality, microbial population and fish yield. The experiment was conducted using 20 L glass aquaria with two sources of carbon, namely, wheat flour and sweet potato flour. Glass aquaria with no carbon source were used as control. The treatments and control were done in duplicate. Biofloc was produced in the glass aquaria for 10 days, followed by stocking of tilapia juveniles (ABW: 15 g) at a density of 4 fish/aquarium and reared for 30 days. During the rearing period, biofloc production was maintained by adding the different flours at a carbon:nitrogen (C:N) ratio of 16. Commercial feeds (35 % crude protein) were also given at 6 % body weight. Moderate aeration was provided in all aquaria to maintain the flocs in suspension. The levels of total suspended solids and biochemical oxygen demand as well as total heterotrophic bacteria and bacilli populations were significantly higher in the treatment groups than the control. However, the levels of nitrogenous wastes (ammonia-N, nitrite-N and nitrate-N) were generally lower in glass aquaria added with sweet potato flour. At the end of the experiment, there were no significant differences in the average weights of tilapias from the biofloc groups and the control, although the tilapias that were reared in the biofloc group with sweet potato flour had the lowest average body weight and growth rate. Tilapias from both biofloc groups had significantly lower feed conversion ratio compared with the control, with the wheat flour biofloc group performing better than the sweet potato flour biofloc group. Taken together, the results of this study showed that the bioflocs that are produced using sweet potato flour as a carbon source could lower nitrogenous waste in the rearing water and to a certain degree serve as food for tilapias in freshwater culture. Efficient feeding management is needed to ensure good growth of the fish if sweet potato flour is to be utilized for biofloc production.

Keywords Aquaculture · Biofloc technology · BFT · Carbon sources · Water quality

Introduction

High-yield, high-density aquaculture is a rapidly growing industry and is accompanied with increased pollution of the aquatic environments (De Schryver and Verstraete 2009). Aquaculture animals usually absorb 20–25 % of protein in the feed, with the rest discharged into the water in the form of ammonia nitrogen, food remains and excrement, resulting in reduced water quality (Crab et al. 2007; Piedrahita 2003). High con-

C. M. A. Caipang (✉) · H. X. Choo · Z. Bai · H. Huang · C. M. Lay-yag
School of Applied Science, Temasek Polytechnic, Singapore 529757, Singapore
e-mail: cmacaipang@yahoo.com

centrations of ammonia–nitrogen and nitrite in aquaculture water also cause disease or even death of the farmed animals, which is a serious impediment to the development of the aquaculture industry (Yang and Qiu 2006).

Several new methods have been proposed to solve the problems associated with high concentrations of ammonia–nitrogen and nitrite in the water. Biofloc technology (BFT) is one of these methods, which have shown numerous advantages (Wang et al. 2015). The use of BFT in aquaculture has received considerable attention because the adoption of this technique results in high production yields, feed protein recycling, better water quality, and bacterial infection control (Avnimelech 2006; Crab et al. 2007; Little et al. 2008; Choo and Caipang 2015). The ideal candidates for BFT include shrimp, which are detritus feeders and tilapia, which are generally filter feeders (Azim and Little 2008). The basic principle of this technology is the generation of a nitrogen cycle in a stagnant aquaculture system by enhancing the growth of heterotrophic bacteria. This bacterial population utilizes nitrogenous wastes and subsequently serves as food for the cultured species (Avnimelech 1999; De Schryver et al. 2008). Biofloc systems have not only been shown to reduce nitrogenous wastes but they also improve the feed nutrient utilization efficiency and immunity of the cultured animals (Avnimelech 1999; Crab et al. 2009, 2012; Azim and Little 2008; Wang et al. 2015).

Maintaining an optimum carbon:nitrogen (C:N) ratio is crucial to the formation and production of bioflocs (Avnimelech 2006). High C:N ratio ranging 10–20 could promote the formation of biofloc as shown in previous studies (Avnimelech 1999; Hargreaves 2006; Asaduzzaman et al. 2008; Ballester et al. 2010). However, the C:N ratio in most commercial feeds is approximately 10; thus, additional carbon is required to increase the ratio and maintain biofloc production (Avnimelech 1999; De Schryver et al. 2008). The C:N ratio in feed could be increased by the addition of extra carbohydrate to the basic feed, or by the reduction of the crude protein content in feed (Wang et al. 2015). Although bioflocs have the capacity to augment the nutritional requirements of fish and shrimp in aquaculture, a number of studies have indicated that the ability of bioflocs to control the water quality in the culture system and their nutritional properties are largely influenced by the type of carbon source that was used to produce the flocs (Crab 2010; Crab et al. 2010). There are several organic carbon sources that could favor the growth of specific bacteria, protozoa and algae, and hence influence the microbial composition of the bioflocs (Crab 2010). These differences in the composition of the bioflocs also affect their nutritional properties (Crab 2010; Ekasari et al. 2010, 2014). Feeding experiments also demonstrated that the type of carbon source affected the availability, palatability and digestibility of these bioflocs (Crab 2010; Crab et al. 2010; Ekasari et al. 2010). For example, Ekasari et al. (2010) showed that regardless of the salinity, bioflocs that are produced using glycerol as a carbon source had higher total n-6 PUFAs than that of using glucose during the culture of tilapia. This was also shown by the studies done by Crab (2010). In shrimp culture, molasses is the most widely used carbon source for biofloc production (Baloi et al. 2013; Schveitzer et al. 2013; Correia et al. 2014). Previously, we compared wheat flour and corn flour as carbon sources for biofloc production in freshwater tilapia and showed that there were no significant effects on water quality between the biofloc groups and the control (Caipang et al. 2015). However, in terms of weight gain and feed conversion ratio (FCR), wheat flour performed better than corn flour, indicating that the former can be a good candidate as a source of carbon during biofloc production in freshwater tilapia culture. In spite of the popularity of molasses and wheat flour as carbon sources for biofloc production, further research should focus on the use of low-cost non-conventional agro-industrial products as carbon source. Different carbon sources stimulate the growth of the indigenous microbiota in various ways and would greatly influence the water quality, in situ feed production and utilization of the flocs by the cultured organisms. Various products or residues of the local industry can provide a low-cost external carbon source for application in biofloc technology in the culture of fish, but will need extensive research before implementation on a larger scale. Hence, this study evaluated the use of sweet potato (*Ipomoea batatas*) flour, an abundant and cheap source of carbon for biofloc production in terms of its effects on water quality, microbial population and yield during culture of tilapia in freshwater. The effects of sweet potato flour were compared against the performance of wheat flour to determine whether the former can be used as an alternative source of carbon for biofloc production in tilapia culture.

Materials and methods

Glass aquaria and experimental design

The experiment was carried out at the Aquaculture Workstation of Temasek Polytechnic, Singapore. Glass aquaria (20 L capacity) were cleaned, disinfected and sundried prior to the experiment. Each aquarium was filled with 1 L of water taken from Bedok Reservoir, Singapore, scaled up to 15 L using de-chlorinated freshwater and provided with aerators. After the addition of water, 100 mL of *Chlorella* (density of 10^6–10^8 cells mL^{-1}) was added to each aquarium and supplemented with 8.5 mL of the algal nutrient. The treatments consisted of A: biofloc production using wheat flour as carbon source; B: biofloc production using sweet potato flour as carbon source and C: no biofloc production with no addition of flour. Each treatment was done in duplicate.

Biofloc production and fish stocking

For biofloc production in glass aquaria, the corresponding flour and commercial feed (crude protein of 35 %) combinations were added following a previously described procedure (Caipang et al. 2015). The amount of flour that was added together with the commercial feeds had a C:N ratio of 16:1 following the calculations of Crab et al. (2012). The use of the commercial feed for biofloc production in glass aquaria was to trigger the production of nitrogenous wastes and to be subsequently utilized by the heterotrophic bacteria that reside in the rearing water. Feeds and/or flour were added to each aquarium every morning at 9 AM. For the control group, 20 % of the total water volume was replaced every 2 days, while for the treatment groups, the water was replaced at the same volume as the control every 10 days or when the total suspended solids (TSS) exceeded 500 mg L^{-1}. At the 11th day of the experiment, tilapia fingerlings with average weight of 15 g were stocked at a density of 4 fish in each aquarium (4 g L^{-1}). The experiment lasted for 41 days.

Water quality monitoring

Water samples for determination of selected physico-chemical parameters were collected at the initial, 7th, 30th and 40th day of the experiment. Total ammonia–nitrogen (TAN), nitrite (NO_2-N) and nitrate (NO_3-N) from the water samples were determined using commercial water quality kits following the procedures of the manufacturer (Hach Company, Colorado, USA). TSS and biochemical oxygen demand (BOD) were analyzed using the procedures described in APHA (1989). The water samples for TSS and BOD determination were obtained and processed following previously described procedures (Caipang et al. 2015).

Bacterial counts

Total heterotrophic bacteria and bacilli from the water in all glass aquaria were enumerated on the 20th and 40th day of the experiment. The enumeration of total heterotrophic bacteria provided an estimate of the culturable bacterial population in the rearing water during biofloc production, while the bacilli counts would likely indicate the presence of potential probiotic candidates that could be utilized in further studies. The water samples were serially diluted using 1 × PBS and 100 μL of the various dilutions was plated onto both nutrient agar plates and MRS agar plates for the enumeration of total heterotrophic bacteria and bacilli, respectively. Each dilution was performed in duplicate. The plates were then incubated at 28 °C for 2 days and bacterial counts were expressed as log10 colony-forming units (CFU) mL^{-1}.

Assessment of fish yield

At the end of the experiment, the weights of the tilapia were taken and recorded. The average body weight (ABW), average daily gain (ADG), specific growth rate (SGR) and feed conversion rate (FCR) were calculated from each group.

Data analyses

The data are represented as mean ± SD. One-way ANOVA was used to determine significant differences in the fish yield parameters among the different groups; thereafter, the differences between the groups were checked by Tukey's Multiple Comparison Test. All significance levels were set at $p < 0.05$.

Results and discussion

Physico-chemical parameters of the rearing water

The total ammonia–nitrogen concentration in the wheat flour treatment group was higher compared with the control and the sweet potato treatment group (Fig. 1a). There was a large degree of fluctuation in the NO_2-N concentration throughout the experiment for all the treatment groups with no observable trend among the treatment groups and the control (Fig. 1b). However, at the last sampling date, the NO_2-N concentration in the sweet potato flour group was significantly lower compared to the wheat flour group and the control. The levels of NO_3-N were generally lower in the sweet potato group than the wheat flour group and the control during the duration of the experiment (Fig. 1c).

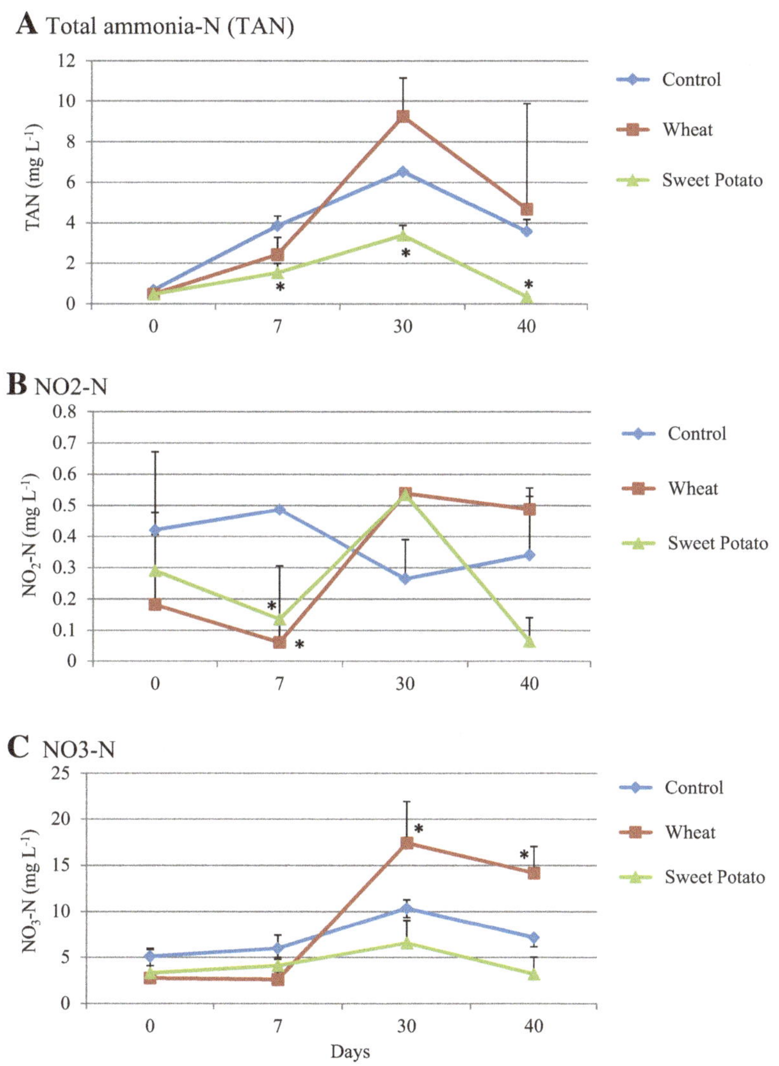

Fig. 1 Levels of **a** total ammonia–nitrogen (TAN), **b** nitrite-N (NO_2-N) and **c** nitrate-N (NO_3-N) in the rearing water of freshwater tilapia with and without biofloc. *Asterisk* at each time point indicates significant difference at $p < 0.05$

These results suggest that sweet potato as a carbon source for biofloc production in a small-scale level might result in better water quality due to a reduction in the levels of the various forms of nitrogenous wastes within the system when this particular flour was used as a carbon source. In a previous study using wheat flour and corn flour as sources of carbon for small-scale biofloc production in freshwater tilapia showed variability in the levels of ammonia, nitrate and nitrite following addition of these carbon sources (Caipang et al. 2015). How these differences in the fluctuations of the nitrogenous wastes following the addition of these carbon sources related to the type of flour remain to be supported by further studies. Until such thorough experimentation is done, no firm conclusion can be advanced with regard to the best carbon source that could help mitigate the occurrence of high levels of nitrogenous wastes in freshwater BFT system for tilapia. However, the present study showed that the use of sweet potato in small-scale production of biofloc in freshwater tilapia culture seemed to be beneficial in lowering nitrogenous wastes in comparison with using either wheat or corn flour.

The total suspended solids (TSS) in both the biofloc treatment groups were higher compared to the control (Fig. 2a). The levels of the biochemical oxygen demand (BOD) as shown in Fig. 2b corresponded to the TSS concentration, where higher readings for both parameters were obtained in both treatment groups compared to the control. The results obtained were similar to the findings of Azim and Little (2008), where TSS concentrations and BOD levels are higher in the BFT tanks compared to the control. Caipang et al. (2015) also obtained similar trends for the levels of TSS and BOD when they used wheat and corn flour as carbon sources for small-scale biofloc production in freshwater tilapia culture. Regardless of the type of carbon source, addition of such will result in higher BOD and TSS in BFT systems.

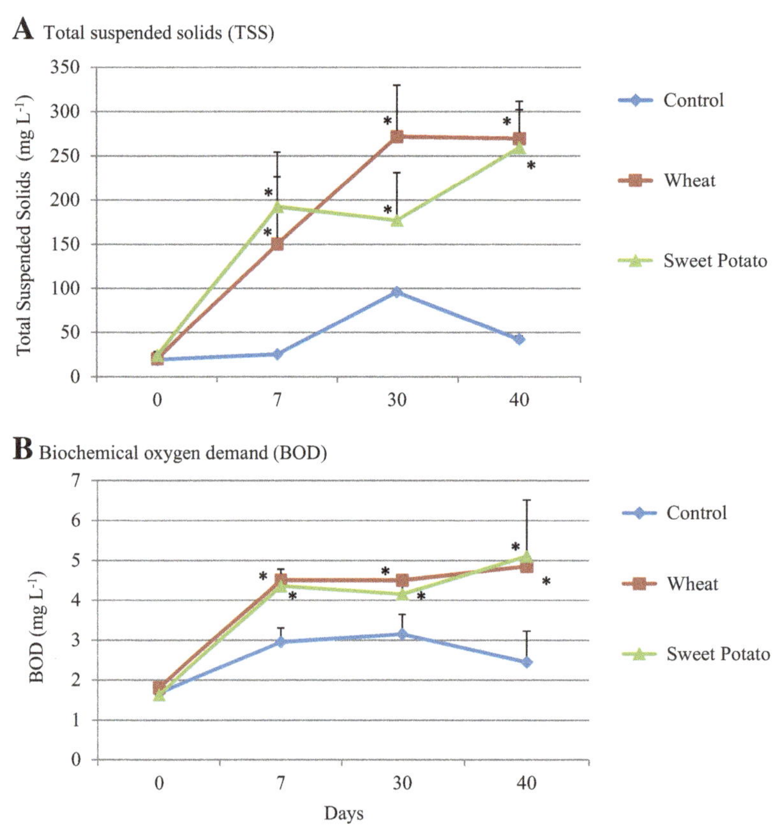

Fig. 2 Levels of **a** total suspended solids (TSS) and **b** biochemical oxygen demand (BOD) in the rearing water of freshwater tilapia with and without biofloc. *Asterisk* at each time point indicates significant difference at $p < 0.05$

Bacterial load

In Fig. 3a, it can be observed that total bacteria counts in the BFT groups were higher than the control throughout the duration of the experiment. This could be credited to the increased in bacteria population growth from the addition of carbon source in the BFT tanks (Avnimelech 1999). Similar results were also obtained in our previous study using wheat and corn flour as carbon sources for biofloc production in freshwater tilapia (Caipang et al. 2015). For the bacilli count, the wheat flour biofloc group has a consistently high bacilli population throughout the experiment period as shown in Fig. 3b. However, bacilli count was on a constant decrease in the sweet potato treatment, reaching a point where bacilli count was close to those of the control group at the last sampling date. An earlier study on the use of various carbon sources for biofloc production revealed that the organic carbon source of choice determines the composition of the flocs produced (De Schryver et al. 2008). This could have an effect on the bacterial populations that are closely associated with the flocs. The decreased population of bacilli in the group added with sweet potato flour could be an indication that this particular type of flour may not be a good source of carbon to maintain a stable population of bacilli compared to wheat flour. Additional research is needed to support the present study.

The presence of higher bacteria population in the BFT group as a result of the addition of carbon source could have resulted in the increased volume of the flocs. Higher biofloc production triggered the increase in TSS and a subsequent rise in respiration, which led to higher BOD levels. Previous studies on the use of bioflocs in tilapia culture are in agreement with the findings of the present study (Azim and Little 2008; Crab et al. 2010; Caipang et al. 2015). It is interesting to find out whether a correlation exists between bacterial population and the levels of nitrogenous wastes in the rearing water as this relationship could provide insight on the possible role of these bacteria in the dynamics of nitrogenous waste in the water. Such relationship was not established in the present study because there were differences in the sampling dates for determination of

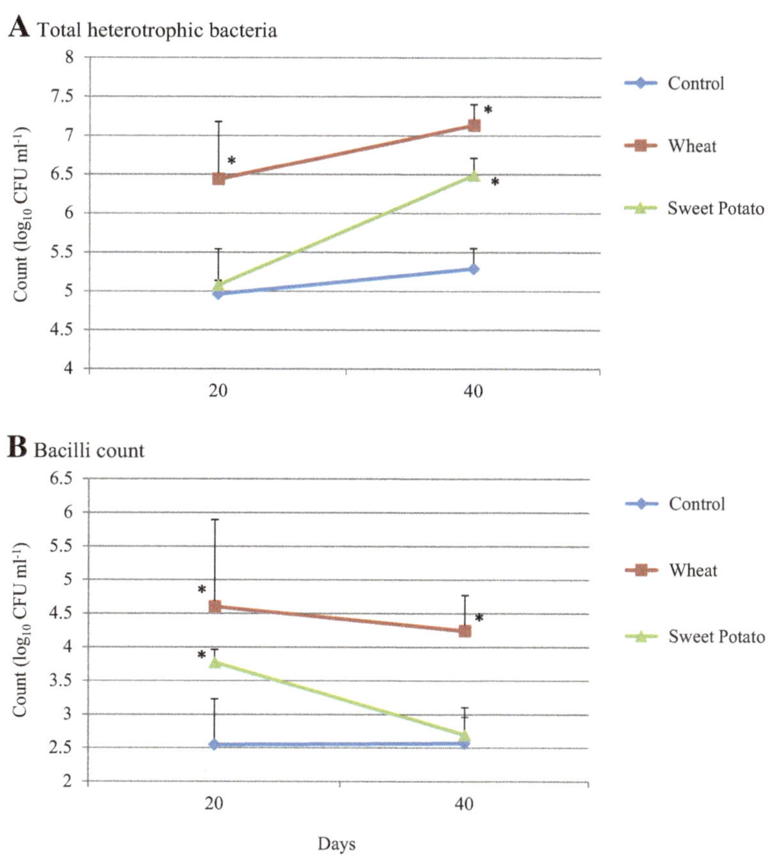

Fig. 3 Total counts for **a** heterotrophic bacteria and **b** bacilli population in the rearing water of freshwater tilapia with and without biofloc. *Asterisk* at each time point indicates significant difference at $p < 0.05$

Table 1 Production of tilapia from biofloc and non-biofloc aquaria

Parameters	Control	Wheat	Sweet potato
Biomass (g)	75.8 ± 2.55	71.0 ± 2.55	54.2 ± 4.24*
Final weight (g)	33.5 ± 0.05	32.7 ± 0.66	28.5 ± 1.07*
FCR	0.79 ± 0.03	0.34 ± 0.01*	0.44 ± 0.03*
ADG (g day^{-1})	0.62 ± 0.01	0.59 ± 0.03	0.46 ± 0.04*
SGR (% d^{-1})	2.68 ± 0.01	2.61 ± 0.06	2.20 ± 0.10*

*Significant difference from the control at $p < 0.05$

nitrogenous wastes and the enumeration of bacterial counts. However, this link between the two parameters will be the point of focus in future studies involving the use of BFT in tilapia culture.

Fish yield and growth parameters

Feed conversion rate (FCR) was significantly lower in the treatments compared to the control, with the FCR in the wheat flour treatment being lower compared to the sweet potato flour treatment (Table 1). The average daily gain (ADG; g day^{-1}) in the wheat flour BFT treatment was almost similar to the control, while the ADG in the sweet potato flour was lower compared to both the control and wheat flour BFT treatment. The specific growth rate (SGR) of tilapia in the sweet potato treatment was lower than the control and wheat flour BFT treatment. No fish mortality was observed in both the BFT and non-BFT groups during the monitoring period.

The results obtained indicate that sweet potato flour as a carbon source could result in lower FCR compared to conventional tanks fed with commercial feed, suggesting the increased utilization of protein in these treatments. However, FCR in the sweet potato treatment is higher compared to the wheat flour treatment. Relatively slow growth rate was also observed in the sweet potato treatment compared to both control and wheat flour treatment. These could be credited to nutritional differences in flocs produced between the wheat flour treatment and sweet potato flour treatment, resulting in slower growth in fish feeding on flocs produced with sweet potato flour. This is supported by the study of Ekasari et al. (2010), which shows that there is a difference in nutritional content of biofloc produced by different carbon source. Further research on the nutritional difference in flocs between the two treatments is needed in this aspect.

Conclusion

In conclusion, small-scale studies showed that the bioflocs produced during the addition of sweet potato flour as a carbon source could lower nitrogenous waste and to a certain degree serve as food for tilapias in freshwater culture. However, to ensure that the fish are able to attain the desired growth and weight gain, efficient feeding management is needed when rearing fish in this particular type of system using sweet flour as carbon source for biofloc production.

Acknowledgments The data included in this paper are part of the Major Student's Project (Optimization Studies of Biofloc Technology (BFT) in Freshwater Tilapia Culture: Effects on Water Quality and Fish Growth) of H.X. Choo. The authors gratefully acknowledge the funding support provided by the School of Applied Science, Temasek Polytechnic.

Compliance with ethical standards

Conflict of interest On behalf of the co-authors, I declare that there is no conflict of interest regarding the manuscript.

References

APHA (American Public Health Association), American Water Works Association, and Water Pollution Control Association (1989) Standard methods for the examination of water and wastewater, 17th edn. American Public Health Association, Washington, DC

Asaduzzaman M, Wahab MA, Verdegem MCJ, Huque S, Salam MA, Azim ME (2008) C/N ratio control and substrate addition for periphyton development jointly enhance freshwater prawn *Macrobrachium rosenbergii* production in ponds. Aquaculture 280:117–123

Avnimelech Y (1999) Carbon and nitrogen ratio as a control element in aquaculture systems. Aquaculture 176:227–235

Avnimelech Y (2006) Bio-filters: the need for an new comprehensive approach. Aquac Eng 34:172–178

Azim ME, Little DC (2008) The biofloc technology (BFT) in indoor tanks: water quality, biofloc composition, and growth and welfare of Nile tilapia (*Oreochromis niloticus*). Aquaculture 283:29–35

Ballester ELC, Abreu PC, Cavalli RO et al (2010) Effect of practical diets with different protein levels on the performance of *Farfantepenaeus paulensis* juveniles nursed in a zero exchange suspended microbial flocs intensive system. Aquac Nutr 16:163–172

Baloi M, Arantes R, Schveitzer R, Magnotti C, Vinatea L (2013) Performance of Pacific white shrimp *Litopenaeus vannamei* raised in biofloc systems with varying levels of light exposure. Aquacult Eng 52:39–44

Caipang CMA, Choo HX, Bai Z, Huang H, Lay-Yag CM, Lim J (2015) Small-scale production of biofloc using various carbon sources for the freshwater culture of tilapia, *Oreochromis* sp. ABAH Bioflux 7:103–111

Choo HX, Caipang CMA (2015) Biofloc technology (BFT) and its application towards improved production in freshwater tilapia culture. AACL Bioflux 8:362–366

Correia ES, Wilkenfeld JS, Morris TC, Wei L, Prangnell DI, Samocha TM (2014) Intensive nursery production of the Pacific white shrimp *Litopenaeus vannamei* using two commercial feeds with high and low protein content in a biofloc-dominated system. Aquacult Eng 59:48–54

Crab R (2010) Bioflocs technology: an integrated system for the removal of nutrients and simultaneous production of feed in aquaculture. PhD thesis, Ghent University

Crab R, Avnimelech Y, Defoirdt T, Bossier P, Verstraete W (2007) Nitrogen removal techniques in aquaculture for a sustainable production. Aquaculture 270:1–14

Crab R, Kochva M, Verstraete W, Avnimelech Y (2009) Bio-flocs technology application in over-wintering of tilapia. Aquacult Eng 40:105–112

Crab R, Chielens B, Wille M, Bossier P, Verstraete W (2010) The effect of different carbon sources on the nutritional value of bioflocs, a feed for *Macrobrachium rosenbergii* postlarvae. Aquac Eng 41:559–567

Crab R, Defoirdt T, Bossier P, Verstraete W (2012) Biofloc technology in aquaculture: beneficial effects and future challenges. Aquaculture 365–357:351–356

De Schryver P, Verstraete W (2009) Nitrogen removal from aquaculture pond water by heterotrophic nitrogen assimilation in lab-scale sequencing batch reactors. Bioresour Technol 100:1162–1167

De Schryver P, Crab R, Defoirdt T, Boon N, Verstraete W (2008) The basics of bio-flocs technology: the added value for aquaculture. Aquaculture 277:125–137

Ekasari J, Crab R, Verstraete W (2010) Primary nutritional content of bio-flocs cultured with different organic carbon sources and salinity. HAYATI J Biosci 17:125–130

Ekasari J, Angela D, Waluyo SH, Bachtiar T, Surawidjaja EH, Bosier P, De Schryver P (2014) The size of biofloc determines the nutritional composition and the nitrogen recovery by aquaculture animals. Aquaculture 426–427:105–111

Hargreaves JA (2006) Photosynthetic suspended-growth systems in aquaculture. Aquac Eng 34:344–363

Little DC, Murray JF, Azim ME, LeschenW Grady K, Young J, Watterson A (2008) Warm-water fish production in the UK: limits to green growth? Trends Food Sci Technol 19:255–264

Piedrahita RH (2003) Reducing the potential environmental impact of tank aquaculture effluents through intensification and recirculation. Aquaculture 226:35–44

Schveitzer R, Arantes R, Costodio PFS, do Espirito Santo CM, Arana LV, Seiffert WQ, Andreatta ER (2013) Effect of different biofloc levels on microbial activity, water quality and performance of *Litopenaeus vannamei* in a tank system operated with no water exchange. Aquacult Eng 56:59–70

Wang G, Yu E, Xie J, Yu D, Li Z, Luo W, Qiu L, Zheng Z (2015) Effect of C/N ratio on water quality in zero-water exchange tanks and the biofloc supplementation in feed on the growth performance of crucian carp, *Carassius auratus*. Aquaculture 443:98–104

Yang SP, Qiu DQ (2006) Water quality in the high-density shrimp culturing ponds. Fisheries Sci 25:459–462

13

Mesozooplankton of the Omani shelf: taxonomy, seasonality, and spatial distribution

Sergey A. Piontkovski · Asyla Al-Mawali · Ahlam Al-Kharusi · WardalMuna Al-Manthri · Sharon Smith · Elena Popova

Abstract The total zooplankton biomass was determined for 216 samples collected during seasonal surveys onboard a research vessel. Some of these samples were processed to the level of species. In 2007–2008, the Omani shelf was populated by a highly productive epipelagic plankton community. The chlorophyll-*a* concentration was high throughout the seasonal cycle and likewise the zooplankton biomass, where seasonal values varied from 543 to 723 mg m^{-3}. Spatial distribution of zooplankton biomass over shelf waters was highly heterogeneous, with maximal heterogeneity observed during the South-west Monsoon. The mean biomass and size structure of the zooplankton community did not exhibit statistically significant seasonal changes. On the species level, seasonal changes dealt with a massive appearance of the copepod *Calanoides carinatus s.f.* in shelf waters, to which organisms migrated from the deep, during the South-west Monsoon. The population represented mainly by c4 and c5 copepodite stages ascending to upper layers during its ontogenetic migration has occupied the entire Omani shelf area. However, this migration did not contribute markedly to seasonal variation of the total zooplankton biomass.

Keywords Zooplankton · Arabian Sea · Coastal upwelling · Ontogenetic migration

Introduction

The issue of spatial–temporal structure in the zooplankton community inhabiting the Omani shelf was basically skipped by basin-scaled International expeditions devoted to the Arabian Sea. For instance, a major part of JGOFS oceanographic stations of 1990s was oriented toward the pelagic processes in oceanic regions (Smith 2005). The earliest expedition that covered coastal Oman, was carried out by R/V "Fridtjof Nansen," from March to June 1976 and dealt with zooplankton studies (Anonymous 1975). From December 1989 to

S. A. Piontkovski (✉) · W. Al-Manthri
CAMS, Sultan Qaboos University, P.O. Box 34, Al-Khod 123, Muscat, Sultanate of Oman
e-mail: spiontkovski@gmail.com

A. Al-Mawali · A. Al-Kharusi
Marine Fisheries Research Center, Muscat, Sultanate of Oman

S. Smith
RSMAS, University of Miami, 4600 Rickenbacker Causeway, Miami, FL 33149, USA

E. Popova
Institute of Biology of the Southern Seas, 2 Nakhimov Prospect, Sevastopol, Russia

November 1990, a zooplankton survey of the Omani shelf was carried out by R/V "Rastrelliger," focusing on the assessment of spatial–temporal variations of total zooplankton biomass (Thangaraja et al. 2011). Later on in August 1995, the US R/V "Malcolm Baldrige" made a part of its sampling along the Omani shelf, with an accent on acoustical measurements of diel vertical migrations (Hitchcock et al. 2002). In 2002, the R/V "Charles Darwin" carried out a 28-day cruise along the Omani shelf, with the main objective of producing a bathymetry map and sampling the benthic community (Jacobs 2003). The survey was accompanied by CTD casts which were important for further insights into the thermo-haline structure of water masses.

A comprehensive study of zooplankton ecology is of great importance for analyzing the states of the northern Indian Ocean like Oman, because many commercial fish species are planktivorous; a challenge which points out the necessity of studying the details of their food resources. An academic interest in zooplankton research is associated in part with seasonal changes, in particular with the phenomenon entitled the "paradox" of the Arabian Sea zooplankton, stating that biomass is invariant despite seasonally changing primary production (Baars 1999; Madhupratap et al. 1996). Magnitudes of seasonal changes in zooplankton biomass are valuable characteristics in understanding the limitation of algal bloom magnitudes by zooplankton grazing (Smith 2001). It is believed that about 40 % of primary production could be ingested by mesozooplankton, which is an indicative of the importance of this group in trophic relationships in epipelagic plankton communities (Roman et al. 2000).

The Omani shelf is the region notorious for its intensive seasonal upwelling, which is footprinted in ontogenetic vertical migrations of abundant zooplankton species. The copepod *Calanoides carinatus s.f.* is one of these. It has been shown that the population migrates to the upper layers rich in diatoms which it grazes on. However, the data establishing this concept came from a single local sampling site (Smith et al. 1998). In fact, the Oman shelf area was largely missed in all the JGOFS sampling. Schemes of standard transect of the JGOFS cruises point out only one station out of eight, located on the shelf (Lane et al. 1998). In order to estimate the spatial scale of the ontogenetic migration phenomenon, a field survey of the whole shelf is inevitable.

Here we present results of five seasonal surveys carried out over the Omani shelf overlooking the western Arabian Sea. Data come from the expedition sponsored by the Oman Ministry of Fisheries and implemented in 2007–2008 by New Zealand Consultant Bruce Shallard and Associates who subcontracted the New Zealand Institute of Water and Atmospheric Research, Deliotte New Zealand, Independent Fisheries New Zealand, and Al Riyami Group (McKoy et al. 2009). We discuss spatial distribution and seasonal changes in the abundance of species, as well as the total zooplankton biomass.

Data and methods

Data on the wind speed were retrieved from the NCAR/NCEP reanalysis database (Kistler et al. 2001), in which the daily averages were extracted for the western Arabian Sea region incorporating the Omani shelf (15–23°N; 57–60°E). Maps for sea surface height anomalies were acquired from the CCAR Global Near Real-Time SSH Anomaly/Ocean Color Data Viewer (http://eddy.colorado.edu/ccar/data_viewer/index).

Satellite-derived MODIS Aqua weekly Level-3 products for the chlorophyll-*a* concentration (with a 4-km spatial resolution) were obtained from the National Aeronautics and Space Administration (NASA) database (http://gdata1.sci.gsfc.nasa.gov/daac-bin/G3/gui.cgi?instance_id=ocean_8day). Monthly time series were acquired using the GES-DISC Interactive Online Visualization and Analysis Infrastructure software, as part of the NASA's Goddard Earth Sciences Data and Information Services Center.

Zooplankton sampling was a part of seasonal surveys consisting of five voyages of the R/V "Mustaqila1" carried out along the shelf in the 20–250 m depth range, from the Ras al Hadd region to the southernmost part of the Omani shelf (to the Yemen border). Sampling was accompanied by the RBD CTD probe casts (with conductivity, temperature, density, and dissolved oxygen concentration measurements), echo-soundings, and fish trawls (Table 1).

Zooplankton were collected around the clock at 216 stations along the offshore transects (Fig. 1). A net with an opening of 60 cm equipped with 500 μm mesh net was used for oblique vertical hauls. Samples were preserved in 4 % buffered formaldehyde. The wet weight (mass) of 216 samples was estimated in the lab by using an electronic balance, after blotting, and expressed in mg m^{-3}. Subsamples from 20 samples collected

Table 1 General characteristics of the "Mustaqila1" voyages

Voyage code	Time range	Demersal stations/trawls	CTD stations	Zooplankton samples (stations)
OMA0701	Sep–Oct 2007	101	386	0
OMA0702	Nov–Dec 2007	104	459	0
OMA0801	Jan–Mar 2008	120	644	55
OMA0802	Apr–Jun 2008	123	603	79
OMA0803	Aug–Sep 2008	129	520	82

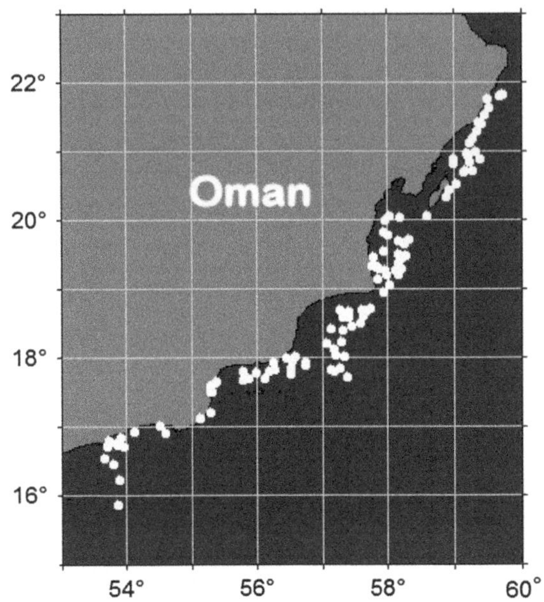

Fig. 1 Zooplankton sampling sites along the Omani shelf from September 2007 to September 2008

during winter, spring, and summer seasons were subjected to taxonomic analysis to the species level and size measurements.

A Simrad EK60 acoustic instrument with 38, 120, and 200 kHz transducers was used for routine acoustic surveys. Transects across the shelf began at a 200 m depth and extended out to 8 km offshore. When deep-scattering layers were identified on the sounder, pelagic trawls were used to catch the organisms. Later on, catches were sorted by species (McKoy et al. 2009). The P55 midwater trawl with a 10 mm liner in the cod end was used to sample mesopelagic organisms.

Results

In terms of time span, a major part of the zooplankton sampling was carried out from January to September 2008. We begin the analysis of seasonal changes from the Winter North-east Monsoon, through the Spring Inter-monsoon and finally, to the Summer South-west Monsoon period. During winter and summer monsoons, the prevailing wind direction is parallel to the coast, but it has opposite directions (Böhm et al. 1999). This does not allow a coastal current (the East Arabian Current) to set up upwelling during winter, but maximizes the offshore Ekman transport and thus the upwelling intensity during the summer.

The year 2008 was notorious for its unusually calm winter monsoon, in which the wind speed in February was gradually decreasing, compared to the previous years, whereas the winds featured in the South-west

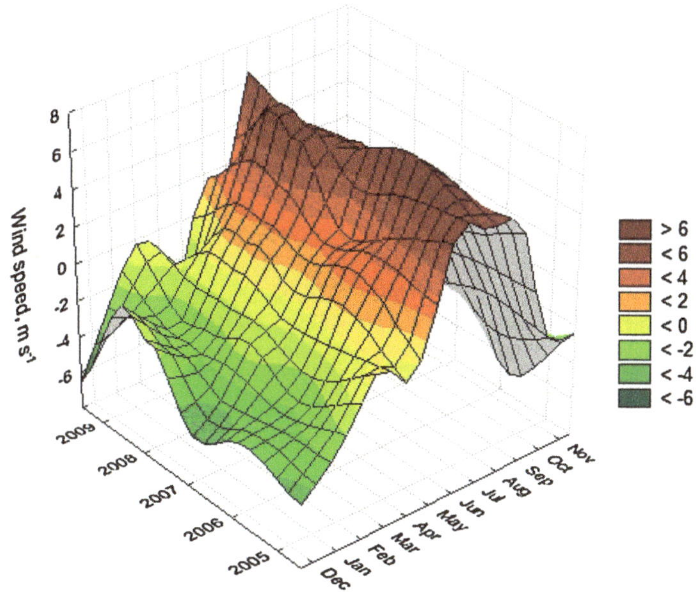

Fig. 2 Monthly changes of meridional component of the wind speed over the western Arabian Sea including the Omani shelf region (15–23°N; 57–60°E). Values are positive when the wind blows from west to east

Monsoon (in August) were gradually increasing (Fig. 2). Overall, summer winds possessed high meridional speeds compared to the winter monsoon winds.

The series of CTD casts carried out during zooplankton sampling have captured the seasonal variability of the thermo-haline structure over the Omani shelf (Fig. 3).

This variability is largely based on a reversing of the East Arabian Current, changing its direction from north-eastward during the South-west Monsoon to the opposite (south-westward) during the North-east Monsoon. Figure 3 exemplifies vertical structure corresponding with the monsoonal and inter-monsoonal seasons. Visible is a deepened mixed layer established by an evaporatively driven convective overturning characteristic for the North-east Monsoon. Salinity in the deeper portion of this layer was in the region of 36.55 ppm.

During the Spring Inter-monsoon, the mixed layer was much less pronounced. A relaxation and subsequent collapse of the East Arabian Current has favored the formation of layered structures, which, in the given case, was reflected by vertical distribution of salinity (Fig. 3). The South-west Monsoon is notorious for its wind-driven coastal upwelling, shallow mixed layer, and subsequent low temperatures in the seasonal thermocline entrained from beneath. For instance, temperature was 21 °C at a 50 m depth, versus 21.5 and 23.5 °C during the Spring Inter-monsoon and North-east Monsoon periods.

Mesoscale variability contributed to the vertical thermo-haline structure as well. Cyclonic and anticyclonic eddies were observed across the shelf throughout the year (Fig. 4). A 3-year weekly animation of the sea surface height anomalies (with a fragment exemplified by Fig. 4) has pointed out three major types of eddies observed in the western Arabian Sea.

Numerous eddies move from the Arabian Sea interior westward toward the Omani shelf, throughout the entire year. During the North-east Monsoon, eddies in the western Arabian Sea were poorly developed and the sea surface height anomalies associated with eddies were quite weak. These eddies were observed throughout the Spring Inter-monsoon season as well. During the South-west Monsoon, parts of eddies were generated by the East Arabian Current and its confluence with the current headed out from the Gulf of Oman. In terms of kinetic energy, these are the most powerful eddies in this region. Some of them moved southward when the confluence of currents collapsed, approaching the Fall Inter-monsoon season.

Footprints of monsoon and inter-monsoon periods were pronounced in the chlorophyll-*a* concentration remotely sensed by the MODIS scanner (Fig. 5). The seasonality of chlorophyll changes was pronounced over shelf waters, but most clearly it was observed over the open sea regions. During the South-west Monsoon, shelf concentrations exceeded the Spring Inter-monsoon period nearly fourfold.

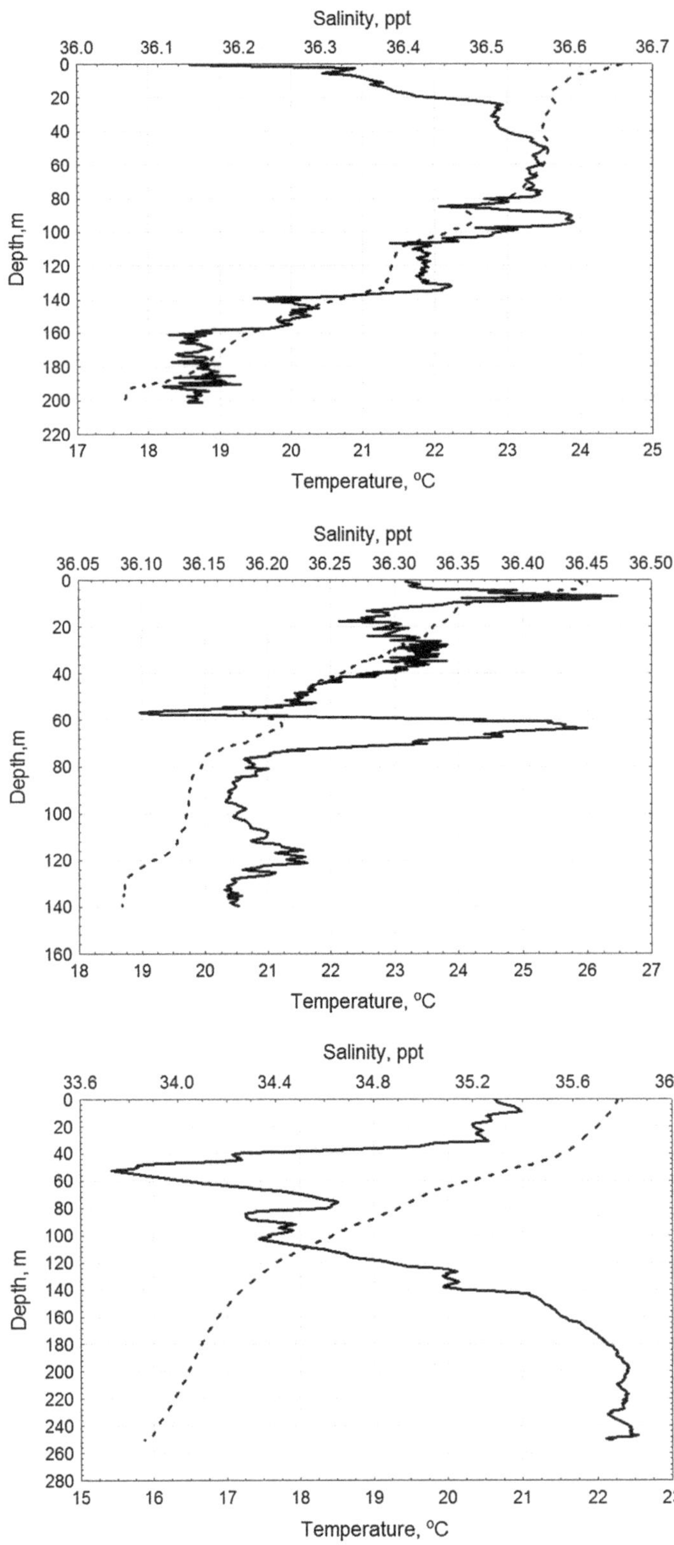

Fig. 3 Vertical distribution of temperature and salinity over the Omani shelf during North-east Monsoon, Spring Inter-monsoon, and South-west Monsoon. *Upper panel* St.316, 18 February, 2008; 10°01.73°N, 57°21.09°E. *Middle panel* St.210, 6 May, 2008; 19°11.92°N, 58°08.82°E, and *low panel* St.173, 18 August, 2008; 18°19.78°N, 57°33.43°E

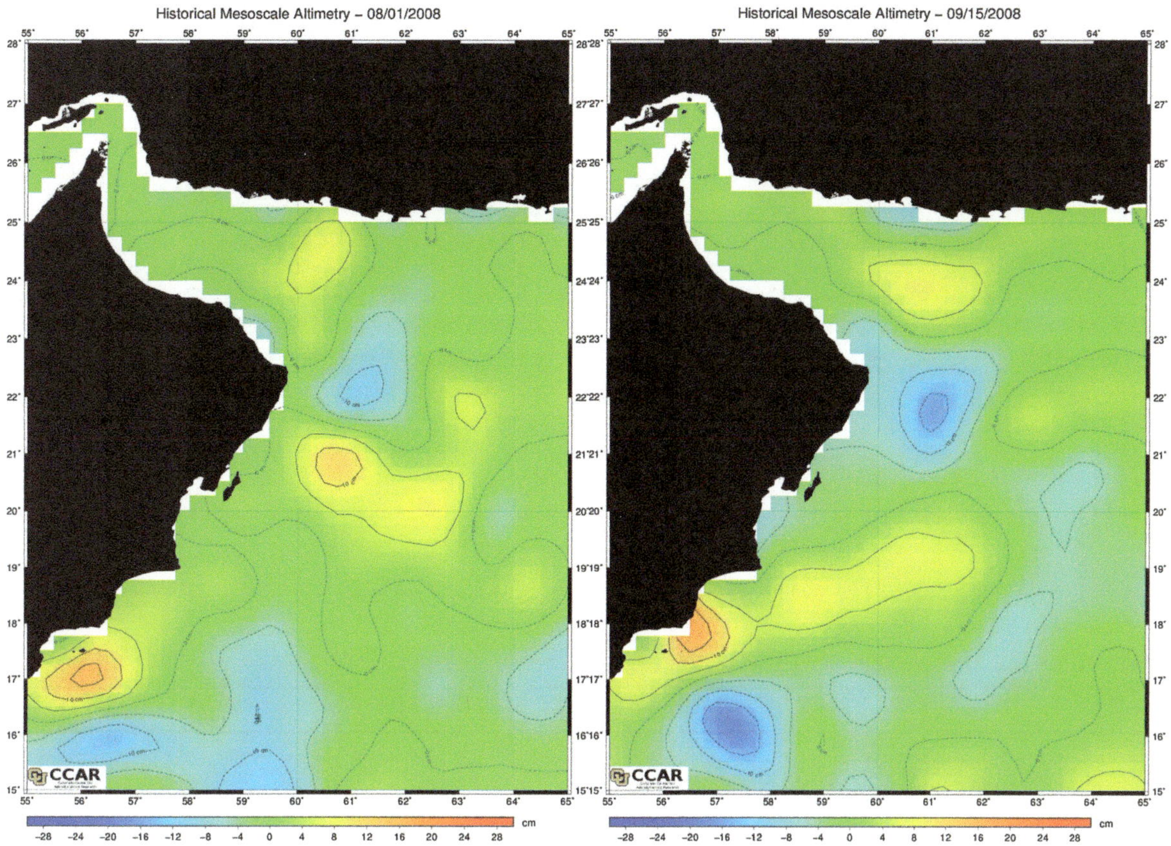

Fig. 4 Maps of sea surface height anomalies (in cm) featuring the South-west Monsoon season. *Left* to *right* 1 August, 2008, and 15 September, 2008

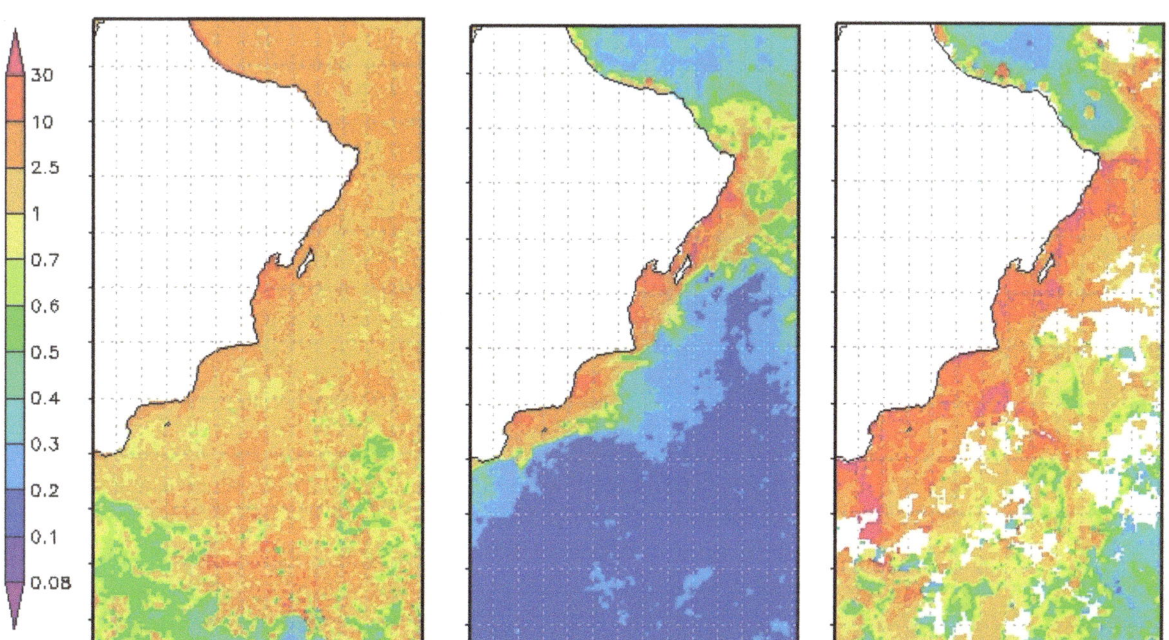

Fig. 5 Seasonal changes of the chlorophyll-*a* concentration in the western Arabian Sea (including the Omani shelf). The vertical color bar stands for the chlorophyll-*a* concentration (mg m^{-3}). From *left* to *right* North-east Monsoon (January–February, 2008), Spring Inter-monsoon (April–May, 2008), and South-west Monsoon (August–September, 2008). White zones in the right figure stand for a cloud mask

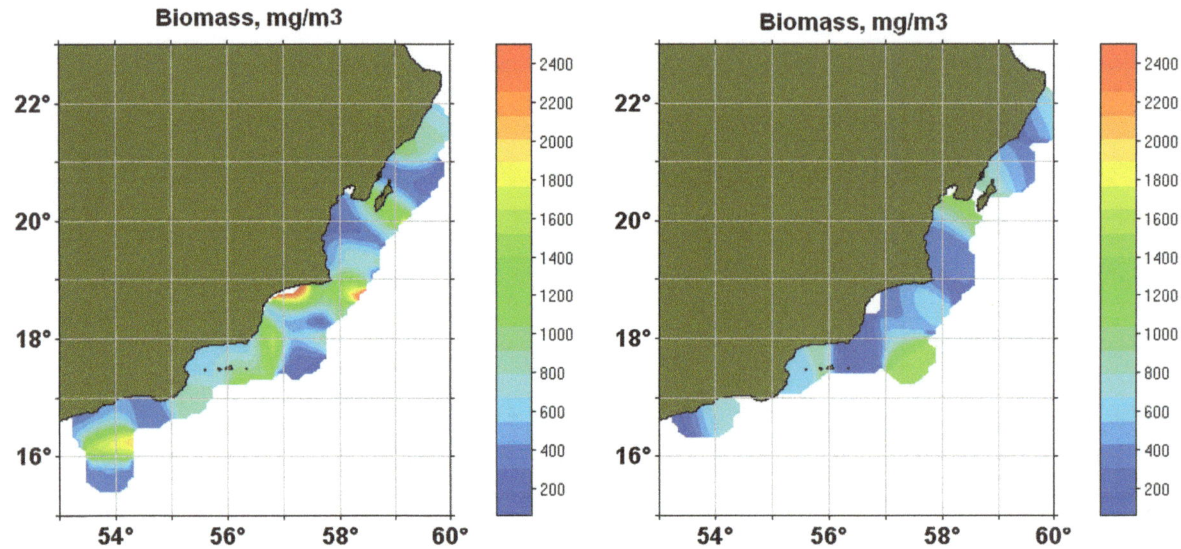

Fig. 6 Spatial distribution of zooplankton biomass during Spring Inter-monsoon (*left figure*) and South-west Monsoon periods

Table 2 Some basic characteristics of the Omani shelf (2008)

Seasons	NEM	SIM	SWM	FIM
Euphotic depth (m)	39 ± 6	66 ± 5	22 ± 8	No data
Dissolved oxygen (mg L^{-1}) (1–10 m)	4.6 ± 0.7	4.5 ± 0.2	4.3 ± 0.9	No data
Temperature (°C) (1–10 m)	24.2 ± 0.9	26.8 ± 1.3	24.5 ± 3.9	No data
Salinity (ppt) (1–10 m)	36.4 ± 0.1	36.4 ± 0.2	36.2 ± 0.2	No data
Chlorophyll-*a* concentration (mg m^{-3}) (surface)	2.0 ± 1.8	2.4 ± 5.2	8.8 ± 10.8	No data
Number of daily zooplankton stations	25	53	42	3
Zooplankton biomass (mg m^{-3})	689 ± 672	723 ± 690	543 ± 705	619
Variance/mean ratio of zooplankton biomass (mg m^{-3})	656	659	915	No data
Mean prosome length of copepods (mm)	1.9 ± 0.8	2.0 ± 1.0	1.8 ± 0.9	No data
Myctophids (biomass) (t km^{-2})	No data	155	1,200,000	No data

NEM North-east Monsoon, *SIM* Spring Inter-monsoon, *SWM* South-west Monsoon, *FIM* Fall Inter-monsoon

In analyzing spatial and seasonal distribution of zooplankton, we excluded samples collected at night, in order to eliminate the impact of diel vertical migrations, which could cause from a two- to a fivefold increase of zooplankton abundance and biomass (Goswami and Shrivastava 1996; Luo et al. 2000; Smith et al. 1998). Only samples collected during day time (from 6 am to 6 pm) were used to construct the maps. This procedure has limited our subsequent assessments of spatial distribution, making them statistically reliable for the Spring Inter-monsoon and the South-west Monsoon periods only (Fig. 6).

Maps revealed the biomass distributed mosaically over the shelf area during both seasons. Local maxima with values of approximately 1400–2000 mg m^{-3} were observed across the shelf and near the shelf edge during both monsoon periods. No seasonal changes were pronounced, regarding the total zooplankton biomass. The mean seasonal values were 689, 723, 543, and 619 mg m^{-3} for the North-east Monsoon, Spring Inter-monsoon, South-west Monsoon, and Fall Inter-monsoon periods (Table 2). No statistical differences among mean biomass values were observed since the variation was large, which is reflected by the standard deviations.

In spatial distribution, the variance-to-mean ratio has been used as the spatial heterogeneity index (Okubo 1980; Piontkovski and Melnik 2006). In the present study, the ratio was treated as a simple way to assess spatial heterogeneity of zooplankton biomass distribution. As far as the Omani shelf waters are concerned, patchiness in biomass distribution was maximal during the South-west Monsoon (Table 2).

An absent seasonality on the level of total zooplankton biomass was accompanied by a relatively stable species composition of the dominant (by abundance) groups—copepods; 40 species were found during the North-east Monsoon, 38 species during the Spring Inter-monsoon, and 34 species were present in the South-west Monsoon. Of those, 11 species (about 30 %) were present throughout the year (Table 3).

As for the groups of dominant zooplankton species (including copepods), the composition changed throughout the seasons. In the North-east Monsoon, the top five dominant groups were represented by families *Oikopleuridae*, *Sagitta*, and copepod species *Cosmocalanus darwinii*, *Nannocalanus minor*, and *Euchaeta sp.*, which comprised 15, 10, 10, 5, and 10 % of the total zooplankton abundance, respectively.

In the Spring Inter-monsoon season, fish eggs, decapod larvae, and females of the copepod *Centropages tenuiremis* dominated the total abundance of zooplankton, and contributed 8 % to total numbers. In the South-west Monsoon, the zooplankton community was dominated by c4 and c5 developmental stages of the copepod *Calanoides carinatus s.f.*, followed by *Centropages tenuiremis* and *Oikopleuridae. C. carinatus s.f.* comprising 51 % of the total zooplankton abundance. The other relatively abundant copepod species was *Rhincalanus nasutus*—with a maximal total prosome length of about 3.5 mm (Table 4).

The dominant group exhibited its presence all across the shelf, although the abundance of c4 and c5 copepodite stages of *Calanoides carinatus s.f.*, varied markedly, from 19 to ~6800 ind. m^{-3} (Fig. 7). In comparison to the North-east Monsoon and Spring Inter-monsoon seasons, the South-west Monsoon was notorious for the dominance of the late copepodite stages and adults of *Calanoides carinatus s.f., Centropages tenuiremis,* and *Rhincalanus nasutus* in samples.

In order to analyze seasonal changes in the size structure of the entire zooplankton community, organisms were distributed across 16 size classes with the resolution of 0.4 mm. The medians of these classes were used to approximate size distributions (Fig. 8). No prominent changes were observed over seasons. The group with the mean size of 1.8 mm was the dominant one throughout all three seasons. Overall, the community of the North-east Monsoon was characterized by a broad spectrum, whereas the spectrum of the South-west Monsoon community was narrow. The contribution of the dominant group increased, from 28 % during the North-east Monsoon to 59 % during the South-west Monsoon. The twofold increase was driven by the appearance of *Calanoides carinatus s.f.* No statistical differences were observed among seasons change in the organisms' mean size. Intra-seasonal variation in the mean size was large, which is reflected in the range of standard deviation (Table 2).

Discussion

Seasonal surveys of the zooplankton community across the Omani shelf have enabled us to characterize it as highly productive, with the mean mesozooplankton biomass varying from 543 to 723 mg m^{-3} throughout the seasonal cycle. These values exceeded the average biomass of the open Arabian Sea regions by three times, which was in a range of 100–200 mg m^{-3} according to the map of the International Indian Ocean Expedition (Bogorov et al. 1968; Prasad 1968).

Processing of samples has allowed us to identify three general spatial–temporal patterns: (1) the dominance of copepods in the total zooplankton abundance over the shelf waters, (2) the absence of seasonal changes in the level of total zooplankton biomass, and (3) the absence of seasonal changes in the size structure of the zooplankton community.

A leading contribution of copepods to the total abundance of zooplankton has been reported for various regions of the Arabian Sea (Padmavati et al. 1998; Smith and Madhupratap 2005) including Omani coastal waters (Piontkovski et al. 2013). This taxonomic group dominates the zooplankton abundance regardless of seasons.

As far as the seasonality of mesozooplankton biomass is concerned, published data are controversial. The "paradox" of zooplankton biomass reported for the western and eastern Arabian Sea in the early 1990's (Baars 1999; Madhupratap et al. 1992, 1996) was challenged by subsequent investigators stressing the presence of seasonality (Jyothibabu et al. 2010).

On the level of abundance, lack of seasonal changes of copepod abundance was reported for the central and eastern Arabian Sea (Madhupratap et al. 1996). Nonetheless, it was denoted that the North-east Monsoon season "is always characterized by an increase in the abundance of the cyclopoid copepods, particularly

Table 3 List of the copepod species found in samples collected during the North-east Monsoon (NEM), Spring Inter-monsoon (SIM), and the South-west Monsoon (SWM)

NEM	SIM	SWM
Acartia plumosa	*Acartia amboinensis*	*Acartia amboinensis*
Acrocalanus gracilis	*Acartia plumosa*	*Acartia plumosa*
Acrocalanus longicornis	*Acrocalanus longicornis*	*Acrocalanus longicornis*
Calanopia elliptica	*Calocalanus pavo*	*Bestiolina arabica*
Calanopia parathompsoni	*Candacia bradyi*	*Calanoides carinatus s.f.*
Candacia catula	*Candacia curta*	*Calocalanus plumulosus*
Candacia curta	*Canthocalanus pauper*	*Centropages orsini*
Candacia discaudata	*Centropages furcatus*	*Centropages tenuiremis*
Candacia truncata	*Centropages tenuiremis*	*Copilia mirabilis*
Canthocalanus pauper	*Copilia mirabilis*	*Corycaeus pacificus*
Centropages orsini	*Copilia quadrata*	*Eucalanus elongatus*
Centropages tenuiremis	*Corycaeus crassiusculus*	*Euterpina acutifrons*
Clausocalanus furcatus	*Corycaeus pacificus*	*Farranula gibbula*
Copilia mirabilis	*Corycaeus speciosus*	*Haloptilus mucronatus*
Corycaeus crassiusculus	*Cosmocalanus darwinii*	*Heterorhabdus papilliger*
Corycaeus pacificus	*Eucalanus elongatus*	*Mesocalanus tenuicornis*
Cosmocalanus darwinii	*Euchaeta indica*	*Nannocalanus minor*
Euchaeta indica	*Euchaeta rimana*	*Oithona brevicornis*
Euchaeta rimana	*Labidocera acuta*	*Oithona plumifera*
Labidocera acuta	*Lucicutia gaussae*	*Oithona setigera*
Nannocalanus minor	*Nannocalanus minor*	*Oncaea mediterranea*
Oithona plumifera	*Oncaea venusta*	*Oncaea venella*
Oncaea venusta	*Paracalanus aculeatus*	*Paracalanus aculeatus*
Paracalanus aculeatus	*Paracalanus tropicus*	*Paracalanus aculeatus minor*
Pareucalanus attenuatus s. l.	*Pontellina plumata*	*Paracalanus indicus*
Pontellina plumata	*Rhincalanus cf. nasutus*	*Pleuromamma indica*
Pontellopsis regalis	*Sapphirina metallina*	*Pseudodiaptomus serricaudatus*
Rhincalanus rostrifrons	*Sapphirina nigromaculata*	*Rhincalanus cf. nasutus*
Sapphirina nigromaculata	*Sapphirina opalina*	*Scolecitrichopsis ctenopus*
Sapphirina opalina	*Sapphirina scarlata*	*Subeucalanus crassus*
Sapphirina scarlata	*Subeucalanus crassus*	*Subeucalanus mucronatus*
Sapphirina stellata	*Subeucalanus mucronatus*	*Subeucalanus pileatus*
Scolecithrix danae	*Subeucalanus pileatus*	*Subeucalanus subtenuis*
Subeucalanus crassus	*Subeucalanus subcrassus*	*Temora turbinata*
Subeucalanus pileatus	*Subeucalanus subtenuis*	
Subeucalanus subcrassus	*Temora discaudata*	
Subeucalanus subtenuis	*Temora turbinata*	
Temora discaudata	*Undinula vulgaris*	
Temora turbinata		
Undinula vulgaris		

Oithona species, regardless of location in the Arabian Sea" (Smith and Madhupratap 2005). We have not observed this phenomenon, presumably because of the coarse mesh size used; the *Oithona* species were extremely rare in winter samples, in which females of *Oithona plumifera* were observed.

Thangaraja et al. (2011), who analyzed the total zooplankton biomass collected by a 200 μm net across the Omani shelf, have reported a 1.8-fold increase during the North-east Monsoon in comparison to the Fall Inter-monsoon period.

Table 4 List of the most abundant copepod species on the Omani shelf

	Ontogenetic stage	Body length (mm)
NEM		
Cosmocalanus darwinii	f	2.3–2.4
Nannocalanus minor	f	1.8–2.0
Subeucalanus subtenuis	f	3.3–3.6
Euchaeta indica	f	2.4–2.7
Euchaeta rimana	f	3.5
Euchaeta spp.	c5	1.0–3.0
Nannocalanus minor	f	1.8
Acrocalanus longicornis	f	1.2–1.4
Subeucalanus crassus	c5	2.5–2.7
Temora discaudata	f	1.8–2.0
SIM		
Centropages tenuiremis	f	1.6–2.0
Centropages tenuiremis	m	1.5–1.7
Subeucalanus subtenuis	f	3.0–3.2
Subeucalanus subtenuis	c4	2.0–2.2
Subeucalanus pileatus	f	2.0–2.3
Subeucalanus pileatus	c5	1.8–2.0
Subeucalanus pileatus	c4	1.6–1.7
Subeucalanus crassus	c2	1.4–1.5
Subeucalanus crassus	c3	1.9–2.0
Acartia amboinensis	f	1.5–1.6
Subeucalanus crassus	f	3.0–3.3
Subeucalanus crassus	c2	1.5–1.7
Subeucalanus pileatus	c3	1.4–1.5
Temora discaudata	c2	1.2–1.3
SWM		
Calanoides carinatus s.f.	f	2.3–2.7
Calanoides carinatus s.f.	c4	1.7–2.0
Calanoides carinatus s.f.	c5	2.3–2.5
Centropages tenuiremis	f	1.7–1.9
Centropages tenuiremis	m	1.5–1.7
Rhincalanus cf. nasutus	c5	3.5

NEM the North-east Monsoon, *SIM* Spring inter-monsoon, *SWM* the South-west Monsoon, *f* females, *m* males, *c2–c5* ontogenetic stages (copepodites)

In 1994–1996, the absolute maximum in zooplankton biomass was observed during the South-west Monsoon, in the area of coastal upwelling (Smith et al. 1998). The most abundant copepod species in the upwelling area in the upper 25 m layer were the small-bodied copepods *Paracalanus aculeatus*, *Paracalanus denudatus*, and *Parvocalanus dubia*. Four species (*P. denudatus, P. dubia, P. aculeatus* and *Acartia longicornis*) accounted for 36 % of the total abundance. With two other species added (*Subeucalanus crassus* and *Calanoides carinatus s.f.*), the group accounted for 37 % of total abundance (Smith and Madhupratap 2005). In our case, the group accounting for a major part of total abundance was notoriously different (Table 3). However, the mesh size used in our studies was larger (500 vs 150 μm), which shifted size characteristics and species composition within the community.

During the 1994–1995 cruises, high-frequency acoustical measurements carried out off the Omani shelf pointed out the lowest zooplankton biovolumes during the summer South-West Monsoon, intermediated during fall inter-monsoon, and the highest during the winter North-east Monsoon (Pieper et al. 2001).

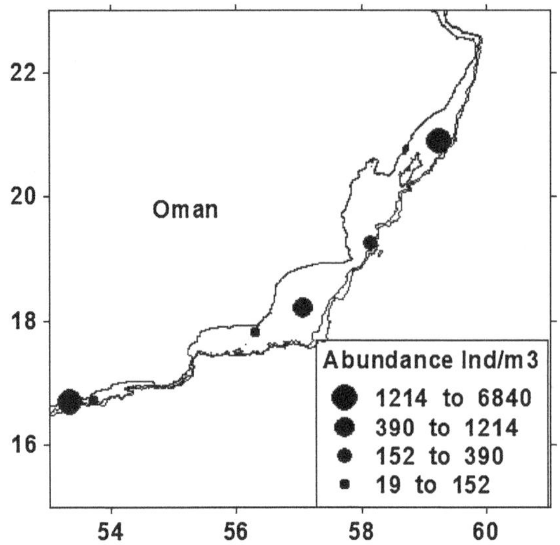

Fig. 7 Spatial distribution of the abundance of *Calanoides carinatus s.f.* during the South-west Monsoon

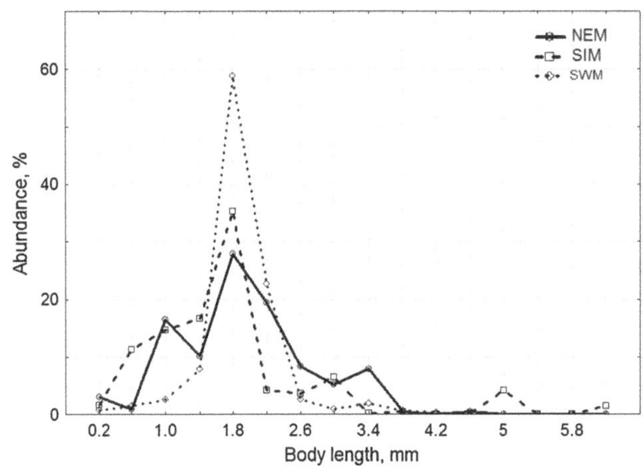

Fig. 8 Size structure of zooplankton community over seasons. The abundance of size groups is shown by % of the total zooplankton abundance. *NEM* North-east Monsoon, *SIM* spring inter-monsoon, and *SWM* South-west Monsoon

The UK "Arabesque" expedition sampled mesozooplankton during and after the 1994 South-west Monsoon (Stelfox et al. 1999), but almost all their "zooplankton stations" (1 out of 8) were outside the shelf area, in which the zooplankton biomass showed a decline from the North-east Monsoon to the fall inter-monsoon period.

One of the factors contributing to seasonal changes and heterogeneous distribution of zooplankton biomass should be a trophic pressure imposed by myctophids. These fishes actively ingest zooplankton, in particular at night when myctophids migrate to upper layers from the deep. Fish trawls and echo-soundings carried out during five seasonal cruises have implied dense layers of myctophids, 20–40 m thick, at about a 200–450 m depth, extending offshore from the shelf during the day and ascending to the upper layer at night, along the shelf edge. The myctophid biomass varied gradually over seasons and was composed mainly of *Benthosema pterotum*. For instance in the Ras al Hadd region, total myctophid catches accounted for 155 t km^{-2} during the Spring Inter-monsoon, with a subsequent increase of up to 1.2 M t km^{-2} by the end of the South-west Monsoon (Devine and Macaulay 2009). Interestingly, these huge changes did not induce an immediate reduction of zooplankton biomass in summer, although the mean value was minimal during the subsequent Fall Inter-monsoon season (Table 2).

On the species level, the most striking feature about seasonal changes in the zooplankton community was the massive appearance of *Calanoides carinatus s.f.* in the upper mixed layer, to which the population migrated in September, ascending from the depth of about 400 m. This species ingests predominantly diatoms in upwelling areas of the western Indian Ocean (Smith 2001). Interestingly, the appearance of *Calanoides carinatus s.f.* does not induce prominent seasonal changes in the total zooplankton biomass (Table 2).

We do not have the phytoplankton samples collected at the same place and time as the zooplankton sampling, so remotely sensed chlorophyll-*a* concentration was the only source used to estimate the magnitude of phytoplankton biomass change over seasons. The chlorophyll maps implied a fairly high biomass in all seasons. A peculiarity of 2008, which was unusual in many aspects including the distortion of a common well-developed seasonal cycle of chlorophyll-*a*, was the appearance of a massive harmful algal bloom of *Cochlodinium polykrikoides* during the Fall inter-monsoon, which lasted until spring-2009 (Richlen et al. 2010; Al-Azri et al. 2014).

In terms of spatial distribution, our data point out that the population of *Calanoides carinatus s.f.* occupies the whole shelf area—from Salalah to Ras al Hadd, although the spatial pattern has exhibited a prominent heterogeneity of biomass distribution—the feature reported by the previous survey carried out in August-1995, when the threefold variance of the dry weight of the biomass was reported along the southern part of the Omani shelf (Hitchcock et al. 2002).

Once the population of *Calanoides carinatus s.f.* occupies shelf waters, the question is how it gets there (over the continental slope, to depths less than 200 m), if the core of the population is located seaward, in oceanic waters, at the depth of 400 m? At least two aspects of the on-shelf advection could be proposed; the first one is transport by the East Arabian Current from the south. The second one is transport by cyclonic and anticyclonic eddies induced by Rossby waves propagating westward toward the Omani coast (Chelton et al. 2007). In the western Arabian Sea, spatial heterogeneity of zooplankton distribution could be largely explained as being caused by eddies and the mesoscale frontal zones between them (Hitchcock et al. 2002; Piontkovski and Melnik 2006).

On the scale of the Arabian Sea, a statistical relationship between the available potential energy of an eddy field and spatial heterogeneity of zooplankton biomass was reported (Piontkovski et al. 1995). The Omani shelf is a small part, but there, mesoscale eddies are common elements of a regional circulation (Flagg and Kim 1998; Piontkovski and Al-Jufaili 2013; Smith et al. 1991). Eddies entering Omani shelf waters from the central Arabian Sea have been the subject of a specialized expedition aimed at studying the role of eddies in the structure and functioning of the pelagic ecosystem of the western Arabian Sea (Piontkovski and Banse 2006). Zooplankton were collected in the upper 100 m layer at 109 stations in February–April 1990. Biomass maxima usually occurred at the edges of eddies and in frontal zones. The difference between biomass in these maxima was in a region of two- to fourfold.

The variance-to-mean ratio pointed out that zooplankton biomass possessed the most heterogeneous distribution throughout the shelf during the South-west Monsoon period (Table 2). This is the season when eddies are most common over a seasonal cycle of their occurrence in the western Arabian Sea (Piontkovski and Al-Jufaili 2013).

Conclusion

In 2007–2008, the Omani shelf was populated by a highly productive epipelagic plankton community. The chlorophyll-*a* concentration was high throughout a seasonal cycle and so was the mesozooplankton biomass, values of which nearly threefold exceeded the average biomass of the open Arabian Sea regions. Spatial distribution of zooplankton biomass over shelf waters was highly heterogeneous. The biomass and size structure of the zooplankton community did not exhibit prominent seasonal changes. However, spatial distribution of the zooplankton biomass throughout the shelf was most heterogeneous during the South-west monsoon season.

On the species level, the most striking feature in seasonal changes of the zooplankton community was the massive appearance of the copepod *Calanoides carinatus s.f.* in shelf waters, to which the population migrated in September. The population, ascending to upper layers during its ontogenetic migration, has occupied the entire Omani shelf area from Salalah to Ras al Hadd.

Acknowledgments This research was supported by the Ministry of Agriculture and Fisheries Wealth (Oman), the Research Council Grant No. RC/AGR/FISH/12/01, U.S. National Science Foundation Grants Nos. 0825598 and 1259255. L. Galkovskaya is thanked for technical assistance with data quality control and maps.

References

Al-Azri AR, Piontkovski SA, Al-Hashmi KA, Goes JI, Gomes HR, Glibert P (2014) Mesoscale and nutrient conditions associated with the massive 2008 *Cochlodinium polykrikoides* bloom in the Sea of Oman/Arabian Gulf. Estuar Coast 37:325–338

Anonymous (1975) Reports on cruises Nos. 1 and 2 of R.V. "Dr. Fridtjof Nansen". Indian Ocean Fishery and Development Programme. Pelagic Fish Assessment Survey, North Arabian Sea. FAO, Rome

Baars MA (1999) On the paradox of high mesozooplankton biomass, throughout the year in the western Arabian Sea: re-analysis of IIOE data and comparison with newer data. Ind J Mar Sci 28:125–137

Bogorov VG, Vinogradov ME, Voronina NM, Kanaeva IP, Syetova IA (1968) The distribution of zooplankton biomass in the upper layer of the world's ocean. Rep USSR Acad Sci (Dokladyi Academii Nayk SSSR) 182:232–245 **(in Russian)**

Böhm E, Morrison J, Manghnani V, Kim HS, Flagg C (1999) The Ras al Hadd Jet: remotely sensed and acoustic Doppler current profiler observations in 1994-1995. Deep-Sea Res II 46:1531–1549

Chelton DB, Schlax MG, Samelson RM, Szoeke RA (2007) Global observations of large oceanic eddies. Geophys Res Let 34:L15606. doi:10.1029/207GL030812

Devine J, Macaulay G (2009) Fish resources assessment survey of the Arabian Sea Coast of Oman: Habitat. Technical Report 4. Mesopelagic Fisheriy Resources of the Arabian Sea Coast of Oman. Bruce Shallard and Associates, Muscat

Flagg CN, Kim HS (1998) Upper ocean currents in the northern Arabian Sea from shipboard ADCP measurements collected during the 1994-1996 JGOFS and ONR programs. Deep-Sea Res II 45:1917–1959

Goswami SC, Shrivastava Y (1996) Zooplankton standing stock, community structure and diversity in the northern Arabian Sea. In: Proceedings of the Second Workshop-Scientific Results. FORV Sagar Sampada, pp. 127–137

Hitchcock GL, Lane P, Smith S, Luo J, Ortner PB (2002) Zooplankton spatial distributions in coastal waters of the northern Arabian Sea, August 1995. Deep Sea Res II 49:2403–2423

Jacobs CL (2003) RRS Charles Darwin Cruise 143. Sheherezade II: Geological and biological surveys of the Arabian Sea and the continental slope of Oman. Southhampton Oceanography Centre Cruise Report, No. 42

Jyothibabu R, Madhu NV, Habeebrehman H, Jayalakshmy KV, Nair KKC, Achuthankutty CT (2010) Re-evaluation of "paradox" in the eastern Arabian Sea based on ship and satellite observations. J Mar Syst 81:235–251

Kistler RE, Kalnay W, Collins S, Saha G, White J, Woollen M, Chelliah W, Ebisuzaki M, Kanamitsu V, Kousky H, van den Dool RJ, Fiorino M (2001) The NCEP-NCAR 50-year reanalysis: monthly means CD-ROM and documentation. Bull Am Met Soc 82:247–268

Lane P, Smith S, Zavagoza J, Prusova I, Roman M (1998) U.S. JGOFS Technical Report: Copepod taxonomy, abundance and biomass in the upper 300 meters of the Arabian Sea during the Southwest Monsoon of 1995. RSMAS Technical Report 98-007. University of Miami, Miami, p 409

Luo J, Ortner PB, Forcucci D, Cummings SR (2000) Diel vertical migration of zooplankton and mesopelagic fish in the Arabian Sea. Deep Sea Res II 47:1451–1473

Madhupratap M, Haridas P, Ramaiah N, Achuthankutty CT (1992) Zooplankton of southwest coast of India: abundance, composition, temporal and spatial variability in 1987. In: Desai BN (ed) Oceanography of Indian Ocean. Oxford & IBH, New Delhi, pp 99–112

Madhupratap M, Gopalakrishnan TC, Haridas P, Nair KKC, Aravindakshan PN, Padmavati G, Paul S (1996) Lack of seasonal and geographic variation in mesozooplankton biomass in the Arabian Sea and its structure in the mixed layer. Cur Sci 71:863–868

McKoy J, Bagley N, Gauthier S, Devine J (2009) Fish Resources of the Arabian Sea coast of Oman: Project summary. In Technical Report 1. Bruce Shallard and Associates, Muscat

Okubo A (1980) Diffusion and ecological problems: mathematical models. Biomathematics, vol 10. Springer, New York

Padmavati G, Harids P, Nair KK, Gopalakrishnan TC, Shiney P, Madhupratap M (1998) Vertical distribution of mesozooplankton in the central and eastern Arabian Sea during the winter monsoon. J Plankton Res 20:343–354

Pieper RE, McGehee DE, Greenlaw CF, Holliday DV (2001) Acoustically measured seasonal patterns of zooplankton in the Arabian Sea. Deep-Sea Res II 48:1325–1343

Piontkovski SA, Al-Jufaili S (2013) Coastal upwellings and mesoscale eddies of the western Arabian Sea: some biological implications. Int J Ocean Oceanogr 7:93–115

Piontkovski SA, Banse K (2006) Overview of results. In: Banse K, Piontkovski SA (eds) The mesoscale structure of the epipelagic ecosystem of the open northern Arabian Sea. Universities Press, Hyderabad

Piontkovski SA, Melnik TA (2006) Mesozooplankton of the north-west Arabian Sea. In: Banse K, Piontkovski SA (eds) The mesoscale structure of the epipelagic ecosystem of the open northern Arabian Sea. Universities Press, Hyderabad

Piontkovski SA, Williams R, Peterson W, Kosnirev VK (1995) Relationship between oceanic zooplankton and energy of eddy fields. Mar Ecol Progr Ser 128:35–41

Piontkovski SA, Al-Maawali A, Al-Mantheri W, Al-Hashmi K, Popova E (2013) Zooplankton of Oman coastal waters. Agr Mar Sci 18:37–50

Prasad RR (1968) Maps on total zooplankton biomass in the Arabian Sea and the Bay of Bengal. International Indian Ocean Expedition Plankton Atlas, vol. 1, Fasc. 1. Indian Biological Centre, National Institute of Oceanography, CSIR, New Delhi

Richlen ML, Morton SL, Jamali EA, Rajan A, Anderson DM (2010) The catastrophic 2008-2009 red tide in the Arabian Gulf region, with observations on the identification and phylogeny of the fish-killing dinoflagellate *Cochlodinium polykrikoides*. Harmful Algae 9:163–172

Roman M, Smith S, Wishner K, Zhang X, Gowing M (2000) Mesozooplankton production and grazing in the Arabian Sea. Deep-Sea Res II 47:1423–1450

Smith S (2001) Understanding of the Arabian Sea: reflections on the 1994-1996 Arabian Sea expedition. Deep-Sea Res II 48:1385–1402

Smith S (2005) The Arabian Sea of the 1990s: new biogeochemical understanding. Progr Oceanogr 65:113–115

Smith SL, Madhupratap M (2005) Mesozooplankton of the Arabian Sea: patterns influenced by seasons, upwelling, and oxygen concentrations. Progr Oceanogr 65:214–239

Smith SL, Banse K, Cochran JK, Codispoti LA, Ducklow HW, Luther ME, Olson DB, Peterson WT, Prell WL, Surgi N, Swallow JC, Wishner K (1991) U.S.JGOFS: Arabian Sea Process Study. U.S. JGOFS Planning Report No. 13. Woods Hole Oceanographic Institution, Woods Hole

Smith S, Roman M, Prusova I, Wishner K, Gowing M, Codispoti L, Barber R, Marra J, Flagg C (1998) Seasonal response of zooplankton to monsoonal reversals in the Arabian Sea. Deep-Sea Res II 45:2369–2404

Stelfox CE, Burkill PH, Edwards ES, Harris RP, Sleigh MA (1999) The structure of zooplankton communities, in the 2 to 200 μm size range, in the Arabian Sea during and after the SW monsoon, 1994. Deep-Sea Res II 46:815–842

Thangaraja M, Al-Aisry A, Al-Kharusi L (2011) Ecology, abundance and spatial distribution of zooplankton biomass in the waters of Sultanate of Oman. South Ind Coast Mar Bull 3:5–13

14

Effect of water de-stratification on dissolved oxygen and ammonia in tilapia ponds in Northern Thailand

Patcharawalai Sriyasak · Chanagun Chitmanat · Niwooti Whangchai · Jongkon Promya · Louis Lebel

Abstract Episodes of low concentrations of dissolved oxygen and high concentrations of ammonia are major causes of fish stress, which in turn, reduces growth and increases mortality rates in aquaculture ponds. This study measured the effects of water de-stratification on dissolved oxygen and ammonia concentrations in tilapia ponds in Northern Thailand. Fifteen ponds in five provinces in Northern Thailand were sampled on multiple dates in the hot, wet, and dry seasons. Thermal water stratification peaked around 14:00–16:00 h each day; wherein the differences between surface and lower water temperatures in the 0.8–2.0 m deep ponds reached 1.3–4.0 °C. Thermal de-stratification in the hot season and dry season usually occurred late at night; in the wet season however, it occurred early in the evening due to the cooling effects of rain. The mixing of surface and bottom waters decreased dissolved oxygen levels in water near the surface, and, increased concentration levels in the bottom water layer. Mean DO concentrations of integrated and commercial ponds were lower than 1 mg/L from 02:00 to 06:00 h, in 68 % of the observations. Repeated measures ANOVA analysis showed that water depth and fish culture systems have significant effects on total ammonia nitrogen (TAN). TAN concentrations near the surface were lower than at the bottom, and increased after water de-stratification. TAN in the integrated culture system was significantly higher than in commercial and subsistence systems. Toxic, un-ionized ammonia (NH_3) concentrations in all culture systems, however, were still lower than standard toxicity level for tilapia. The improved understanding of the seasonal and culture-system specific effects of de-stratification on pond water quality provided by this study, are useful for improving pond management practices to reduce the risks of exceeding DO thresholds. In particular, fish farmers should: (1) take care to avoid over-feeding and manage water and sediments to prevent excessive accumulation of organic matter and waste at the bottom of ponds, which can influence other water quality parameters; and, (2) use aeration and mechanical mixing interventions at critical times to reduce stress on fish from low DO concentrations, and thus avoid risks of mass mortality events.

Keywords Thermal de-stratification · Dissolved oxygen · Ammonia · Fish culture

P. Sriyasak · C. Chitmanat · N. Whangchai · J. Promya
Faculty of Fisheries Technology and Aquatic Resources, Maejo University, Chiang Mai, Thailand

P. Sriyasak (✉) · L. Lebel
Unit for Social and Environmental Research, Chiang Mai University, Chiang Mai, Thailand
e-mail: kpatchara@hotmail.com

Introduction

Nile tilapia culture in earthen ponds in Thailand is expanding: between 2003 and 2013 culture area and production increased by 12 and 46 %, respectively (DOF 2013). This is due to high consumer demand, improvements in seed production, including, the introduction of sex-reversal technologies (Bhujel 2013), and the availability of pellet feeds; thus allowing producers to achieve high fish growth rates and yields (Abdel-Tawwab 2012; Bhujel 2013). At the same time, fish farmers are facing increasing problems of maintaining adequate water quality in fish ponds (Sriyasak et al. 2014).

Water quality in a pond influences feeding, growth, disease burdens, and survival rates (Chainark and Boyd 2010; Pandit and Nakamura 2010). Water quality is controlled by a complex interplay of many factors, including weather conditions (Ahmed et al. 2011). For example, dissolved oxygen (DO) is related to phytoplankton production and respiration (Smith and Piedrahita 1988); nitrogen waste such as ammonia is related to the amount of organic matter inputs and ammonium excretion by fish (Gross et al. 1999; Hargreaves 1998); and, water temperature and thermal stratification are controlled by sunlight and air temperature (Egna and Boyd 1997; Losordo and Piedrahita 1991).

Adequate DO concentrations in a fish pond are essential for maintaining optimal growth (Chang and Ouyang 1988). Ponds with high nutrient input have high phytoplankton concentrations, and as a consequence, have high DO concentrations in the afternoon due to phytoplankton photosynthesis (Diana et al. 1991), and low DO concentrations in the early morning due to phytoplankton, water column, and sediment respiration (Chang 1989). Oxygen concentrations lower than 3 mg/L reduce feed intake and growth in Nile tilapia (Duy et al. 2008).

Ammonia is the first form of nitrogen released when organic matter decomposes, and can exist in two forms: un-ionized ammonia (NH_3) or ionized ammonia (NH_4^+) (Durborow et al. 1992; Gandhi 2012; Gross et al. 1999). Un-ionized ammonia (NH_3) is toxic to fish, and becomes even more toxic as water temperature and pH increase (EPA 2013; Hargreaves and Tucker 2004). Acute toxicity of ammonia is due to its effect on the central nervous system (Randall and Tsui 2002); ammonia concentrations of 7.40 mg/L have been shown to cause mass mortality in tilapia fingerling within 24 h (Benli and Koksal 2005). Fish exposed to toxic levels of ammonia cannot excrete ammonia efficiently; as a result, ammonia levels in blood and tissues increase along with pH levels, therefore affecting enzyme activity (Lawson 1995). This can lead to poor feed conversion, slower growth rates, and reduced resistance to diseases (Gandhi 2012; Hargreaves and Tucker 2004).

Shallow aquaculture ponds undergo diel cycles of thermal and chemical vertical stratification (Losordo and Piedrahita 1991). Stratification of water in ponds arises due to the differences in water density, as the absorption of solar radiation decreases with increasing depth of the water column (Chowdhury et al. 2014; Klimaszyk and Tomaz 2007; Losordo and Piedrahita 1991). Warm surface water also has higher DO levels from photosynthesis, while cooler bottom water has lower DO, plus accumulated nitrogenous waste from uneaten feed and feces that drop down to the sediment (Gandhi 2012). De-stratification, or the remixing of water, occurs as air temperature drops and the surface layer cools down, therefore becoming denser than water in the lower layers. The surface water then sinks and forces water at the bottom to rise, thus turning over the layers (Chowdhury et al. 2014; Szyper and Lin 1990). After water mixing, DO in the surface layer falls, and the concentration of nitrogenous waste like ammonia increases. Stratification with depth in ponds, followed by rapid turnover or exchange of surface and bottom water layers, can expose fish to dangerously low dissolved oxygen levels, even near the surface. Conversely, mixing helps to increase oxygen levels near the bottom.

For Nile tilapia cultured in earthen ponds in Thailand, tilapia mortality from disease outbreak frequently occurs when DO levels are low and ammonia concentrations are high (Chitmanat 2013). Fish culture in pellet feed based and poultry or livestock integrated systems appear to be more impacted by periods of prolonged cloud cover and heavy rain than subsistence culture system (Pimolrat et al. 2013). Fish in ponds that are exposed to hot weather for several days, and then treated to a sudden period of heavy rain or cloud cover, also tend to be vulnerable to diseases due to low oxygen condition and high organic wastes (Srisapoome and Areechon 2014); this is because rain encourages water mixing between the surface and bottom layers.

Stratification and water turnover is known to occur in tilapia fish ponds in Northern Thailand (Sriyasak et al. 2013). The objective of this study was to measure the effect of water de-stratification on concentrations of DO and ammonia in fish ponds that are managed under different culture systems during different seasons in Northern Thailand. As tilapias are known to be sensitive to episodes of low concentrations of DO and high concentrations of ammonia, a better understanding of the effects of de-stratification on pond water quality should be useful for managing water- and climate-related risks to aquaculture production.

Methods

Study site

This study made observations and measurements in 15 tilapia earthen ponds located in five provinces in Northern Thailand: Chiangrai, Chiangmai, Phayao, Lampang, and Nakhon Sawan (Fig. 1). Pond sizes ranged

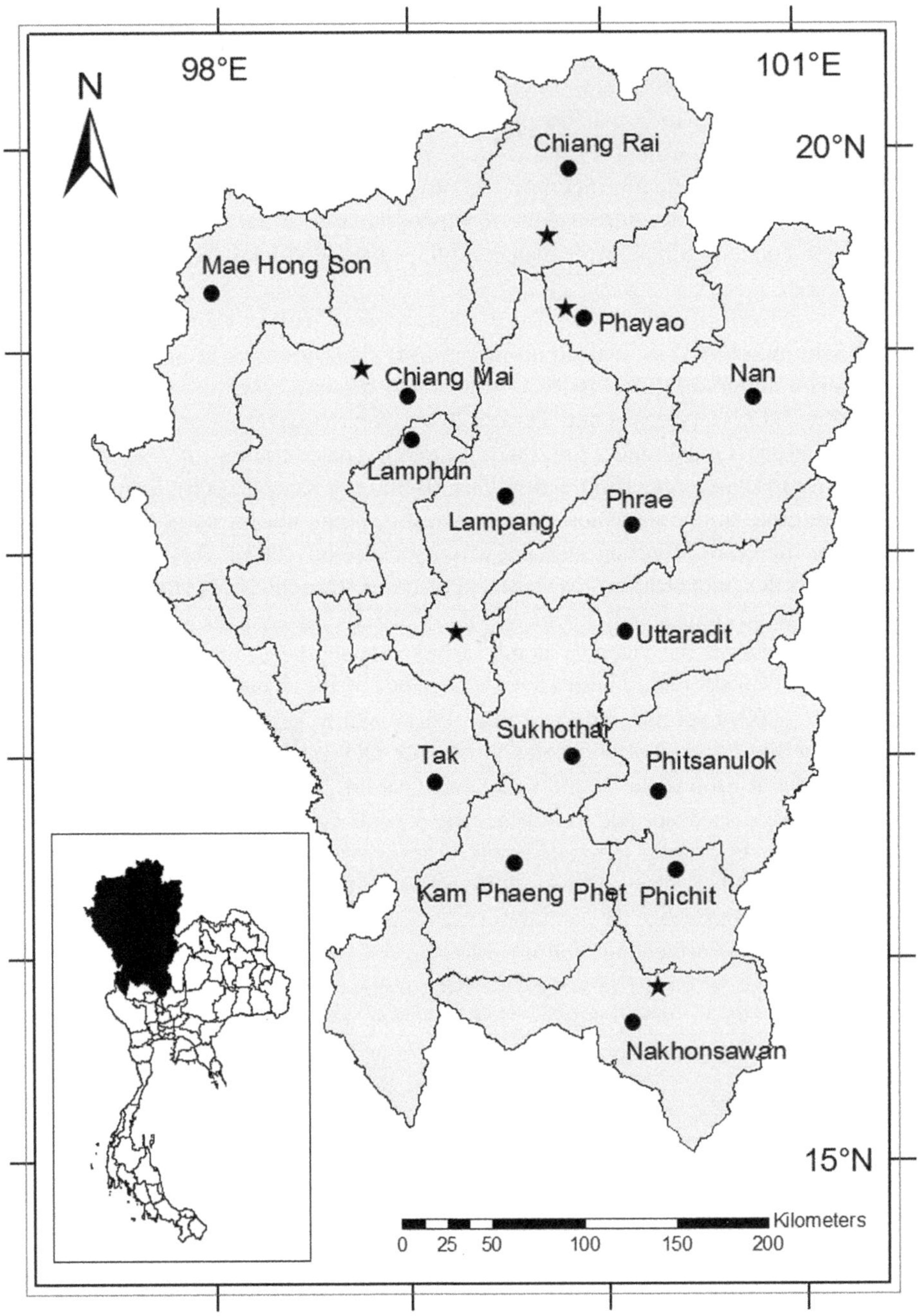

Fig. 1 Map of study area in Northern Thailand. *Stars* indicate farm locations, *dots* the city district in each province

Table 1 Major features of ponds managed under different culture systems that were monitored in the study

Pond feature	Culture system		
	Integrated	Commercial	Subsistence
Number of ponds	3	6	6
Pond area (m^2)	2600 ± 1830	3400 ± 1050	460 ± 340
Pond depth (m)	1.2 ± 0.4	1.2 ± 0.1	1.0 ± 0.2
Stocking rate ($fish/m^2$)	2.7 ± 0.5	2.8 ± 0.4	0.8 ± 0.3
Culture period (months)	5.8 ± 0.7	6.5 ± 0.5	14.0 ± 3.1
Feed type	Pellet feed	Pellet feed	Pellet feed, vegetable waste and food scarps
Manure inputs	Yes	No	No
Water renewal rate	Seldom	10 % per day	Seldom
Production	All harvest sold	All harvest sold	Harvest consumed and sold

Mean ± SD

from 0.16 to 0.64 ha, with depths of 0.8–2.0 m. Ponds were grouped according to culture system: commercial (where prepared pellet feed was regularly provided and crops tended to be harvested once); integrated (where fish and livestock were raised in the same area and the wastes from the latter were dumped into ponds); and subsistence (where feeding was sporadic and fish were harvested continuously for consumption and market). The characteristics of each culture system are summarized in Table 1.

Water parameters

Altogether, 660 samples of fish pond water were collected for water quality analysis (15 ponds × 11 - months × 4 samples) between May 2013 and 2014. Samples were classified into three seasons for analysis: the "dry" season (November–February) when air temperatures and precipitation are lowest; the pre-monsoon "hot" season (March–May) when temperatures are highest; and the monsoon "wet" season (June–October) when precipitation is highest (Thai Meteorological Department 2012). Ambient air temperature, water temperature, pH, turbidity and DO were monitored at 2 h intervals, over a 24 h period, in each season, at every 20 cm depth, with a multimeter (TOA DKK WQC-22A model, Japan). Moreover, water samples were collected 20 cm below the surface and 20 cm above the pond bottom during stratified and de-stratified conditions. Chemical analyses were carried out for total ammonia-nitrogen (TAN) and chlorophyll-a according to standard methods (APHA 1980).

Toxic un-ionized ammonia (NH_3) concentration was calculated from the equation:

$$NH_3 = (a)\mathrm{TAN}$$

where, (a) is the mole fraction of un-ionized ammonia in aqueous solution at different pH values and temperatures (Emerson et al. 1975).

The maximum difference between surface and bottom temperatures of the pond was used as an indicator of stratification (Stefan et al. 1996). The de-stratification time was defined as when there was no detectable difference (less than 0.1 °C) between surface and bottom water temperatures. Data on precipitation were taken from the nearest weather station to each farm (Thai Meteorological Department 2014).

Statistical analysis

A factor analysis using principal component analysis (PCA) with a varimax rotation was first carried out to help understand the patterns of co-variation among the many water quality variables. Separate ANOVA models were then used to analyze the combined effects of culture system, season, and sampling depth factors on maximum water temperature stratification, minimum DO, DO stratification, and mean TAN. TAN values were log-transformed prior to analysis to homogenize variances, and then analyzed using repeated measures ANOVA for concentrations before and after stratification. Pond depth (<1 and >1 m) was included as a separate predictor as it could potentially influence thermal stratification; but as it was later found not to be significant it was thus excluded from the final models.

Results

Principal component analysis

Logically related individual water quality measures were strongly correlated with the same factor (Table 2). Thus, temperature variables covaried and all loaded on Factor 1, whereas TAN variables were correlated with Factor 3, and DO variables with Factor 4. Factor 2 could be interpreted as high plankton and organic input conditions, as both chlorophyll-a and turbidity variables were positively correlated with the factor, while pH was negatively correlated. Based on the clear separation of outcome variables of interest in this exploratory analysis, it seems justified to proceed to analyzing the effects of season, culture system, and sampling depth on the selected temperature, DO and TAN variables separately.

Weather, water stratification and de-stratification in fish ponds

Averaged over 24 h, air temperatures, as well as surface and bottom water temperatures, at pond sites on sampling dates, were significantly lower in the dry season than in the hot or wet seasons as expected (Table 3). Maximum water surface temperature usually occurred at 16:00 h, and minimum at 06:00 h. The highest water temperature observed was 38.0 °C in the hot season, and the lowest was 18.3 °C during the dry season. Daily air temperature range was lower in the wet season than in the other two seasons. Average mean precipitation also varied among seasons as expected (Table 3).

The maximum difference between surface and bottom water temperatures always occurred in measurements made in the afternoon, and ranged between 0 and 7.1 °C. The degree of stratification varied among seasons and culture systems with a significant interaction between the two factors ($P < 0.05$, Fig. 2). In the wet season, culture systems behaved alike, whereas in the hot and dry seasons, integrated ponds showed relatively larger thermal stratification (Fig. 2).

Stratification usually began from 14:00 to 16:00 h, and became less apparent during the night, as the upper layer cooled and water began to mix. Thermal de-stratification in the wet season occurred between 18:00 and 22:00 h; in the dry season between 20:00 and 24:00 h; and in the hot season it occurred between 22:00 and 04:00 h. Water de-stratification in the wet season in all culture system types occurred earlier than in the dry and hot seasons; possibly because cold water from rain decreased surface water temperature, in turn,

Table 2 Factor loading, eigenvalues, and percentage of variance from the principal component analysis

Variables	Factor 1	Factor 2	Factor 3	Factor 4
Air temperature	**0.91**	0.12		−0.15
Surface water temperature	**0.97**	0.13		
Bottom water temperature	**0.92**		0.18	
DO surface		0.22		**0.83**
DO bottom	−0.19	−0.36		**0.73**
TAN before stratification			**0.96**	
TAN after stratification		0.21	**0.92**	
pH		−0.72	−0.17	0.43
Turbidity		**0.88**		0.14
Chlorophyll-a	0.20	**0.85**	0.14	
Eigenvalue	3.54	2.00	1.46	1.34
Percentage of variance	35.4	20.0	14.6	13.4
Cumulative (%)	35.4	55.3	70.0	83.4

Values larger than 0.7 are in bold

Table 3 Air and water temperatures as well as daily precipitation (Mean ± SD) at fish ponds in three different seasons; different superscript letters in the same rows indicate a significant difference between means at $P < 0.05$ (Tukey's HSD)

Variables	Season		
	Hot (March–May)	Wet (June–October)	Dry (November–February)
Mean maximum air temperature (°C)	34.3 ± 6.9^a	30.5 ± 8.8^b	30.8 ± 6.6^b
Daily air temperature range (°C)	12.5 ± 3.7^a	8.1 ± 2.8^b	14.4 ± 4.2^a
Air temperature (°C)	28.2 ± 2.9^a	27.7 ± 2.7^a	22.8 ± 3.9^b
Surface water temperature (°C)	29.6 ± 2.6^a	28.7 ± 2.3^a	24.2 ± 3.4^b
Bottom water temperature (°C)	29.1 ± 2.7^a	28.3 ± 2.1^a	23.7 ± 3.4^b
Daily precipitation (mm/day)	2.9 ± 1.7^b	6.5 ± 2.2^a	0.8 ± 1.0^c

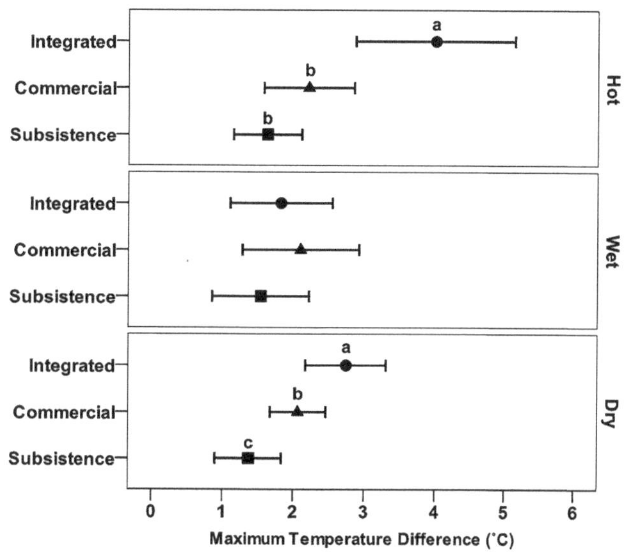

Fig. 2 Maximum difference between surface and bottom water temperatures in different culture systems and seasons. *Plotted symbols* are means and bars 95 % confidence intervals (CI) for those means

encouraging de-stratification. To support this argument, Fig. 3 shows the thermal stratification on three separate observation days, one of which happen to rain heavily at about 16:00 h (Fig. 3c); 2 h later, complete stratification could be observed (Fig. 3c), two or more hours earlier than usual. Thermal stratification was sometimes substantial (Fig. 3b), and at other times less so (Fig. 3a).

In a few instances however, no stratification was observed: 11 occasions in the dry season, 7 occasions in the hot season, and 4 occasions in the wet season. The reasons for these instances include: heavy rain during the day ($n = 6$); high cloud cover ($n = 8$); exchanging of pond water ($n = 4$); and the use of aerators ($n = 4$). These events were too rare for powerful statistical comparisons to be made.

Effect of water de-stratification on dissolved oxygen in fish ponds

Thermal stratification in ponds (Fig. 3) was accompanied by DO stratification (Fig. 4). The highest DO was found in the surface water layer during the afternoon, which began to decrease after sunset. The maximum DO recorded at the surface layer in the afternoon was 22.7 mg/L; this was in an integrated pond system in May. Oxygen oversaturation in the late afternoon was found to occur in both integrated and commercial ponds. On the other hand, under-saturation of oxygen in the late afternoon was observed in some subsistence ponds.

These differences in extremes among culture systems are consistent with measures of chlorophyll-a and turbidity. The mean chlorophyll-a concentration of integrated ponds (551 ± 300 μg/L) was significantly higher than commercial ponds (227 ± 208 μg/L) or subsistence ponds (36 ± 42 μg/L). Likewise, the

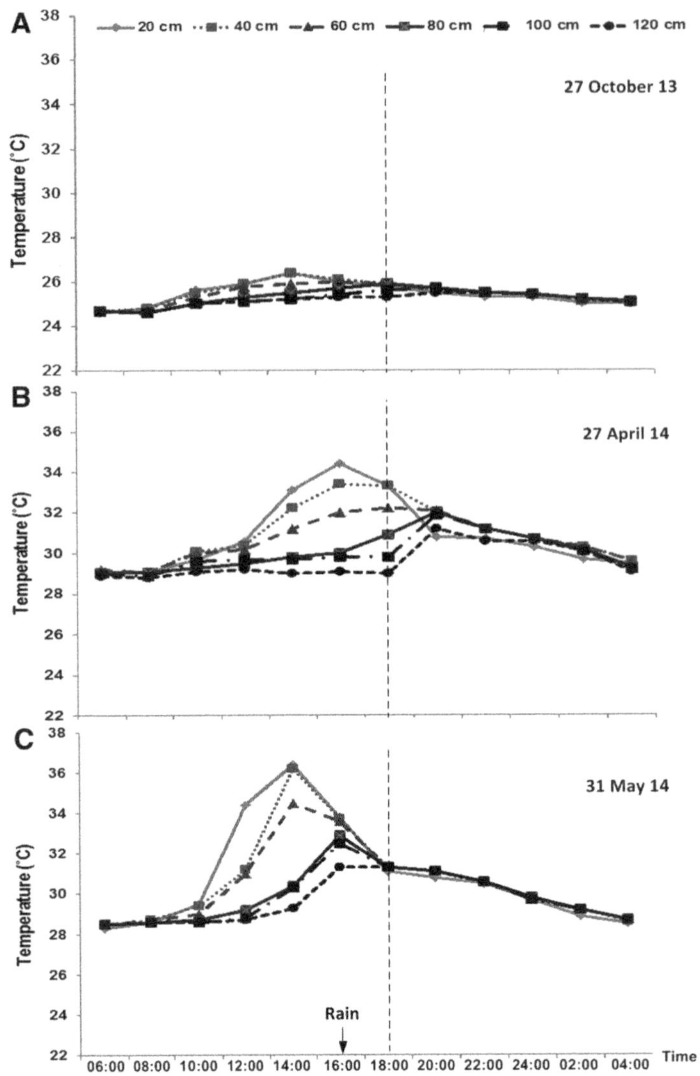

Fig. 3 Thermal stratification of pond waters on three illustrative dates with small maximum temperature difference (**a**); large maximum temperature difference (**b**); and large temperature difference and heavy afternoon rainfall (**c**)

turbidity in integrated ponds (153 ± 82 NTU) was significantly higher than commercial ponds (88 ± 52 NTU) or subsistence ponds (36 ± 26 NTU).

Water de-stratification resulted in decreased DO concentration at the surface, and increased DO at the bottom of the pond (Fig. 4). Mean DO concentration of integrated and commercial ponds were lower than 1 mg/L, from 02:00 to 06:00 h, in 68 % of observations; whereas the subsistence systems were always over 1 mg/L during the same periods.

ANOVA was used to analyze the influence season and culture system had on average minimum DO concentrations over a 24-h period. Average minimum DO concentration was low in integrated ponds, and was especially lower than the other two culture systems during the dry season (Fig. 5). Average 24-h DO concentrations were lower at the bottom of the pond (2.8 ± 2.3 mg/L) than at the surface (5.6 ± 2.3 mg/L, $P < 0.001$). Average 24-h DO concentrations were higher in integrated (6.5 ± 6.0 mg/L) and commercial ponds (6.1 ± 4.5 mg/L), than in subsistence ponds (4.6 ± 3.2 mg/L, $P < 0.05$).

Mean maximum DO difference between top and bottom water layers varied significantly among culture systems, but was unaffected by season (Fig. 6). Integrated culture system showed the greatest DO stratification, whereas subsistence system showed the least (Fig. 6).

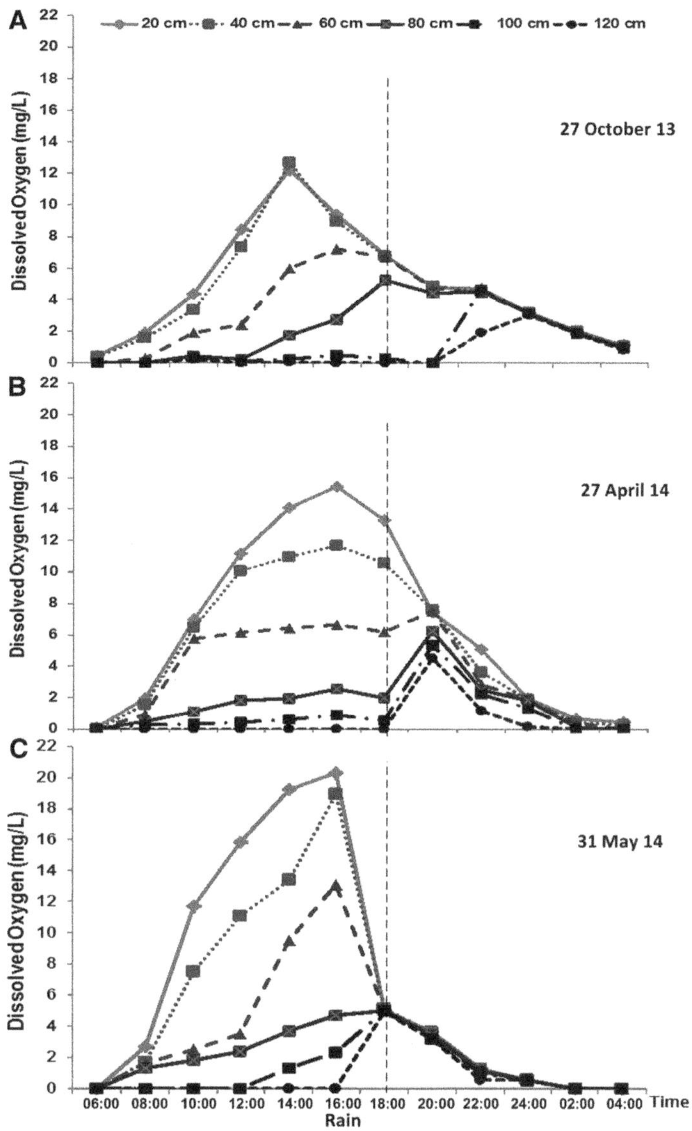

Fig. 4 Diel stratification of dissolved oxygen of pond waters on three illustrative dates (as in Fig. 3) with small maximum temperature difference (**a**); large maximum temperature difference (**b**); and large temperature difference and heavy afternoon rainfall (**c**)

Effect of water de-stratification on total ammonia in fish ponds

Repeated measures, multi-factor ANOVA was used to analyze the influence of depth of water (surface and bottom), season, and fish culture system, on TAN before and after de-stratification. The analysis showed that fish culture systems and depth of pond had significant effects on TAN (Fig. 7). TAN concentrations in tilapia ponds at the surface were lower than at the bottom of the pond ($P < 0.05$). After de-stratification, TAN at the surface water layer clearly increased (Fig. 7). Water de-stratification resulted in increased TAN at the surface layer. TAN of the integrated system was significantly higher than in commercial and subsistence systems (Tukey's HSD, $P < 0.05$). However, season had no significant effects on TAN ($P > 0.2$). The average levels of un-ionized ammonia (NH_3) concentration of integrated and commercial pond systems were 0.012 ± 0.01 mg/L, whilst in subsistence system it was 0.005 ± 0.01 mg/L.

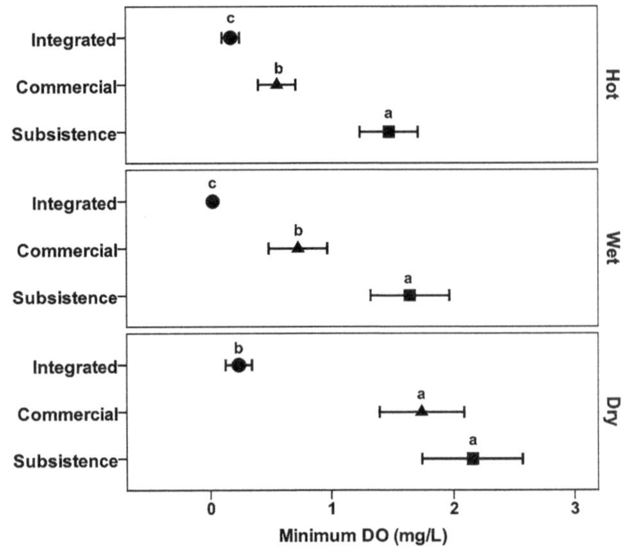

Fig. 5 Minimum DO concentration in different culture systems and seasons. *Plotted symbols* are means and bars 95 % confidence intervals (CI) for those means

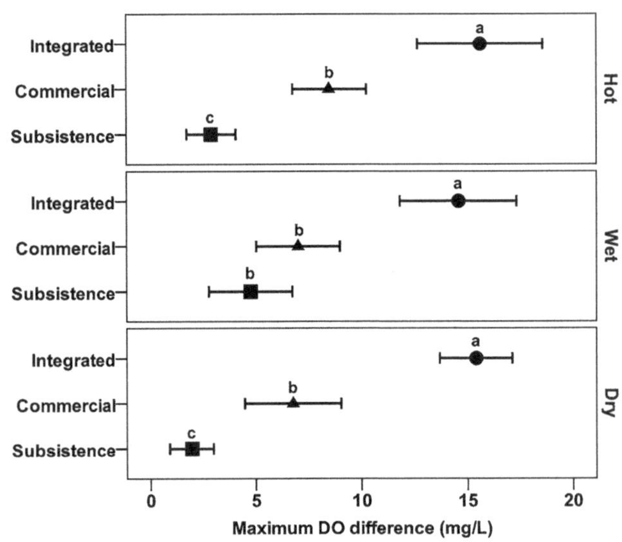

Fig. 6 Difference between surface and bottom DO concentrations in the afternoon in different culture systems and seasons. *Plotted symbols* are means and bars 95 % confidence intervals (CI) for those means

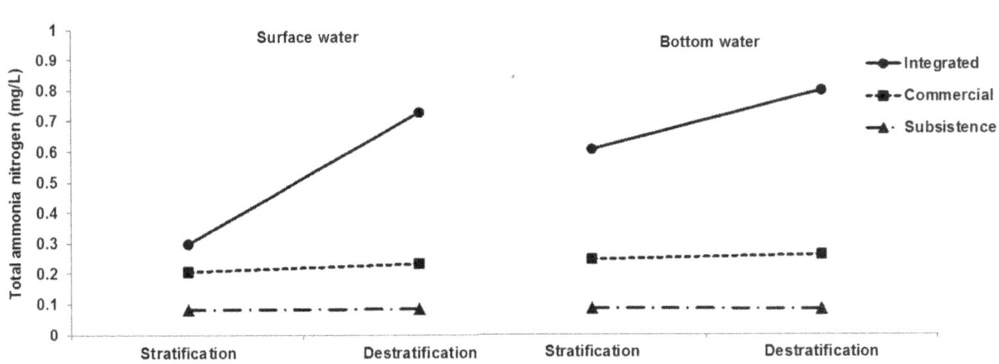

Fig. 7 Total ammonia nitrogen of different culture systems at surface and bottom pond before and after de-stratification

Discussion

The largest differences in temperature between surface and bottom layer were observed in integrated ponds during the hot season; whereas differences in the dry season were modest and in the wet season largely absent (Fig. 2). Four factors help explain these finding. First, the high load of nutrient and fertilizer inputs in integrated ponds caused phytoplankton blooms (Little and Edwards 2004; Szyper and Lin 1990). In this study, concentrations of Chlorophyll-a and overall turbidity, were much higher in integrated ponds than the other two systems. Consequently, high biomass of plankton or suspended particles in the surface layer limits light penetration to the lower water layers (Litchman et al. 2004; Tadesse et al. 2004). Second, is that solar radiation in the hot season tends to be higher than in the other seasons of Thailand (Janjai and Wattan 2011). Factors one and two explain the high potential for thermal stratification to occur during the hot season. A third factor, is that the air temperature range within a 24-h period tended to be greater in the hot and dry seasons than in the wet season (Table 3), which, if wind-mixing was limited, would generate thermal gradients (Losordo and Piedrahita 1991). A fourth factor, is heavy day time rainfall in the wet season; which could help explain why thermal stratification sometimes does not occur in this season (Fig. 2).

Thermal stratification usually peaked in the afternoon and collapsed in the evening (Fig. 3). Water de-stratification in the wet season occurred earlier than in the dry and hot season, probably due to mixing effects from rain (Diana et al. 1991). Seasonal differences in typical wind conditions might also have played a role (Diana et al. 1991; Culberson and Piedrahita 1996). Thermal de-stratification effects DO concentrations; decreasing them in surface waters while increasing them in the bottom layers (Boyd et al. 1978; Chang 1989). Water mixing is an important factor to consider therefore, as it effects fish respiration. On the other hand, the increase of DO that occurs at the bottom water layers as a result of mixing may be significant to improving the aerobic conditions at the bottom of ponds (Chang and Ouyang 1988). The use of aerators or pond mixers to assist in water mixing during the afternoons can help to reduce stratification in fish ponds; in addition, introduce more oxygen that can be used during the night, when DO concentrations otherwise might reach critically low levels (Chowdhury et al. 2014; Szyper and Lin 1990).

Oxygen stratification had a similar diel pattern with thermal stratification. The highest DO level was found in surface water level in the afternoon, and decreased in the evening (Fig. 4). Photosynthetically active solar radiation (PAR) drives photosynthesis (Romaire and Boyd 1979). Therefore, DO concentration generally increases with increasing PAR in the morning (Tadesse et al. 2004). In the afternoon, DO concentrations continue to increase even after PAR decreases; this observation is consistent with previous studies that suggest there is a time lag in photosynthesis response to changes in light intensity (Cloot 1994; Szyper et al. 1992).

The degree of DO stratification varied significantly among culture systems: integrated culture systems showed the greatest DO stratification, and subsistence systems the least (Fig. 6). In integrated and commercial ponds, oxygen oversaturation occurred during the late afternoon, while the bottom layer was often depleted of any oxygen. A likely explanation is that high input nutrient ponds have much higher phytoplankton biomass than the lower input systems, and this is thus reflected in DO concentrations. The bottom layers had low photosynthesis activity due to the abundance of phytoplankton in the upper layers, which prevented light from penetrating into the deeper layers (Diehl 2002). Respiration by phytoplankton, settling detritus and the sediments may have also contributed to the depletion of oxygen in the bottom layers (Chang 1989; Hargreaves 1998).

In two-thirds of the early morning observations in integrated and commercial ponds, DO levels were lower than 1 mg/L. The critical threshold value for respiration for Nile tilapia is a DO concentration of 0.8 mg/L, at 26 °C (Duy et al. 2008). Be that as it may, tilapia can survive under DO concentrations of lower than 1 mg/L, by drawing air directly from the atmosphere at the pond surface. Mortality can occur when DO concentrations become lower than 2 mg/L for more than 2–3 days consecutively (Chervinski 1982). On days with dense cloud cover, low light intensities affect the oxygen budget of a pond; as a result, the oxygen production and the oxygen reserve for use at night declines (Kepenyes and Varadi 1984). One commercial pond in Nakhon Sawan experienced a mass mortality event, losing about 400 kg of full-sized fish (each weighing 700–800 g) during a week of cloudy and rainy days, when DO levels remained consistently low (DO concentration at night to early morning was about 0.1 mg/L). The average minimum DO concentrations in integrated and commercial culture systems were lower than 1 mg/L (Fig. 5), confirming that both types of culture systems are prone to risk of

DO depletion (Sriyasak et al. 2013). The reason for this is because these systems have high fish stocking densities and high nutrient inputs (Table 1) which stimulate high phytoplankton biomass.

Fish culture system and depth of pond had significant effects on TAN. In the bottom layer, TAN concentration was higher than surface water. This was likely due to the accumulation of nitrogen in the sediments (Hargreaves 1998; Kaggwa et al. 2010). Water de-stratification resulted in TAN concentration increasing at the surface layer; consequently, the diffusion of ammonia into the water column created a nutrient source to support the proliferation of phytoplankton (Hargreaves and Tucker 2004). TAN levels of integrated ponds were higher than commercial and subsistence systems. Ammonia of integrated and commercial culture systems, which had high feeding and fish stocking densities, were mainly from uneaten feed, feces, and fish excretes (Durborow et al. 1992; Gross et al. 1999). Average un-ionized ammonia concentration of all culture systems was lower than the proposed threshold toxicity levels for tilapia (less than 0.5 mg/L) (Hargreaves and Tucker 2004). The level of un-ionized ammonia which have an impact on growth rate and can kill fish over a few days ranged between 0.2 and 0.6 mg/L (Durborow et al. 1992; El-Shebly and Gad 2011; Joel and Amajuoyi 2010), while the acute toxicity (24 h) of Nile tilapia fingerling is 7.4 mg/L (Benli and Koksal 2005). The suggested safe level for chronic exposure (over 2–3 week) to un-ionized ammonia concentration is lower than 0.05 mg/L (Lawson 1995; MacIntyre et al. 2008). Ammonia levels measured in this present study were therefore not considered toxic to the fish; and with continued good water management practices in place, it should not be a major problem in future.

The complex interactions among factors as found in this study, suggest that there should be value in adapting existing pond aquaculture models (e.g. Culberson and Piedrahita 1996) to explore in more detail the dynamic consequences of weather and climate in different culture systems on thermal and DO stratification.

Conclusions

Thermal water stratification results in low DO levels in fish ponds. De-stratification reduces DO near the surface and increases DO at the bottom layer. Stratification interventions like mixing water or using an aerator should be employed to reduce stratification, and further increase oxygen levels at the bottom of the pond. Integrated and commercial culture systems are more prone to risk from very low DO (<1 mg/L), at around 02:00–06:00 h, than the subsistence system. In both systems, organic matter inputs from feed and waste are high, resulting in high concentrations of phytoplankton; which consume a lot of oxygen at night. De-stratification also causes ammonia concentrations in the surface water layer to increase. Ammonia concentrations in all ponds observed in this study however, were lower than the toxic threshold levels for tilapia. This study found widespread evidence indicating that integrated and commercial culture systems are especially prone to low levels of DO concentrations at night; approaching or exceeding thresholds likely to impact fish health and survival. Pond management practices should be adjusted to reduce the risks of exceeding DO thresholds. In particular, fish farmers should: (1) take care to avoid over-feeding; manage water and sediments to prevent excessive accumulation of organic matter and waste at the bottom of ponds, which can influence other water quality parameters; and, (2) use aeration and mechanical mixing interventions at critical times to reduce stress on fish from low DO concentrations, and thus avoid risks of mass mortality events.

Compliance with ethical standards

Conflict of interest All authors declare that they have no conflicts of interest.

References

Abdel-Tawwab M (2012) Effects of dietary protein levels and rearing density on growth performance and stress response of Nile tilapia, *Oreochromis niloticus* (L.). Int Aquat Res 4:3

Ahmed Z, Hisham A, Rahman A (2011) Ecomonitoring of climate impact on earthen pond water quality in El-Fayoum, Egypt. Int Res J Microbiol 2:442–454

APHA (1980) Standard method for the examination of water and wastewater. American public health association, Washington DC

Benli KA, Koksal G (2005) The acute toxicity of ammonia on tilapia (*Oreochromis niloticus* L.) larvae and fingerlings. Turk J Vet Anim Sci 29:339–344

Bhujel RC (2013) On-farm feed management practices for Nile tilapia (*Oreochromis niloticus*) in Thailand. In: Hasan MR, New MB (ed) On-farm feeding and feed management in aquaculture. FAO Fisheries and Aquaculture Technical Paper No. 583. Rome, FAO

Boyd CE, Romaire RP, Johnson E (1978) Predicting early morning dissolved oxygen concentration in channel catfish ponds. Trans Am Fish Soc 107:484–492. doi:10.1577/1548-8659(1978)107<484:PEMDOC>2.0.CO;2

Chainark S, Boyd CE (2010) Water and sediment quality, phytoplankton communities and channel catfish production in sodium nitrate-treated ponds. J Appl Aquac 22:171–185. doi:10.1080/10454431003736524

Chang WYB (1989) Estimates of hypolimnetic oxygen deficits in ponds. Aquac Fish Manag 20:167–172

Chang WYB, Ouyang H (1988) Dynamics of dissolved oxygen and vertical circulation in fish ponds. Aquaculture 74:263–276. doi:10.1016/0044-8486(88)90370-5

Chervinski J (1982) Environmental physiology of tilapias. In: Roger S, Pullin RSV, Lowe-McConell RH (ed) ICLARM Conference: topic in the Biology and Culture of Tilapias, Manila, September 1980. Worldfish, Bellagio, pp 119–128

Chitmanat C (2013) Nile tilapia disease. Chiangmai Vet J 11:75–86

Chowdhury AS, Hasan K, Alam K (2014) The use of an aeration system to prevent thermal stratification of water bodies: pond, lake and water supply reservoir. Appl Ecol Environ Sci 2:1–7. doi:10.12691/aees-2-1-1

Cloot A (1994) Effect of light intensity variations on the rate of photosynthesis of algae: a dynamical approach. Math Comput Model 19(9):23–33. doi:10.1016/0895-7177(94)90038-8

Culberson SD, Piedrahita RH (1996) Aquaculture pond ecosystem model: temperature and dissolved oxygen prediction-mechanism and application. Ecol Model 89:231–258. doi:10.1016/0304-3800(95)00140-9

Diana SJ, Lin KC, Schneeberger JP (1991) Relationships among nutrient inputs, water nutrient concentrations, primary production, and yield of *Oreochromis niloticus* in ponds. Aquculture 92:323–341. doi:10.1016/0044-8486(91)90038-9

Diehl S (2002) Phytoplankton, light, and nutrients in a gradient of mixing depth: theory. Ecology 83:386–398. doi:10.1890/0012-9658(2002)083

DOF (2013) Report of tilapia production in the first quarter of 2013 [in Thai]. http://fishco.fisheries.go.th/fisheconomic/Doc/fishnews_164.pdf. Accessed 10 Jan 2014

Durborow MR, Crosby MD, Brunson WM (1992) Ammonia in fish ponds. Southern Regional Aquaculture Center, Stoneville

Duy AT, Scharama J, Dam AV, Verreth AJ (2008) Effects of oxygen concentration and body weight on maximum feed intake, growth and hematological parameters of Nile tilapia, *Oreochromis niloticus*. Aquculture 275:152–162. doi:10.1016/j.aquaculture.2007.12.024

Egna HS, Boyd CE (1997) Dynamics of pond aquaculture. CRC Press, New York

El-Shebly AA, Gad MH (2011) Effect of chronic ammonia exposure on growth performance, serum growth hormone (GH) levels and gill histology of Nile tilapia (*Oreochromis niloticus*). J Microbiol Biotechnol Res 1(4):183–197

Emerson K, Russo RC, Lund RE, Thurston RV (1975) Aqueous ammonia equilibrium calculations: effect of pH and temperature. J Fish Res Board Can 32:2379–2383. doi:10.1139/f75-274

EPA (2013) Aquatic life ambient water quality criteria for ammonia—freshwater 2013. Office of Science and Technology, Washington, DC

Gandhi KT (2012) A study of water quality parameters to better manage our ponds or lakes. Int J Late Res Sci Technol 1:359–363

Gross A, Boyd CE, Wood CW (1999) Ammonia volatilization from freshwater fish ponds. J Environ Qual 28:793–797. doi:10.2134/jeq1999.00472425002800030009x

Hargreaves J (1998) Nitrogen biogeochemistry of aquaculture ponds. Aquaculture 166:181–212. doi:10.1016/S0044-8486(98)00298-1

Hargreaves JA, Tucker CS (2004) Managing ammonia in fish ponds. Southern Regional Aquaculture Center, Stoneville

Janjai S, Wattan R (2011) Development of a model for the estimation of photosynthetically active radiation from geostationary satellite data in a tropical environment. Rem Sens Environ 115:1680–1693. doi:10.1016/j.rse.2011.02.026

Joel O, Amajuoyi CS (2010) Determination of the concentration of ammonia that could have lethal effect on fish pond. J Eng Appl Sci 5:1–4

Kaggwa R, Dam AA, Kipkemboi J, Denny P (2010) Evaluation of nitrogen cycling and fish production in seasonal ponds ('Fingerpond') in lake victoria wetlands, East Africa using a dynamic simulation model. Aquac Res 42:74–90. doi:10.1111/j.1365-2109.2010.02563.x

Kepenyes J, Varadi L (1984) Aeration and oxygenation in aquaculture. In: FAO (ed) Inland aquaculture engineering. FAO, Rome

Klimaszyk P, Tomaz J (2007) Seasonal changeability of thermal and oxygen conditions in polyhumic lakes: Glodne Jeziorko III and Czarny Dol (northern Poland). Limnol Rev 7:199–203

Lawson TB (1995) Fundamentals of aquaculture engineering. Chapman & Hall, New York

Litchman E, Klausmeier CA, Bossard P (2004) Phytoplankton nutrient competition under dynamic light regimes. Limnol Oceanogr 49((4, part 2)):1457–1462. doi:10.4319/lo.2004.49.4_part_2.1457

Little DC, Edwards P (2004) Impact of nutrition and season on pond culture performance of mono-sex and mixed-sex Nile tilapia (*Oreochromis niloticus*). Aquaculture 232:279–292. doi:10.1016/S0044-8486(03)00482-4

Losordo TM, Piedrahita RH (1991) Modelling temperature variation and thermal stratification in shallow aquaculture ponds. Ecol Model 54:189–226. doi:10.1016/0304-3800(91)90076-D

MacIntyre CM, Ellis T, North BP, Turnbull JF (2008) The influences of water quality on the welfare of farmed rainbow trout: a review. In: Branson EJ (ed) Fish welfare. Blackwell Publishing Ltd., Oxford

Pandit NP, Nakamura M (2010) Effect of high temperature on survival, growth and feed conversion ratio of Nile Tilapia, *Oreochromis niloticus*. Our Nat 8:219–224. doi:10.3126/on.v8i1.4331

Pimolrat P, Whangchai N, Chitmanat C, Promya J, Lebel L (2013) Survey of climate-related risks to Tilapia pond farms in northern Thailand. Int J Geosci 4:54–59. doi:10.4236/ijg.2013.45B004

Randall DJ, Tsui TKN (2002) Ammonia toxicity in fish. Mar Pollut Bull 45:17–23. doi:10.1016/S0025-326X(02)00227-8

Romaire PR, Boyd EC (1979) Effects of solar radiation on the dynamics of dissolved oxygen in channel catfish ponds. Trans Am Fish Soc 108(5):473–478. doi:10.1577/1548-8659(1979)108<473:EOSROT>2.0

Smith DW, Piedrahita RH (1988) The relation between phytoplankton and dissolved oxygen in fish ponds. Aquculture 68:249–265. doi:10.1016/0044-8486(88)90357-2

Srisapoome P, Areechon N (2014) Serious disease outbreaks in Nile tilapia cultured in Thailand. http://www.rdi.ku.ac.th/kasetresearch52/05-animal/prapansak_Nile/animal_00.html. Accessed 24 Dec 2014

Sriyasak P, Chitmanat C, Whangchai N, Promya J, Lebel L (2013) Effects of temperature upon water turnover in fish ponds in northern Thailand. Int J Geosci 4:18–23. doi:10.4236/ijg.2013.45B004

Sriyasak P, Whangchai N, Chitmanat C, Promya J, Lebel L (2014) Impacts of climate and season on water quality in aquaculture ponds. KKU Res J 19(5):743–751

Stefan HG, Hondzo M, Fang X, Eaton JG, McCormick JH (1996) Simulated long-term temperature and dissolved oxygen characteristics of lakes in the north-central United States and associated fish habitat limits. Limnol Oceanogr 41:1124–1135. doi:10.4319/lo.1996.41.5.1124

Szyper PJ, Lin CK (1990) Techniques for assessment of stratification and effects of mechanical mixing in tropical fish ponds. Aquac Eng 9:151–165. doi:10.1016/0144-8609(90)90002-H

Szyper PJ, Rosenfeld ZJ, Piedrahita HR, Giovannini P (1992) Diel cycles of planktonic respiration rates in briefly incubated water samples from a fertile earthen pond. Limnol Oceanogr 37(6):1193–1201. doi:10.4319/lo.1992.37.6.1193

Tadesse I, Green FB, Puhakka JA (2004) Seasonal and diurnal variations of temperature, pH and dissolved oxygen in advanced integrated wastewater pond systems treating tannery effluent. Water Res 38:645–654. doi:10.1016/j.watres.2003.10.006

Thai Meteorological Department (2012) The weather of Thailand. http://www.tmd.go.th/info/info.php?FileID=22. Accessed 9 Dec 2014

Thai Meteorological Department (2014) The weather data of Thailand. http://119.46.126.50/web/aws/aws_alphanumeric.asp. Accessed 12 Feb 2015

15

Seasonal dynamics of phytoplankton communities residing in different types of shallow waters in the Kuibyshev Reservoir (Russia)

L. Yu. Khaliullina · G. V. Demina

Abstract Unstable water level regime in the Kuibyshev Reservoir affects coastline formation as well as allocation and development of aquatic vegetation on the coast. These factors determine the composition and structure of biocenosis in shallow waters of the reservoir. The aim of this study was to reveal the patterns of formation, distribution and dynamics in phytoplankton communities in the shallow coastal waters on two reaches of the Kuibyshev Reservoir (Volga and Volga-Kama, Russia). These reaches differ by the extent of anthropogenic influence, protection from wind and waves and other environmental conditions. The research was done during the growing season of 2002 in the thickets of *Typha angustifolia* L. and *Phragmites australis* (Cav.) Trin. Ex Steud., as well as in the open water areas. Seasonal changes in the total biomass and abundance of phytoplankton in the thickets of macrophytes and in the open areas of the reservoir differed little. On the outer edge of the thickets, where intensive contact with the open water occurs, the highest algal species diversity and abundance were revealed, which is known as the "edge effect". Two peaks of phytoplankton development with maximums in June–July and late August were observed. By the end of the summer, a decrease in water level led to the autumn outbreak in volvocine algae abundance and biomass. The maintaining of an optimal water level in the reservoir is recommended for controlling of water "blooming" and thus maintaining high water quality.

Keywords Phytoplankton diversity · Phytoplankton dynamics · Algae · Water reservoir · Thickets of macrophytes

Introduction

The Kuibyshev Reservoir is the sixth stage in the Volga cascade reservoirs, the filling of which took place from 1955 to 1957 after damming the Volga by hydroengineering constructions near the Zhiguli Mountains (Gidrometeorologicheskii rezhim 1978). The total area of the reservoir is 5900 km^2. Seasonal water level fluctuations of high amplitude are characteristic to the Kuibyshev Reservoir. The annual water level varies due to significant water inflow and outflow. Spring floods fill the reservoir to maximum levels. Then, for 2–3 months, the level stays close to the normal headwater level of 53 m. By the beginning of the freezing-

L. Yu. Khaliullina · G. V. Demina (✉)
Department of Botany and Plant Physiology, Institute of Fundamental Medicine and Biology, Kazan Federal University, Kremlyovskaya 18, Kazan 420008, Russia
e-mail: deminagv@mail.ru

L. Yu. Khaliullina
e-mail: liliya-kh@yandex.ru

over, the level gradually lowers to 48–49 m, causing significant changes in hydrological and biological parameters of the reservoir, especially in the extensive shallow-water areas (Gidrometeorologicheskii rezhim 1978). At the normal headwater level, shallow waters (not deeper than 2 m) occupy ~ 15 % of the reservoir's total area (Kujbyshevskoe vodohranilishhe 1983, 2008).

The main sources of primary production in shallow waters are macrophytes, coastal phytoplankton and periphyton (Komarkova et al. 1983; Qin et al. 2006; Sayer et al. 2010). The ratio of each structural element in the total vegetation primary production is different depending on the type of the reservoir and may vary significantly throughout the year. The dynamics in the hydrological regime of the reservoir determines the cenotic structure of higher aquatic vegetation (Guseva and Jekzercev 1966; Koreljakova 1977; Kutova 1974a, 1974b; Potapov 1958) and primarily affects the development of planktonic algae communities. Thickets of higher aquatic vegetation are the areas where biocenotic relationships between hydrobionts are more complex than in the pelagic (Gaevskaja 1966; Hilt 2006; Mulderij et al. 2007; Sand-Jensen and Borum 1991; Schriver et al. 1995; Zimbalevskaja 1972, 1973). These relations include not only food networks and toxic effects, but also the roles of plants as substrates or shelters. Phytophilic fauna and algae development in the thickets occur along with the seasonal formation of plant biotopes.

The area of overgrown shallow waters in the Kuibyshev Reservoir is 8500 ha, with the most extensive areas in the Volga and the Volga-Kama reaches. The phytoplankton communities in the Volga cascade artificial water reservoirs are crucial for reservoirs' ecosystems, in particular for the trophic status and biological productivity of the reservoir (Khaliullina et al. 2009; Konovalov and Pautova 1989; Lavrent'eva 1977; Okhapkin 1994; Pautova and Nomokonova 1994, 2001). Since the 1970s, long-term monitoring of seasonal phytoplankton dynamics in the Kuibyshev water reservoir has been carried out (Gidrometeorologicheskii rezhim 1978; Khaliullina et al. 2009; Konovalov and Pautova 1989; Pautova and Nomokonova 1994). However, some open questions still remain. In particular, the mechanisms of phytoplankton succession in the reservoirs are not well understood. Here, we provide more insights into the patterns of formation, distribution and dynamics of phytoplankton in the coastal shallows in the two reaches of the Kuibyshev Reservoir. The comparison of phytoplankton dynamics in these sites is of special interest, as the sites differ in the extent of anthropogenic impact, the degree of protection from wind and waves and other environmental conditions. Samples collected from both macrophytes thickets and areas of open water were analyzed for the phytoplankton abundance and biomass.

Materials and methods

Water samples were taken from two shallow bays of the Volga (site 1) and the Volga-Kama (site 2) reaches of the Kuibyshev Reservoir (Russian Federation, Republic of Tatarstan) (Fig. 1) in 2002.

Site 1 is located in Pobedilovo township (southern part of Kazan city) and is substantially isolated from the other parts of the reservoir. In the studied area, the water flow is slow, and the bank is protected from wind.

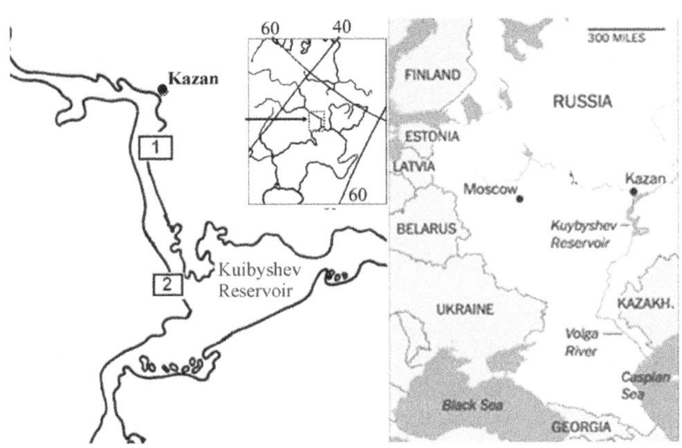

Fig. 1 Layout of the studied coastal shallow water areas in the Volga and the Volga-Kama reaches of the Kuibyshev Reservoir: site 1 (*1*) (Pobedilovo, Kazan) and site 2 (*2*) (Saralinsky, Volga-Kama State Nature Biosphere Reserve) (Khaliullina 2015)

The site is influenced by waste waters produced by sewage disposal plants and suburban areas. The right bank of the bay, where the observing stations were located, is low-laying, overgrown with single willow bushes and riverside vegetation. The bottom is flat and covered with sticky mud. Next to willow thickets, higher aquatic vegetation is represented mainly by thickets of narrow-leaved cat's-tail (*Typha angustifolia* L.) and common reed [*Phragmites australis* (Cav.) Trin. Ex Steud.]. In July, the general projective covering by plants reaches 100 %, 50 % of which are narrow-leaved cat's-tail and common reed, and the other 50 % are free-swimming plants (*Spirodela* sp., *Lemna minor* L., *L. trisulca* L.). At the end of August, the ratio becomes 90 and 10 %, respectively. In the thickets area, argillaceous sand grounds with high content of organic residues dominate. This area is characterized by a relatively high trophic status. In the open shallow water, sludgy argillaceous sand grounds also dominate.

Site 2 is located near the settlement Tatar Sarali (Laishevski district) and is a part of Saralinsky section of the Volga-Kama State Nature Biosphere Reserve. In site 2, along with seasonal fluctuations in water level, the impact of transit water flows (generally characteristic for this water area of the reservoir) is significant. This site is characterized by slightly irregular coastline, and is located in the zone of high wind and wave activity. The sampling stations were located on the peninsula, the coastal zone of which is covered with willow beds and thickets of narrow-leaved cat's-tail. The latter ones extend along the coastline as continuous zones (~60 m wide) with small clearings of open littoral. Their projective cover is 100 %. In mid-summer, the dominant narrow-leaved cat's-tail occupies ~60 % of the territory, and 30–40 % is occupied by *Spirodela* sp. and *Lemna minor* L. In coastal thickets, the bottom is covered mainly with sandy-mud and argillo-arenaceous grounds. Large amounts of decaying plant residues are accumulating in the bottom, which leads to siltation and oxygen deficiency in the thickets. In shallow waters of open areas, slightly muddy argillo-arenaceous grounds dominate. Below the area of thickets, open littoral is represented by fine brown (homogeneous) clayey silts.

The studied areas of both sites included the thickets of narrow-leaved cat's-tail and common reed, as well as open water areas. Water samples were taken from three stations with different water levels: from the top of macrophyte thickets to the deepest water level in the bottom of the thickets (Fig. 2). For comparison, we laid three control points in the areas free of vegetation at the same water levels.

At each site, the studies were carried out starting from the time of flooding in the shallow waters with higher aquatic vegetation (June) until the time of drainage (mid-October). The samples were taken once every 2 weeks. In total, more than 240 quantitative and over 300 qualitative samples of phytoplankton were handled. During the study period, weather conditions and hydrological characteristics of the reservoir were recorded.

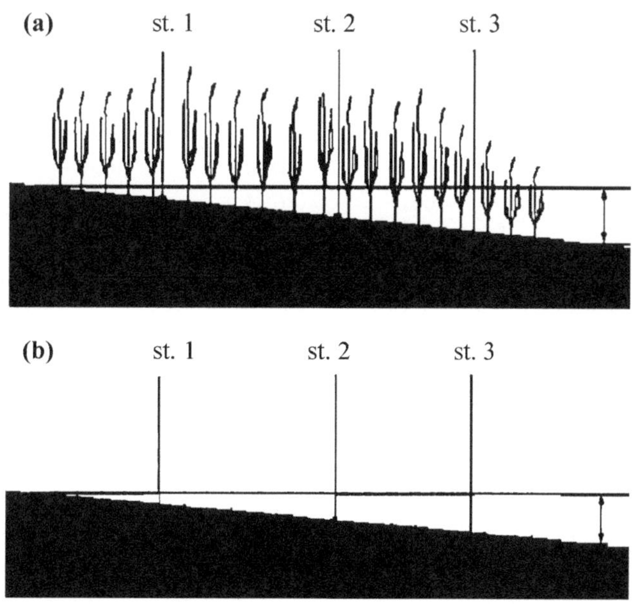

Fig. 2 The schematic location of the stations in the thickets (**a**) and in the open shallow waters (**b**) of two sites in the Volga and the Volga-Kama reaches shown as cross-sections. Station 1 (st. 1) was located near the bank, at the water depth of 0.5–0.7 m, station 2 (st. 2) at 0.7–1.0 m, and station 3 (st. 3) at 1.0–1.2 m. *Arrows* indicate the amplitude of seasonal water level fluctuations

The sampling and laboratory processing of phytoplankton samples was performed according to conventional methods (Vodorosli spravochnik 1989; Morduhaj-Boltovskoj 1975). The samples were taken with Molchanov bathometer. All quantitative samples (0.5 l) were fixed with 4 % (v/v) formalin. Qualitative samples were taken with Apshtein small network (mill sieve No. 73) by filtering 50–100 l of water. The fixed samples were concentrated to 20–30 ml. The present organisms were counted in hemocytometer. For each algae species, the individual cell size was calculated by stereometric method. To determine the biomass, the counting-volumetric method was used. In each sample, an average volume of cells was determined for each algae species (not less than 20 cells inspected). For rare organisms, their individual volumes were used, measured in every sample. Statistical data processing was performed using parametric (*t* test, mean and its standard error) and non-parametric tests (Wilcoxon test, Spearman rank correlation) (Borovikov and Borovikov 1998).

Results and discussion

Hydrochemical regime of the studied sites

According to the data provided by the hydrochemistry laboratory of the Institute of Natural Systems Ecology (Tatarstan Academy of Sciences), as well as our own data, it was found that site 1 belongs to the α-

Table 1 Average hydrochemical parameters of the studied sites of shallow waters in the Kuibyshev Reservoir in 2002–2003, adapted from Ratushnyak et al. (2006a, b)

Parameter	Site 1			Site 2		
	M ± SE	Min.	Max.	M ± SE	Min.	Max.
O_2 (mg/l)	10.7 ± 0.9	2.2	19.6	7.6 ± 0.8	0.4	16.3
Eh (mV)	148.5 ± 10.2	57.0	223.0	130.4 ± 21.4	201.0	313.0
pH	7.3 ± 0.1	6.0	8.7	7.2 ± 0.1	6.4	9.1
Σmin. (mg/l)	468.2 ± 15.9	307.6	854.0	248.6 ± 6.2	171.0	414.9
Ca^{2+} (mg/l)	84.7 ± 3.9	40.0	178.0	44.1 ± 1.4	20.8	96.0
Mg^{2+} (mg/l)	20.1 ± 1.7	1.8	40.9	9.3 ± 0.7	1.2	29.2
HCO_3^- (mg/l)	170.6 ± 3.8	122.0	263.6	136.8 ± 3.5	110.0	276.1
SO_4^{2-} (mg/l)	165.4 ± 8.3	21.0	377.0	54.3 ± 2.6	15.8	99.9
Cl^- (mg/l)	39.9 ± 1.2	24.8	59.6	22.7 ± 0.9	15.6	46.8
BO (mg O_2/l)	33.9 ± 2.6	14.3	92.8	24.0 ± 2.3	3.1	78.0
PO (mg O_2/l)	12.2 ± 0.9	1.7	28.8	11.5 ± 0.5	6.7	20.4
BOD_5 (mg O_2/l)	5.9 ± 0.6	0.9	11.5	4.8 ± 0.8	0.5	18.6
P_{total} (mg/l)	0.28 ± 0.03	0.07	0.59	0.29 ± 0.04	0.05	0.74
PO_4^{3-} (mg/l)	0.09 ± 0.01	0.003	0.46	0.10 ± 0.02	0.001	0.80
NO_3^- (mg/l)	1.88 ± 0.22	0.10	7.86	1.15 ± 0.11	0.11	3.60
NO_2^- (mg/l)	0.09 ± 0.02	0.003	0.69	0.07 ± 0.02	0.001	1.25
NH_4^+ (mg/l)	0.67 ± 0.10	0.02	5.34	0.49 ± 0.04	0.02	1.48
Cd (μg/l)	1.15 ± 0.27	0.01	7.50	1.0 ± 0.2	0.1	4.6
Pb (μg/l)	11.6 ± 2.1	0.1	48.0	9.3 ± 1.6	0.2	31.5
Cu (μg/l)	6.4 ± 0.7	0.5	29.5	5.8 ± 0.5	0.5	16.1
Co (μg/l)	7.9 ± 0.6	0.7	14.3	6.3 ± 0.5	0.9	14.2
Ni (μg/l)	12.5 ± 0.9	1.0	19.7	10.8 ± 0.8	0.5	19.9
Zn (μg/l)	17.8 ± 1.7	1.1	42.6	17.9 ± 2.5	0.9	67.1
Cr (μg/l)	4.9 ± 0.7	1.5	9.5	3.4 ± 0.5	1.7	8.1
Mn (μg/l)	89.4 ± 29.3	3.5	1198.0	114.6 ± 35.7	6.0	1176.0
Fe (μg/l)	239.0 ± 47.4	1.0	1344.0	209.5 ± 42.7	22.5	1555.0

M mean, *SE* standard error, *Min.* the minimal value, *Max.* the maximum value, *Σmin.* total mineralization, *BO* bichromate oxidizability, *PO* permanganate oxidizability, BOD_5 biological oxygen demand for 5 days

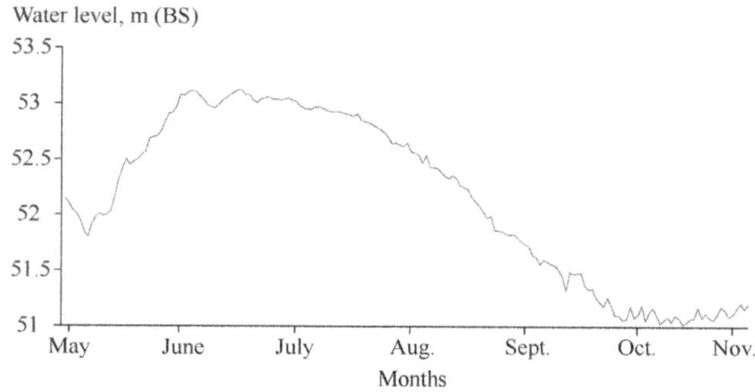

Fig. 3 Seasonal dynamics of the water level in the Kuibyshev Reservoir in 2002. (Kujbyshevskoe vodohranilishhe 2008). Water level is in Baltic system (BS)

mesosaprobic zone, 3–4 class of water quality, and is assessed as "polluted". Site 2 belongs to β-mesosaprobic zone, 3 class of water quality and is characterized as "of satisfactory purity" (Kazda et al. 2004; Ratushnyak et al. 2006a, b). The studied shallow water sites were significantly different in chemical composition. Integrated assessment of water quality in sites 1 and 2 based on the average ranking index is shown in Table 1.

Identified excess of mineral content, organic and biogenic elements in site 1 compared to site 2 is explained by the inflow of contaminated industrial and domestic wastewaters from Kazan sewage disposal plants, located upstream. Their typical ingredients are sulfates, chlorides, and heavy metals (Stepanova 1999).The concentration of dissolved oxygen in the two studied sites is characterized by a negative correlation with the content of organic compounds. The oxygen concentration at site 1 remained consistently higher than that at the cleaner site 2. Significant correlations between the oxygen, organic matter contents (BO, ON, BOD_5), and phytoplankton indicators in the surface layers have not been identified. Probably, the reason for such distribution is to be found in production–destruction features of these sites.

Seasonal dynamics of water level and weather conditions in 2002

In 2002, the water level rose constantly starting with the spring filling of the reservoir until June 5 when it reached 53.11 m in the Baltic System (BS) and stayed at this level with small fluctuations caused by wind-driven waves until the June 17 (Kujbyshevskoe vodohranilishhe 2008). Later, the water level gradually decreased, and in September–October, it was 51.02 m (Fig. 3).

May 2002 was cold and rainy. Monthly rainfall exceeded the norm by 18 mm (Gosudarstvennyj doklad 2003, 2005). Monthly average temperatures were 3–4 °C lower than long-term averages. At the soil surface frosts were recorded (−3 to −11 °C). Especially cold weather was observed in the third decade of May. The average air temperature of the decade was the lowest for the observation period since 1924. In the northern regions of Tatarstan Republic the snowfalls were registered on May 21, and in some areas, the formation of a temporary snow cover was observed. The average monthly temperature in June was 1–2 °C lower than the long-term average. Monthly rainfall exceeded the long-term "norm", except for the most part of Predvolzhje, where the rainfall of only 28–70 % of the "norm" was registered. In July, average temperatures exceeded the long-term average by 2–4 °C, and the weather was hot and dry. In August, cool weather (average temperature of 2–3 °C below the norm) with a deficit of rainfall was observed. September and October were characterized by unstable weather and significant variations in temperature (in the first decade of September, the maximum air temperature rose to 25–29 °C). The average monthly air temperature was 1–2 °C above the norm. Precipitation in October amounted to 1–2 norms.

Seasonal changes in phytoplankton species diversity at different biotopes in 2002

During the observation period, 336 taxa, including 323 species of algae belonging to eight divisions, were found in phytoplankton of the studied sites. The greatest number of order-level taxa was identified in the divisions of chlorophytes and diatoms. In species-level diversity diatoms (26.2 %) and *Chlorococcales* (36.8 %) also dominated. Other groups were less diverse: cyanobacteria (12.1 %), euglenophytes (7.8 %), chrysophytes (6.9 %), xanthophytes (4.4 %), dinoflagellates (3.7 %), and cryptophytes (2.2 %).

At the beginning of the growing season, phytoplankton of both sites differed little in composition and was represented by diatoms. In mid-summer, when the thickets of cat's-tail and reed were grown, a complex of phytoplankton algae formed at the station 2 was different from that formed in the open shallow water, which was particularly noticeable at site 1 (Pobedilovo). This algocoenosis included several ecological groups of algae (true phytoplankton, epipelon, epiphyton, free-floating strands and clumps of filamentous algae), which differed in species composition, habitat conditions and development dynamics. In the species diversity diatoms, cyanobacteria, *Desmidiales*, and euglenophytes dominated.

The highest algae abundance and species diversity was typical for station 3 (at the outer edge of the thickets at both sites), where an intensive contact with open water occurred ($p < 0.05$). The border area between algae communities of thickets and the open pelagic zone of the reservoir is a sort of ecotone. For such zones an "edge effect" is known, i.e. there is an increase in biodiversity and abundance in the mixing zone (Harchenko

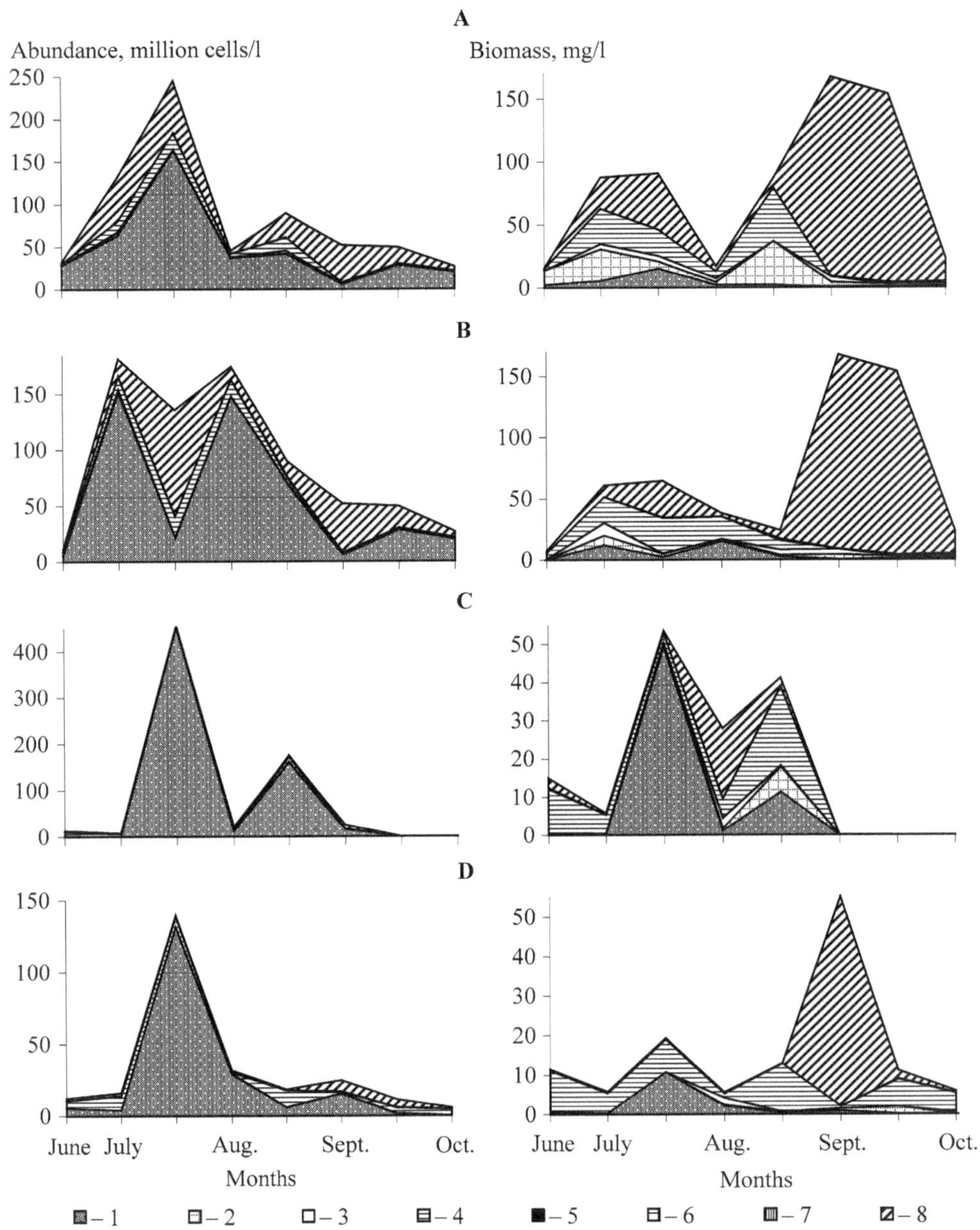

Fig. 4 Dynamics of phytoplankton total abundance and biomass at different biotopes of site 1 (**a**, **b**) and site 2 (**c**, **d**) in cats'-tail thickets (**a**, **c**), common reed thickets (**b**), open shallow water (**d**). Cyanobacteria (*1*), euglenophytes (*2*), dinoflagellates (*3*), diatoms (*4*), xanthophytes (*5*), cryptophytes (*6*), chrysophytes (*7*), chlorophytes (*8*)

1991; Solov'eva and Rozenberg 2006). The effect was observed in both studied sites. Under the conditions of mixing with water masses coming from the open pelagic, the role of true phytoplankton significantly increased. Here, the above-mentioned groups of station 2 were joined by algae washed to the thickets by wind, as these algae can grow only in the open waters of reservoirs and under conditions of wind-mixing water (Vodorosli. lishajniki 1977). Epiphytic species of diatoms and chlorophytes at station 3 maintained a high abundance and species diversity throughout the whole growing season, withstanding even strong shading by macrophytes starting from mid-summer. No significant differences in species composition of phytoplankton communities in cat's-tail and reed thickets were observed during the study period. In open shallow water, phytoplankton was distributed evenly and consisted mainly of truly planktonic species and less of epipelon species.

Thus, the floristic list of phytoplankton in shallow coastal water of the reservoir is "saturated" during the entire growing season. The total number of recorded algae taxa approaches the maximum during the summer–autumn period. The "saturation" of the diatoms list occurs already in early summer, while the list of chlorophytes is constantly increasing by the beginning of autumn. The highest species diversity is observed in genera *Aulacoseira*, *Navicula*, *Nitzschia*, and *Scenedesmus.* In formed algocoenoses, relatively large-cell species dominate during spring, and species with small-sized cells are more prevalent during the summer. The observed decrease in the average quantities of diatoms and chlorophytes along the formation of biocenosis in shallow water zone and during the summer seems to be the result of competitive relationships between algae. At the same time, the advantage is given to rapidly proliferating algae with larger relative cell surface.

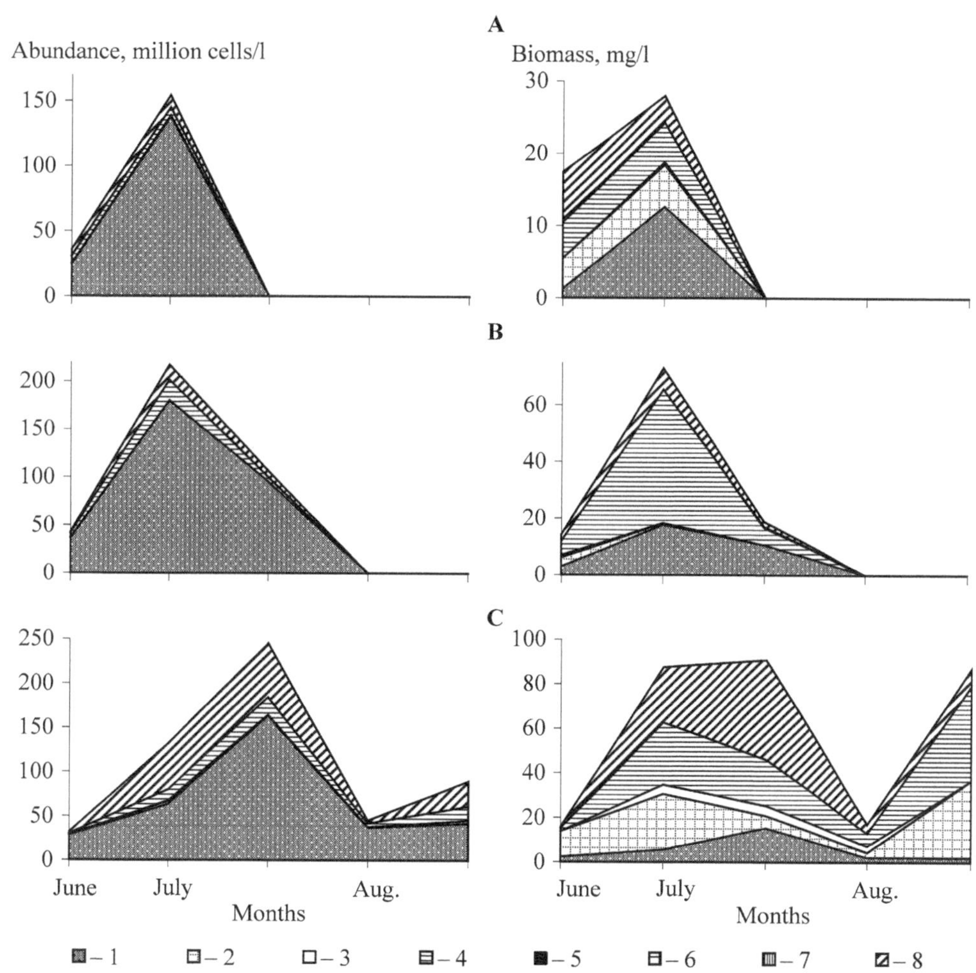

Fig. 5 Seasonal dynamics of phytoplankton abundance and biomass in cats'-tail thickets at station 1 (**a**), station 2 (**b**), and station 3 (**c**) of site 1 (Pobedilovo). Cyanobacteria (*1*), euglenophytes (*2*), dinoflagellates (*3*), diatoms (*4*), xanthophytes (*5*), cryptophytes (*6*), chrysophytes (*7*), chlorophytes (*8*)

Seasonal changes in phytoplankton abundance and biomass at different biotopes in 2002

During the period of maximal water level (in the first half of summer), higher aquatic vegetation was poorly developed, and algocoenoses associated with thickets had not been formed yet. The water in the reservoir was homogeneous, and the quantitative parameters of phytoplankton in macrophytes thickets and in open shallow water differed slightly. In this period, the total abundance and biomass of phytoplankton at both sites were low and were associated mainly with the development of cyanobacteria, diatoms, and volvocine algae (Fig. 4).

At both sites, all the biotopes were dominated by cyanobacteria *Aphanizomenon flos-aquae* (l.) Ralfs. (up to 1.99×10^8 cells/l), *Anabaena flos-aquae* Breb., *Anabaena scheremetievi* Elenc., diatoms *Stephanodiscus hantzschii* Crun., genera *Aulacoseira*, *Melosira*, *Cyclotella*, *Nitzschia*, dinoflagellates of genus *Peridinium*, chlorophytes of genus *Chlamydomonas*. In associations with narrow-leaved cats'-tail and common reed, the dominant complex also included euglenophytes of genera *Euglena* and *Trachelomonas*. The highest biomass of these algae was observed at site 1. Spearman correlation analysis revealed a negative relationship between the abundance of this group and water clarity ($p < 0.01$). Quantitative parameters of phytoplankton in the thickets of cats'-tail and reed differed little.

In macrophytes thickets at site 1 (Pobedilovo), the number of chlorophytes was higher in the areas closer to the open water ($p < 0.03$). The distribution of other algae other groups at three stations was even, and no significant differences between the coastal zone (station 1) and the outer edge of the thickets (station 3) were observed (Figs. 5, 6).

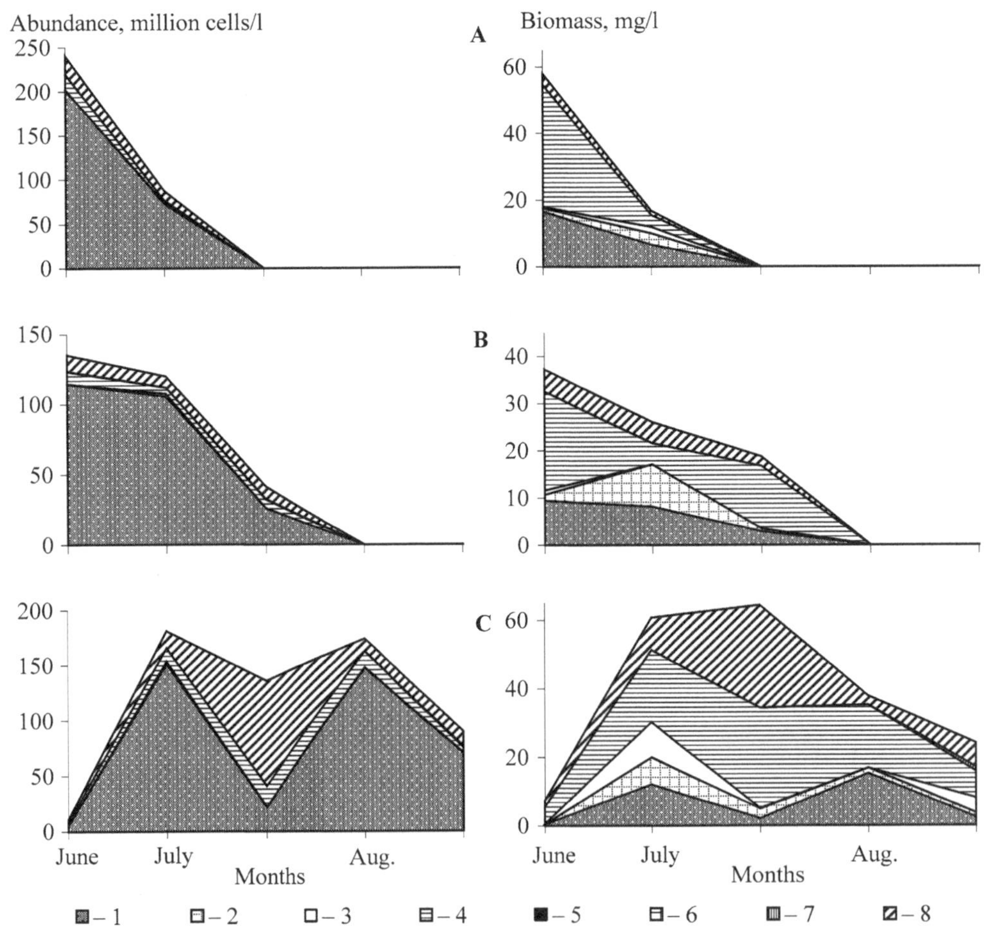

Fig. 6 Seasonal dynamics of phytoplankton abundance and biomass in common reed thickets at station 1 (**a**), station 2 (**b**), and station 3 (**c**) of site 1 (Pobedilovo). Cyanobacteria (*1*), euglenophytes (*2*), dinoflagellates (*3*), diatoms (*4*), xanthophytes (*5*), cryptophytes (*6*), chrysophytes (*7*), chlorophytes (*8*)

At site 2 (Saralinsky), in cats'-tail thickets, the abundance of diatoms increased closer to the coast (Figs. 7, 8). At the stations located close to the water's edge and rich with detritus (stations 1, 2), high biomass of periphyton and benthic-planktonic diatoms with large cells and low abundance was noticed. These algae included *Nitzschia sigmoidea* (Nitzsch.) W.Sm., *N. vermicularis* (Kiitz.) Grun., *Gyrosigma acuminatum* (Kiitz.) Rabenh., *Cymatopleura elliptica* (Breb.) W.Cm., *C. solea* (Breb.)W.Sm., *Cymbella* sp. In the open shallow water, no differences in phytoplankton were detected at the stations.

In May–June 2002, large fluctuations in water level were observed. After that, the water level was gradually decreasing, which led to the drying up of some shallow coastal waters. Moderate dynamics of water masses and high warming up of the water due to hot and dry weather led to intensive development of phytoplankton. During this period, the spring algocoenosis was completely replaced by the summer one. By that time, stations 1 of all studied biotopes almost remained above the water's edge.

The first maximum in phytoplankton abundance and biomass in the thickets and open shallow water was observed in mid-July. Green–blue algae *A. flos-aquae*, *An. flos-aquae*, *An. scheremetiev*, *Microcystis aeruginosa* f. *flos-aquae* (Wittr.) Elenk. were the most abundant, causing water "blooming". Moreover, their high abundance was observed at all stations of both sites, except for stations 1 and 2 in the open shallow water of site 2, where diatoms dominated. In addition, by that time, the abundance of chlorophytes and euglenophytes had increased at site 1 (Pobedilovo).

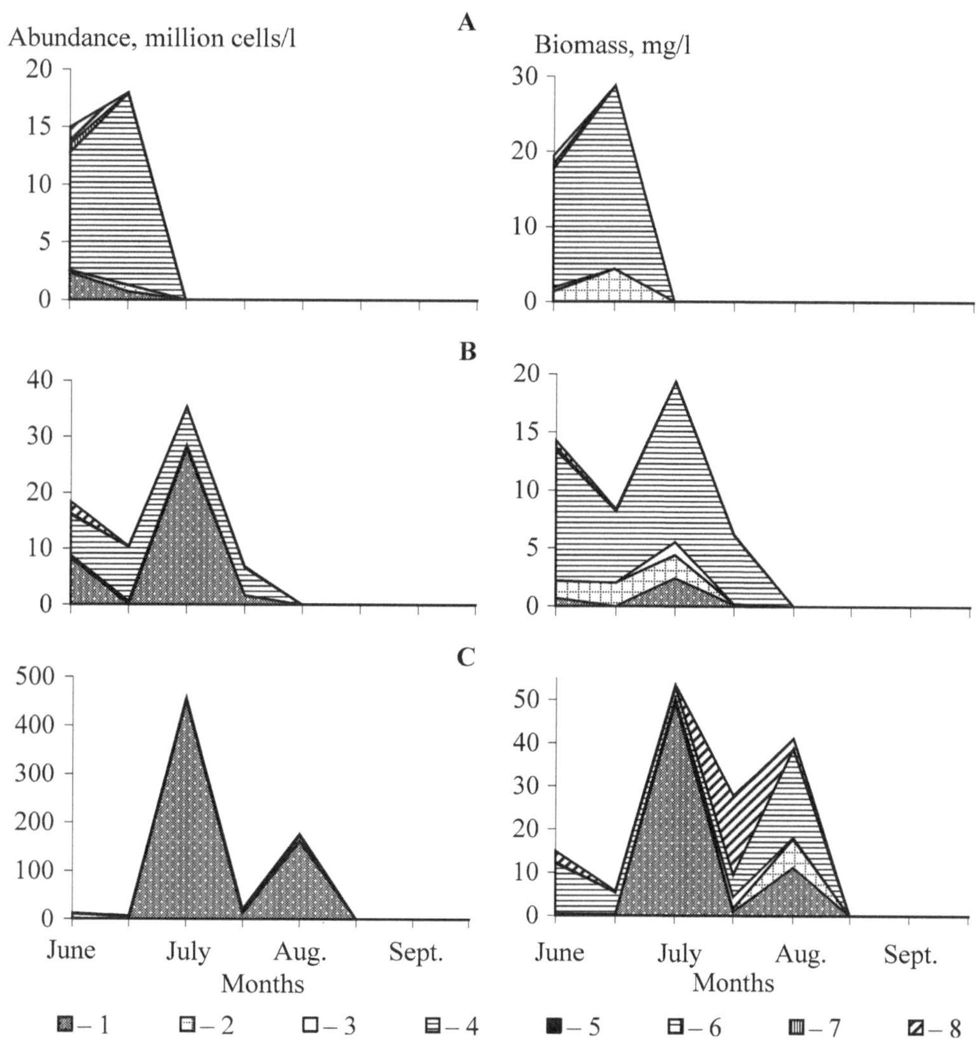

Fig. 7 Seasonal dynamics of phytoplankton abundance and biomass in cats'-tail thickets at station 1 (**a**), station 2 (**b**), and station 3 (**c**) of site 2 (Saralinsky). Cyanobacteria (*1*), euglenophytes (*2*), dinoflagellates (*3*), diatoms (*4*), xanthophytes (*5*), cryptophytes (*6*), chrysophytes (*7*), chlorophytes (*8*)

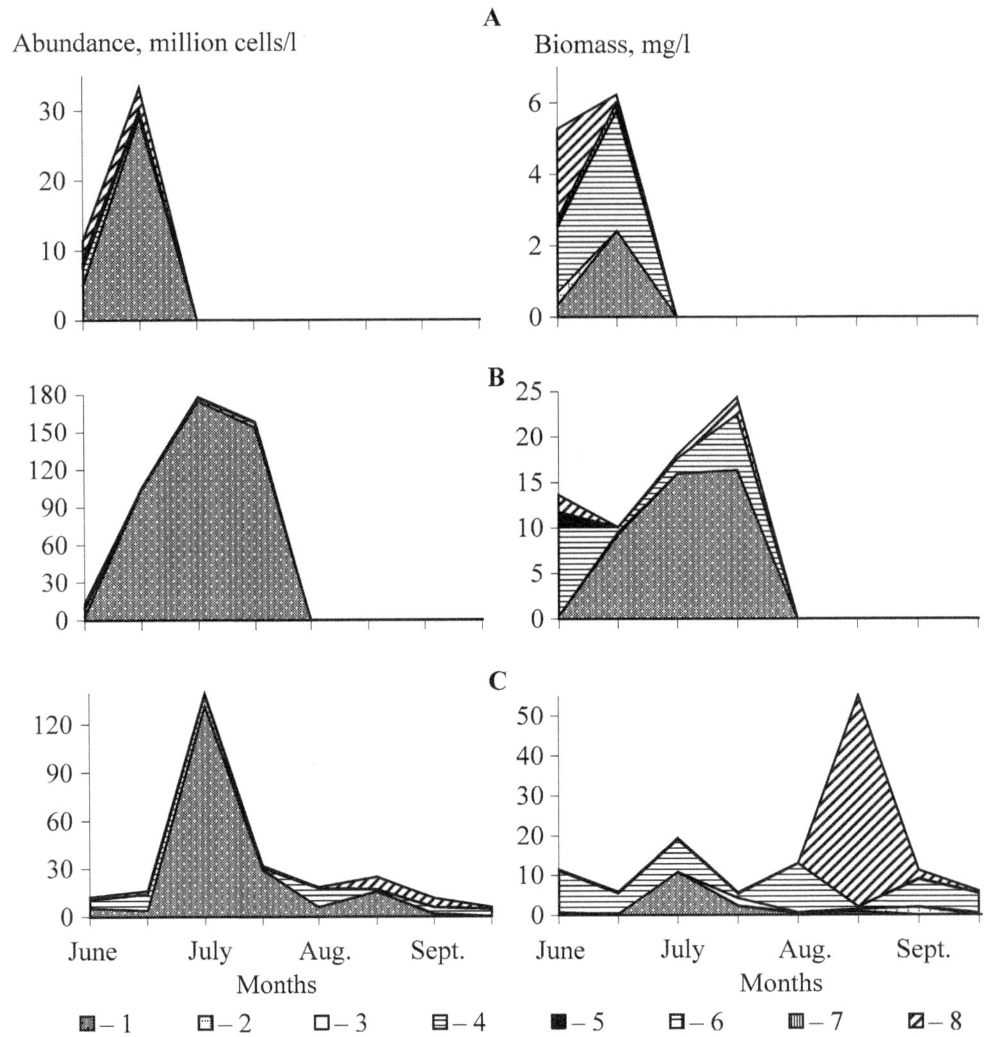

Fig. 8 Seasonal dynamics of phytoplankton abundance and biomass in open shallow water at station 1 (**a**), station 2 (**b**), and station 3 (**c**) of site 2 (Saralinsky). Cyanobacteria (*1*), euglenophytes (*2*), dinoflagellates (*3*), diatoms (*4*), xanthophytes (*5*), cryptophytes (*6*), chrysophytes (*7*), chlorophytes (*8*)

Cyanobacteria were developing rapidly first in weakly flowing shallow waters which were protected from waves and wind. Then these algae spread to the entire area of the reservoir. Under the meteorological conditions observed during that time period, open shallow water was found to be the most favorable area for cyanobacteria vegetation. At site 2 (Saralinsky) where hydrodynamic activity was intensive, cyanobacteria also occasionally accumulated at the outer edge of the thicket, reaching very high concentrations (up to 4.52×10^8 cells/l). Since the hydrological conditions of the site were not favorable for further development of these algae, such outbreaks in their abundance were short-termed. Due to intensive water dynamics, cyanobacteria could not settle even in the dense thickets of macrophytes.

When the biotopes of site 1 (Pobedilovo) are compared, in this period, no significant differences in quantitative indicators of phytoplankton seasonal changes in the thickets of cats'-tail and reed were observed. High concentration of cyanobacteria was noticed at all stations of this site, with the increase in the thickets closer to the coast. The biomass and abundance of diatoms increased slightly, and the most abundant species were *S. hantzschii*, *Cyclotella meneghiniana* Kiitz., *Melosira varians* Ag., *Aulacosira granulata* (Ehr.) Sim., *A. Italic* (Kiitz.) Sim., *A. islandica* (O. Mull.) Sim., *Fragilaria construens* (Ehr.) Grun., *Nitzschia palea* (Kiitz).W.Sm.

With the growth of macrophytes in the thickets, a reduction in the concentration of cyanobacteria and diatoms was detected. In ditch-water, phytoplankton development in the thickets is usually limited by shortage of nutrients in water and low lightening. In the dense thickets of reeds and cats'-tail, sometimes less than 1 % of photosynthetically active radiation reaches the surface of the water, resulting in a low number of phytoplankton (Komarkova et al. 1983). In this period, at station 2, the decrease in the total quantity indicators of phyton was also observed. Conversely, the number of euglenophytes *Trachelomonas hispida* (Perty.) Stein emend. Defl., *T. intermedia* Dang., *Euglena viridis* Ehr., *Phacus pleuronectes* (Ehr.) Daj. and dinoflagellates of the genus *Peridinium*, i.e. more mobile and mixotrophic forms, increased. The dynamics of algal biomass of these algae groups were the same.

At station 3, in the contact zone of algal communities residing in the thickets and open water, with the development of macrophytes, the abundance and biomass of algae, on the contrary, increased. By the end of July, chlorophytes reached the seasonal peak in abundance, while the numbers of other algae in the water began to decline. The dominant chlorophytes complex consisted of volvocine algae *Chlamydomonas* sp., *Pandorina morum* (Mill.) Bory. and *Chlorococcales Coelastrum proboscideum* Bohl., *Crucigenia rectangularis* (A.Br.) Gay., *C. Tetrapedia* (Kirchn.) W.et.W., *Oocystis natans* Wille., *Pediastrum duplex* Meyen., *Scenedesmus guadricauda* (Turp.) Breb. Spatially, the maximal abundance and biomass of chlorophytes were observed at station 3.

At site 2 (Saralinsky), the peaks of quantitative phytoplankton indicators were observed later compared with site 1. In the thickets, the phytoplankton abundance and biomass was 2.5–3 times higher than those in the open shallow water. In all biotopes, the accumulation of cyanobacteria causing "blooming" of water was observed. In open areas more away from the coast (station 3), diatoms, mainly represented by *M. varians*, *Cocconeis placentula* Ehr., *Cyclotella comta* (Ehr.) Kiitz., *Gomphonema olivaceum* (Lyngb.) Kiitz., *A. italica*, *A. islandica*, *F. construens*, *S. hantzschii*, *Amphora ovalis* Kiitz., *Tabellaria fenestrata* (Lyngb.) Kiitz., *Nitzschia palea* (Kiitz). W.Sm., *Navicula* sp. were found in high numbers.

In cats'-tail thickets, closer to the water's edge (stations 1, 2), the proportion of diatoms was significantly higher than that of green-blue algae. Also, the number of euglenophytes *Euglena viridis* Ehr., *P. triquetrus* (Ehr.) Daj. increased. By the end of July, the concentration of volvocine algae of the genus *Chlamydomonas* reached 7.5×10^6 cells/l and 27.8 mg/l, which accounted for 34.8 % of the total population and 65.8 % of the total phytoplankton biomass. At the same time, the abundance of other algal groups in the thickets started to decrease.

Since July, the water level in the reservoir has been steadily lowering. By the end of September–beginning of October the level was 51 m, which is well below the long-term averages. Due to the lowering of the water level, some shallow water areas were drained in the middle of the summer, and at stations 1 and 2, only one maximum in biomass occurred. It was possible to follow the dynamics of phytoplankton growth throughout the summer and autumn only at station 3. At the beginning of August, a reduction in the number of all algae groups in the water was detected. The second maximum in phytoplankton abundance and biomass was observed in mid-August due to the intensive development of cyanobacteria and diatoms in the studied biotopes. Main contributors to massive development of cyanobacteria during the summer and autumn period were the species *A. flos-aquae*, *An. flos-aquae*, *An. scheremetievi*, *M. aeruginosa*. The abundance of *M. aeruginosa* has increased dramatically in the summer–autumn complex of algae.

The dynamics of the total phytoplankton biomass in the studied biotopes had significant differences. Firstly, in cats'-tail thickets associations, euglenophytes intensively developed at both sites. The same species dominated elsewhere: *E. viridis*, *P. triquetrus*, *P. pleuronectes*, *Trachelomonas volvocina* Ehr., *Trachelomonas hispida* (Perty.) Stein emend. Defl. The diatoms of the thickets at site 1 were represented by the species: *S. hantzschii*, *C. comta*, *C. placentula*, *A. italica*, *N. vermicularis*, *M. varians*, *F. construens*, *T. fenestrata*, *Gomphonema acuminatum* Ehr., *Nitzschia acicularis* W.Sm., *Navicula cryptocephala* Kiitz., *Diatoma elongatum* (Lyngb.) Ag.

In the open shallow water of site 2, diatoms dominated in biomass. The most frequently found species were *N. cryptocephala*, *C. placentula*, *M. granulata*, *S. hantzschii*, *A. ovalis*, *M. varians*, *G. olivaceum*, *Diatoma vulgare* Bory., *Nitzschia paleacea* Grun., *Synedra ulna* (Nitzsch.) Ehr. The ratio of quantitative indicators of dominant species in the thickets was changing, and the dominant forms (listed in descending order) were *M. varians*, *G. acuminatum*, *G. olivaceum*, *A. ovalis*, *C. comta*, *T. fenestrata*, and *N. palea*.

In the summer–autumn mean-water, the average water level in the reservoir remained at the levels well below the long-term average (51.0 m) until the middle of November. By the beginning of September, shallow water areas of both sites have dried up below the outer edge of the thickets, and further research continued in the shallows below the thickets. During that period, the quantitative indicators of previously dominating algae decreased across the whole reservoir. By the end of water "blooming" caused by cyanobacteria, in all parts of the reservoir, the outbreak of volvocine algae (*Carteria multifilis* (Fres.) Dill. и *Chlamydomonas* sp.) abundance occurred and continued in spring 2003 (unpublished observations, Khaliullina).

When volvocine algae abundance started to rise, an increase of biogenic elements in the water was detected (nitrate, ammonium and phosphorus compounds) (Ratushnyak et al. 2006a, b). Aging and death of cyanobacteria (especially *M. aeruginosa* which biomass was high) cell colonies was coupled with their mineralization. As a result, cellular content was released into water and enriched it with organic compounds. Since then, the water level in the reservoir stayed abnormally low for some time, and the concentration of nutrients in the water remained high. Because of this, fast-reproducing volvocine algae could grow in mass and cause the autumn "blooming" of water. It is known that volvocine algae of the genus *Chlamydomonas* can develop in large numbers at fairly low temperatures (Vodorosli. lishajniki 1977). Similar cases of water "blooming" causing the development of volvocine algae were described for Cheboksary, Volgograd and Saratov water reservoirs in different years (Gerasimova 1996; Okhapkin 1994).

Conclusion

Seasonal changes in the total biomass and abundance of phytoplankton in the thickets of macrophytes and in the open areas of the reservoir differed little. During the growing season, two peaks with maximums in June–July and late August were observed at all studied sites.

The development of planktonic algae in the thickets occurs in parallel with the seasonal formation of plant biotopes. In the beginning of summer, when higher aquatic vegetation is developed little, qualitative and quantitative composition of phytoplankton in the thickets and in open water areas differed slightly. With the growth of macrophytes, chlorophytes production is limited by low lightening. In contrast, the content of euglenophytes, dinoflagellates, and diatoms, which are mixotrophic and mobile (i.e. they can find the most favorable location in the water), increases. In mid-summer, the algocoenosis formed in the thickets is different from that residing in the open water areas. This was particularly noticeable during the periods of water mass stability and at the site 1 (Pobedilovo), which is relatively isolated from the pelagic zone of the reservoir and is characterized by more sustained runoff.

In areas of more intense water masses dynamics (site 2, Saralinsky), the effect of macrophyte vegetation on cyanobacteria is less pronounced. Due to wind-driven waves, cyanobacteria occasionally accumulate in high numbers at the outer edge of the thickets. Also, during the periods of sharp increase of the water level, the intensive dynamics of water masses contributes to the widespread distribution of heterotrophic flagellar heterotrophic euglenophytes, dinoflagellates, and cryptomonads, which developed mainly in the thickets of macrophytes.

The zone on the outer edge of the thickets, where an intensive contact with the open water occurs and where the role of true phytoplankton significantly increases, is characterized with the highest species diversity and abundance of algae. For this mixing zone, the "edge effect" is noticed, i.e. there is a tendency for increase in biodiversity and quantitative phytoplankton indicators.

By the end of the summer of 2002, a decrease in water level lower than 51.5 m led to the autumn outbreak in volvocine algae abundance and biomass, which continued during the spring of the next year (unpublished observations, Khaliullina). Thus, the water level lowering in the reservoir leads to massive development of phytoplankton, cyanobacteria or other algae, representatives of which can be indicators of higher saprobity zones. The maintaining of an optimal water level in the reservoir is recommended for controlling of water "blooming" and thus maintaining high water quality.

Under the conditions of the water reservoir, in the biological communities, all the development processes, having a potential to create a structural and functional order, are regularly interrupted by various factors (flood, rain floods, drought, sedimentation, etc.). Formed communities are also constantly transforming mainly due to water level fluctuations. Therefore, it is impossible to identify clear boundaries of the communities in different biotopes.

Water level fluctuations in the Kuibyshev Reservoir (seasonal or caused by other reasons) primarily affect the shallow coastal waters phytoplankton communities, both in the macrophytes thickets and in the open water. Despite the fact that in this work, we found no direct correlations between phytoplankton community structure formation and such factors as concentration of nutrients, relief, dynamics of water masses, and the degree of higher aquatic vegetation overgrowth in the coastal shallow-water littoral zone, apparently these factors together with climatic conditions determine the development of phytoplankton. Lack of pair correlations between algal abundance and abiotic factors is not an indication that the latter ones do not affect the development of phytoplankton. The effects of such factors are often indirect and may appear with delay (Harris 1987; Reynolds 1990). In addition, multiple factors may overlap, which makes it impossible to reveal direct causal connections.

Acknowledgments The studies were done as a part of an international program INTAS: "Metabolic Biodiversity and Monitoring of Benthic forms: Strategies of Adaptation for Different Oxygen regimes controlled with Natural and Anthropogenic factors" in 2002–2004: "Influence of water level fluctuation on *Typha*-dominated littoral communities of the Kuibyshev Water Reservoir (Tatarstan Republic, Russia)", funded by the Volkswagen Stiftung; Ref. No. I/77 616. The authors are sincerely grateful to Prof. V.A. Yakovlev for the help with the study design and valuable advice throughout the whole study. The work is performed according to the Russian Government Program of Competitive Growth of Kazan Federal University.

Authors' contributions LYu collected and analyzed the water samples and drafted the manuscript. GV helped in analyzing the obtained data and edited the manuscript. All authors read and approved the final manuscript.

Compliance with ethical standards

Conflict of interest The authors declare that they have no competing interests.

References

Borovikov VP, Borovikov IP (1998) STATISTICA. statisticheskij analiz i obrabotka dannyh v srede windows [STATISTICA. Statistical analysis and data processing in Windows] (2nd edn). Filin

Gaevskaja NS (1966) Rol' vysshih vodnyh rastenij v pitanii zhivotnyh presnyh vodoemov [The role of higher aquatic plants in animals nutrition in freshwaters]. Nauka

Gerasimova NA (1996) Fitoplankton saratovskogo i volgogradskogo vodokhranilishch [Phytoplankton of the Saratov and Volgograd Reservoirs]. IEVB RAN, Tolyatti

Gidrometeorologicheskii rezhim ozer i vodokhranilishch SSSR. Kuibyshevskoe i Saratovskoe vodokhranilishcha (1978) [Hydrometeorological Regime of Lakes and Reservoirs in the USSR: Kuibyshev and Saratov Reservoirs]. Gidrometeoizdat, Leningrad

Gosudarstvennyj doklad o sostojanii prirodnyh resursov i ob ohrane okruzhajushhej sredy Respubliki Tatarstan v 2002 godu (2003) [National report on the state of natural resources and environmental protection in the Republic of Tatarstan in 2002]. Minprirody RT, Kazan

Gosudarstvennyj doklad o sostojanii prirodnyh resursov i ob ohrane okruzhajushhej sredy Respubliki Tatarstan v 2004 godu (2005) [National report on the state of natural resources and environmental protection in the Republic of Tatarstan in 2004]. Minprirody RT, Kazan

Guseva KA, Jekzercev VI (1966) Formirovanie fitoplanktona i vysshej vodnoj rastitel'nosti v ravninnyh vodohranilishhah [Formation of phytoplankton and higher aquatic vegetation in the plain reservoirs]. In: Jekologija vodnyh organizmov [Ecology of aquatic organisms], Nauka

Harchenko TA (1991) Koncepcija jekotonov v gidrobiologii [The concept of ecotones in Hydrobiology]. Gidrobiologicheskij Zhurnal 27(4):3–9

Harris GP (1987) Phytoplankton ecology: structure, function and fluctuation. Chapman and Hall, New-York, p 384

Hilt S (2006) Allelopathic inhibition of epiphytes by submerged macrophytes. Aquat Bot 85(3):252–256

Kazda M, Yakovlev V, Ivanov D, Gang N, Leffler S, Amenitskij S (2004) The importance of *Typha* for littoral communities of the Kuybishev water reservoir—overview of the project financed by the Volkswagen foundation. Ecological problems of littoral in flat water reservoirs. International scientific conference, Kazan, pp 45–48

Khaliullina LI (2015) Features of the structural organization of phytoplankton of coastal shallow water of Volga and Volga-Kama reaches of the Kuibyshev Reservoir. Res J Pharm Biol Chem Sci (RJPBCS) 6(4):105–116

Khaliullina LY, Yakovlev V, Khaliullin I (2009) Seasonal and year-to-year dynamics of phytoplankton in connection with the level regime of the Kuibyshev Reservoir. Water Resour 36(4):459–465

Komarkova IJ, Marvan P, Rychkova MA (1983) Pervichnaja produkcija i rol' vodoroslej v litoral'noj zone vodoemov razlichnogo tipa [Primary production and the role of algae in the littoral zone of various types of water bodies]. In: Gidrobiologicheskie processy v vodoemah [Hydrobiological processes in water reservoirs]. Nauka, Leningrad

Konovalov SM, Pautova VN (1989) Jekologija fitoplanktona kujbyshevskogo vodohranilishha [Ecology of phytoplankton of the Kuibyshev reservoir]. Nauka, Leningrad

Koreljakova IL (1977) Rastitel'nost' kremenchugskogo vodohranilishha [Vegetation of the Kremenchugskoe Reservoir] Naukova dumka, Kiev

Kujbyshevskoe vodohranilishhe (1983) [The Kuibyshev Reservoir]. Nauka, Leningrad

Kujbyshevskoe vodohranilishhe (2008) [The Kuibyshev reservoir]. IEVB RAN, Tolyatti

Kutova TN (1974a) Rastitel'nost' melkovodij gor'kovskogo vodohranilishha [Vegetation of the shallows of the Gorky Reservoir]. Izvestija GosNIORH 89:30–36

Kutova TN (1974b) Rastitel'nost' mshichinskogo zaliva [Vegetation of the Mshihinskogo Bay]. In: Prirodnye Resursy Mologo-Sheksninskoj Niziny. Rybinskoe Vodohranilishhe [Natural Resources Mologa-Sheksna Lowland. Rybinsk Reservoir], pp 98–110

Lavrent'eva GM (1977) Fitoplankton vodohranilishh volzhskogo kaskada. [Phytoplankton of the Volga cascade reservoirs] Izvestija GosNIORH, 114(6):1937–1938

Morduhaj-Boltovskoj FD (1975) Metodika izuchenija biogeocenozov vnutrennih vodoemov [The methods of studying the biogeocenosis in inland waters], Nauka, Moscow

Mulderij G, Van Nes EH, Van Donk E (2007) Macrophyte–phytoplankton interactions: the relative importance of allelopathy versus other factors. Ecol Model 204(1):85–92

Okhapkin AG (1994) Fitoplankton Cheboksarskogo Vodohranilishha [Phytoplankton in the Cheboksary Reservoir]. IEVB RAN, Tolyatti

Pautova VN, Nomokonova VI (2001) Dinamika fitoplanktona nizhnej volgi–ot reki k kaskadu vodohranilishh [The phytoplankton dynamics in the Lower Volga, from the river to the cascade of reservoirs]. IEVB RAN, Tolyatti

Pautova VN, Nomokonova VI (1994) Produktivnost' fitoplanktona kujbyshevskogo vodohranilishha [Phytoplankton productivity in the Kuibyshev reservoir] IEVB RAN, Tolyatti

Potapov AA (1958) Kormovoe znachenie vodnoj i pribrezhnoj rastitel'nosti vodohranilishh. [Forage value of aquatic and riparian vegetation of water reservoirs] Vestnik Sel'Skohozjajstvennoj Nauki (6):139–144

Qin B, Song Y, Gao G (2006) The role of periphytes in the shift between macrophyte and phytoplankton dominated systems in a shallow, eutrophic lake (lake Taihu, China). Sci China Ser C Life Sci 49(6):597–602

Ratushnyak AA, Borisovich MG, Valeev VS, Ivanov DV, Andreeva MG, Trushin MV (2006a) The hydrochemical and hydrobiological analysis of the condition of the Kuibyshev Reservoir littorals (republic of Tatarstan, Russia). Ekoloji 15(61):22–28

Ratushnyak AA, Borisovich MG, Valeev VS, Ivanov DV, Andreeva MG, Trushin MV (2006b) Diagnostics of the water quality of the Kuibyshev Reservoir littorals with various anthropogenic load: hydrobiological and multifractal analysis. Fresenius Environ Bull 15(7):626–632

Reynolds CS (1990) Temporal scales of variability in pelagic environments and the response of phytoplankton. Freshw Biol 21(1):23–53

Sand-Jensen K, Borum J (1991) Interactions among phytoplankton, periphyton, and macrophytes in temperate freshwaters and estuaries. Aquat Bot 41(1):137–175

Sayer CD, Davidson TA, Jones JI (2010) Seasonal dynamics of macrophytes and phytoplankton in shallow lakes: a eutrophication-driven pathway from plants to plankton? Freshw Biol 55(3):500–513

Schriver P, Bøgestrand J, Jeppesen E, Søndergaard M (1995) Impact of submerged macrophytes on fish-zooplankton-phytoplankton interactions: large-scale enclosure experiments in a shallow eutrophic lake. Freshw Biol 33(2):255–270

Solov'eva VV, Rozenberg GS (2006) Sovremennoe predstavlenie ob jekotonah ili teorija jekotonov [The theory of ecotones]. Uspehi sovremennoj biologii 126(6):531–549

Stepanova NY (1999) Jekologicheskie kriterii normirovanija nagruzki na vodoem v uslovijah ego zagrjaznenija vozvratnymi vodami [Environmental criteria of burden valuation on the pond under the conditions of its pollution with return waters]. Dissertation, Kazan State University

Vodorosli. lishajniki [Algae. Lichens.] (1977) In: Gollerbah AA (ed) Zhizn' rastenij [Plants life]. Prosveshhenie, Moscow

Vodorosli spravochnik (1989) In: Vasser SP (ed) Algae directory. Naukova dumka, Kiev

Zimbalevskaja LN (1972) Raspredelenie fitofil'nyh bespozvonochnyh i metody ih kolichestvennogo ucheta [Distribution of phytophilic invertebrates and the methods for their quantitative count]. Gidrobiologicheski Zhurnal 8(2):49–55

Zimbalevskaja LN (1973) Raspredelenie fitofil'nyh bespozvonochnyh i metody ih kolichestvennogo ucheta [Distribution of phytophilic invertebrates and the methods for their quantitative count]. Gidrobiologicheski Zhurnal 6(9):38–48

16

Antioxidative gelatin hydrolysate from unicorn leatherjacket skin as affected by prior autolysis

Supatra Karnjanapratum · Soottawat Benjakul

Abstract Gelatin hydrolysates from autolysed non-swollen and swollen unicorn leatherjacket skin prepared using partially purified glycyl endopeptidase (GE) from papaya latex were examined for their antioxidative activities. Autolysed swollen skin was more hydrolysed by GE as indicated by higher α-amino group content, compared with autolysed non-swollen skin. ABTS radical scavenging activity and ferric reducing antioxidant power (FRAP) of hydrolysates were increased with increasing levels of GE used. Antioxidative gelatin hydrolysate from autolysed skin, both non-swollen and swollen skins, using 8 % GE termed 'NS-8GE' and 'SS-8GE' exhibited different modes of action. When both hydrolysates were tested in lecithin liposome system, the efficiency in retardation of lipid oxidation was in a dose-dependent manner. Antioxidative activity of hydrolysates at 5.0 g/L was comparable to that of 0.1 g/L Trolox, in which the oxidation was almost completely inhibited. NS-8GE and SS-8GE also showed their antioxidative activities in gastrointestinal tract model system (GIMs). ABTS radical scavenging activity of both hydrolysates increased, but FRAP decreased in a duodenal condition. Based on size exclusion chromatography, major antioxidative peptides in NS-8GE and SS-8GE had molecular weight of 1,170 and 750 Da, respectively. Therefore, GE could enhance antioxidative activity of autolysed skin. Additionally, swelling process directly determined the modes of actions of resulting gelatin hydrolysates. NS-8GE had high ABTS radical scavenging activity, whereas SS-8GE showed high FRAP.

Keywords Unicorn leatherjacket · Gelatin hydrolysate · Autolysis · Gylcyl endopeptidase · Antioxidative activity

Abbreviations

FRAP	Ferric reducing antioxidant power
PV	Peroxide value
TBARS	Thiobarbituric acid reactive substances
GIMs	Gastrointestinal tract model system

S. Karnjanapratum · S. Benjakul (✉)
Department of Food Technology, Faculty of Agro-Industry, Prince of Songkla University, Hat Yai, Songkhla 90112, Thailand
e-mail: soottawat.b@psu.ac.th

S. Karnjanapratum
e-mail: agro_bow@hotmail.com

Introduction

Lipid oxidation has been known as a serious cause of food deterioration as well as various human diseases, such as cardiovascular diseases and cancers (Lobo et al. 2010). The reactive oxygen species (ROS) such as hydroxyl radical ($^{\bullet}OH$), superoxide radical ($O_2^{\bullet -}$), hydrogen peroxide (H_2O_2), hydroperoxyl radical ($HO_2^{\bullet -}$), lipid peroxyl radical ($LOO^{\bullet}$), alkoxyl radical ($LO^{\bullet}$), and singlet oxygen ($^1O_2$) are the main initiators of lipid oxidation through free radical chain reaction (Min and Ahn 2005). Antioxidants are substances that can inhibit the generation of free radicals in a system and prevent or delay the lipid oxidation reaction. Generally, the most common synthetic antioxidants including butylated hydroxytoluene (BHT), butylated hydroxyanisole (BHA), tert-butylhydroquinone (TBHQ) and propyl gallate (PG) have been used to enhance the oxidative stability and inhibit lipid oxidation in lipid containing foods (Min and Ahn 2005; Xu 2008).

Nowadays, natural additives involving antioxidants have gained attention from consumers. Protein hydrolysates, especially from marine processing byproducts, have attracted interest due to their safety as well as bioactivities (Kim and Wijesekara 2010; Gómez-Guillén et al. 2011). Enzymatic hydrolysis has been intensively employed to produce antioxidative gelatin hydrolysates (Gómez-Guillén et al. 2011). However, the high cost of production is mainly associated with enzyme, energy consumed and production process. Recently, autolysis mediated by indigenous protease was used as the aid for production of gelatin hydrolysates with antioxidative activity from unicorn leatherjacket skin that could reduce the amount of commercial proteases and energy used (Karnjanapratum and Benjakul 2014a). The selection of proper enzyme to substrate was crucial for production of antioxidative gelatin hydrolysate from Alaska pollack skin (Kim et al. 2001). Kittiphattanabawon et al. (2012) found that glycyl endopeptidase, the major cysteine protease in papaya latex, showed the effectiveness in cleaving peptide bonds with Gly at P_1 of fish gelatin from blacktip shark skin, yielding antioxidative gelatin hydrolysate with high degree of hydrolysis. Additionally, the active fraction of glycyl endopeptidase from papaya latex was prepared using aqueous two-phase system to remove offensive odour compounds (Karnjanapratum and Benjakul 2014b).

To reduce the amount of enzyme and energy used, the autolysis-assisted process mediated by indigenous protease in unicorn leatherjacket skin could be employed prior to further hydrolysis by glycyl endopeptidase from papaya latex. Additionally, pretreatment of skin via swelling process might affect the hydrolysis by protease used. As a consequence, gelatin hydrolysate with different antioxidative activities could be obtained. The objective of this study was to investigate the impact of autolysis-assisted process of non-swollen and swollen skin of unicorn leatherjacket and effect of partially purified glycyl endopeptidase at different levels on antioxidative activities of resulting gelatin hydrolysates.

Materials and methods

Chemicals

Polyethylene glycol (PEG) 6000 was obtained from Fluka (Buchs, Switzerland). 2,4,6-trinitrobenzenesulphonic acid (TNBS), 2,2′-azinobis (3- thylbenzothiazoline-6-sulfonic acid) (ABTS), 6-hydroxy-2,5,7,8-tetramethyl-chroman-2-carboxylic acid (Trolox), 3-(2-pyridyl)-5,6-diphenyl-1,2,4-triazine-4′,4″-disulfonic acid sodium salt (ferrozine), pepsin from porcine gastric mucosa (EC 3.4.23.1), pancreatin from porcine pancreas, trypsin from bovine pancreas (EC 3.4.21.4) and bile extract porcine were purchased from Sigma-Aldrich Chemical Co. (St. Louis, MO, USA). Trichloroacetic acid (TCA), ferrous chloride and iron standard solution were obtained from Merck (Darmstadt, Germany). 2-thiobarbituric acid and cumene hydroperoxide were procured from Fluka (Buchs, Switzerland). Ammonium thiocyanate was purchased from Lab-Scan (Bangkok, Thailand). All chemicals were of analytical grade.

Preparation of skin and autolysed skin

Preparation of skins

The skins of unicorn leatherjacket (*Aluterus monoceros*) were obtained from a dock, Songkhla, Thailand. The fish with the size of 1.0–1.5 kg/fish (30–35 cm of body length) were caught from the Gulf of Thailand, stored in ice and offloaded approximately 2–3 days after capture. Three different lots of skins were collected. For each lot, skins were pooled and used as the composite sample. Skins were stored in ice with a skin/ice ratio of 1:2 (w/w) and transported to the Department of Food Technology, Prince of Songkla University, Hat Yai within 1 h. Upon arrival, the skins were washed with iced tap water (0–2 °C) and cut into small pieces (0.5×0.5 cm^2), placed in polyethylene bags and stored at −20 °C until use. The storage time was less than 2 weeks.

Pretreatment of skins

The skins were pretreated to remove non-collagenous proteins following the method of Kaewruang et al. (2013). Fish skins (0.5×0.5 cm^2) were soaked in 0.05 M NaOH with a skin/alkaline solution ratio of 1:10 (w/v) to remove non-collagenous proteins. The mixture was stirred continuously at room temperature using an overhead stirrer equipped with a propeller (RW 20.n, IKA Labortechnik, Staufen, Germany) at a speed of 150 rpm. The alkaline solution was changed after 2 h and total pretreatment time was 4 h. Pretreated skins were washed with tap water until neutral or faintly basic pH of wash water was obtained.

Preparation of autolysed skin

Use of non-swollen skin

Autolysis was conducted using pretreated skin (non-swollen skin) following the method of Karnjanapratum and Benjakul (2014a). The pretreated skins were mixed with deionised water at a ratio of 1:5 (w/v). The autolysis was performed by incubating the mixture in a water bath (Model W350, Memmert, Schwabach, Germany) at 55 °C for 12 h and terminated by heating at 90 °C for 15 min. The mixture was centrifuged at 5,000×*g* at 4 °C using a refrigerated centrifuge model Avanti J-E (Beckman Coulter, Inc., Palo Alto, CA, USA) for 10 min to remove the debris. Autolysed skin was collected and referred to as 'NS'.

Use of swollen skin

To prepare swollen-skin, the pretreated skin was soaked in 0.1 M phosphoric acid with a skin/solution ratio of 1:10 (w/v) for 6 h with a gentle stirring at room temperature as per the method of Kaewruang et al. (2013). The acidic solution was changed every 3 h. Acid-treated skin was washed thoroughly with tap water until wash water became neutral or faintly acidic. To prepare autolysed skin, the swollen skin was mixed with deionised water at a ratio of 1:5 (w/v) and subjected to autolysis as previously described. Autolysed skin obtained was referred to as 'SS'.

Production of gelatin hydrolysate from autolysed skin using partially purified glycyl endopeptidase (GE)

Preparation of crude extract from papaya (C. papaya) latex

Fresh papaya latex was collected from green papaya fruit cultivated in Songkhla, Thailand. Four to six longitudinal incisions were made on the green papaya fruit using a stainless steel knife. The exuded latex was collected using a receiving container. The latex was then transferred to a beaker and stored below 10 °C and used within 3 h.

The crude extract was prepared using the method of Kittiphattanabawon et al. (2012). The latex was mixed with cold distilled water (≤4 °C) with a latex-to-water ratio of 1:3 (w/v). The mixture was gently stirred at 4 °C for 1 h. Then, the mixture was centrifuged at 9,000×*g* at 4 °C for 20 min using a refrigerated centrifuge.

The supernatant was filtered using a Whatman No. 1 filter paper (Whatman International, Maidstone, Kent, UK), followed by lyophilisation (Scanvac Model Coolsafe 55 freeze dryer, Coolsafe, Lynge, Denmark). The powder obtained was referred to as 'crude extract'.

Fractionation of glycyl endopeptidase using aqueous two-phase system (ATPS) in combination with ammonium sulphate precipitation

Crude extract powder (1 g) was dissolved in 8 mL of distilled water. The pH of solution was adjusted to 6.0 using 6 M HCl and the volume was made up to 10 mL by distilled water to obtain a concentration of 100 mg/mL prior to fractionation using ATPS.

The glycyl endopeptidase was fractionated using the method of Karnjanapratum and Benjakul (2014b). ATPS with 10 % PEG 6000 and 10 % ammonium sulphate $(NH_4)_2SO_4$ was used for fractionation of glycyl endopeptidase. The salt-rich bottom phase was collected and further precipitated using ammonium sulphate at 60 % saturation. After centrifugation at 9,000×g at 4 °C for 20 min, the pellet was re-dissolved in distilled water and dialysed against 20 volumes of distilled water, six times. After lyophilisation, the powder referred to as 'partially purified glycyl endopeptidase, GE' was stored at −40 °C until use.

Production of gelatin hydrolysates using GE

Autolysed fish skins, both NS and SS, were used as substrates. The solutions (3 %, w/v) were then added with GE at different concentrations (2, 4, 6 and 8 %, based on solid matter). The reaction proceeded at 40 °C for 60 min. After enzyme inactivation by heating at 90 °C for 15 min, the resulting gelatin hydrolysate was centrifuged at 9,000×g at 4 °C for 20 min. The supernatant was collected and lyophilised. The gelatin hydrolysate powders were placed in polyethylene bag and stored at −40 °C. Hydrolysate powders were subjected to analyses.

Analyses

Determination of α–amino group content

The α–amino group content was determined according to the method of Benjakul and Morrissey (1997). To diluted samples (125 μL), 2.0 mL of 0.2 M phosphate buffer, pH 8.2 and 1.0 mL of 0.01 % TNBS solution was added. The solution was mixed thoroughly and placed in a temperature-controlled water bath at 50 °C for 30 min in the dark. The reaction was terminated by adding 2.0 mL of 0.1 M sodium sulphite. The mixture was cooled at room temperature for 15 min. L-leucine standard solutions with concentrations ranging from 0.5 to 5.0 mM were used. The absorbance was read at 420 nm and α–amino group content was expressed in terms of L-leucine.

Determination of in vitro antioxidative activities

ABTS radical scavenging activity

ABTS radical scavenging activity of gelatin hydrolysates was determined as described by Binsan et al. (2008). The stock solutions included 7.4 mM ABTS solution and 2.6 mM potassium persulfate solution. The working solution was prepared by mixing the two stock solutions in equal quantities. The mixture was allowed to react for 12 h at room temperature in the dark. The solution obtained (1 mL) was then diluted with 50 mL distilled water in order to obtain an absorbance of 1.1 ± 0.02 units at 734 nm using a UV-1601 spectrophotometer (Shimadzu, Kyoto, Japan). Fresh ABTS solution was prepared for each assay. Sample (150 μL) was mixed with 2,850 μL of ABTS solution and the mixture was allowed to stand at room temperature for 2 h in the dark. The absorbance was then measured at 734 nm using a spectrophotometer. The blank was prepared in the same manner, except that distilled water was used instead of the sample. A standard curve of Trolox ranging from 50 to 600 μM was prepared. The activity was expressed as μmol Trolox equivalent (TE)/g sample.

Hydrogen peroxide scavenging activity

Hydrogen peroxide scavenging activity was assayed according to the method of Wettasinghe and Shahidi (2000) with a slight modification. Briefly, the sample (1 mL) was mixed with 83 µl of 100 mM hydrogen peroxide (prepared in 0.1 M phosphate buffer, pH 7.4). The mixture was allowed to react for 40 min at room temperature. The absorbance at 230 nm of the reaction mixture was read. The blank, in which 0.1 M phosphate buffer was used instead of hydrogen peroxide, was prepared. Trolox (0–10 mM) was used as the standard. The hydrogen peroxide scavenging activity was expressed as mmol Trolox equivalent (TE)/g sample.

Ferric reducing antioxidant power (FRAP)

FRAP was determined according to the method of Benzie and Strain (1996). FRAP reagent was prepared by mixing acetate buffer (30 mM, pH 3.6), 10 mM TPTZ solution in 40 mM HCl and 20 mM iron (III) chloride solution in proportions of 10:1:1 (v/v). The sample solution (100 µL) was mixed with 3 mL of working FRAP reagent and incubated in dark condition at room temperature for 30 min. The absorbance of the reaction mixture was read at 593 nm using a spectrophotometer. The standard curve was prepared using Trolox ranging from 0 to 500 µM. The activity was expressed as µmol Trolox equivalent (TE)/g sample.

Ferrous chelating activity

Chelating activity of gelatin hydrolysates towards ferrous ion (Fe^{2+}) was measured by the method of Thiansilakul et al. (2007) with a slight modification. Sample (200 µL) was mixed with 800 µL of distilled water. Thereafter, 0.1 mL of 2.0 mM $FeCl_2$ and 0.2 mL of 5 mM ferrozine were added. The mixture was allowed to react for 20 min at room temperature. The absorbance was then read at 562 nm. The standard curve of EDTA (0–1.0 mM) was prepared. The control was prepared in the same manner except that distilled water was used instead of the sample. Ferrous chelating activity was expressed as µmol EDTA equivalents (EE)/g sample.

Antioxidative activity of gelatin hydrolysate in lecithin liposome model system

Gelatin hydrolysates prepared using 8 % GE rendering the highest in vitro antioxidative activity were tested in a lecithin liposome model system according to the method of Thiansilakul et al. (2007). Both gelatin hydrolysates were named 'NS-8GE' and 'SS-8GE' where NS and SS were used as substrates, respectively.

Lecithin liposome system was prepared by suspending lecithin in deionised water at a concentration of 8 mg/mL. The mixture was stirred with a glass rod, followed by sonification for 30 min in a sonicating bath (Elma Model S30H, Singen, Germany). Gelatin hydrolysate (3 mL) was added to the lecithin liposome system (15 mL) to obtain the final concentrations of 2.5 and 5.0 g/L. The mixture was sonicated for 2 min. To initiate the reaction, 40 µL of 0.15 M cupric acetate was added. The mixture was shaken in the dark at 120 rpm using a shaker (Unimax 1010, Heidolph Model, Schwabach, Germany) at 37 °C. The system containing 0.10 g/L Trolox was also prepared. The control was prepared in the same manner, except that distilled water was used instead of gelatin hydrolysates or Trolox. Oxidation in lecithin liposome systems was monitored during 24 h of incubation by determining PV and TBARS values.

Determination of peroxide value (PV)

PV was measured according to the method of Wu et al. (2003). Liposome sample (0.1 mL) was mixed with 4.7 mL of ethanol/distilled water (3:1), 0.1 mL of 30 % ammonium thiocyanate, and 0.1 mL of 20 mM ferrous chloride solution in 3.5 % HCl. After incubation at 40 °C for 3 min, the absorbance was read at 500 nm using a spectrophotometer. PV was calculated from a standard curve of cumene hydroperoxide (0–100 mg/L) and expressed as mg cumene hydroperoxide/L lecithin liposome.

Determination of thiobarbituric acid reactive substances (TBARS)

TBARS was determined as described by Buege and Aust (1978) with a slight modification. Liposome sample (0.5 mL) was homogenised with 2.5 mL of TBARS solution (0.375 % thiobarbituric acid, 15 % TCA and 0.25 M HCl). The mixture was heated in boiling water for 10 min to develop the pink colour. The mixture was then cooled with running water and centrifuged at 5000xg for 10 min at room temperature using Hettich centrifuge (Hettich Model MIKRO-20, Tuttlingen, Germany). The supernatant was collected and the absorbance at 532 nm was read using a spectrophotometer. TBARS value was calculated from a standard curve of malondialdehyde (MDA) (0–10 mg/L) and expressed as mg MDA/L lecithin liposome.

Changes in antioxidative activities of gelatin hydrolysate in gastrointestinal tract model system

Gastrointestinal tract model system mimicking the conditions of the digestive organs was prepared according to the method of Enari et al. (2008) with slight modifications. Gelatin hydrolysate solution (10 mg/mL in distilled water 100 mL) was mixed with 0.5 ml of 1 M HCl-KCl buffer (pH 1.5), followed by adding 32 U/mL of pepsin solution in 1 M HCl-KCl buffer (pH 1.5) (5 mL) and incubating for 60 min at 37 °C (stomach condition) with a continuous shaking using a Unimax 1010 shaker (Heidolph Model, Schwabach, Germany). Thereafter, the pH of the reaction mixture was raised to 6.8 with 1 M $NaHCO_3$ (1 mL), and the enzyme mixture of bile and pancreatic juice (1 mL) that contained pancreatin (10 mg/mL), trypsin (14,600 U/mL) and bile extract (13.5 mg/mL) in 10 mM phosphate buffer (pH 8.2), was added to the solution, followed by incubation at 37 °C for 3 h (duodenal condition) with a continuous shaking. The digestion was terminated by placing in a boiling water bath for 10 min. During the simulated gastrointestinal digestion, the mixture was randomly taken at 0, 0.5, 1, 2, 3 and 4 h for determination of ABTS radical scavenging activity and FRAP.

Molecular weight distribution

Gelatin hydrolysates (NS-8GE and SS-8GE) were separated by size exclusion chromatography using a Sephadex G-25 gel filtration column (2.5 × 50 cm) (17-0032-01, GE Healthcare Bio-Science AB, Uppsala, Sweden). The experiment was conducted in a walk-in cold room (4 °C) and 2 mL of sample solution (60 mg gelatin/mL) was loaded. After being loaded, the elution was performed using a low pressure chromatography system (BioLogic LP system, Bio-Rad Laboratories Ltd., Hercules, CA, USA) coupled with a fraction collector (Model 2128, Bio-Rad Laboratories Ltd., Hercules, CA, USA). Fifty mM sodium phosphate buffer (pH 7.0) was used as the elution buffer at a flow rate of 0.5 mL/min. The fractions of 3 mL were collected. The absorbance was recorded at 220 and 280 nm. ABTS radical scavenging activity and ferric reducing antioxidant power (FRAP) of each fraction were determined. Blue dextran (2,000,000 Da) was used for void volume measurement. The molecular weight markers included insulin chain B (3,496 Da), vitamin B_{12} (1,355 Da), glycine-tyrosine (238 Da) and tyrosine (181 Da). MW of fraction with the highest ABTS radical scavenging activity and FRAP was estimated from the plot between available partition coefficient (K_{av}) and the logarithm of the molecular weight of the protein standards.

Statistical analysis

All experiments were run in triplicate using the three different lots of skin. Data were subjected to analysis of variance (ANOVA) and mean comparisons were carried out using Duncan's multiple range test (Steel and Torrie 1980). Statistical analysis was performed using the Statistical Package for Social Sciences (SPSS for Windows SPSS Inc., Chicago, IL, USA). The data with $P < 0.05$ were considered to be statistically significant.

Results and discussion

Effect of GE on hydrolysis of autolysed skin

The hydrolysis of autolysed non-swollen and swollen skin by GE at different levels as monitored by α-amino group content is shown in Fig. 1. Without hydrolysis using GE, NS and SS had the similar α-amino group

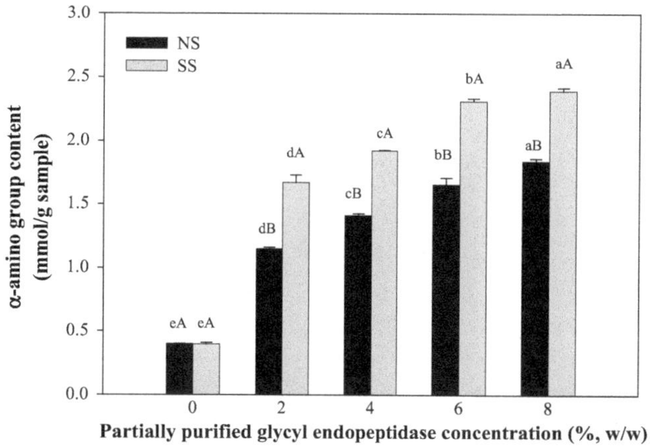

Fig. 1 α-Amino group content of gelatin hydrolysates prepared using partially purified glycyl endopeptidase at different concentrations (2, 4, 6 and 8 % of solid matter). NS autolysed non-swollen skin; SS autolysed swollen skin. Different *uppercase* and *lowercase letters* on the *bars* within the same enzyme concentration and the same substrate, respectively, indicated significant difference ($P < 0.05$). *Bars* represent standard deviation ($n = 3$)

content ($P > 0.05$). The α-amino group content of resulting gelatin hydrolysates increased with increasing GE levels for both substrates (NS and SS) used. This result was in accordance with Karnjanapratum and Benjakul (2014a) who reported that the increase in enzyme/substrate (0.5, 1.0 and 2.0 %) resulted in the increase in degree of hydrolysis of gelatin hydrolysate from unicorn leatherjacket skin prepared using papain. When NS was used as substrate, α-amino group contents in resulting hydrolysates were lower than those found in hydrolysate from SS at all enzyme levels used ($P < 0.05$). The result indicated that peptides in SS were more likely the preferable substrate for GE. Autolysis of gelatin in unicorn leatherjacket skin at 55 °C caused by indigenous protease (Kaewruang et al. 2013) yielded available peptides for further hydrolysis by GE. For swollen skin (SS), α-chains or their cross-link, e.g. β- or γ-chains, could be more repulsed. As a result, those proteins could be cleaved by indigenous protease easily. The resulting peptides might have free exposed glycine along the chain to a higher extent, in comparison with peptides produced from non-swollen skin. When the skin matrix was still compact, cleavage sites might be limited and the resulting peptides were plausibly present as bundles, in which cleavage site, especially glycine, could be hindered. This might result in the lower hydrolysis as indicated by lower α-amino group content.

In vitro antioxidative activities of gelatin hydrolysates

Figure 2 shows the antioxidative activities of gelatin hydrolysates prepared using GE at different levels (2–8 % of solid matter) as examined by different assays. ABTS radical scavenging activity of NS (52.72 μmol/g sample) was higher than SS (29.50 μmol/g sample) ($P < 0.05$). Notably, NS and SS had similar α-amino group content (Fig. 1). Different peptides with various chain length and amino acid sequence might be obtained as influenced by prior swelling. Therefore, the difference in antioxidative activity of autolysed skin (NS and SS) could be observed. The increase in ABTS radical scavenging activity of gelatin hydrolysates was obtained as GE was applied for the second step of hydrolysis. The higher activity was found when GE levels used increased ($P < 0.05$), regardless of substrate. At the same level of GE, gelatin hydrolysates obtained from SS possessed the lower ABTS radical scavenging activity, compared with those from NS ($P < 0.05$). It was noted that ABTS radical scavenging activity of resulting gelatin hydrolysates was not in accordance with α-amino group content (Fig. 1). This might reflect the difference in antioxidative peptides in hydrolysate when different substrates were used. Wu et al. (2003) reported that size, level and composition of free amino acids of peptides affected the antioxidative activity. ABTS radical scavenging assay is based on biamperometric measurements using $ABTS^+$/ABTS redox couple. The ability of antioxidants to scavenge the preformed $ABTS^+$ was estimated by the decolourisation (Re et al. 1999). Therefore, hydrolysate prepared from NS using GE more likely showed higher ability of scavenging ABTS radical, compared with those from SS.

H_2O_2 scavenging activity of gelatin hydrolsates prepared using GE at different concentrations (2–8 % of solid matter) is depicted in Fig. 2b. NS (1.17 mmol/g sample) and SS (1.16 mmol/g sample) had a similar H_2O_2 scavenging activity ($P > 0.05$). The decrease in H_2O_2 scavenging activity of gelatin hydrolysates was observed after further hydrolysis by GE, regardless of substrates. However, the resulting gelatin hydrolysate

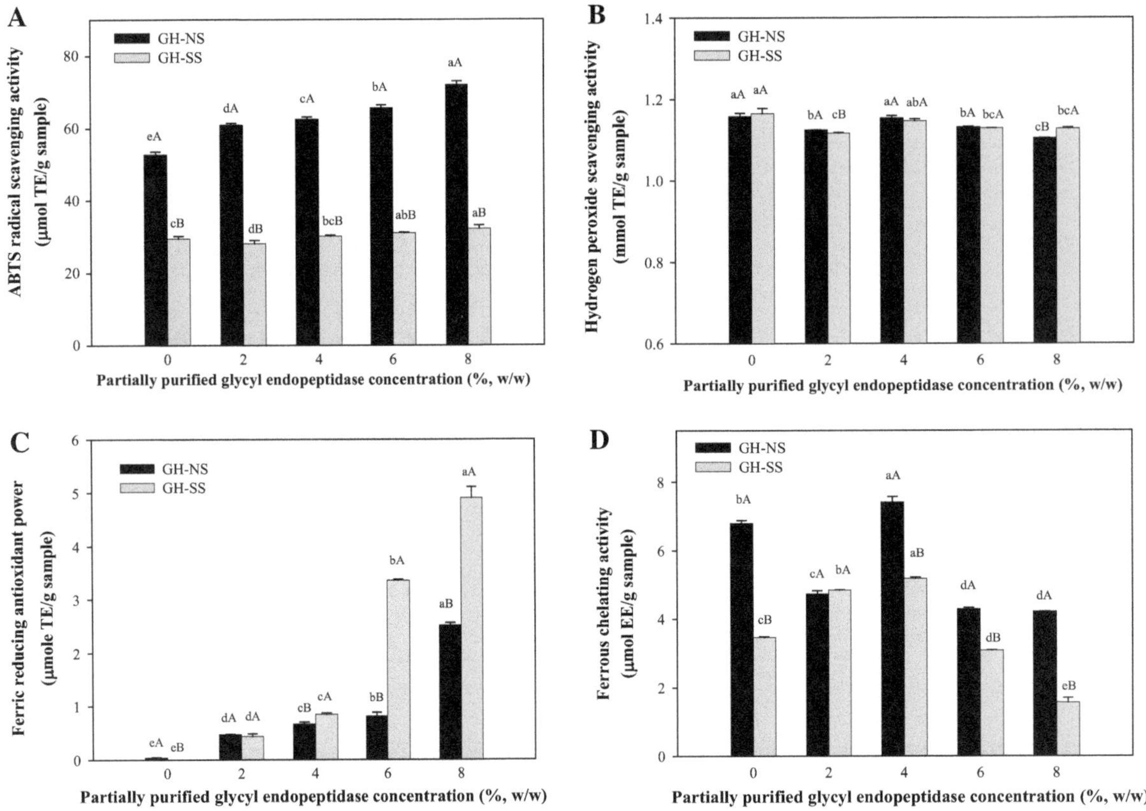

Fig. 2 ABTS radical scavenging activity (**a**), hydrogen peroxide scavenging activity (**b**), ferric reducing antioxidant power (**c**) and ferrous chelating activity of gelatin hydrolysates prepared using partially purified glycyl endopeptidase at different concentrations (2, 4, 6 and 8 % of solid matter). NS autolysed non-swollen skin; SS autolysed swollen skin. Different *uppercase* and *lowercase letters* on the *bars* within the same enzyme concentration and the same substrate, respectively, indicated significant difference ($P < 0.05$). *Bars* represent standard deviation ($n = 3$)

prepared with 4 % GE had a similar H_2O_2 scavenging activity, compared with the substrates, NS and SS ($P > 0.05$). These results suggested that the peptides generated after hydrolysis by GE had low ability in scavenging H_2O_2. Karnjanapratum and Benjakul (2014a) also reported that H_2O_2 scavenging activity of gelatin hydrolysates from unicorn leatherjacket prepared using papain decreased when the level of papain increased. It was reported that degree of hydrolysis had no relationship with antioxidative activity of protein hydrolysate (Chen et al. 1995). The extensive hydrolysis probably caused the lower antioxidative activity, due to the generation of free amino acids which had lower antioxidative activity, compared to peptides (Hernandez-Ledesma et al. 2005). H_2O_2, which is a weak oxidising agent, is not directly involved in the initiation of lipid oxidation because its reduction potential is lower than that of unsaturated fatty acid (Choe and Min 2005). However, H_2O_2 shows indirect contribution to lipid oxidation. H_2O_2 is the precursor for the generation of hydroxyl radical, which is a strong initiator of lipid oxidation (Choe and Min 2005). Thus, further hydrolysis of autolysed skin by GE slightly lowered H_2O_2 scavenging activity of resulting hydrolysates.

Ferric reducing antioxidant power (FRAP) of gelatin hydrolsates prepared using GE at different concentrations is shown in Fig. 2c. NS had low FRAP (0.04 μmol/g sample). On the other hand, SS had no FRAP. The activity of gelatin hydrolysates increased when GE levels used increased ($P < 0.05$), regardless of substrates. Generally, gelatin hydrolysate obtained from SS exhibited the higher FRAP than those from NS, especially when GE at levels higher than 2 % was used. The increases in FRAP were in accordance with increasing α-amino group content of gelatin hydrolysate (Fig. 1). It was suggested that smaller peptides in hydrolysate obtained from SS more likely had the higher ability in reducing TPTZ-FE(III) complex. FRAP is generally used to measure the capacity of a substance in reducing TPTZ-FE(III) complex to TPTZ-Fe(II) complex (Binsan et al. 2008). This was in agreement with Kittiphattanabawon et al. (2012) who reported that

FRAP of hydrolysate from blackstip shark skin prepared using papaya latex enzyme increased with increasing degree of hydrolysis. After hydrolysis using GE, the resulting gelatin hydrolysates from both substrates, NS and SS, possibly contained high amounts of peptides which donated electron to free radicals, thereby terminating the chain reaction.

Ferrous chelating activity of gelatin hydrolsates is shown in Fig. 2d. NS (6.79 μmol/g sample) showed the higher ferrous chelating activity than SS (3.46 μmol/g sample) ($P < 0.05$). After further hydrolysis by GE, the resulting gelatin hydrolysates from NS had the decreases in chelating activity, especially when GE levels increased ($P < 0.05$). However, the increase in ferrous chelating activity was obtained when 4 % GE was used for hydrolysis. When SS was used as substrate, the ferrous chelating activity of resulting hydrolysates increased as GE levels increased up to 4 % ($P < 0.05$). Nevertheless, the decrease in chelating activity was observed in hydrolysate prepared using GE level higher than 4 % ($P < 0.05$). Karnjanapratum and Benjakul (2014a) also reported that ferrous chelating activity of gelatin hydrolysate from unicorn leatherjacket skin decreased with increasing papain levels. The result indicated that different peptides in hydrolysates might have varying ability in sequestering Fe^{2+}. It was noted that gelatin hydrolysates from NS after further hydrolysis using GE showed the higher chelating activity, compared with those from SS. It is well-known that transition metal ions such as iron or copper may catalyse the formation of reactive oxygen species that accelerate lipid oxidation. Thus, gelatin hydrolysates obtained in the present study were able to prevent lipid oxidation via metal chelating ability. Nevertheless, their ability varied, depending upon the level of GE used for hydrolysis.

From the above measurements, gelatin hydrolysate prepared with 8 % GE from both NS (NS-8GE) and SS (SS-8GE) showed the highest radical scavenging activity and electron donating activity, which are crucial for termination of chain reaction. Therefore, NS-8GE and SS-8GE were selected for determination of antioxidative activity in lecithin liposome model system in comparison with NS and SS.

Antioxidative activity of selected gelatin hydrolysates in lecithin liposome model system

Oxidation of lecithin liposome system added with gelatin hydrolysates at different concentrations (2.5 and 5.0 g/L) was monitored by determining peroxide value (PV) (Fig. 3a) and TBARS values (Fig. 3b) during incubation of 48 h at 37 °C. For the control (without hydrolysate or trolox), the sharp increase in PV was found within the first 42 h of incubation ($P < 0.05$), in which the highest value (138.74 ± 3.64 mg cumene hydroperoxide/L lecithin liposome) was found at 42 h of incubation. Thereafter, the sharp decrease in PV was observed up to 48 h. The sharp decrease in PV after 42 h of incubation was plausibly caused by the decomposition of hydroperoxide to the secondary products (Frankel et al. 1997). The system added with NS and SS showed the lower PV than the control throughout 48 h of incubation ($P < 0.05$). In general, the rate of increase varied with the concentration of gelatin hydrolysate used. On the other hand, the system added with NS-8GE and SS-8GE showed very low increase in PV throughout the incubation of 48 h. Ability in prevention of hydroperoxide formation was comparable to that of 0.1 g/L trolox, especially for NS-8GE and SS-8GE at high concentration tested (5.0 g/L). NS-8GE and SS-8GE possessed the higher antioxidative activity in lecithin liposome system than their substrates (NS and SS). This was in accordance with higher antioxidative activities of hydrolysates, especially ABTS radical scavenging activity and FRAP, as shown in Fig. 2a, c, respectively. NS-8GE and SS-8GE showed both radical scavenging activity and reducing antioxidant power. Gelatin hydrolysates from seabass and bigeye snapper skins displayed antioxidative activity in lecithin liposome systems (Senphan and Benjakul 2014, Phanturat et al. 2010).

Changes in TBARS of liposome system during incubation of 48 h are depicted in Fig. 3b. The sharp increase in TBARS was observed after 18 h of incubation in the control and systems added with NS and SS at all levels ($P < 0.05$), indicating the formation of the secondary lipid oxidation products. The systems containing gelatin hydrolysate at 5.0 g/L had lower TBARS than those added with hydrolysate at a lower level ($P < 0.05$). There were no marked changes in TBARS of lecithin liposome system added with 1.0 g/L trolox. Slight increases in TBARS were observed in the system containing 5.0 g/L NS-8GE and SS-8GE throughout 48 h of incubation. The result suggested that NS-8GE and SS-8GE, especially at the high concentration (5.0 g/L) could retard the formation of the secondary oxidation products effectively. After 36 h of incubation time, the decreased TBARS was noticeable, suggesting the loss of oxidation products. The secondary products with

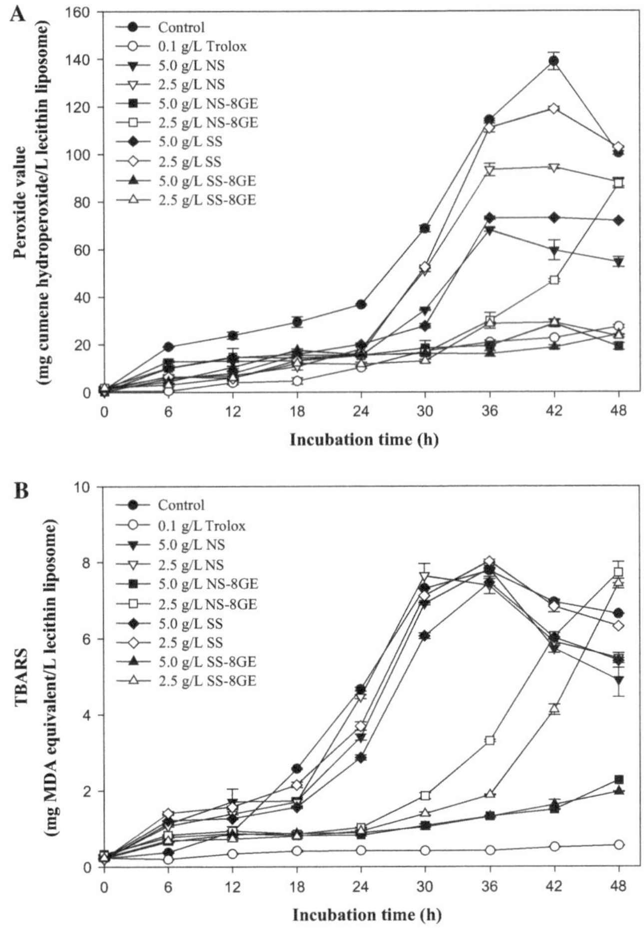

Fig. 3 The formation of peroxide value (**a**) and TBARS (**b**) in lecithin liposome system containing gelatin hydrolysates from unicorn leatherjacket skin. NS autolysed non-swollen skin, NS-8GE autolysed non-swollen skin hydrolysed using 8 % partially purified glycyl endopeptidase, SS autolysed swollen skin. SS-8GE autolysed swollen skin hydrolysed using 8 % partially purified glycyl endopeptidase. *Bars* represent standard deviation ($n = 3$)

low molecular weight could be lost easily (Stahnke 1995). Thus, NS-8GE and SS-8GE were able to retard the early stages as well as the advanced stage of oxidation in lecithin liposome system.

Changes in antioxidative activities of gelatin hydrolysate in gastrointestinal tract model system

The gastrointestinal tract model system (GIMs) was used to mimic the digestive system containing several proteases. The changes in antioxidative activity of NS-8GE and SS-8GE, as monitored by ABTS radical scavenging activity and FRAP, throughout the system are presented in Fig. 4. ABTS radical scavenging activity decreased after pepsin digestion ($P < 0.05$) but increased under duodenal condition during 1 and 6 h. After digestion in duodenum, proteases in pancreatin juice might cleave the peptides and release some new peptides with higher antioxidative activities, especially radical scavenging activity as evidenced by the increased ABTS radical scavenging activity. It was found that ABTS radical scavenging activity was higher for SS-8GE than NS-8GE throughout ingestion system. The results suggested that NS-8GE and SS-8GE had different peptides, which could be further hydrolysed at varying degrees. As a result, the generated peptides more likely had different chain length as well as amino acid sequence. This directly determined ABTS radical scavenging activity. Kittiphattanabawon et al. (2012) also found the increased antioxidative activity of gelatin hydrolysate from blacktip shark skin prepared using papaya latex after being ingested in the simulated model system. The gelatin hydrolysate from seabass skin prepared using protease from hepatopancreas of Pacific white shrimp showed the increase of antioxidative activity with further hydrolysis in intestinal simulated system (Senphan and Benjakul 2014). On the other hand, FRAP of both NS-8GE and SS-8GE decreased after ingestion in the GIMs over digestion time of 0–4 h. The extensive hydrolysis during digestion might provide small peptides with low ability on donation of electron to free radicals. The result suggested that radical scavenging activity of NS-8GE and SS-8GE could be increased in the gastrointestinal tract.

Fig. 4 ABTS radical scavenging activity (**a**) and ferric reducing antioxidant power (**b**) of gelatin hydrolysates from unicorn leatherjacket skin in gastrointestinal tract model system. NS-8GE autolysed non-swollen skin hydrolysed using 8 % partially purified glycyl endopeptidase; SS-8GE autolysed swollen skin hydrolysed using 8 % partially purified glycyl endopeptidase. *Bars* represent standard deviation ($n = 3$)

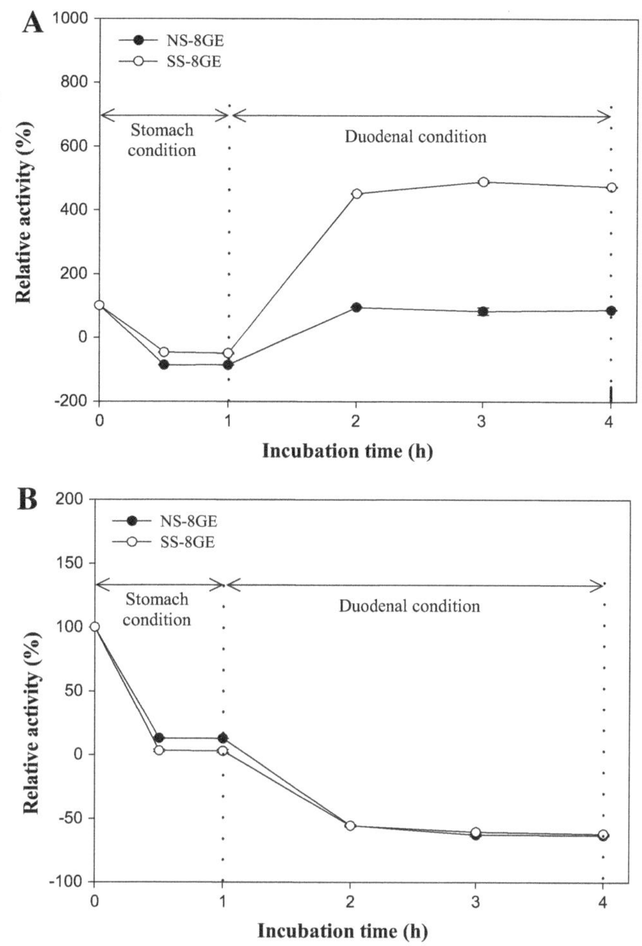

Molecular weight distribution

Elution profiles of NS-8GE and SS-8GE on the Sephadex™-G15 gel filtration chromatography are shown in Figs. 5 and 6, respectively. Based on A_{280}, gelatin hydrolysates showed three major peaks, indicating the presence of peptides containing aromatic amino acid with varying MW (Figs. 5a and 6a). Different elution profiles between NS-8GE and SS-8GE were generally obtained. SS-8GE showed the higher peak of peptides containing aromatic amino acids with low MW than NS-8GE. Based on A_{220} representing peptides, it was noted that NS-8GE showed higher content of larger peptides, in comparison with SS-8GE. The result indicated that prior autolysis and swelling process used for preparing substrate might enhance hydrolysis by GE. Moreover, this result was in accordance with free α-amino group content of SS-8GE which was higher than that of NS-8GE as shown in Fig. 1. ABTS radical scavenging activity and FRAP of fractions of NS-8GE and SS-8GE are depicted in Figs. 5b and 6b, respectively. NS-8GE fraction with MW of 1,170 Da exhibited the highest ABTS radical scavenging activity, followed by the peptide with MW of 3,380 Da (Fig. 5b). Additionally, NS-8GE contained peptides (2,150 Da, 1,010 Da and 220 Da) which possessed the ability for donating electrons to free radicals. For SS-8GE, the fraction with MW of 750 exhibited the highest FRAP, followed by that with MW of 1,590 Da. ABTS radical scavenging activity was also found in fractions with MW of 1,850 and 1,010 Da. Both gelatin hydrolysates showed ABTS radical scavenging activity and FRAP, in which NS-8GE and SS-8GE possessed the ability to retard the oxidation mainly via their radical scavenging and electron donating activities, respectively. The small peptides generally possess high antioxidative activity (Kittiphattanabawon et al. 2012; Phanturat et al. 2010). However, the level and amino acid composition of peptide also affected the antioxidative activity (Wu et al. 2003). Gelatin hydrolysate from skin of unicorn leatherjacket prepared using autolysis-assisted process with MW of 2,200 Da exhibited higher antioxidative activity than those with MW of 870 Da (Karnjanapratum and Benjakul 2014a). Yarnpakdee et al. (2014)

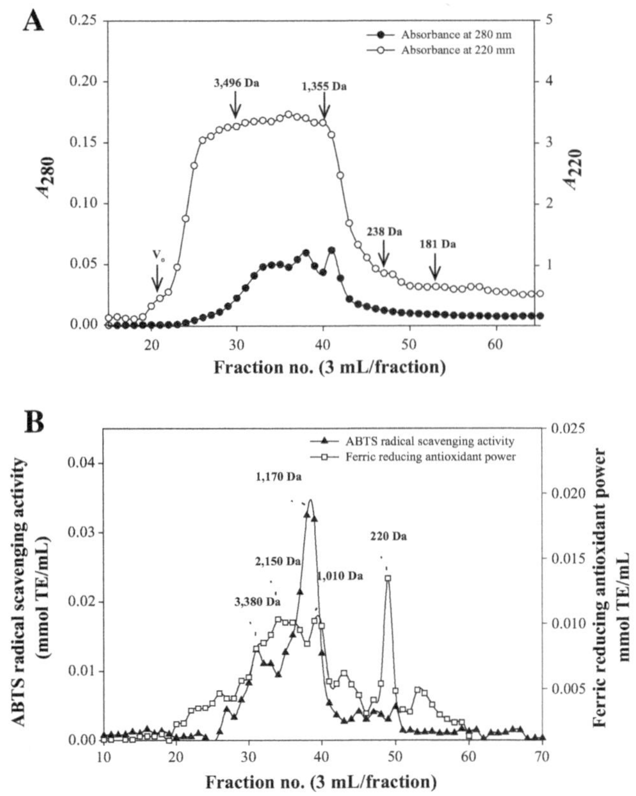

Fig. 5 Elution profile of gelatin hydrolysate (NS-8GE) from unicorn leatherjacket skin subjected to Sephadex G-25 size exclusion chromatography. Absorbance at 280 nm (*filled circle*), 220 nm (*opened circle*) (**a**) ABTS radical scavenging activity (*opened triangle*) and ferric reducing antioxidant power (*opened box*) (**b**) molecular weight of markers (*right arrow*) molecular weight of selected fractions (*dashed right arrow*)

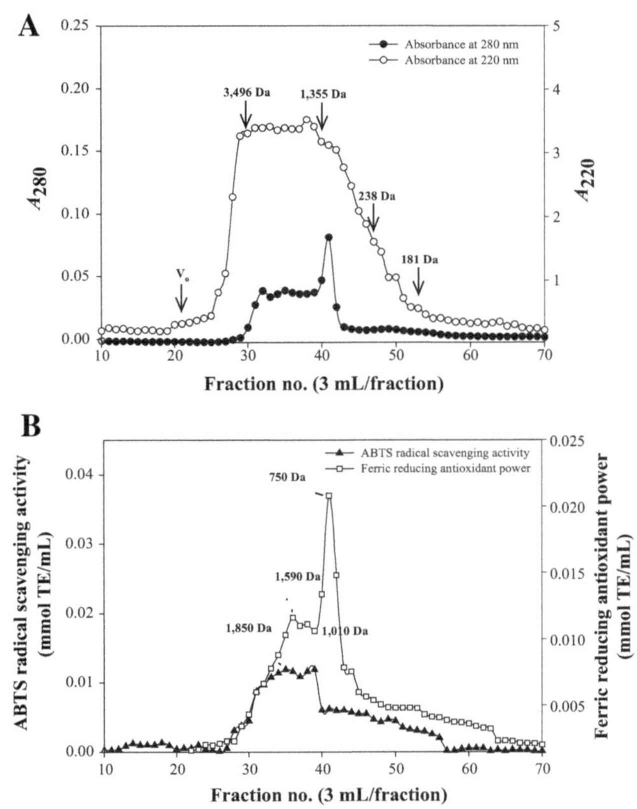

Fig. 6 Elution profile of gelatin hydrolysate (SS-8GE) from unicorn leatherjacket skin subjected to Sephadex G-25 size exclusion chromatography. Absorbance at 280 nm (*filled circle*), 220 nm (*opened circle*) (**a**) ABTS radical scavenging activity (*opened triangle*) and ferric reducing antioxidant power (*opened box*) (**b**) molecular weight of markers (*right arrow*) molecular weight of selected fractions (*dashed right arrow*)

found that Nile tilapia protein hydrolysates contained antioxidative peptides with different modes of action. Fraction containing peptides with MW of 513 Da and 1,484 Da showed the strongest ABTS radical scavenging activity and metal chelating activity, respectively (Yarnpakdee et al. 2014). The result suggested that NS-8GE and SS-8GE contained peptides with different sizes and varying modes of action.

Conclusion

Antioxidative gelatin hydrolysate could be produced from both autolysed non-swollen and swollen skin of unicorn leatherjacket skin with 8 % partially purified glycyl endopeptidase from papaya latex. Gelatin hydrolysates possessed antioxidative activities and showed the effectiveness in retarding lipid oxidation. However, modes of action of both hydrolysates were different, more likely governed by swelling process. Peptide having MW of 1,170 Da with the dominant radical scavenging activity was found in NS-8GE, whereas that with MW of 750 Da having the profound FRAP was observed in SS-8GE. Therefore, autolysis mediated by indigenous protease in the skin could be applied for production of gelatin hydrolysate with antioxidative activity.

Acknowledgments The authors would like to express their sincere thanks to Thailand Research Fund under the Royal Golden Jubilee Ph.D. Program to Supatra Karnjanapratum (PHD/0019/2554) and the Grant-in-Aid for dissertation from Graduate School, Prince of Songkla University, Thailand for financial support. The TRF Distinguished Research Professor Grant was also acknowledged.

Conflict of interest The authors declare that they have no competing interests.

Authors' contributions SB developed the initial idea and designed the study. SK was responsible for conducting experiments and analysis of data. SB and SK carried out the analytical work. SK wrote the manuscript with assistance from SB. SB read and approved the final manuscript.

References

Benjakul S, Morrissey MT (1997) Protein hydrolysates from Pacific whiting solid wastes. J Agric Food Chem 45:3423–3430

Benzie IFF, Strain JJ (1996) The ferric reducing ability of plasma (FRAP) as a measure of ''antioxidant power'': the FRAP assay. Anal Biochem 239:70–76

Binsan W, Benjakul S, Visessanguan W, Roytrakul S, Tanaka M, Kishimura H (2008) Antioxidative activity of Mungoong, an extract paste, from the cephalothorax of white shrimp (*Litopenaeus vannamei*). Food Chem 106:185–193

Buege JA, Aust SD (1978) Microsomal lipid peroxidation. Meth enzymol 52:302–310

Chen HM, Muramoto K, Yamauchi F (1995) Structural analysis of antioxidative peptides from soybean β-conglycinin. J Agric Food Chem 43:574–578

Choe E, Min DB (2005) Chemistry and reactions of reactive oxygen species in foods. J Food Sci 70:R142–R159

Enari H, Takahashi Y, Kawarasaki M, Tada M, Tatsuta K (2008) Identification of angiotensin I-converting enzyme inhibitory peptides derived from salmon muscle and their antihypertensive effect. Fisheries Sci 74:911–920

Frankel E, Huang SW, Aeschbach R (1997) Antioxidant activity of green teas in different lipid systems. J Am Oil Chem Soc 74:1309–1315

Gómez-Guillén MC, Giménez B, López-Caballero ME, Montero MP (2011) Functional and bioactive properties of collagen and gelatin from alternative sources: a review. Food Hydrocolloid 25:813–1827

Hernandez-Ledesma B, Davalos A, Bartolome B, Amigo L (2005) Preparation of antioxidant enzymatic hydrolysates from α-lactalbumin and β-lactoglobulin. Identification of active peptides by HPLC-MS/MS. J Agric Food Chem 53:588–593

Kaewruang P, Benjakul S, Prodpran T (2013) Molecular and functional properties of gelatin from the skin of unicorn leatherjacket as affected by extracting temperatures. Food Chem 138:1431–1437

Karnjanapratum S, Benjakul S (2014a) Characteristics and antioxidative activity of gelatin hydrolysates from unicorn leatherjacket skin as affected by autolysis-assisted process. J Food Process Preserv (in press)

Karnjanapratum S, Benjakul S (2014b) Glycyl endopeptidase from papaya latex: partial purification and the use for production of fish gelatin hydrolysate. Food Chem (in press)

Kim SK, Wijesekara I (2010) Development and biological activities of marine-derived bioactive peptides: a review. J Funct Foods 2:1–9

Kim SK, Kim YT, Byun HG, Nam KS, Joo DS, Shahidi F (2001) Isolation and characterization of antioxidative peptides from gelatin hydrolysate of Alaska pollack skin. J Agric Food Chem 49:1984–1989

Kittiphattanabawon P, Benjakul S, Visessanguan W, Shahidi F (2012) Gelatin hydrolysate from blacktip shark skin prepared using papaya latex enzyme: antioxidant activity and its potential in model systems. Food Chem 135:1118–1126

Lobo V, Patil A, Phatak A, Chandra N (2010) Free radicals, antioxidants and functional foods: impact on human health. Pharmacogn Rev 4:118–126

Min B, Ahn DU (2005) Mechanism of lipid peroxidation in meat and meat products-a review. Food Sci Biotechnol 14:152–163

Phanturat P, Benjakul S, Visessanguan W, Roytrakul S (2010) Use of pyloric caeca extract from bigeye snapper (*Priacanthus macracanthus*) for the production of gelatin hydrolysate with antioxidative activity. LWT: food. Sci Technol 43:86–97

Re R, Pellegrini N, Proteggente A, Pannala A, Yang M, Rice-Evans C (1999) Antioxidant activity applying an improved ABTS radical cation decolorization assay. Free Radic Biol Med 26:1231–1237

Senphan T, Benjakul S (2014) Antioxidative activities of hydrolysates from seabass skin prepared using protease from hepatopancreas of Pacific white shrimp. J Funct Foods 6:147–156

Stahnke L (1995) Dried sausages fermented with Staphylococcus xylosus at different temperatures and with different ingredient levels-Part II. Volatile components. Meat Sci 41:193–209

Steel RGD, Torrie JH (1980) Principles and procedures of statistics: a biometrical approach, 2nd edn. McGraw-Hill, New York

Thiansilakul Y, Benjakul S, Shahidi F (2007) Antioxidative activity of protein hydrolysate from round scad muscle using alcalase and flavourzyme. J Food Biochem 31:266–287

Wettasinghe M, Shahidi F (2000) Scavenging of reactive-oxygen species and DPPH free radicals by extracts of borage and evening primrose meals. Food Chem 70:17–26

Wu HC, Chen HM, Shiau CY (2003) Free amino acids and peptides as related to antioxidant properties in protein hydrolysates of mackerel (*Scomber austriasicus*). Food Res Int 36:949–957

Xu Z (2008) Antioxidants in grains, vegetables and fruits. Food Sci Tech Bull Funct Foods 5:61–70

Yarnpakdee S, Benjakul S, Kristinsson HG, Kishimura H (2014) Antioxidant and sensory properties of protein hydrolysate derived from Nile tilapia (*Oreochromis niloticus*) by one- and two-step hydrolysis. J Food Sci Technol (in press)

17

Preliminary assessment of the Gut weed *Ulva intestinalis* as food for herbivorous fish

Muhammad Abu Bakar Siddik · Nguyen Thi Ngoc Anh

Abstract A set of feeding trials was carried out to assess the potential use of gut weed (GW) *Ulva intestinalis* as a direct food to replace commercial feed (CF) in an alternative approach for feeding Spotted scat (*Scatophagus argus*), Red tilapia (*Oreochromis niloticus*), and Giant gourami (*Osphronemus goramy*) juveniles for 60, 45, and 56 days, respectively. Four feeding regimes were assigned to triplicate tanks. The different groups were fed CF every day, GW every day, and one day of CF followed by one (1CF_1GW) or two (1CF_2GW) days of GW, respectively. The results indicated that growth performance of the *S. argus* fed on single GW was not significantly different from the control group but the performance of *O. niloticus* and *O. goramy* in the alternative feeding treatments were comparable to the control. The feed conversion ratio could be reduced by 26.1–57.8 % in the combined feeding regimes to control will minimize the production cost of fish. Water quality in terms of NO2 and NH4/NH3 in the rearing tanks was better in the combined feeding than in the single CF. These results indicated that GW can be used as a food to partially substitute CF for herbivorous fish.

Keywords Gut weed · Macroalgae · Water quality · Herbivorous · FCR · *Ulva intestinalis*

Background

Gut weed (GW) *Ulva intestinalis* (Linnaeus) a macroalga distributed in a wide variety of environments from freshwater to seawater including ocean coast, brackish waters, and inland freshwater all over the world (Ergun et al. 2009; Siddik et al. 2014a). It has high nutritional values such as 9–14 % protein, 2.0–3.6 % lipids, and 32–36 % ash on a dry weight basis. In addition, the n−3 and n−6 fatty acids pose 10.4 and 10.9/100 g of the total fatty acids. It is also rich in essential amino acids, pigments, and minerals (Haroon et al. 2000; Aguilera-Morales et al. 2005; Shanmugan and Palpandi 2010). It can therefore be used as an ingredient in diets or as direct feed for fish/shrimp and for integrated aquaculture system (FAO 2011). Macroalgae have been reported in several studies as a booster for keeping water quality better in intensive culture of fish (Valente et al. 2006;

M. A. B. Siddik (✉)
Sado Marine Biological Station, Faculty of Science, Niigata University, Niigata, Japan
e-mail: siddik@pstu.ac.bd

M. A. B. Siddik
Department of Fisheries Biology and Genetics, Patuakhali Science and Technology University, Patuakhali, Bangladesh

N. T. N. Anh
College of Aquaculture and Fisheries, Can Tho University, Can Tho, Vietnam

Ergun et al. 2009; Siddik et al. 2014a). Since tropical fish are generally sensitive to poor water quality and a large number of fish is kept in a small tank during the culture period, the water quality should be given prime importance for keeping fish in tranquil. Using GW as food for fish can play a dual role to maintain good condition of water for better production in one hand and on the other hand, its incorporation may reduce feed cost for the fish production (Anh et al. 2012).

The long-term sustainability of aquaculture may be threatened by its present over-dependence on fish meal and fish oil (FAO 2011). Moreover, fish feeding represents over 50 % of operating costs in intensive aquaculture, with protein being the most expensive dietary source (Aguilera-Morales et al. 2005; Ergun et al. 2009). An intensive effort during the last decades has been made in order to evaluate the potential of alternative protein sources in aquaculture (El-Sayed 1999; Rad et al. 2013). Several studies have been conducted to evaluate the replacement of fish meal by plant ingredients in diets of European sea bass and Tilapia (Valente et al. 2006; Pereira et al. 2012; Kaushik et al. 2004). However, research data about the potential use of macroalgae in herbivorous fish diets is scarce (El-Sayed, 1999; Pereira et al. 2012). Therefore, the aim of this study was to assess the effect of commercial feed (CF) replacement with GW as direct feed on water quality, growth, and feeding efficiency of the herbivorous fishes like Spotted scat (*Scatophagus argus*), Red tilapia (*Oreochromis niloticus*), and Giant gourami (*Osphronemus goramy*). This work could encourage farmers using local availability of GW as food source for fish, and it could contribute to reduce feed costs and provide more profits for farmers.

Materials and methods

Experimental design

Three separate experiments were performed in the experimental hatchery complex of the College of Aquaculture and Fisheries, Can Tho University, Vietnam in the period from January to September, 2012. The fresh and dried GWs were used as a direct feed to replace CF in alternative regimes for feeding *S. argus*, *O. niloticus,* and *O. goramy* juveniles for 60, 45, and 56 days, respectively.

Four feeding regimes were assigned to triplicate tanks. The different groups were fed CF every day, GW every day (GW), and one day of CF followed by one (1CF_1GW) or two (1CF_2GW) days of GW, respectively.

Experiential diets

Fresh GW *U. intestinalis* was collected from the extensive farm of Bac Lieu province, Vietnam and cleaned with 5 g/L saline water to maintain the quality. Dried GW was obtained by air drying a thin layer of biomass until reaching a moisture content of 12–14 %. Both fresh and dried GWs were cut with scissors into small pieces (±2 mm) for feeding the fish. The CF (GROBEST-GB640) used in this study had the crude protein ≥30 %, lipid ≥6 %, ash ≥14 %, fiber ≥6 %, and moisture ≥11 % (information from manufacturer). Proximate composition (% of dry matter) of GW was analyzed by the standard methods of AOAC (2005), giving a protein content of 12.7 ± 2.6 %, a lipid content of 2.4 ± 0.2 %, an ash content of 32.4 ± 7.6, a fiber content of 4.5 ± 1.1 %, and a carbohydrate content of 47.9 ± 6.7 %.

Culture system

The 100-L plastic tanks were filled with 80 L water with salinity of 5 g/L for *S. argus* and with freshwater for *O. niloticus* and *O. goramy*. 30 fish were introduced in each tank, provided with continuous aeration and were fed to satiation twice a day at 8:00 a.m. and 4:00 pm. Uneaten feed was collected after 1.5 h of feeding. The water exchange rate was approximately 50 % of the tank volume every third day.

Data collection

Water temperature, dissolved oxygen, and pH were recorded twice a day at 7:00 and 14:00 h. Temperature and pH were determined using a thermo pH meter (YSI 60 Model, HANNA instruments, Mauritius). The dissolved oxygen (DO) was determined using a DO meter (YSI, Model-58, USA), and the concentrations of NO2 and NH4/NH3 (TAN) were determined weekly using DREL/2 HACH kits (HACH Co., Loveland, CO, USA). For the estimation of growth performance of fishes, weights were recorded and average growth calculated throughout the experimental period. Samples for this purpose were collected at the beginning and end of the study, along with every 14th or 15th day of the experimental period. Specific growth rate (SGR), feed conversion ratio (FCR), and survival rate were calculated using the following equations:

SGR (%/day) = (Final weight−Initial weight)/culture period × 100.

Survival (%) = Final number of fish/Initial number of fish × 100.

FCR = Feed intake (dry weight)/Weight gain (wet weight).

Statistical analysis

Statistical analysis of all measured variables was performed using SPSS for windows, Version 14.0. Variations of dietary treatments were compared by one-way ANOVA. The Tukey HSD post hoc analysis was used to detect differences between means.

Results and discussion

Water quality parameters

Daily water temperature, dissolved oxygen, and pH over three experiments fluctuated in the ranges of 26.5–30.2 °C, 4.90–5.66 mg/L, and 7.1–8.3, respectively. These factors were generally within the suitable range for tropical fish growth (Boyd 2003; Siddik et al. 2014b; Ahsan et al. 2013). However, the mean concentrations of dissolved inorganic nitrogen such as NO2 and TAN ranged from 0.04 to 1.85 mg/L and 0.02–0.97 mg/L, respectively. These values were highest at the end of the experiment and higher in the control treatment (commercial feed) compared to other treatments (Fig. 1).

The results showed that feeding treatments had significant effects on water quality. The mean concentration of NO2 and TAN tended to increase during the culture period in all groups. The highest concentrations of NO2 (1.5 mg/L) and TAN (0.90 mg/L) were observed in the control group, followed by the 1CF_1GW and the 1CF_2GW groups. The lowest values (NO2 of 0.33 mg/L and TAN of 0.28 mg/L, respectively) were found in the group fed only GW. This indicates that the water quality is much better in the rearing tanks with GW than in the tanks with only CF. Hence, the concentration of N-nutrients in the culture tanks was increased with regard to the amount of pellet feeds used, which could be reduced by water exchange. For aquaculture production, suitable level of NO2 and TAN for aquatic animals should be less than 0.5 and 0.2–2 mg/L, respectively (Rahman et al. 2008; Zhao et al. 2014). However, in the present study, because of the relatively high rate of water exchange (about 50 % of volume for every 3 days interval), these parameters might not be harmful to the experimental fish.

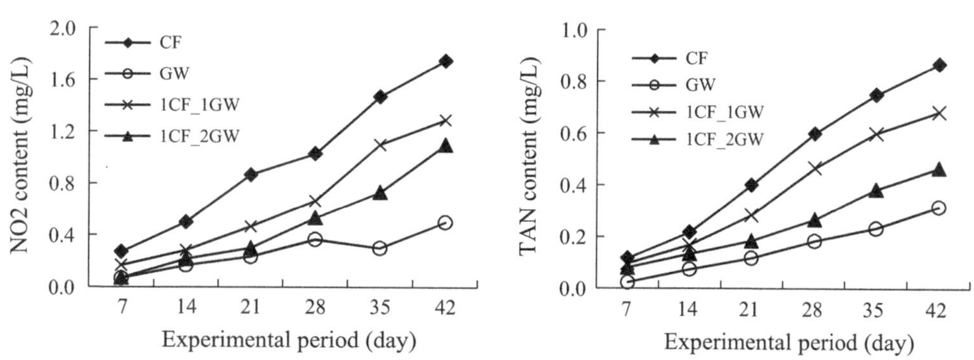

Fig. 1 Variation in the mean concentrations of NO2 and TAN during experimental period

Survival rate

The survival rates of the experimental fishes were not significantly different among treatments, varying in the ranges of 88.3–93.3, 82.7–85.3, and 93.3–100 % for Spotted scat, Red tilapia, and Giant gourami, respectively (Table 1). Similar findings have been reported where utilization of seaweeds as fish feed did not affect survival rate of experimental fishes (Swain and Padhi 2011; Guroy et al. 2011).

Growth performance

There was high variability in the final weight and SGR of experimental fish in the different treatments (Table 1).

For the Spotted scat, alternate feeding 1CF_1GW and 1CF-2GW resulted in significantly higher growth compared to the control (commercial feed) and the group fed solely GW. However, fish fed on only GW had similar growth to the control group. This could be related to fish feeding habit of the species. Gandhi (2002) studied on the food and feeding habits of Spotted scat, and indicated that GW dominated the gut contents in fishes of size above 50–100 mm in total length. On the other hand, better water quality in the group received solely GW may also support the fish growth (Siddik et al. 2014a). For Red tilapia, growth of fish receiving alternate feeding regimes of CF and GW (1CF_1GW and 1CF_2GW) was also comparable to the control group, while the lowest growth rates were observed in the group fed single GW which was significantly different from other treatments.

Growth performances of Giant gourami followed similar pattern as observed for the red tilapia. Growth rate of the fish fed 1CF_1GW was equivalent to those receiving solely CF. When fish received 1CF_2GW, their growth was significantly lower than the control and the 1CF_1GW treatments. Fish fed only on GW showed the lowest growth compared to other treatments. In Vietnam, herbivorous fish are preferred in household-scale

Table 1 Growth, survival rate, and food conversion ratio (dry weight) of three herbivorous fishes in four different feeding regimes (mean values ± SD)

Parameters	Final weight (g)	SGR (% day^{-1})	Survival (%)	FCR		CF reduced to control (%)
				Commercial feed	Gut weed	
Spotted scat (*S. argus*) reared for 60 days						
CF (control)	9.66 ± 2.27^{c}	2.66 ± 0.42^{b}	93.3 ± 6.7	2.07 ± 0.17	–	–
1CF-1GW	8.13 ± 2.06^{a}	2.36 ± 0.46^{a}	86.7 ± 3.3	1.53 ± 0.04	0.97 ± 0.06	−26.1
1CF-2GW	8.26 ± 1.89^{ab}	2.40 ± 0.38^{a}	88.9 ± 1.9	1.18 ± 0.14	1.46 ± 0.04	−43.0
GW	9.03 ± 1.35^{bc}	2.58 ± 0.26^{b}	88.3 ± 7.1	–	2.82 ± 0.03	−100
Red tilapia (*O. niloticus*) reared for 45 days						
CF (control)	10.76 ± 2.33^{b}	2.31 ± 0.51b	84.0 ± 0.0	1.56 ± 0.12	–	–
1CF-1GW	10.63 ± 2.59^{b}	2.28 ± 0.51^{b}	82.7 ± 2.3	1.05 ± 0.17	0.27 ± 0.28	−32.7
1CF-2GW	10.25 ± 2.13^{b}	2.20 ± 0.48^{b}	85.3 ± 2.3	0.87 ± 0.09	0.41 ± 0.05	−44.2
GW	7.10 ± 1.29^{a}	1.40 ± 0.40^{a}	84.0 ± 4.0	–	2.47 ± 0.35	−100
Giant gourami (*O. goramy*) reared for 56 days						
CF (control)	6.10 ± 0.88^{c}	1.74 ± 0.26^{c}	98.3 ± 2.9	1.99 ± 0.27	–	–
1CF-1GW	5.97 ± 0.83^{c}	1.71 ± 0.26^{c}	100 ± 0.0	1.03 ± 0.07	0.95 ± 0.06	−48.2
1CF-2GW	5.28 ± 0.86^{b}	1.48 ± 0.28^{b}	93.3 ± 2.9	0.84 ± 0.06	1.41 ± 0.04	−57.8
GW	4.60 ± 0.70^{a}	1.24 ± 0.27^{a}	95.0 ± 5.0	–	3.29 ± 0.12	−100

Values are mean ± standard deviation. Mean values in each column bearing different superscripts are significantly different ($P < 0.05$). Average initial weights of experimental fishes were 1.90 ± 0.19, 3.72 ± 0.34, and 2.27 ± 0.22 g for the Spotted scat, Red tilapia, and Giant gourami, respectively

CF commercial feed, *FGW* fresh gut weed, *DGW* dried gut weed culture systems that primarily produce fish for home or local consumption (Anh et al. 2012; Siddik et al. 2014a). In particular Spotted scat, Red tilapia, Giant gourami, and other species are fed on aquatic plants, macroalgae, and grass, which are abundant in the Mekong delta. However, a combination of natural food (macroalgae, aquatic plants) with pellet feed can reduce the production cycle (Ergun et al. 2009; Guroy et al. 2013).

Feed conversion ratio

Since fresh GW is not available all the year round, dried GW was obtained by air drying of biomass till reaching a moisture content of 12–14 % for using as fish diet. In the present study, the highest FCR was observed in the fish fed on single CF in three fish species. With respect to the use of GW as single feed, FCRs of GW were much higher (2.47–3.29) than using solely CF (1.56–2.07). This could be due to different proximate composition between CF and GW as CF contains ≥30 % crude protein and ≥6 % lipid, while GW had average crude protein and lipid of 12.7 and 2.4 %, respectively. However, when combined feeding of CF and fresh or dried GW were applied, the FCR of CF was considerably reduced in all species. The reduction was from 26.1 to 43.0 % in Spotted scat, from 32.7 to 44.2 % in Red tilapia, and from 48.2 to 57.8 % in Giant gourami, respectively. This approach could help farmers reduce the feed cost and enhance profit.

Conclusions

The findings of this study confirmed that GW may be combined with CF in order to reduce the feed costs and improve the water quality in the culture system. This result also may encourage feed manufacturers and fish farmers generating low-cost aquafeed production.

Acknowledgments This study was funded by the Belgian Government (VLIR-UOS) scholarship. The authors would like to express their sincere thanks to College of Aquaculture and Fisheries, Can Tho University (CTU), Vietnam for their logistics and manpower support.

Conflict of interest The authors declare that they have no competing interests.

Authors' contributions MABS was responsible for conducting experiments, analysis of data, and wrote the manuscript. NTNA developed the initial idea, designed the study, and finalized the manuscript. Both authors read and approved the final manuscript.

References

Aguilera-Morales M, Casas-Valdez M, Carrillo-Dominguez S, Gonzalez-Acosta B, Perez-Gil F (2005) Chemical composition and microbiological assays of marine algae *Enteromorpha spp.* as a potential food source. J Food Compos Anal 18:79–88

Ahsan ME, Wahab MA, Siddik MAB, Alam MA, Sharker MR, Nahar A (2013) Impacts of inclusion of column feeder rohu (*Labeo rohita*) at different stocking densities on growth, production and environment in freshwater prawn-carp-mola polyculture system. Int J Biol Res 1(2):48–54

Anh NTN, Tien NM, Hai TN (2012) Proceedings of the 12th International Fisheries on sharing knowledge for sustainable aquaculture and fisheries in the South-East Asia, Can Tho University, Vietnam, 6–8th December 2012

AOAC (2005) Official methods of analysis of chemist, 18th edn. AOAC International, Maryland

Boyd CE (2003) Guidelines for aquaculture effluent management at the farm-level. Aquaculture 226:101–112

El-Sayed AFM (1999) Alternative dietary protein sources for farmed tilapia, *Oreochromis* spp. Aquaculture 179:149–168

Ergun S, Soyuturk M, Guroy B, Guroy D, Merrifield D (2009) Influence of Ulva meal on growth, feed utilization, and body composition of juvenile Nile tilapia (*Oreochromis niloticus*) at two levels of dietary lipid. Aquacult Int 17:355–361

FAO (2011) Demand and supply of feed ingredients for farmed fish and crustaceans. FAO Fisheries and Aquaculture Technical paper No. 564, ISBN 2070-7010, Rome, Italy

Gandhi V (2002) Studies on the food and feeding habits of cultivable butterfish *Scatophagus argus*). J Mar Biol Ass India 44:115–121

Guroy D, Guroy B, Merrifield DL, Ergun S, Tekinay AA, Yigit M (2011) Effect of dietary Ulva and Spirulina on weight loss and body composition of rainbow trout, *Oncorhynchus mykiss* (Walbaum), during a starvation period. J Anim Physiol Anim Nutr 95:320–327

Guroy B, Ergun S, Daniel LM, Guroy D (2013) Effect of autoclaved Ulva meal on growth performance, nutrient utilization and fatty acid profile of rainbow trout, *Oncorhynchus mykiss*. Aquacult Int 21:605–615

Haroon AM, Szaniawska A, Normant M, Janas U (2000) The biochemical composition of *Enteromorpha spp.* from the Gulf of Gdansk coast on the Southern Baltic Sea. Oceanologia 42:19–28

Kaushik SJ, Covès D, Dutto G, Blanc D (2004) Almost total replacement of fish meal by plant protein sources in the diet of a marine teleost, the European seabass, *Dicentrarchus labrax*. Aquaculture 230:391–404

Pereira R, Valente LMP, Sousa-Pinto I, Rema P (2012) Apparent nutrient digestibility of seaweeds by rainbow trout (*Oncorhynchus mykiss*) and Nile tilapia (*Oreochromis niloticus*). Algal Res 1:77–82

Rad A, Zakeri MM, Yavari V, Mosavi SM (2013) Effect of different levels of dietary supplementation of *Saccharomyces cerevisiae* on growth performance, feed utilization and body biochemical composition of Nile Tilapia (*Oreochromis niloticus*) fingerlings. World J Fish Mar Sci 5(1):88–95

Rahman MM, Jo Q, Gong YG, Miller SA, Hossain MY (2008) A comparative study of common carp (*Cyprinus carpio* L.) and calbasu (*Labeo calbasu* Hamilton) on bottom soil resuspension, water quality, nutrient accumulations, food intake and growth of fish in simulated rohu (*Labeo rohita* Hamilton) ponds. Aquaculture 285:78–83

Shanmugan A, Palpandi C (2010) Biochemical composition and fatty acid profile of the green alga *Ulva reticulata*. Asian J Biochem 5(3):188–193

Siddik MAB, Nahar A, Rahman NM, Anh NTN, Nevejan N, Bossier P (2014a) Gut Weed, *Enteromorpha* sp. as a partial replacement for commercial feed in Nile Tilapia (*Oreochromis niloticus*) culture. World J Fish Mar Sci 6(3):267–274

Siddik MAB, Nahar A, Ahamed F, Hossain MY (2014b) Over-wintering growth performance of mixed-sex and mono-sex Nile tilapia *Oreochromis niloticus* in northeastern Bangladesh. Croat J Fish 72:70–76

Swain PK, Padhi SB (2011) Utilization of seaweeds as fish feed in aquaculture. J Biol Sci Biohelica 2:35–46

Valente LMP, Gouveia A, Rema P, Matos J, Gomes EF, Pinto IS (2006) Evaluation of three seaweeds *Gracilaria bursa, Ulva rigida* and *Gracilaria cornea* as dietary ingredients in European sea bass (*Dicentrarchus labrax*) juvenile. Aquaculture 252:85–91

Zhao Z, Xu Q, Luo L, Wang C, Li J, Wang L (2014) Effect of feed C/N ratio promoted bioflocs on water quality and production performance of bottom and filter feeder carp in minimum-water exchanged pond polyculture system. Aquaculture 434:442–448

18

Effect of initial stocking size of the predatory African sharptooth catfish (*Clarias gariepinus*) on recruits, growth performance, survival and yield of mixed-sex Nile tilapia (*Oreochromis niloticus*) in concrete tank culture system

Samwel Mchele Limbu · Amon Paul Shoko · Hieromin Amon Lamtane · Evantuce Dominick Shirima · Mary Alphonce Kishe-Machumu · Huruma Fredrick Mgana · Yunus Daud Mgaya

Abstract Prolific breeding and production of high percentage of recruits are main problems in mixed-sex Nile tilapia (*Oreochromis niloticus*) culture in earthen ponds. The current study assessed the efficiency of different sizes of African sharptooth catfish (*Clarias gariepinus*) in controlling recruitment in mixed-sex Nile tilapia cultured in tanks and its effect on growth performance, percentage survival and yield. Three treatments consisting of triplicate ponds stocked with mixed-sex Nile tilapia in monoculture and in polyculture with large (62.50 ± 3.26 g) and small (40.00 ± 2.68 g) African sharptooth catfish were fed on a 297.50 g kg^{-1} crude protein diet for 126 days. Results showed that, the number of recruits was significantly lower in larger African sharptooth catfish predator than smaller ones and monoculture of Nile tilapia. Polyculture with larger African sharptooth catfish resulted in significantly higher growth performance of Nile tilapia. Large African sharptooth catfish in polyculture reduced the amount of small, low-value recruits, while the yield of large and high value Nile tilapia was increased. This study revealed that fish farmers can reduce prolific breeding, obtain higher growth performance and produce larger size of marketable Nile tilapia by predominantly stocking ponds with large African sharptooth catfish predator of at least 60 g.

Keywords Predator · Recruitment · Polyculture · Monoculture · Prolific breeding · Production

Introduction

Nile tilapia (*Oreochromis niloticus*, Linnaeus) is one of the most cultured species in the world (FAO 2012) providing major sources of animal protein and income (Chakraborty et al. 2011; Yakubu et al. 2012) in the majority of tropical countries (Lowe et al. 2012). Production of Nile tilapia is carried out in ponds, cages, tanks and raceways through mixed-sex monoculture, all-male monoculture and polyculture with predators

S. M. Limbu · Y. D. Mgaya (✉)
Department of Aquatic Sciences and Fisheries, University of Dar es Salaam, P. O. Box 35064, Dar es Salaam, Tanzania
e-mail: yunusmgaya@gmail.com

A. P. Shoko · M. A. Kishe-Machumu
Tanzania Fisheries Research Institute, Headquarters, P. O. Box 9750, Dar es Salaam, Tanzania

H. A. Lamtane · E. D. Shirima
Department of Animal Science and Production, Sokoine University of Agriculture, P. O. Box 3004, Morogoro, Tanzania

H. F. Mgana
Tanzania Fisheries Research Institute, Kigoma Centre, P. O. Box 90, Kigoma, Tanzania

(Hernández et al. 2014). Its production in developing countries occurs primarily in semi-intensive ponds with fertilization and/or supplementary feeding. In Tanzania, more than 95 % of fish farmers culture Nile tilapia in earthen ponds under mixed-sex culture (Kaliba et al. 2006).

One of the main problems facing the culture of mixed-sex Nile tilapia in earthen ponds is early maturation leading to prolific breeding and production of high percentage of recruits (Offem et al. 2009; Chakraborty et al. 2011; Yakubu et al. 2012; Dagne et al. 2013). As a result, the total biomass at harvest comprises 28–70 % of small-size, low-value fish originating from spontaneous breeding (de Graaf et al. 1996; Ofor et al. 2011). The increased number of recruits in mixed-sex culture causes stunted growth due to shortage of space and competition for food (Yi et al. 2002; Abdel-Tawwab 2005; Offem et al. 2009).

Various methods for controlling recruitment in mixed-sex Nile tilapia have been proposed. The methods and their limitations have been reviewed by Mair and Little (1991). The methods include culture with predators, culture in cages, intermittent harvesting, as well as hybridization, induction of sterility and production of mono-sex progenies. However, most of these methods are not only expensive to the local fish farmers in Tanzania, but also increase the production cost and limit the expansion of the aquaculture industry (Carlberg et al. 2000). Using a predatory fish to control recruitment in mixed-sex Nile tilapia cultured in ponds is perceived as one of the most appropriate, safe and affordable method in developing countries (Abdel-Tawwab 2005; Offem et al. 2009). However, the efficiency of a predator fish is determined by its capacity to prey upon the recruits of Nile tilapia (de Graaf et al. 1996; Hernández et al. 2014).

African sharptooth catfish (*Clarias gariepinus*, Burchell) is one of the commonly used predators in controlling recruits in mixed-sex Nile tilapia culture (Lin and Diana 1995; de Graaf et al. 1996). Stocking African sharptooth catfish at a proper size reduces the number of recruits in mixed-sex culture of Nile tilapia, enhances growth performance and produces large marketable Nile tilapia sizes due to adequate food and space (Offem et al. 2009). It has been shown that, recruitment of Nile tilapia can be completely controlled by large (6.8–130 g) rather than small-size (<3.65 g) African sharptooth catfish (de Graaf et al. 1996). However, in certain circumstances, African sharptooth catfish has been criticized as a lazy predator with a low predatory efficiency in controlling recruits in mixed-sex Nile tilapia (de Graaf et al. 1996; Yi et al. 2002; Ofor et al. 2011). This is due to the limited availability of appropriate large-sized African sharptooth catfish with optimum predation efficiency (Abdel-Tawwab 2005; El-Naggar 2007; Offem et al. 2009). Low predatory efficiency in certain stocks of African sharptooth catfish has resulted in similar yields between polyculture with Nile tilapia and monoculture systems (Offem et al. 2009). Choosing an efficient African sharptooth catfish predator of a specific size capable of completely controlling recruits in mixed-sex Nile tilapia represents a limitation to the success of this technique (Fagbenro 2004). To limit the proliferation of small-size Nile tilapia, it is imperative to stock ponds with predators of an adequate size, as predation is size dependent (Shrestha et al. 2011).

The objective of this study was therefore to assess the efficiency of two sizes of African sharptooth catfish in controlling recruitment in mixed-sex Nile tilapia cultured in concrete tanks. The study also evaluated the growth performance, survival and production of Nile tilapia in monoculture and polyculture with African sharptooth catfish.

Materials and methods

Experimental design

The study was conducted at the Sokoine University of Agriculture (SUA) in Morogoro, Tanzania. Nine outdoor circular concrete tanks each having 7 m^2 and mean depth of 1 m were used with fortnightly complete renewal rate of water in each tank. The experiment consisted of three replicates of mixed-sex Nile tilapia in monoculture (as a control) and a polyculture system with African sharptooth catfish. Both the mixed-sex population of Nile tilapia brooders and African sharptooth catfish predators were collected from grow-out ponds at the site and acclimatized in separate tanks for 2 weeks prior to the start of the experiment. The sex ratio of Nile tilapia broodstock was determined by looking at genital papilla prior to stocking to produce homogenous sex ratios across treatments and replicates. This was necessary because the sex ratio has definite impact on the production of offspring (Mair and Van Dam 1996). A homogenous sex ratio of 3:1 equivalent to 12:4 females to males, respectively, was used in all treatments and replicates. In both systems, mixed-sex Nile tilapia brooders and African sharptooth catfish were stocked at a density of 3 fish m^{-2}. The stocking density of 3 fish m^{-2} used in this study was done intentionally to reflect the practice by rural fish farmers in Tanzania. In the monoculture system, 21 mixed-sex Nile tilapia brooders with an initial average weight (mean ± standard deviation) of 51.80 ± 2.30 g were stocked alone to act as a control of the experiment. In the polyculture system, a stocking ratio of about 3:1 for Nile tilapia and African sharptooth catfish, respectively, was used. Sixteen mixed-sex Nile tilapia brooders with an initial average weight (mean ± standard deviation) of 53.50 ± 4.12 g and five large (62.50 ± 3.26 g) African sharptooth catfish were stocked in the first polyculture tanks. A similar number of mixed-sex Nile tilapia brooders (16) with an initial average weight (mean ± standard deviation) of 52.30 ± 3.58 g and five small African sharptooth catfish (40.00 ± 2.68 g) were stocked in the second polyculture tank. The choice of the two sizes of African sharptooth catfish was based on ease of availability. The experiment lasted for 126 days.

Fish in both culture systems were fed on a 297.50 g kg^{-1} crude protein diet made using locally available cotton seed cake (683.40 g kg^{-1} of dry feed), maize bran (315.60 g kg^{-1}) and vitamin premix (1.00 g kg^{-1}) (Table 1) for the entire period of the study. A daily ration of 5 % average body weight of Nile tilapia in monoculture or of the combined body weights of Nile tilapia and African sharptooth catfish in the case of polyculture was adopted. The ration was divided into two portions and fed twice daily in the morning (between 0900 and 1000 hours) and evening (between 1500 and 1600 hours).

Data collection

Before the experiment, all African sharptooth catfish individuals were weighed to the nearest 0.01 g by using a sensitive weighing balance (Electronic Precision Balance Model EJB-KD-3000 g, Endel Global Weighing Company). Before weighing, the specimens were blot dried to ensure accuracy. Thereafter all fishes were weighed every 14 days from each tank. Feeding allowances were calculated on the basis of the weights of the African sharptooth catfish and large Nile tilapia. Feeding regimes were re-computed fortnightly after each sampling using the average weights.

At the end of the experiment, water was completely drained out of each tank and all fishes including the recruits were collected, sorted according to species and size (fingerlings and adults) and counted. Nile tilapia adults and their recruits were weighed individually to determine the final mean weight (FMW, g) and weight classes (g), respectively, at the time of harvesting. African sharptooth catfish, large Nile tilapia and their recruits were then weighed for determination of yield. The yield (g) and tank area (m^2) were extrapolated into kilogram (kg) and hectare (ha), respectively, for reporting purposes. Daily weight gain (DWG, g day^{-1}),

Table 1 Formulation and proximate composition (dry matter basis) of feed used in the study

Ingredients	Composition (g kg^{-1}) dry feed
Cotton seed cake	683.40
Maize bran	315.60
Vitamin premix[a]	1.00
Total	1000.00
Proximate composition of feed	
Dry matter	924.70
Crude protein	297.50
Crude lipid	136.70
Crude fibre	109.60
Ash	68.10
Carbohydrate	388.10

[a] Vitamin premix (per kg of premix): thiamine, 2.50 g; riboflavin, 2.50 g; pyridoxine, 2.00 g; inositol, 100.00 g; biotin, 0.30 g; pantothenic acid, 100.00 g; folic acid, 0.75 g; para-aminobenzoic acid, 2.50 g; choline, 200.00 g; nicotinic acid, 10.00 g; cyanocobalamine, 0.005 g; α-tocopherol acetate, 20.10 g; menadione, 2.00 g; retinol palmitate, 100,000 IU; cholecalciferol, 500,000 IU

specific growth rate (SGR, %), percentage survival (SR, %), net fish yield (NFY) and net annual yield (NAY) were calculated using the following formulae:

$$\text{DWG} = \frac{W_f - W_i}{t} \tag{1}$$

where W_f is the final mean weight (g), W_i the initial mean weight (g) and t is the growth period (days).

$$\text{SGR} = \frac{\ell n W_f - \ell n W_i}{t} \times 100 \tag{2}$$

$$\text{SR} = \frac{N_h}{N_s} \times 100 \tag{3}$$

where N_h is the total number of fish harvested and N_s is the total number of fish stocked.

$$\text{NFY} = \frac{W_h - W_s}{A} \tag{4}$$

where W_h is the total weight of fish harvested (kg), W_s the total weight of fish stocked (kg) and A is the tank area (ha)

$$\text{NAY} = \frac{\text{NFY} \times 365}{t} \tag{5}$$

Water quality parameters were measured twice daily in the morning (0830–0900 hours) and in the afternoon (1630–1700 hours) to gain an insight into the prevailing environmental conditions in fish tanks. Temperature, pH and dissolved oxygen (DO) were measured using a multi-probe kit (model KTO, HQ, 40D PHC 101-LD 101-01 by Hach Company Ltd, Loveland, CO, USA).

Statistical analyses

Data are presented as mean ± standard error (SE) of mean values and tested for homogeneity of variance using Levene's test based on mean values. After confirming homogeneity of variances, one-way analysis of variance (ANOVA) was used to test if there were any significant differences in the number of recruits, their mean weights and their percentage contribution on yield among treatments. The contributions (%) to yield of adult Nile tilapia and African sharptooth catfish were compared using one-way ANOVA and independent-samples t test, respectively. ANOVA was similarly used to identify any significant differences in FMW, NFY and NAY among the treatments. The differences among treatments for DWG and SGR were assessed using repeated measures ANOVA as the same groups of individuals were measured repeatedly over a period of time. The individuals are one factor (usually considered as random and serving as replication) and the time dimension is the second factor, a fixed treatment effect (Winer 1971). When significant differences were detected, Tukey post hoc test was performed to determine specific significant differences among treatments. Percentage data were arcsine-transformed prior to ANOVA and back-transformed afterwards for reporting purposes (Zar 2010). All statistical analyses were performed using SPSS version 20 for Windows. Significant differences were judged at a probability level of $p \leq 0.05$ for ANOVA, t test and Tukey tests (Zar 2010).

Results

Number of recruits and their weights

Results on the effect of size of predator on controlling the number of Nile tilapia recruits are shown in Fig. 1. The initial size of the predator African sharptooth catfish had a significant effect on reducing the number of

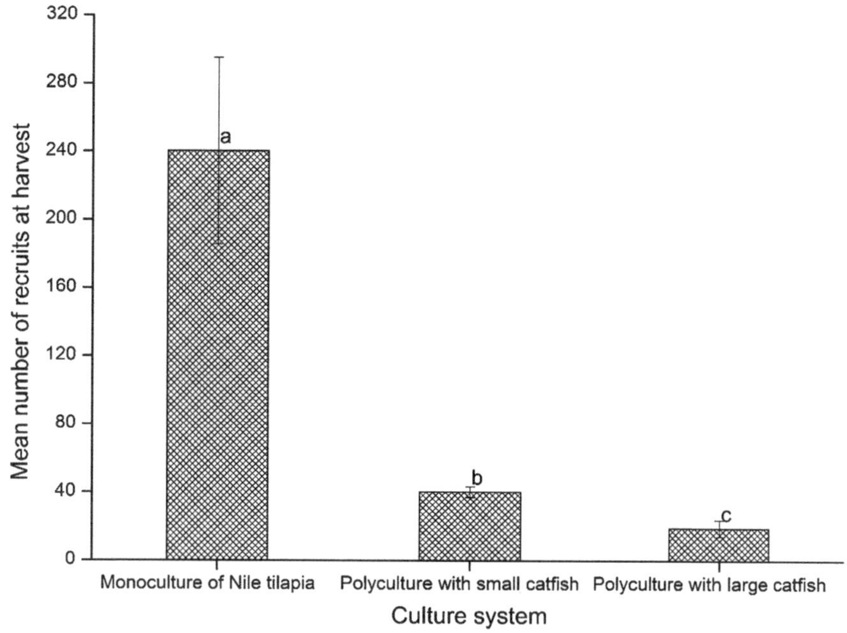

Fig. 1 The mean number of Nile tilapia recruits in the three experimental treatments at the end of the experiment. *Different letters above the bars* show significant differences ($p \leq 0.05$) and *vertical bars* indicate standard error (SE) of the mean

recruits in mixed-sex Nile tilapia culture ($F_{2,6} = 14.712$; $p = 0.005$). The multiple comparisons test showed significantly lower mean number of Nile tilapia recruits in larger (19.33 ± 3.18) compared to smaller African sharptooth catfish predators (40.33 ± 3.18; $p = 0.046$) and monoculture of Nile tilapia (240.33 ± 54.77; $p = 0.006$). Likewise, significantly lower number of Nile tilapia recruits were recorded in polyculture with smaller African sharptooth catfish compared to monoculture of Nile tilapia ($p = 0.010$). Large African sharptooth catfish predators controlled the number of Nile tilapia recruits to less than ten percent (only 6 % remained) at the end of the experiment. Comparatively, this represents about 12 (80 %) and 2 (13 %) times the number of recruits that remained in monoculture of Nile tilapia and polyculture with smaller African sharptooth catfish, respectively.

At the end of the 126-day rearing period, the body weight classes of tilapia recruits ranged from 0.10 to 30.00 g (Fig. 2). In all situations, with or without African sharptooth catfish predators, very small tilapia recruits (<3 g) were abundant. In the ponds stocked with small catfish, recruits >20 g were rare, whereas in those stocked with large African sharptooth catfish, they were more abundant. These variations in weight classes contributed to the significant difference in the mean body weight of Nile tilapia recruits among the treatments ($F_{2,387} = 8.470, p < 0.001$; Fig. 3). The mean body weight of Nile tilapia recruits was significantly higher in the tanks stocked with larger (14.78 ± 1.39 g) than smaller African sharptooth catfish predator (8.71 ± 0.59 g, $p < 0.001$) and monoculture of Nile tilapia (11.52 ± 0.70 g; $p = 0.050$). Nile tilapia recruits stocked in monoculture tanks attained significantly higher mean body weights than those stocked in tanks with smaller African sharptooth catfish predator ($p = 0.024$).

Growth performance and percentage survival

The mean weight increase in Nile tilapia in the three experimental treatments is shown in Fig. 4. The size of predator African sharptooth catfish affected significantly the final weight of Nile tilapia ($F_{2,42} = 278.283$, $p < 0.001$). The multiple comparisons test showed significantly higher FMW (198.74 ± 0.39 g) of Nile tilapia in larger than smaller African sharptooth catfish predators (140.52 ± 3.27 g) and monoculture of Nile tilapia (128.39 ± 2.10 g) ($p < 0.001$; Table 2). Similarly, Nile tilapia reared together with smaller African sharptooth catfish revealed significantly higher FMW than those reared in monoculture ($p = 0.01$; Table 2). Nile tilapia stocked with large African sharptooth catfish had comparatively higher FMW (43 %) than those in polyculture with small African sharptooth catfish (30 %) and monoculture system (27 %).

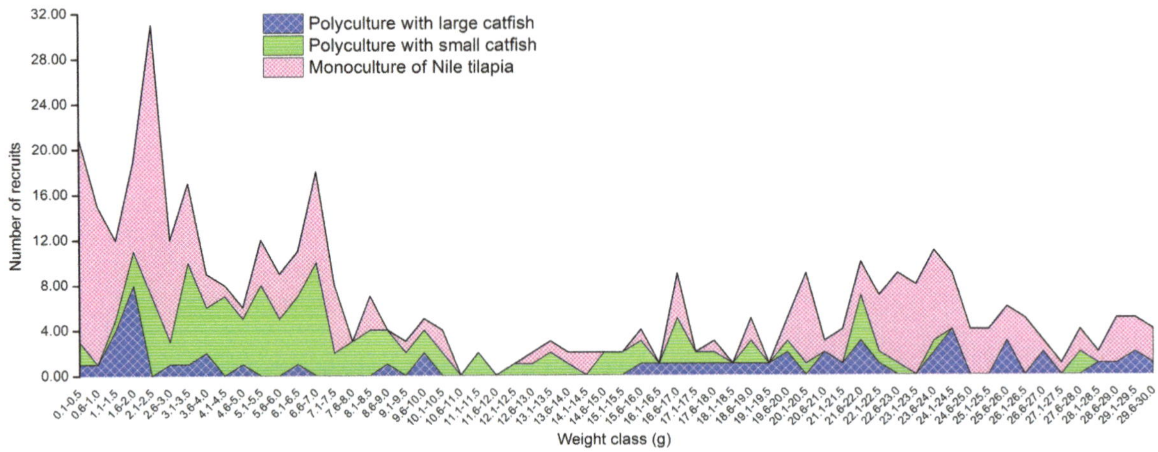

Fig. 2 The number of Nile tilapia recruits at various weight classes obtained at the end of the experiment

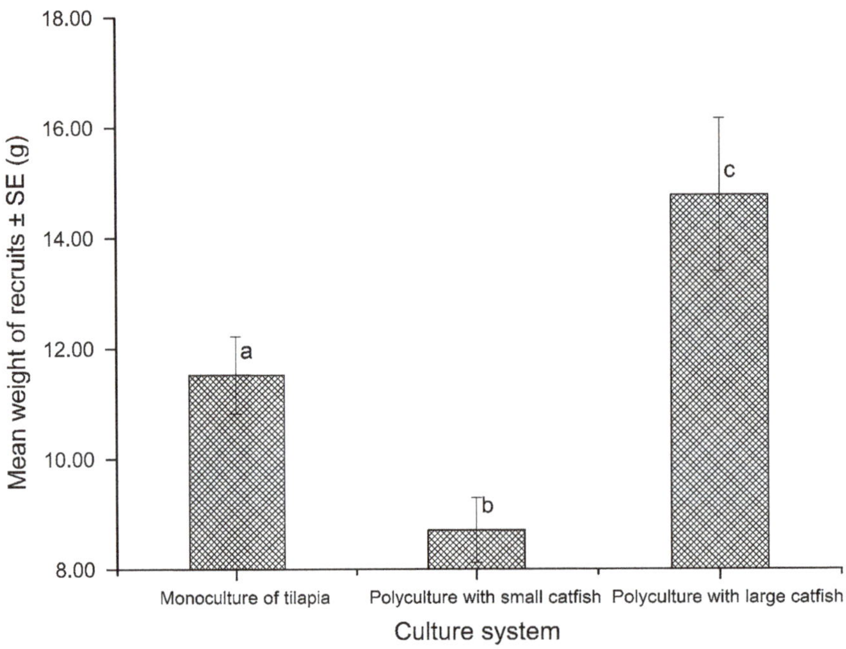

Fig. 3 The mean weight of Nile tilapia recruits in the three experimental treatments at the end of the experiment. *Different letters above the bars* show significant difference ($p \leq 0.05$) and *error bars* indicate standard error (SE) of the mean

Furthermore, DWG of Nile tilapia was also significantly affected by the size of the predator ($F_{8,64} = 2.829$, $p = 0.012$). The polyculture of Nile tilapia with larger African sharptooth catfish resulted in significantly higher DWG of Nile tilapia compared with those reared with smaller African sharptooth catfish and Nile tilapia alone ($p < 0.001$; Table 2). Furthermore, the DWG for Nile tilapia polycultured with small African sharptooth catfish was significantly higher than those reared alone ($p < 0.001$). Similarly, SGR of Nile tilapia was significantly affected by the size of the predator African sharptooth catfish ($F_{8,64} = 161.071$, $p < 0.001$). Nile tilapia reared with larger African sharptooth catfish had significantly higher SGR than those raised with small African sharptooth catfish and in monoculture ($p < 0.001$; Table 2). Likewise, Nile tilapia reared with smaller African sharptooth catfish had a significant SGR than those reared alone in monoculture ($p < 0.001$; Table 2). Percentage survival of Nile tilapia was not affected in any case by the size of the predator ($p > 0.05$).

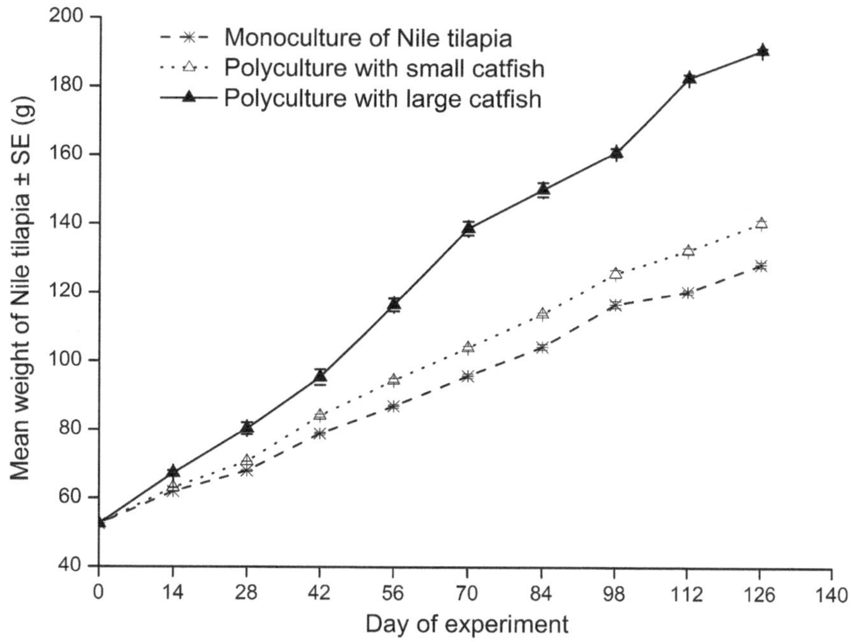

Fig. 4 Mean weight increase of Nile tilapia cultured in monoculture and polyculture with smaller and larger African sharptooth catfish predator. *Vertical bars* indicate standard error (SE) of the mean

Yield

The size of the predator affected significantly the yield performance of the different treatments ($F_{2,6} = 8.179$, $p = 0.019$). The NFY and NAY were significantly lower in polyculture with large African sharptooth catfish than in monoculture of Nile tilapia ($p = 0.016$; Table 3). However, NFY and NAY were not significantly different between monoculture of Nile tilapia and polyculture with small African sharptooth catfish ($p = 0.115$).

The percentage weight of Nile tilapia recruits reared in monoculture contributed almost 60 % to the yield, while those in polyculture contributed less than 20 % (Table 3). The contribution differed significantly among the culture systems ($F_{2,6} = 28.663$, $p < 0.001$). Nile tilapia recruits reared in monoculture contributed significantly more to the yield than those in polyculture with smaller ($p = 0.001$) and larger African sharptooth catfish ($p = 0.002$).

In the polyculture setup, the 21 stocked adult fishes (Nile tilapia and African sharptooth catfish) contributed over 80 % to the total yield, while in the tilapia monoculture the contribution was only 42 % of the total yield. The contribution of adult Nile tilapia to the yield differed significantly among the experimental setup ($F_{2,6} = 12.174$, $p = 0.008$). The percentage weight of adult Nile tilapia cultured in polyculture with smaller African sharptooth catfish contributed significantly higher to the yield compared to those cultured in monoculture ($p = 0.011$) and polyculture with larger African sharptooth catfish ($p = 0.014$; Table 3). Larger African sharptooth catfish exhibited significantly higher percentage weight contribution to yield than smaller ones ($t_{(4)} = -13.270$, $p < 0.001$; Table 3).

Water quality parameters

The water quality parameters (mean ± standard deviation) were stable in all treatments with minor fluctuations. Water temperature ranged from 23.90 to 28.10 °C with mean values between 26.23 ± 1.13 and 26.51 ± 0.96 °C recorded in Nile tilapia alone and in polyculture with small African sharptooth catfish, respectively. Dissolved oxygen ranged from 2.90 to 10.19 mg L^{-1} with mean values between 5.99 ± 1.65 and 6.24 ± 1.54 mg L^{-1} obtained in polyculture with small and large African sharptooth catfish, respectively. pH values ranged from 5.59 to 7.20 with a mean of 6.55 ± 0.66 in all treatments. Results of all water quality parameters measured did not show any significant differences among treatments ($p > 0.05$).

Table 2 Growth performance parameters of Nile tilapia during the study period

Growth parameter	Monoculture of Nile tilapia	Polyculture with small African catfish	Polyculture with large African catfish
Final mean weight (g)	128.39 ± 2.10^{a}	140.52 ± 3.27^{b}	198.74 ± 0.39^{c}
Daily weight gain (g day^{-1})	0.60 ± 0.01^{a}	0.70 ± 0.01^{b}	1.16 ± 0.02^{c}
Specific growth rate (%)	0.71 ± 0.04^{a}	0.78 ± 0.05^{b}	1.06 ± 0.07^{c}

Values in the same row with different superscripts are significantly different ($p \leq 0.05$). Values are mean ± standard error for the replicates

Table 3 Yield parameters and percentage contribution of recruits and adult Nile tilapia and African catfish during the study

Yield parameter	Monoculture of Nile tilapia	Polyculture with small African catfish	Polyculture with large African catfish
Net fish yield (kg ha^{-1})	$6{,}444.59 \pm 1213.94^{a}$	$4{,}028.05 \pm 89.69^{a,b}$	$2{,}406.79 \pm 180.87^{b}$
Net annual yield (kg ha^{-1} $year^{-1}$)	$18{,}668.85 \pm 3516.58^{a}$	$11{,}668.57 \pm 259.82^{a,b}$	$6{,}972.05 \pm 523.95^{b}$
% contribution on yield			
Recruits	57.70 ± 11.85^{a}	12.36 ± 2.45^{b}	$16.51 \pm 6.24^{b,c}$
Adult Nile tilapia	42.30 ± 6.84^{a}	70.49 ± 1.47^{b}	$43.94 \pm 3.57^{a,c}$
Adult African catfish	NA	17.15 ± 0.98^{a}	39.55 ± 1.37^{b}

Values in the same row with different superscripts are significantly different ($p \leq 0.05$). Values are mean ± standard error for the three replicates

NA not applicable

Discussion

The results have shown that, large-sized African sharptooth catfish have a significant effect on reducing the number of recruits in mixed-sex Nile tilapia culture. A similar efficiency of larger compared to smaller size of catfish predation on recruits of Nile tilapia has been reported by Abdel-Tawwab (2005). Moreover, Offem et al. (2009) found highest percentage (71 %) of sub-marketable Nile tilapia recruits in monoculture followed by polyculture stocked with small (49 %) and large (7 %) Vundu catfish (*Heterobranchus longifilis*) predator.

The efficiency of larger compared to smaller catfish predators on recruitment control of Nile tilapia is related to body size. Body size of a predator has been shown as a critical factor in predator–prey relationships (Fessehaye et al. 2006). Variation in size between them is considered the most significant factor in predation efficiency owing to mouth width limitations (Wysujack and Mehner 2005; Offem et al. 2009), because prey recruits are generally swallowed whole by predators (Fessehaye et al. 2006). In the present study, African sharptooth catfish were able to swallow Nile tilapia recruits depending on initial size of the predator and this ability was two times in larger African sharptooth catfish than the smaller ones. Thus, based on the present experimental design, it is recommended to stock African sharptooth catfish of at least 60 g to control mixed-sex Nile tilapia recruitment in the culture systems depending on initial size of the prey.

The existence of 6 and 13 % in the number of recruits at the end of the experiment from polyculture with large and small African sharptooth catfish, respectively, indicates that the recruitment of Nile tilapia was only partially controlled. These findings are in agreement with those of de Graaf et al. (1996), Abdel-Tawwab (2005) and Fessehaye et al. (2006) who reported failure of large-sized African sharptooth catfish to control completely the recruitment of Nile tilapia. This indicates that other factors such as supplementary feeding in the culture systems also play a role in affecting the predation efficiency. In their study, de Graaf et al. (1996) reported increasing efficiency in controlling Nile tilapia recruitment by African sharptooth catfish in low food systems. Equally, El-Gamel et al. (1998) observed reduced predatory efficiency by African sharptooth catfish in earthen ponds when artificial feed was offered. Moreover, Abdel-Tawwab (2005) using different sizes of African sharptooth catfish (35, 75, 180, 275, 400 and 650 g) indicated that artificial feeding reduced the predatory rate of all African sharptooth catfish weights. Furthermore, Oyelese (2007) reported reduced predation efficiency on Nile tilapia recruits by African sharptooth catfish fed to satiation compared to those fed sparingly. These findings imply that, feeding the fishes in polyculture reduced the ability of African sharptooth catfish to completely control recruitment because of nutritional satisfaction. Consequently, the numbers of recruits at the end of the experiment from polyculture with large and small African sharptooth catfish were 6 and 13 %, respectively.

Polyculture of Nile tilapia with large African sharptooth catfish depicted significantly higher mean weights of the number of recruits remaining at the end of the experiment. These results are corroborated by the findings of Offem et al. (2009) using Vundu catfish. The predatory ability of African sharptooth catfish depends on growth rate of the prey population. Oyelese (2007) showed that, predation of African sharptooth catfish on Nile tilapia recruits stopped after the sixth week owing to the increased body size of the prey. This shows that, there is a critical period of predation by African sharptooth catfish on Nile tilapia recruits which is controlled by growth rate of the prey. As the prey grew bigger, it became less vulnerable to predation by African sharptooth catfish due to development of fin rays which complicate predation (Abdel-Hadi et al. 2011). This allowed a few Nile tilapia recruits available in polyculture with large African sharptooth catfish to utilize the available resources such as space and food to grow faster resulting into significantly bigger size (weight).

The present study demonstrated that, weight gained by Nile tilapia stocked with large African sharptooth catfish was more than 40 % compared to those stocked with small African sharptooth catfish and in monoculture system. Similarly, all the growth parameters were significantly higher in Nile tilapia stocked with large African sharptooth catfish. These results are similar to those obtained by de Graaf et al. (1996) using African sharptooth catfish and African snakehead (*Ophiocephalus obscuris*), Oyelese (2007) using African sharptooth catfish and Offem et al. (2009) using Vundu catfish and Nile tilapia. Likewise, Rudolf (2006) using spring salamander (*Gyrinophilus porphyriticus*) as a predator and Southern two-lined Salamander *(Eurycea cirrigera)* as a prey found an increased growth rate of prey at large size of the predator.

The increased growth performance of Nile tilapia stocked in polyculture with large African sharptooth catfish is due to predation on recruits by the latter. Predation by African sharptooth catfish reduced competition for space and food caused by recruits derived from the initially stocked adults through reproduction (Oyelese 2007). The availability of food and space influenced increased growth performance because of enhanced feeding (Offem et al. 2009). On the contrary, the presence of 80 and 13 % recruits in monoculture and polyculture with smaller African sharptooth catfish, respectively, interfered with the growth performance through feed competition and space between the stocked adult Nile tilapia and their recruits. Consequently, the growth of adult Nile tilapia was restricted in polyculture with small African sharptooth catfish and in monoculture. The restriction in growth of the adult Nile tilapia by recruits is significant because of the higher metabolic rates of young ones (de Silva et al. 1986). The restriction problem was reduced by removing the recruits through stocking large predatory African sharptooth catfish resulting in increased growth performance of the adult Nile tilapia in the present study.

Percentage survival of Nile tilapia was not affected by the size of the predator African sharptooth catfish. Ultimately, the percentage survival was 100 % in all the treatments. This indicates that the experimental conditions in this study were conducive for Nile tilapia survival in all treatments (Offem et al. 2009). The high percentage survival of Nile tilapia was caused by enough food and adequate space for co-existence with African sharptooth catfish. Adult Nile tilapia has shown to have high survival when cultured in monoculture and polyculture with African sharptooth catfish (Lin and Diana 1995; Veverica et al. 2001). The two species can be grown together to improve the growth of Nile tilapia and ensure diversified products at the end of the growth period to the producer for increased economic benefits. This is important because the two species are known to be good source of protein and income (Chakraborty et al. 2011; FAO 2012; Yakubu et al. 2012). Since large African sharptooth catfish provide conducive environment (optimal predation efficiency) for the rapid rearing of adult Nile tilapia for food and income, this technology should be advocated to fish farmers.

Recruitment control by large African sharptooth catfish predator reduced significantly the NFY and NAY in polyculture than monoculture of Nile tilapia. A reduction in yield in polyculture systems has been observed in earlier experiments with other catfish species such as African snakehead (de Graaf et al. 1996). The reduced yields in polyculture with large than small African sharptooth catfish and in monoculture is contributed by the minimized number of Nile tilapia recruits by the large African sharptooth catfish. Similar observation has been reported by Abdel-Tawwab (2005) who found lower contribution of recruits to the total yield in polyculture with large African sharptooth catfish than in monoculture. Recruits contribution to the yield in monoculture was higher (58 %) than in polyculture with large catfish (17 %), suggesting that, the consumed recruits were converted into a lower weight by the predator. The lost fingerlings weight could not be fully compensated by the observed higher growth rates of Nile tilapia in polyculture with large African sharptooth catfish. However, this is not worrisome because the loss in biomass is economically compensated by the higher price obtained after selling the larger Nile tilapia produced (de Graaf et al. 1996).

Conclusion

Large-sized African sharptooth catfish used in the present study is more efficient in reducing the number of recruits in mixed-sex Nile tilapia culture. The reduced number of recruits by the large African sharptooth catfish predator lowers competition for space and food for adult Nile tilapia resulting in increased growth performance. The reduction in number of Nile tilapia recruits in polyculture with large African sharptooth catfish contributes to low yields. Nevertheless, the low yields are compensated by the higher price obtained from selling the larger Nile tilapia produced. This study revealed that, fish farmers can reduce prolific breeding, obtain higher growth performance and produce larger-sized marketable Nile tilapia by predominantly stocking them with large African sharptooth catfish predator with initial average weight of at least 60 g.

Acknowledgments We express our sincere thanks to Tanzania Commission for Science and Technology (COSTECH) through "*KILIMO KWANZA*" programme for provision of a research grant. Tanzania Fisheries Research Institute (TAFIRI), the University of Dar es Salaam and Sokoine University of Agriculture (SUA) are acknowledged for granting permission for authors to participate in the study. SUA is also acknowledged for provision of research facilities.

Conflict of interest The authors declare that they have no competing interests.

Authors' contributions SML designed the study, collected and analysed data and drafted the manuscript. APS designed the study, collected and analysed data. HAL designed the study, collected and analysed data. EDS designed the study, collected and analysed data. MAK-M designed the study, collected and analysed data. HFM designed the study, collected and analysed data. YDM designed and supervised the study. All authors read and approved the final manuscript.

References

Abdel-Hadi Y, Craig J, Babaluk J, Wassle R (2011) Oxytetracycline marking studies of tilapia; *Oreochromis niloticus*. Better science, better fish, better life. In: Liping L, Fitzsimmons K (eds) Proceedings of the sixth international symposium on Tilapia in Aquaculture: AquaFish Collaborative Research Support Program, pp 60–66

Abdel-Tawwab M (2005) Predation efficiency of Nile catfish, *Clarias gariepinus* (Burchell, 1822) on fry Nile tilapia, *Oreochromis niloticus* (Linnaeus, 1758): effect of prey density, predator size, feed supplementation and submerged vegetation. Turk J Fish Aquat Sci 5:69–74

Carlberg JM, Olst JVC, Massingi MJ (2000) Hybrid striped bass: an important fish in US aquaculture. Aquac Mag 26:26–38

Chakraborty SB, Mazumdar D, Chatterji U, Banerjee S (2011) Growth of mixed-sex and monosex Nile tilapia in different culture systems. Turk J Fish Aquat Sci 11:131–138

Dagne A, Degefu F, Lakew A (2013) Comparative growth performance of mono-sex and mixed-sex Nile tilapia (*Oreochromis niloticus* L.) in pond culture system at Sebeta, Ethiopia. Int J Aquacult 3(7):30–34

de Graaf G, Galemoni F, Banzoussi B (1996) Recruitment control of Nile tilapia, *Oreochromis niloticus*, by the African catfish, *Clarias gariepinus* (Burchell 1822), and the African snakehead, *Ophiocephalus obscuris*. I. A biological analysis. Aquaculture 146(1–2):85–100

de Silva CD, Premawansa S, Keembiyahetty CN (1986) Oxygen consumption in *Oreochromis niloticus* (L.) in relation to development, salinity, temperature and time of day. J Fish Biol 29(2):267–277

El-Gamel AA, Abdel-Halim AMM, Abdel-Razek E, Solomon A (1998) Biological studies on the Nile perch, *Lates niloticus* (L.) and African catfish, *Clarias gariepinus* (T.) in reference to their food habits and predation pattern in culture ponds. Egypt J Agric Res 76:335–349

El-Naggar G (2007) Efficiency of African catfish, *Clarias gariepinus* in controlling unwanted reproduction of Nile Tilapia *Oreochromis niloticus* in low input production system. Egypt Aqua Biol Fish 11(3):105–113

Fagbenro OA (2004) Predator control of overpopulation in cultured tilapias and the alternative uses for stunted tilapias in Nigeria. In: Bolivar RB, Mair GC, Fitzsimmons K (eds) Proceedings from the sixth international symposium on Tilapia Aquaculture (VI ISTA). Bureau of Fisheries & Aquatic Resources, Manila, pp 634–647

FAO (2012) The state of world fisheries and aquaculture. Food and Agriculture Organization of the United Nations, Rome

Fessehaye Y, Kabir A, Bovenhuis H, Komen H (2006) Prediction of cannibalism in juvenile *Oreochromis niloticus* based on predator to prey weight ratio, and effects of age and stocking density. Aquaculture 255:314–322

Hernández M, Gasca-Leyva E, Milstein A (2014) Polyculture of mixed-sex and male populations of Nile tilapia (*Oreochromis niloticus*) with the Mayan cichlid (*Cichlasoma urophthalmus*). Aquaculture 418–419:26–31

Kaliba AR, Osewe KO, Senkondo EM, Mnembuka BV, Quagrainie KK (2006) Economic analysis of Nile tilapia (*Oreochromis niloticus*) production in Tanzania. J World Aquacult Soc 37(4):464–473

Lin CK, Diana JS (1995) Co-culture of catfish (*Clarias macrocephalus* × *C. gariepinus*) and tilapia (*Oreochromis niloticus*) in ponds. Aquat Living Resour 8:449–454

Lowe MR, Wu W, Peterson MS, Brown-Peterson NJ, Slack WT, Schofield PJ (2012) Survival, growth and reproduction of non-native Nile tilapia II: Fundamental niche projections and invasion potential in the Northern Gulf of Mexico. PLoS One 7(7):1–10

Mair GC, Little DC (1991) Population control in farmed tilapias. NAGA, International Center for Living Aquatic Resources Management (ICLARM) Quarterly 4:8–9

Mair GC, Van Dam AA (1996) The effect of sex ratio at stocking on growth and recruitment in Nile tilapia (*Oreochromls niloticus*) ponds. In: Pullin RSV, Lazard J, Legendré M, Kothias Amon JB, Pauly D (eds) Third international symposium on Tilapia in Aquaculture: International Center for Living Aquatic Resources Management (ICLARM)., pp 100–107

Offem BO, Ikpi GU, Ayotunde EO (2009) Effect of stocking size of the predatory African catfish (*Heterobranchus longifilis* V.) on the growth performance of Nile Tilapia (*Oreochromis niloticus* L.) in pond culture. Int J Fish Aquacult 1(3):038–043

Ofor CO, Udo UI, Udoidiong MO (2011) Effect of repeated partial cropping on population dynamics and yield of *Oreochromis niloticus* (L.) during polyculture with *Heterobranchus longifilis* (Val.). Int J Fish Aquacult 3(7):126–135

Oyelese OA (2007) Prey/predator relationship of *Clarias gariepinus* on tilapia (*Oreochromis niloticus*) populations. Res J Biol Sci 2(1):17–24

Rudolf VHW (2006) The influence of size-specific indirect interactions in predator–prey systems. Ecology 87(2):362–371

Shrestha MK, Sharma RL, Gharti K, Diana JS (2011) Polyculture of Sahar (*Tor putitora*) with mixed-sex Nile tilapia. Aquaculture 319(1–2):284–289

Veverica K, Liti D, Were E, Bowman J (2001) Fish yields and economic benefits of Tilapia/Clarias polyculture in fertilized ponds receiving commercial feeds or pelleted agricultural by-products. In: Gupta A, McElwee K, Burke D, Burright J, Cummings X, Egna H (eds) Eighteenth annual technical report pond dynamics/aquaculture CRSP: Oregon State University., pp 27–29

Winer BJ (1971) Statistical principles in experimental design, 2nd edn. McGraw-Hill, New York

Wysujack K, Mehner T (2005) Can feeding of European catfish prevent cyprinids from reaching a sized refuge? Ecol Freshw Fish 14:87–90

Yakubu AF, Obi A, Okonji VA, Ajiboye OO, Adams TE, Olaji ED, Nwogu NA (2012) Growth performance of Nile tilapia (*Oreochromis niloticus*) as affected by stocking density and feed types in water flow through system. World J Fish Mar Sci 4(3):320–324

Yi Y, Lin CK, Diana JS (2002) Culture of mixed-sex Nile tilapia with predatory snakehead. Corvallis, Oregon

Zar JH (2010) Biostatistical analysis, 5th edn. Prentice Hall Inc, New Jersey

19

Outbreak of growth anomalies in coral communities of Qeshm Island, Persian Gulf

Pavriz Tavakoli-Kolour · Javid Kavousi · Hamid Rezai

Abstract Reports on the outbreaks of coral diseases are on the rise, stating that diseases are considered to be one of the main drives in the decline of global coral reefs. In this study, the outbreak of growth anomalies (GAs) in coral communities of the south of Qeshm Island in the Persian Gulf is reported. Despite a previous report on the localized GAs on a few colonies, this is the first report of GA outbreak from the Persian Gulf, in which 28 ± 11 % (±SE) and 21 ± 13 % of *P. daedalea* and *Porites* colonies contracted GAs in July 2012. A year later, we recorded GAs on 37 ± 07 % of *P. daedalea* and 23 ± 12 % of *Porites* colonies. Pearson's correlation test showed a strong positive relationship between coral colony size and the number of GAs ($r = 0.907$, $p < 0.05$). Further investigation is needed to understand the cause of GAs, their progress trend, and prospective impacts on the coral communities at our studied site.

Keywords Disease · Growth anomalies · Massive *Porites* · Persian Gulf

Introduction

Coral diseases are known as one of the most prevalent threats to coral reefs worldwide (Sutherland et al. 2004; Weil et al. 2006). In the recent years, reports on coral diseases and subsequent mass mortalities have increased worldwide (Goreau et al. 1998; Green and Bruckner 2000; Sutherland et al. 2004; Weil et al. 2006). Although the Caribbean Sea is considered the main coral disease hot spot with over 66 % of the world's recorded diseases (Green and Bruckner 2000), Indo-Pacific coral reefs are getting more attention due to increasing instances of coral disease outbreaks recently (e.g., Willis et al. 2004; Weil et al. 2012) and the Persian Gulf is no exception (Riegl et al. 2012).

Although coral disease studies in the southern Persian Gulf started about two decades ago (see Riegl et al. 2012), little information was available on coral diseases of the northern part until recently. Except the Yellow Band Disease, which was reported at two Iranian Islands in 2000 and 2002 (Rezai et al. 2004), the other

P. Tavakoli-Kolour (✉)
Young Researchers and Elite Club, Bandar Abbas Branch, Islamic Azad University, PO Box 79159-14 1311, Bandar Abbas, Iran
e-mail: p.tavakoli@hotmail.com

J. Kavousi
Graduate School of Engineering and Science, University of the Ryukyus, 1 Senbaru, Nishihara, Okinawa 903-0213, Japan

H. Rezai
Iranian National Institute for Oceanography and Atmospheric Sciences, Tehran, Iran

reports on diseases were drawn up only in the last few years (Samimi-Namin et al. 2010; Kavousi and Rezai 2011; Kavousi et al. 2013).

Similarly, serpulid invasions associated with red tides at Qeshm Island (Samimi-Namin et al. 2010) and the White Mat Disease at Hormuz Island (Kavousi et al. 2013) caused mass coral mortalities in the northern Persian Gulf. Other reports included growth anomalies (GAs) only on a few colonies at Larak, Hengam, and southeast of Qeshm Island (Kavousi and Rezai 2011). However, no coral colonies with GAs were observed at the southern part of Qeshm Island. Here, we report the outbreak of GAs in *Platygyra daedalea* and *Porites* spp. colonies as the only coral taxa affected among the southern coral communities of Qeshm Island in 2012 and 2013.

Materials and methods

The study site includes the southern coral communities of Qeshm Island (26°45′N, 55°49′E, Fig. 1) in the Persian Gulf. The present investigation took place during July 2012 and July–August 2013. In total, eight 30 m^{-1} Line-Intercept transects were established at depth 2–5 m^{-1} in each year for preliminary assessment of benthic communities including live coral and coral genera cover. Moreover, to study growth anomalies (GAs) dynamic on the affected corals, eight 30 m^{-2} belt transects were considered. Coral skeletons in GAs and healthy parts of coral colonies were observed under a focal microscope to check out the microstructural differences.

Furthermore, during 2013, 11 colonies of *Platygyra daedalea* with GAs were measured in three dimensions (length, width, and height) and the frequency of GAs were counted followed by the Pearson's correlation test to assess whether there was any relationship between coral colony size and the number of GAs in coral colonies. *Porites* colonies were not considered due to the negligible difference in their size.

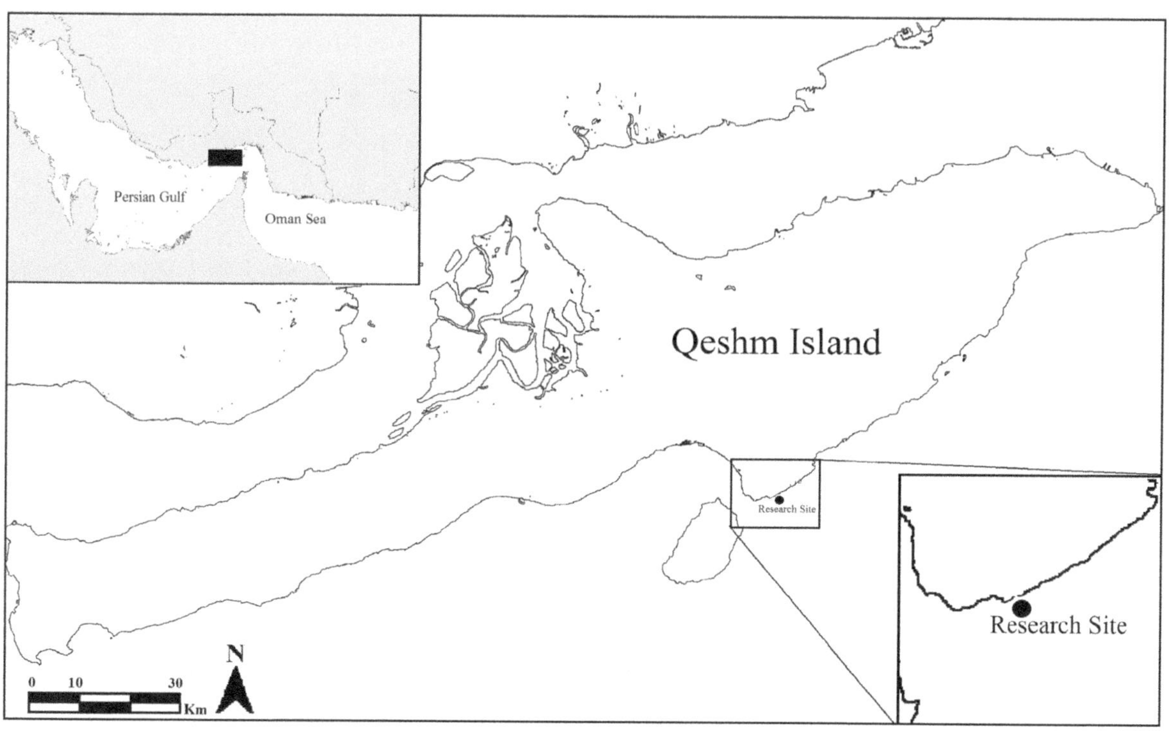

Fig. 1 Map of the site studied in this research

Results and discussion

Live coral cover

Live coral constituted 20 % of the substrate of the studied site at Qeshm Island (Fig. 2), which is higher than that reported for 2009, 14.03 ± 4.8 % (Kavousi et al. 2011). *Platygyra daedalea*, as the dominant coral taxa, accounts for approximately 50 % of the live coral cover in the area (~53 % in 2012 and ~49 % in 2013; Fig. 3). In contrast, *Porites* spp. as dominant genus in the southeast of Qeshm Island constituting 89 ± 12.2 % (Kavousi et al. 2011) accounts for about 6 % of the coral cover at the current studied site (Fig. 3).

Growth anomalies (GAs) on corals

P. daedalea and *Porites* spp. were the only coral taxa with GAs (Fig. 4). GAs accounted for 28 ± 11 % (±SE) and 21 ± 13 % in *P. daedalea* and *Porites* colonies, respectively, during July 2012, whereas they were 37 ± 07 % and 23 ± 12 % in *P. daedalea* and *Porites* spp., respectively (Fig. 5). These results show a 9 % increase in GAs on *P. daedalea* colonies, but the increase in *Porites* spp. is negligible. Although localized GAs on *P. daedalea* and *Dipsastraea* (formerly *Favia*) were recorded at Iranian Islands (Kavousi and Rezai 2011), GA outbreak has never been previously reported for the Persian Gulf.

There is a highly positive correlation between the coral colony size and the number of GAs ($r = 0.907$, $p < 0.05$, Fig. 6). In other words, the bigger the colony, the greater the chances of GAs. Irikawa et al. (2011) also found that *Acropora cytherea* colonies larger than 120 cm in diameter had more GAs at some Japanese reefs. Loya et al. (1984) studying GAs in two species of *Platygyra* at some Australian reefs found that the larger the colony, more likely it is to have GAs.

Microscopic photographs of the coral skeletons show that calices of the affected areas are irregular, destructed, and bigger than those of normal coral skeletons (Fig. 7). Similar results are reported in other studies (McClanahan et al. 2009; Irikawa et al. 2011; Stimson 2011).

Fig. 2 Percentages of biotic and abiotic objects at the studied site. *LC* live coral, *DC* dead coral, *Rb* rubble, *Sd* sand, *Rc* rock, *Al* algae, *Sp* sponge

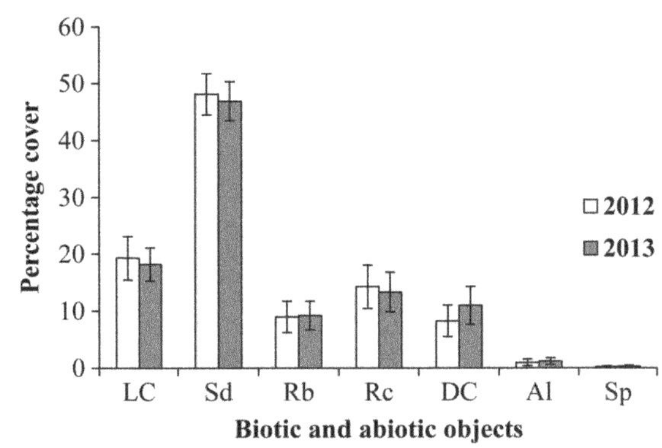

Fig. 3 Coral genera coverage at the studied site

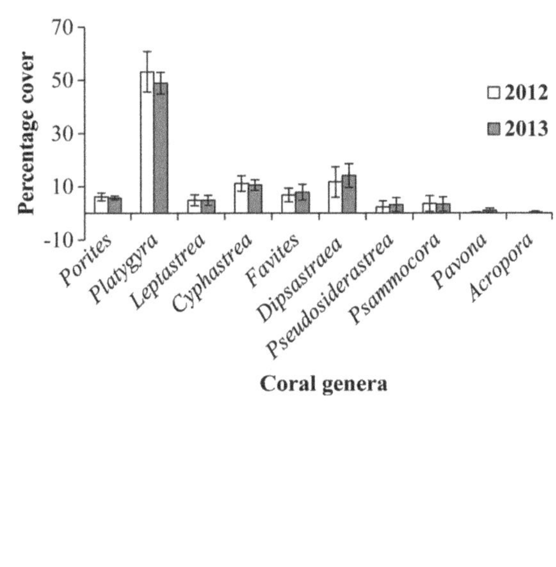

Fig. 4 Growth anomalies on *Platygyra daedalea* (**a**) and on *Porites* spp. (**b**)

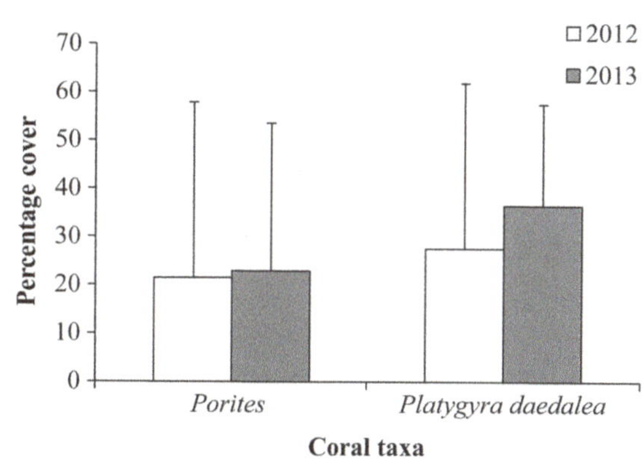

Fig. 5 Percentage of coral colonies contracting growth anomalies in 2012 and 2013

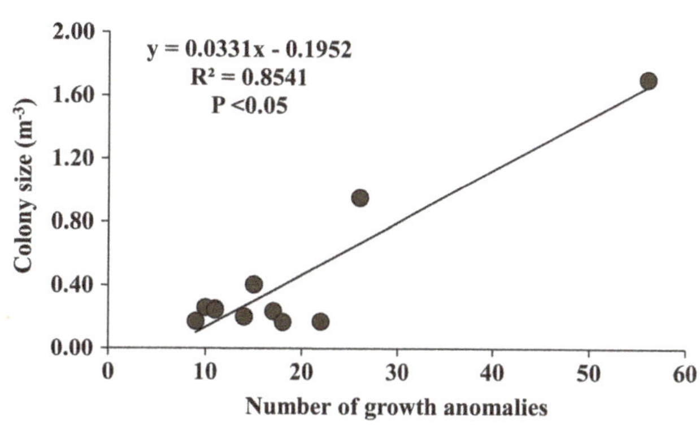

Fig. 6 The relationship between *Platygyra daedalea* colony size and the frequency of growth anomalies in 2013

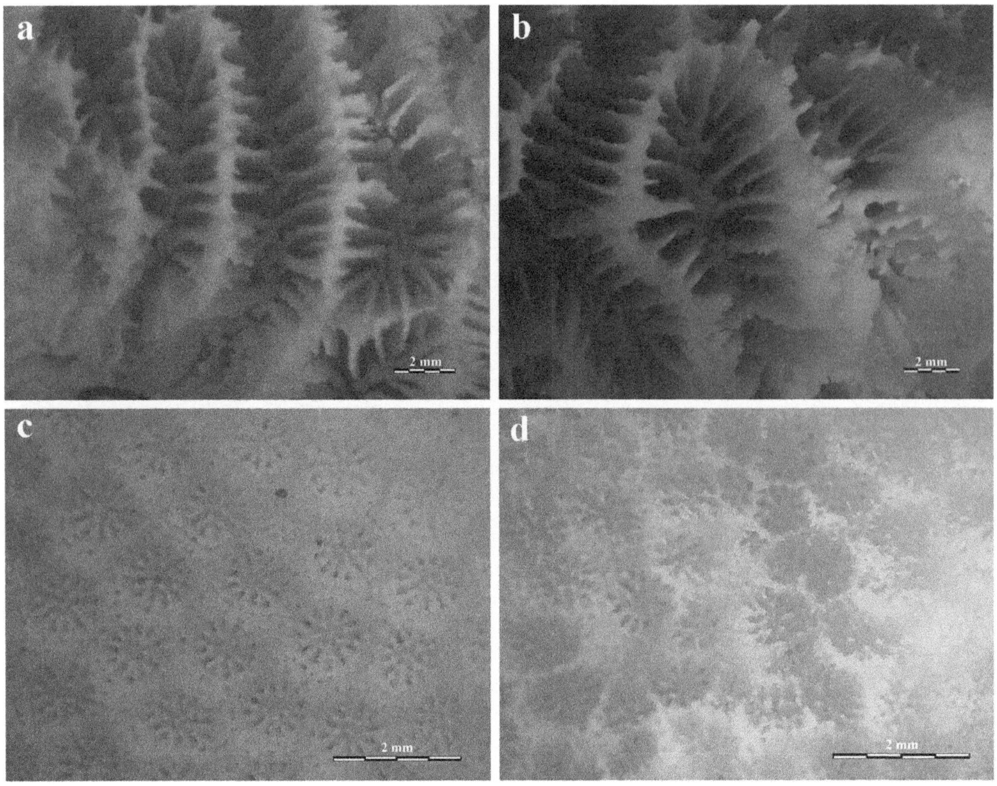

Fig. 7 Microscopic photographs of the coral skeletons of **a** healthy and **b** GA parts of *Platygyra daedalea*, **c** healthy, and **d** GA parts of *Porites* spp.

Although the etiology of GAs, like the majority of coral diseases, is understudied, environmental factors have been suggested as the main cause. For example, McClanahan et al. (2009) found a strong relationship between *Porites* growth anomalies and coral bleaching at Kenyan coral reefs. As a massive bleaching happened in the northern Persian Gulf, extending to our studied site in 2012 (Kavousi et al. 2014), high temperatures might have been a cause of GAs in this area. However, it is unlikely that anomalous temperatures alone could be the cause of the GA outbreak because no GAs were observed at other sites encountering mass bleaching in the area at the same time.

Irikawa et al. (2011) monitoring some Japanese reefs for 10 years suggested that GAs on *Acropora cytherea* were a resultant of aging amplified by temperature anomalies. Some studies also suggested biotic factors such as viruses, fungi, and endolithic algae as the cause of GAs (see Weil et al. 2012). McClanahan et al. (2009) mentioned that altered calice structure in GA areas is due to some microbioeroders including fungi causing the unusual growth.

Another study found that environmental conditions together with the human population size as well as the abundance of host coral communities strongly influence *Acropora* and *Porites* GA prevalence in the Indo-Pacific (Aeby et al. 2011). Our research site, however, encompasses a local community with a few boats, which are used for limited fishing and transportation, it is not close to ports, industrial complexes, or urban areas. The suggestion by Aeby et al. (2011) may not hold true for our studied zone as not only the stressors including human pressure are relatively greater in number at the southeast of Qeshm Island but also *Porites* spp. are dominant ($\sim$89 %) at that site (Kavousi et al. 2011). However, no GA outbreak was discerned in that site (pers. obs.). We hypothesize that a compulsory combination of environmental conditions, host corals' physiological conditions, and microbial invasions may determine occurrence and development of GAs on corals. Partial involvement of the mentioned factors may explain why some studies report transition of GAs to healthy colonies (Kaczmarsky and Richardson 2007), whereas others do not (Irikawa et al. 2011). However, the certain cause of the observed outbreak of the GAs at the south of Qeshm Island remains unknown for the time being.

Although GAs do not kill coral colonies, they can decrease fecundity in the affected parts (Irikawa et al. 2011). Therefore, if an increasing trend in the outbreak of GAs on coral colonies is observed for a long term, so that GAs can cover most of the coral colony surface, it may decline coral recovery after disturbances because of insufficient recruits. The case may be more serious for our studied site as it encounters continuous stressors (Kavousi et al. 2011, 2014). Therefore, it is necessary to conduct field experiments to understand the dynamics and future impacts of GAs on coral communities of the south of Qeshm Island.

References

Aeby GS, Williams GJ, Franklin EC, Haapkyla J, Harvell CD, Neale S, Page CA, Raymundo L, Vargas-Ángel B, Willis B, Work TM, Davy SK (2011) Growth anomalies on the coral genera *Acropora* and *Porites* are strongly associated with host density and human population size across the Indo-Pacific. PLoS One 6:e16887. doi:10.1371/journal.pone.0016887

Goreau TJ, Cervino J, Goreau M, Hayes R, Hayes M, Richardson L, Smith G, DeMeyer K, Nagelkerken I, Garzon-Ferrera J, Gil D, Garrison G, Williams EH, Bunkley-Williams L, Quirolo C, Patterson K, Porter JM, Porter K (1998) Rapid spread of diseases in Caribbean coral reefs. Rev Biol Trop 46:157–171

Green EP, Bruckner AW (2000) The significance of coral disease epizootiology for coral reef conservation. Biol Cons 96:347–361. doi:10.1016/S0006-3207(00)00073-2

Irikawa A, Casareto BE, Suzuki Y, Agostini S, Hidaka M, van Woesik R (2011) Growth anomalies on *Acropora cytherea* corals. Mar Poll Bull 62(8):1702–1707. doi:10.1016/j.marpolbul.2011.05.033

Kaczmarsky L, Richardson LL (2007) Transmission of growth anomalies between Indo-Pacific *Porites* corals. J Invertebr Pathol 94(3):218–221. doi:10.1016/j.jip.2006.11.007

Kavousi J, Rezai H (2011) Coral lesions around some Iranian Islands in the Persian Gulf. Galaxea 13:5–6

Kavousi J, Seyfabadi J, Rezai H, Fenner D (2011) Coral reefs and communities of Qeshm Island, the Persian Gulf. Zool Stud 50:276–283

Kavousi J, Tavakoli-Kolour P, Barkhordari A, Bahrami A (2013) Mass mortality of *Porites* Corals on northern Persian Gulf reefs due to sediment-microbial interactions. Int J Mar Sci 3:306–310. doi:10.5376/ijms.2013.03.0038

Kavousi J, Tavakoli-Kolour P, Mohammadizadeh M, Bahrami A, Barkhordari A (2014) Mass coral bleaching in the northern Persian Gulf, 2012. Sci Mar 78(3):397–404. doi:10.3989/scimar.03914.16A

Loya Y, Bull G, Pichon M (1984) Tumor formations in scleractinian corals. Helgol Mar Res 37:99–112. doi:10.1007/BF01989297

McClanahan TIM, Weil E, Maina J (2009) Strong relationship between coral bleaching and growth anomalies in massive *Porites*. Gobal Change Biol 15:1804–1816

Rezai H, Wilson S, Claereboudt M, Riegl B (2004) Coral reef status in the ROPME sea area, Arabian/Persian Gulf, Gulf of Oman and Arabian Sea. In: Wilkinson C (ed) Status of coral reefs of the world, vol 1. Australian Institute of Marine Science, Washington DC, USA, pp 155–170

Riegl BM, Bruckner AW, Samimi-Namin K, Purkis SJ (2012) Diseases, harmful algae blooms (HABs) and their effects on Gulf coral populations and communities. In: Riegl BM, Purkis SJ (eds) Coral reefs of the Gulf. Springer, Netherlands, pp 107–125

Samimi-Namin K, Risk MJ, Hoeksema BW, Zohari Z, Rezai H (2010) Coral mortality and serpulid infestations associated with red tide, in the Persian Gulf. Coral Reefs 29:509. doi:10.1007/s00338-010-0601-x

Stimson J (2011) Ecological characterization of coral growth anomalies on *Porites compressa* in Hawaii. Coral Reefs 30(1):133–142. doi:10.1007/s00338-010-0672-8

Sutherland KP, Porter JW, Torres C (2004) Disease and immunity in Caribbean and Indo-Pacific zooxanthellate corals. Mar Ecol Prog Ser 266:273–302. doi:10.3354/meps266273

Weil E, Smith GW, Gil-Agudelo DL (2006) Status and progress in coral reef disease research. Dis Aquat Org 69:1–7. doi:10.3354/dao069001

Weil E, Irikawa A, Casareto B, Suzuki Y (2012) Extended geographic distribution of several Indo-Pacific coral reef diseases. Dis Aquat Org 98(2):163–170. doi:10.3354/dao02433

Willis BL, Page CA, Dinsdale EA (2004) Coral disease on the Great Barrier Reef. In: Rosenberg E, Loya Y (eds) Coral health and disease. Springer, Berlin, pp 69–104

20

High tolerance of *Chaetomorpha* sp. to salinity and water temperature enables survival and growth in stagnant waters of central Thailand

Isao Tsutsui · Tatsuo Miyoshi · Dusit Aue-umneoy · Jaruwan Songphatkaew · Chonlada Meeanan · Sirimas Klomkling · Halethichanok Sukchai · Piyarat Pinphoo · Iori Yamaguchi · Monthon Ganmanee · Yukio Maeno · Kaoru Hamano

Abstract The effects of salinity and water temperature on growth and survival of *Chaetomorpha* sp., one of the green tidal filamentous algae, were investigated. Field surveys revealed that this alga was abundant throughout the year in stagnant coastal waters of central Thailand. *Chaetomorpha* sp. was found in salinities of 3.4–90.0 and water temperatures of 20.1–40.9 °C. *Chaetomorpha* sp. biomass fluctuated between 9.2 and 162 kg dry weight in a saline swamp, without significant differences among seasons. The highest mean specific growth rate of approximately, 60 % day^{-1} was observed in laboratory experimental trials at salinities of 20–30 at 30 °C. Salinity and water temperature significantly affected growth of this alga independently and interactively. Based on these results, *Chaetomorpha* sp. can easily grow, survive and monopolize in stagnant waters where the salinity and temperature greatly fluctuate because of their euryhaline and eurythermal nature as well as their high growth rate.

Keywords Cladophoraceae · Euryhalinity · Eurythermy · Algal mat · Mangrove swamps

Introduction

Excessive blooms of ephemeral green seaweeds such as *Ulva*, *Cladophora*, and *Chaetomorpha* are referred to as green tide, which occurs in various regions worldwide. Although it is generally recognized that green tide is negative (Caffrey and Monahan 1999; Lehvo and Bäck 2001; Hiraoka et al. 2004; McAvoy and Klug 2005;

I. Tsutsui · Y. Maeno
Fisheries Division, Japan International Research Center for Agricultural Sciences (JIRCAS), Tsukuba, Ibaraki, Japan

T. Miyoshi · K. Hamano (✉)
National Research Institute of Fisheries and Environment of Inland Sea, Onomichi, Hiroshima, Japan
e-mail: hamano@affrc.go.jp

D. Aue-umneoy · J. Songphatkaew · P. Pinphoo · M. Ganmanee
Department of Animal Production and Fisheries, Faculty of Agricultural Technology, King Mongkut's Institute of Technology Ladkrabang (KMITL), Bangkok, Thailand

C. Meeanan · S. Klomkling · H. Sukchai
Shrimp Co-culture Research Laboratory (SCORL), King Mongkut's Institute of Technology Ladkrabang (KMITL), Bangkok, Thailand

I. Yamaguchi
Faculty of Engineering, Fukuoka Institute of Technology (FIT), Fukuoka, Japan

Morand and Merceron 2005; Liu et al. 2009, 2013; Yamochi 2013), a part of green tide component seaweeds have positive aspects both ecologically and socially (Ohno 1993; Tsutsui et al. 2005, 2010, 2012; Bolton et al. 2009; de Paula Silva et al. 2008; Cruz-Suárez et al. 2010; Maceiras et al. 2011). For example, several species of *Chaetomorpha* were started to investigated for their applicability in integrated aquaculture with marine animals (de Paula Silva et al. 2008; Tsutsui et al. 2010).

Chaetomorpha is a cosmopolitan genus containing 81 species from marine and brackish waters worldwide (Leliaert and Boedeker 2007; Guiry and Guiry 2014). Eight species have been reported in Thailand, most of which occur along the coast (Lewmanomont et al. 1995; Lewmanomont and Ogawa 1995; Coppejans et al. 2010; Hamano et al. 2012; Tsutsui et al. 2012), whereas some occur in stagnant waters, including aquaculture ponds, aqueducts, and reservoirs, in the mangrove hinterlands (Tsutsui et al. 2012). *Chaetomorpha* are often included in the results of floral surveys because the genus is very common in coastal areas; however, very little ecophysiological information is available for this taxon (de Paula Silva et al. 2008; Xu and Lin 2008; Deng et al. 2012).

An unidentified *Chaetomorpha* sp. grows widely and abundantly all through the year in stagnant waters in the mangrove hinterlands of central Thailand. The objectives of the present study were to clarify the ecological conditions of *Chaetomorpha* sp. and evaluate its tolerance to salinity and water temperature. We demonstrate the reason for monopoly of this species in stagnant environments with a wide range of salinity (freshwater to brine water) and water temperature.

Materials and methods

All the research activities in this study including field surveys and laboratory experiments were permitted by National Research Council of Thailand (NRCT) (project ID 2011/005). A part of field survey sites was privately owned land, and the authors got oral approvals for sampling from all private land owners. The field studies did not involve endangered or protected species.

Study organism

An unidentified species *Chaetomorpha* sp. was used for this study. This alga is very thin, light to dark green in color, and soft to the touch; it consists of uniseriate unbranched filaments growing to 50–60 cm when fully developed. Lateral rhizoidal filaments are not observed at any time during the year. The cells are cylindrical, measuring 45–136 μm in length and 27–64 μm in diameter (Fig. 1). The accession numbers of this seaweed at DDBJ were AB759541 and AB819998 for 18s rDNA and 28s rDNA, respectively.

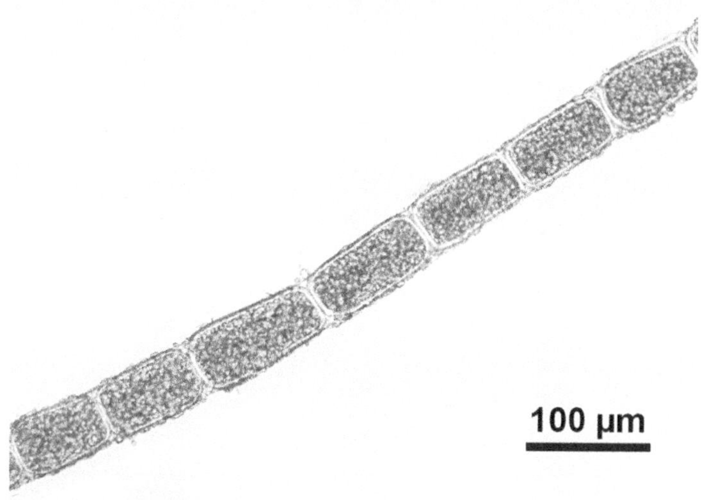

Fig. 1 A microscopic view of *Chaetomorpha* sp. A healthy and *green colored* vegetative filament is shown

Field surveys

Survey for habitat and habit

A field survey to determine the habitat and habit of *Chaetomorpha* sp. was conducted in coastal areas of central Thailand from 2010 to 2013 (Fig. 2; Table 1). During the survey, salinity and water temperature were measured on site using a conductivity meter (Condo 3210, WTW Inc., Weilheim, Germany), and the types of habitats and habits were recorded. A small number of *Chaetomorpha* sp. thalli were directly collected by hand at each survey site and immediately placed in plastic bags with water, after which they were transported to the Shrimp Co-culture Research Laboratory (SCORL), Faculty of Agricultural Technology, King Mongkut's Institute of Technology Ladkrabang (KMITL), Thailand as samples for morphological observation. Sample morphology was observed under an optical microscope (BX 41, Olympus Co. Ltd., Tokyo, Japan) to identify the species. Some samples were dried and pressed and retained as herbarium specimens at SCORL.

Survey for biomass fluctuations

The biomass fluctuations of *Chaetomorpha* sp. were surveyed every 2 weeks from September 2011 to January 2013 in a small salt swamp, which is survey designation number 20 in Table 1. This salt swamp was a semi-closed system with very little water exchange surrounded by a mangrove forest and aqueducts, and had a volume of approximately, 100 m^3 at full capacity (Fig. 3c). A 16 × 16-cm quadrat was prepared with a mesh net to randomly cover a portion of the *Chaetomorpha* sp. community, and all thalli inside the quadrat were collected with scissors. Three replicates were collected during each survey. The total coverage of *Chaetomorpha* sp. was visually estimated, and the water surface area was measured using a laser distance meter (GLM 250 VF, Robert Bosch Inc., Stuttgart, Germany). In addition, salinity and water temperature was measured onsite using a conductivity meter (Condo 3210, WTW Inc.). Water samples were returned to SCORL, where they were analyzed for total inorganic nitrogen (TIN) and phosphate using a portable spectrophotometer (DR2400, Hach Co. Ltd., Loveland, CO, USA). Samples of *Chaetomorpha* sp. were immediately placed in plastic bags with water and transported to SCORL, where they were rinsed with fresh water to remove epiphytes and attached animals. The seaweed samples were then dried in a constant temperature oven (DK300, Yamato Scientific Co. Ltd., Tokyo, Japan) at 85 °C for 5 h, after which the dry weights were measured on an electric balance (EK300H, A&D Co. Ltd., Tokyo, Japan). Total biomass (TB) at the survey station was estimated using the following formula:

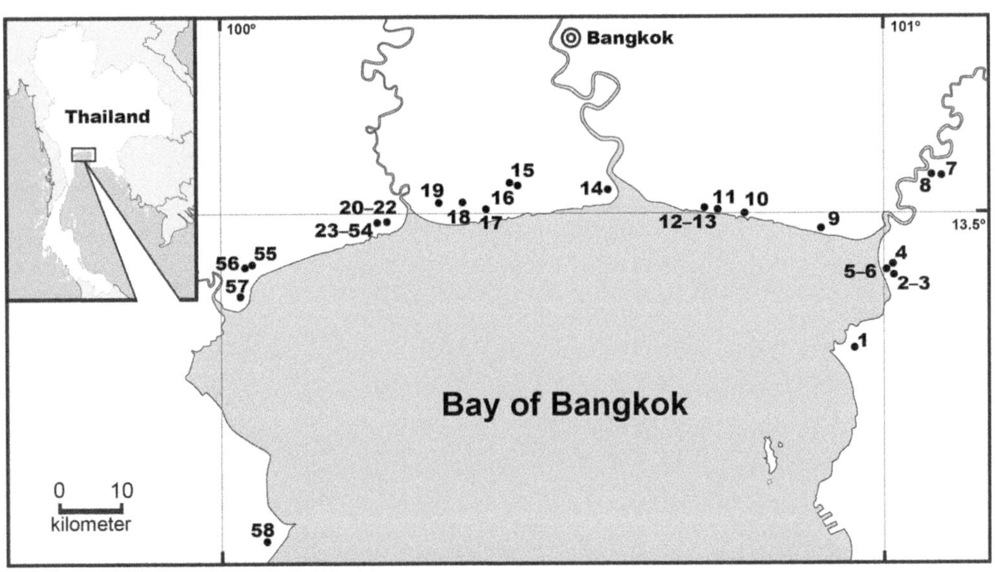

Fig. 2 Map showing locations for the habitat and habit field surveys in central Thailand. *Numbers* refer to the survey designation numbers in Table 1

Table 1 Location, habitats, and habits of *Chaetomorpha* sp. surveyed in central Thailand

No.	Location and date	Habitat	Area (m^2)	Cover (%)	Depth (m)	Habit
01	AngSila, Chonburi 06 July 2011	RE	100	20	0.3	F, B
02	KhlongTamru, Chonburi 06 July 2011	AA	100	60	0.3	F, B
03	KhlongTamru, Chonburi 06 July 2011	SW	30	70	0.2	F, B
04	KhlongTamru, Chonburi 06 July 2011	SW	200	40	0.5	F, B
05	KhlongTamru, Chonburi 09 July 2013	SW	200	10	0.1	F, B
06	KhlongTamru, Chonburi 06 July 2011	RE	1,050	40	0.2	F, B
07	KhlongTamru, Chonburi 09 July 2013	RE	1,050	20	0.4	F, B
08	Bang Pu, Chachengsao 05 July 2013	RE	12,500	60	0.2	F, B
09	Bang Pakong, Chachengsao 05 July 2013	AP	1,450	20	0.2	F, B
10	Khlongdan, SamutPrakan 19 May 2011	AA	100	40	0.5	B
11	Bang Pu Mai, SamutPrakan 19 May 2011	SW	50	40	0.4	F, B
12	Bang Pu Mai, SamutPrakan 9 July 2013	SW	50	10	0.2	F, B
13	Muang, SamutPrakan 05 July 2013	RE	150	40	0.2	F, B
14	LaemFaPha, SamutPrakan 01 June 2011	AA	100	70	0.2	F, B
15	Bang KhunThian, Bangkok 22 July 2013	RE	46,100	5	0.5	F, B
16	Bang KhunThian, Bangkok 22 July 2013	AA	200	60	0.5	F, B
17	Muang, SamutSakhon 22 July 2013	AP	550	50	0.4	F, B
18	Muang, SamutSakhon 22 July 2013	AP	640	40	0.1	F, B
19	Muang, SamutSakhon 22 July 2013	AA	110	60	0.5	F, B
20	Chai Mongkhon, SamutSakhon 03 August 2011	SW	100	75	0.6	F, B
21	Chai Mongkhon, SamutSakhon 23 December 2013	SW	100	75	0.6	F, B
22	Chai Mongkhon, SamutSakhon 29 December 2013	SW	100	75	0.6	F, B
23	Chai Mongkhon, SamutSakhon 25 November 2011	AP	230	80	0.6	F, B
24	Chai Mongkhon, SamutSakhon 14 December 2011	AP	190	80	0.6	F, B

Table 1 continued

No.	Location and date	Habitat	Area (m^2)	Cover (%)	Depth (m)	Habit
25	Chai Mongkhon, SamutSakhon 28 December 2011	AP	75	70	0.6	F, B
26	Chai Mongkhon, SamutSakhon 18 January 2012	AP	230	10	0.7	F, B
27	Chai Mongkhon, SamutSakhon 01 February 2012	AP	230	10	0.7	F, B
28	Chai Mongkhon, SamutSakhon 15 February 2012	AP	180	5	0.6	F, B
29	Chai Mongkhon, SamutSakhon 29 February 2012	AP	150	5	0.5	F, B
30	Chai Mongkhon, SamutSakhon 14 March 2012	AP	230	70	0.7	F
31	Chai Mongkhon, SamutSakhon 28 March 2012	AP	230	20	0.7	F
32	Chai Mongkhon, SamutSakhon 12 April 2012	AP	230	15	0.7	F, B
33	Chai Mongkhon, SamutSakhon 28 April 2012	AP	180	20	0.7	F, B
34	Chai Mongkhon, SamutSakhon 09 May 2012	AP	75	30	0.4	F, B
35	Chai Mongkhon, SamutSakhon 23 May 2012	AP	230	20	0.7	F, B
36	Chai Mongkhon, SamutSakhon 06 June 2012	AP	230	20	0.7	F, B
37	Chai Mongkhon, SamutSakhon 20 June 2012	AP	230	70	0.7	F, B
38	Chai Mongkhon, SamutSakhon 04 July 2012	AP	230	80	0.7	F, B
39	Chai Mongkhon, SamutSakhon 18 July 2012	AP	230	80	0.7	F, B
40	Chai Mongkhon, SamutSakhon 01 August 2012	AP	200	80	0.6	F, B
41	Chai Mongkhon, SamutSakhon 17 August 2012	AP	200	70	0.6	F, B
42	Chai Mongkhon, SamutSakhon 31 August 2012	AP	200	40	0.6	F, B
43	Chai Mongkhon, SamutSakhon 14 September 2012	AP	230	60	0.7	F, B
44	Chai Mongkhon, SamutSakhon 27 September 2012	AP	230	60	0.7	F, B
45	Chai Mongkhon, SamutSakhon 26 Ocober 2012	AP	230	60	0.7	F, B
46	Chai Mongkhon, SamutSakhon 27 September 2012	AP	200	30	0.6	F, B
47	Chai Mongkhon, SamutSakhon 7 November 2012	AP	230	10	0.7	F, B
48	Chai Mongkhon, SamutSakhon 5 December 2012	AP	230	5	0.7	F, B

Table 1 continued

No.	Location and date	Habitat	Area (m^2)	Cover (%)	Depth (m)	Habit
49	Chai Mongkhon, SamutSakhon 19 December 2012	AP	150	5	0.5	F, B
50	Chai Mongkhon, SamutSakhon 2 February 2013	AP	150	50	0.5	F, B
51	Chai Mongkhon, SamutSakhon 16 February 2013	AP	160	20	0.6	F, B
52	Chai Mongkhon, SamutSakhon 2 March 2013	AP	180	50	0.7	F, B
53	Chai Mongkhon, SamutSakhon 15 April 2013	AP	230	5	0.7	F, B
54	Chai Mongkhon, SamutSakhon 11 June 2013	AP	230	5	0.7	F
55	LatYai, SamutSongkhram 29 June 2010	AA	600	40	0.7	F, B
56	LatYai, SamutSongkhram 22 September 2010	AQ	800	70	1.0	F, B
57	LatYai, SamutSongkhram 22 September 2010	AQ	800	40	1.0	B
58	LaemPhakBia, Phetchaburi 19 September 2012	AP	60	50	0.3	F, B

AA aqueduct for aquaculture/salt field, *AP* abandoned aqueduct/pond, *AQ* aquaculture pond, *RE* reservoir, *SW* salt swamp, *F* floating strand from bottom, *B* dense mat on bottom

Fig. 3 *Chaetomorpha* sp. in natural habitats. **a** Plants forming dense and submerged tangled mats on the bottom of an aquaculture pond; **b** plants forming floating strands that cover the water surface in a shrimp culture aqueduct; **c** plants covering the water surface in a salt swamp; **d** plants growing in an abandoned aquaculture pond; **e** plants propagating in a reservoir; **f** close-up view of plants growing in a salt swamp

$$\text{TB (kg DW)} = \text{DW of samples}/(0.16 \times 0.16) \times \text{total area } (\text{m}^2)$$

where DW is dry weight.

Preparation for laboratory experiments

Plants collected at survey designation number 20 in Table 1 and cultured at SCORL were utilized for survival and growth experiments. Prior to the experiments, powdered, artificial seawater (Rhoto Marine II, Rei-Sea Co. Ltd., Tokyo, Japan) was diluted with distilled water and adjusted to salinities of 0.1, 0.2, 0.5, 5, 10, 20, 30, 40, 50, 60, 70, and 80 using a conductivity meter (Condo 3210, WTW Inc.), and distilled water was used for the 0 salinity treatment. These experimental media were sterilized at 121 °C for 20 min using an autoclave (HVE-50, Hirayama Manufacturing Corp., Saitama, Japan), after which the pH was adjusted to 8.0 with NaOH using a pH meter (SevenGo, Mettler Toledo, Schwerzenbach, Switzerland). As a nutrient source, a ratio of 3 mL of stock solution of Provasoli's enriched seawater (PES) in 1 L artificial seawater or freshwater (approximately, 1/6 concentration of PES) was used (Provasoli 1966), and germanium dioxide (3 mg L^{-1}) was added to prevent growth of diatoms (Lewin 1966). Plants cultured at SCORL were cut into small pieces and then acclimated to each of the experimental salinities for 1 week. *Chaetomorpha* sp. was then cut into smaller pieces approximately, 2 mm in length using surgical scissors to prevent maturation (Tanaka 2001). Individual healthy filaments (without epiphytes or damage) measuring exactly 2.0 mm in length were selected using a stereomicroscope (SZX12, Olympus Co. Ltd., Tokyo, Japan) for the experiment.

Laboratory experiments

Experiment for survival

The 2.0-mm filaments were individually placed in each well with 1.8 mL of media of various salinities in multiwell plates. Twenty individuals were used for the experiment involving each combination of salinity (0, 0.1, 0.2, 0.5, 5, 10, 20, 30, 40, 50, 60, 70, and 80) and water temperature (15, 20, 25, 30, 35, and 40 °C). The multiplates were placed in a temperature gradient chamber (TG-180CCFL-3LE, Nippon Medical and Chemical Instruments Co. Ltd., Osaka, Japan) under a 12:12 h light:dark cycle and photon flux density of approximately, 70 μmol m^{-2} s^{-1}. The algae were cultured for 7 days. This culture duration was determined with our preliminary work and previous works (Matsuyama-Serisawa et al. 2004; de Paula Silva et al. 2008). At the end of the experimental period, algae in each of the treatments were observed under a stereomicroscope. The surviving and dead filaments were identified, and survival rates were calculated. We defined green filaments as live filaments (Fig. 1), and clear and empty ones without any pores, from which zoospores or gametes could be released, as dead filaments.

Experiment for growth rate

Twenty filaments, 2.0 mm in length, were individually cultured in each well of the multiwell plates with 1.8 mL of media of various salinities as described for the survival experiment above. The salinities were 0, 0.1, 0.2, 0.5, 5, 10, 20, 30, 40, 50, 60, 70, and 80 and the water temperatures were 15, 20, 25, 30, and 35 °C for the growth experiment. Multiwell plates were set in the same incubation conditions, e.g., light cycle and photon flux density and the algae were cultured for 7 days. At the end of this period, individual algal lengths were measured under a stereomicroscope in each combination of salinity and water temperature. Specific growth rates (SGR) were calculated using the following formula:

$$\text{SGR}\left(\%\ \text{day}^{-1}\right) = (\ln L_t - \ln L_i) \times D^{-1} \times 100$$

where L_t is final total length, L_i is initial length, and D is culture duration.

Experiment for growth patterns

After the survival and SGR experiments, the growth patterns of *Chaetomorpha* sp. were examined under optimum salinities for each water temperature, e.g., salinities of 10, 10, 20, 30, and 30 in water temperatures of 15, 20, 25, 30, and 35 °C, respectively. Filaments of 2.0 mm were cultured in multiwell plates under the same incubation conditions. Algal lengths of each *Chaetomorpha* sp. filament were measured under a stereomicroscope every day for 7 days.

Statistical analysis

Mean values are stated ± their standard error (SE). Statistical analyses were performed using the software JMP10 (SAS Institute, Cary, NC, USA). Differences in mean biomass and mean physical and chemical environmental factors among seasons during the field survey were analyzed by the Tukey–Kramer HSD test ($P < 0.05$). Receiver operating characteristic (ROC) analysis was adopted to determine the cut-off values for survival of the seaweed related to salinity and water temperature. The area under the curve (AUC) of the ROC curve was used to determine adequacy. The best cut-off point was selected as the highest value of sensitivity − (1 − specificity). Logistic regression analysis was then used to consider the relationship among survival of this seaweed in combinations of salinities and water temperatures, which were determined from the result of ROC analysis. A two-way random-effects analysis of variance (ANOVA) was followed by the Tukey–Kramer HSD test (ad hoc test) to compare the effect of salinity and water temperature on growth of *Chaetomorpha* sp.

Results

Habitat and habit

Chaetomorpha sp. was abundant in shallow and stagnant waters with little water exchange, including aquaculture ponds, salt swamps, aqueducts supplying salt fields, and abandoned aquaculture ponds at mangrove hinterlands in coastal areas (Table 1). The stagnant water area ranged from 30 to 46,100 m^2 but usually covered 100–500 m^2. Plants formed entwined spongy mats lying loosely on the bottom and frequently grew upwards as long interwoven strands covering the water surface (Fig. 3). This alga was found in water with salinity and water temperature ranges of 3.4–90.0 and 20.1–40.9 °C, respectively (Fig. 4a). Approximately, 35 % of these sites were within the salinity range of 20–30, and 50 % were between 30 and 35 °C (Fig. 4b, c). No obvious relationship was detected in the distribution between salinity and water temperature.

Seasonal change in biomass

Changes in estimated total biomass and physical and chemical factors at the survey station are shown in Fig. 5. Total biomass fluctuated, rising and falling slightly throughout the year between 9.2 and 162.0 kg DW, and peaks were observed at 2–4-month intervals. Mean total biomass was presented as standard error (SE) and the values were 94.6 ± 8.39, 86.0 ± 20.47, and 58.1 ± 17.50 kg DW during the rainy season (June–November), dry-cold season (December–February), and dry-hot season (March–May), respectively, without significant differences among seasons (Table 2). Salinity fluctuated between 6.0 and 44.0, and tended to stably increase from the late rainy season to the middle dry season with a small rise and fall. Mean salinity (±SE) was lowest at 20.8 ± 2.68 during the rainy season, and this value significantly differed from the salinity of 30.4 ± 2.41 observed during the dry-cold season and 32.2 ± 2.97 observed during the dry-hot season. In addition, water temperature varied between 23.5 and 36.6 °C. The lowest mean (±SE) water temperature was 29.8 °C ± 1.10 °C during the dry-cold season and the highest was 32.6 °C ± 0.96 °C during the dry-hot season. Significant differences were observed between the rainy and dry-hot seasons and dry-cold and dry-hot seasons. Nutrient conditions were highly eutrophicated in the salt swamp of the survey station, as indicated by the TIN values of 1.0–3.6 mg L^{-1} and phosphate values of 2.0–7.6 mg L^{-1}. No significant differences in TIN or phosphate were observed among seasons. *U. intestinalis* and *U. clathrata* opportunistically appeared several times at this survey station; however, their coverage was <5 %.

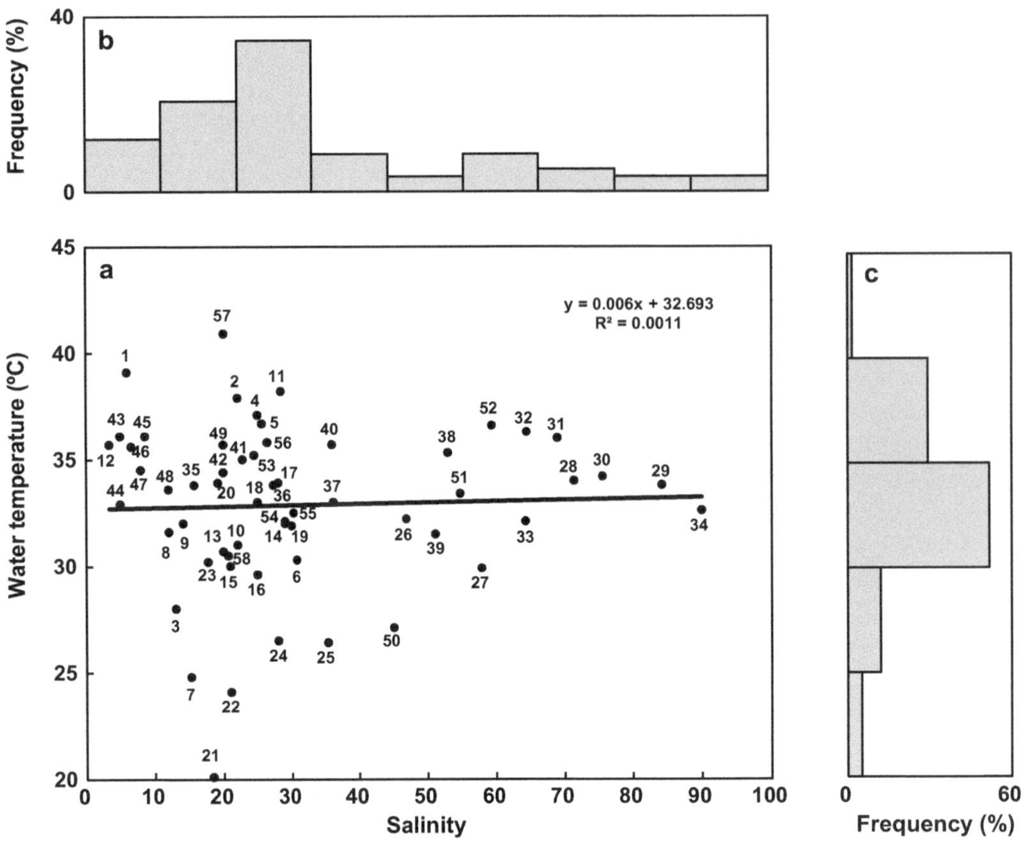

Fig. 4 Salinities and water temperatures in the natural habitats of *Chaetomorpha* sp. in central Thailand. **a** Scatter plot showing combinations at every station. *Numbers* in the plot area refer to the survey designation nos. in Table 1; **b** frequency distributions of salinity; **c** frequency distributions of water temperature

Survival at different salinities and water temperatures

Chaetomorpha sp. exhibited 100 % survival over the salinity range of 0.5–80 and water temperatures of 15–35 °C (Table 3). Even at a salinity of 0, survival rate was 35–70 % at temperatures of 20–30 °C. However, survival was very low in all experimental salinities at 40 °C. Results of the ROC analysis showed that the best cut-off values were observed between 0 and 0.1 for salinity (AUC 0.6718) and between 35 °C and 40 °C for water temperature (AUC 0.8819). These tendencies were statistically analyzed using a logistic regression model. Significant differences were observed between the combination of salinity of 0.1–80 and water temperature of 15–35 °C and at other combinations of salinity and water temperatures (Table 4).

SGR under different salinities and water temperatures

The highest mean (±SE) growth rates of *Chaetomorpha* sp. in 7 days were observed at salinities of 20 and 30 at a water temperature of 30 °C (59.8 ± 4.34 and 61.0 ± 3.24 % day^{-1}, respectively; Figs. 6, 7). The highest individual SGR was 68.2 % in a filament that grew from 2 to 237 mm at a salinity of 30 and water temperature of 30 °C. As water temperature increased, SGR peaks shifted to higher salinities (Fig. 6), with the peak SGR occurring at a salinity of 10 at 15 and 20 °C, at a salinity of 20 at 25 °C, and at a salinity of 30 at 30 and 35 °C. The results of a two-way random-effects ANOVA showed that both salinity and water temperature significantly affected growth of *Chaetomorpha* sp. independently and interactively (Table 5). As the result of Tukey–Kramer HSD test (ad hoc test), significant differences were observed in all pairwise comparisons of water temperature. In addition, significant differences were observed in almost all pairwise comparisons of salinities, except for the combination of salinity 0 vs. 0.1, 0 vs. 80, 0.1 vs. 80, and 0.2 vs. 80.

Fig. 5 Fluctuations in total biomass of *Chaetomorpha* sp., and several physical factors at survey station. Data are mean ± standard error: *R*, rainy season; DC, cold period during dry season; DH, hot period during dry season

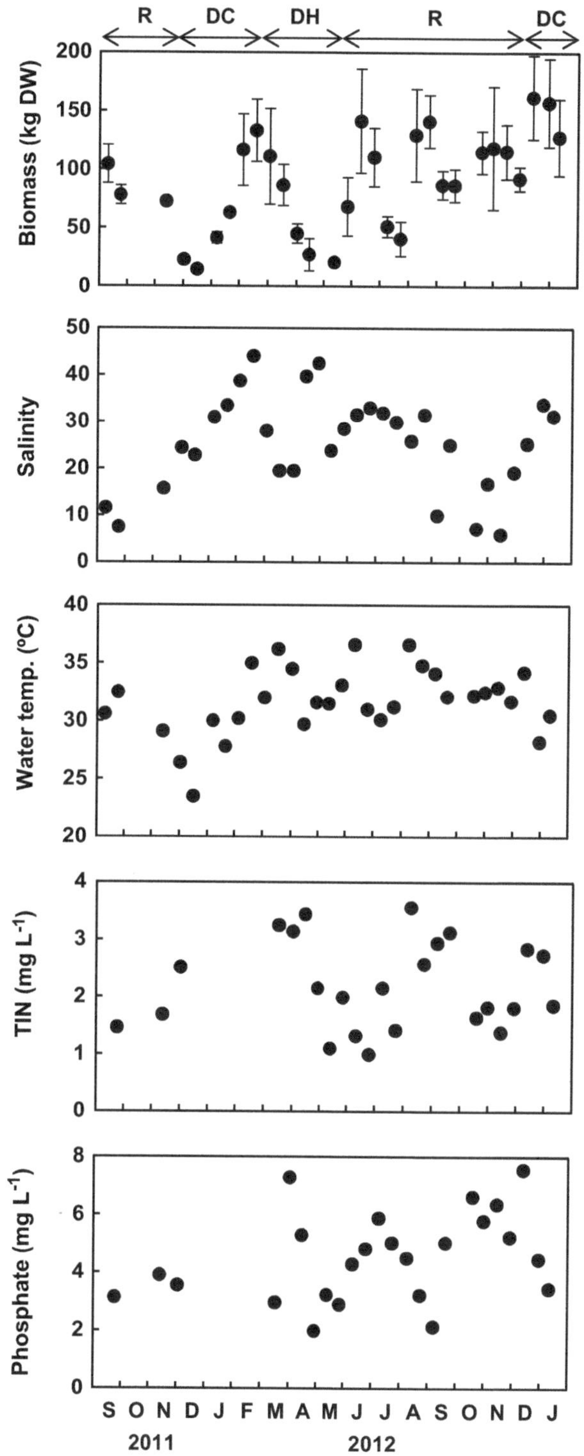

Growth patterns

The increase in lengths of *Chaetomorpha* sp. at optimum salinity under different water temperatures during the final 7 days of the experiment was reflected as quadratic curves (Fig. 8). A maximum mean length (±SE) of 151.9 ± 4.06 mm was observed on the last day in a salinity of 30 at 30 °C, followed by a salinity of 30 at 35 °C, a salinity of 20 at 25 °C, a salinity of 10 at 20 °C, and a salinity of 10 at 15 °C. This growth tendency was the same as that observed in the SGR experiment; however, terminal mean lengths at each water temperature were shorter compared with those of the SGR experiment.

Table 2 Mean biomass of *Chaetomorpha* sp., several physical factors at the survey station during different seasons

	Rainy season	Dry season	
		Cold period	Hot period
Total biomass (kg DW)			
Mean total biomass	94.6 ± 8.39[a]	86.0 ± 20.47[a]	58.1 ± 17.50[a]
Maximum	141.3	162.0	110.0
Minimum	40.8	9.2	20.7
Salinity			
Mean	20.8 ± 2.68[a]	30.4 ± 2.41[b]	32.2 ± 2.97[b]
Maximum	33.0	44.0	42.5
Minimum	6.0	19.2	23.8
Water temperature (°C)			
Mean	32.6 ± 0.56[a]	29.8 ± 1.10[b]	32.6 ± 0.96[a, b]
Maximum	36.6	35.0	36.2
Minimum	29.1	23.5	29.7
Total nitrogen (mg L^{-1})			
Mean	2.0 ± 0.21[a]	2.3 ± 0.21[a]	2.6 ± 0.43[a]
Maximum	3.6	2.8	3.4
Minimum	1.0	1.8	1.1
Dissolved phosphate (mg L^{-1})			
Mean	4.5 ± 0.37[a]	4.8 ± 0.75[a]	4.1 ± 0.95[a]
Maximum	6.6	7.6	7.3
Minimum	2.1	3.4	2.0

Data were collected during 2011 to 2013 in SamutSakhon Province, Thailand. Values are mean ± standard error. Superscripts in the same column indicate homogeneous subsets as determined by the Tukey–Kramer HSD test ($P < 0.05$)

Table 3 Survival rates of *Chaetomorpha* sp. cultured under various salinities and water temperatures for 7 days

Salinity	Water temperature (°C)					
	15	20	25	30	35	40
0	0	35	70	40	0	0
0.1	100	100	100	15	40	0
0.2	100	100	100	100	60	5
0.5	100	100	100	100	100	0
5	100	100	100	100	100	0
10	100	100	100	100	100	0
20	100	100	100	100	100	0
30	100	100	100	100	100	0
40	100	100	100	100	100	0
50	100	100	100	100	100	0
60	100	100	100	100	100	15
70	100	100	100	100	100	10
80	100	100	100	100	100	0

Discussion

Organisms living in estuaries and intertidal zones are subjected to stress from rapid changes in salinity and temperature, particularly when they are exposed at low tides. Organisms living in the intertidal zone are generally able to tolerate wide ranges in salinity and temperature (Macler 1988; Matsuyama-Serisawa et al.

Table 4 Results of logistic regression analysis for survival rates of *Chaetomorpha* sp.

Salinity	Water temperature (°C)	Total number	Survival number	Odds ratio (95 % CI)	Chi squared test
0.1–80	15–35	1,200	1,163	1.000	–
0–80	40	260	6	1,330.637 (555.687–3,186.319)	$P < 0.0001$
0	15–35	100	29	76.955 (44.759–132.311)	$P < 0.0001$

Seaweeds were cultured under different combinations of salinity and water temperatures for 7 days. Combinations were determined from the results of a receiver operating characteristic analysis: CI, confidence interval

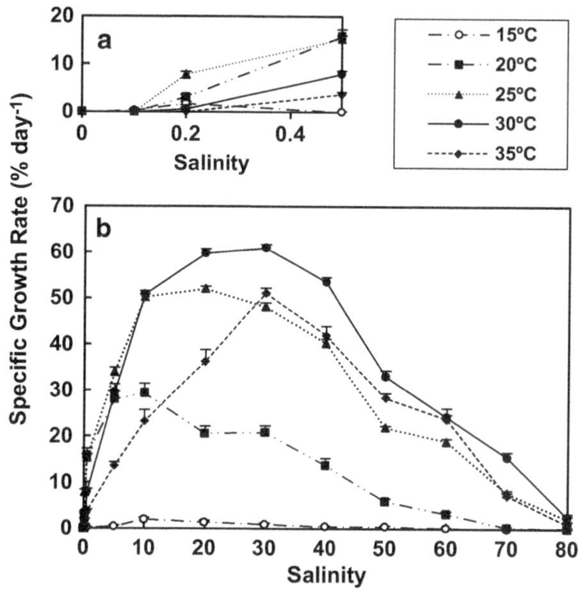

Fig. 6 Mean specific growth rates (SGR) of *Chaetomorpha* sp. under different salinities and water temperatures. Error bars show standard errors for 20 individuals. **a** SGRs at low salinity (0–0.5); **b** SGRs over the full salinity range investigated (0–80)

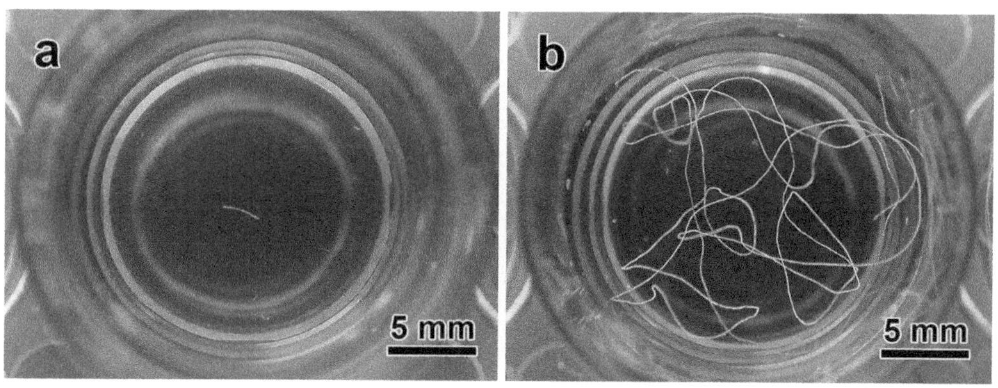

Fig. 7 *Chaetomorpha* sp. cultured in multiwells. **a** Filament of 2 mm at the beginning of the experiment; **b** filaments grew >200 mm by 7 days after starting under a combination of salinity of 30 and water temperature of 30 °C

2004). The salinity ranges in intertidal environments where several green tidal species observed are shown in Table 6. Comparing among green tidal species, it is obvious that the present seaweed, *Chaetomorpha* sp, has wider tolerance to salinity. Flora in the stagnant waters of the present study was almost monopolized by *Chaetomorpha* sp., and intertidal species, such as *U. intestinalis* and *U. clathrata*, were only opportunistically observed. Salinity ranged from 3.4 to 90 (Fig. 4) in this study, which is likely one of the reasons that ordinal intertidal species could not establish their communities because of narrower salinity tolerances of *Ulva*. In contrast, based on our survival experimental result, *Chaetomorpha* sp. could survive in salinities of 0.1–80. This alga is capable of survive and growth in water environments that widely fluctuate from freshwater to brine, whereas other seaweeds are unable to dominate the population and maintain a dense community throughout the year in stagnant waters.

Table 5 Results of two-way random-effect analysis of variance for specific growth rates of *Chaetomorpha* sp.

Source	*df*	Mean square	*F* value	Significance (*P* value)
Salinity				
Hypothesis	4	27,012.948	11.993	0.000
Error	48	2,252.441		
Water temp.				
Hypothesis	12	18,769.030	8.333	0.000
Error	48	2,252.441		
Interaction (S × W)				
Hypothesis	48	2,252.441	118.688	0.000
Error	1,235	18.978		

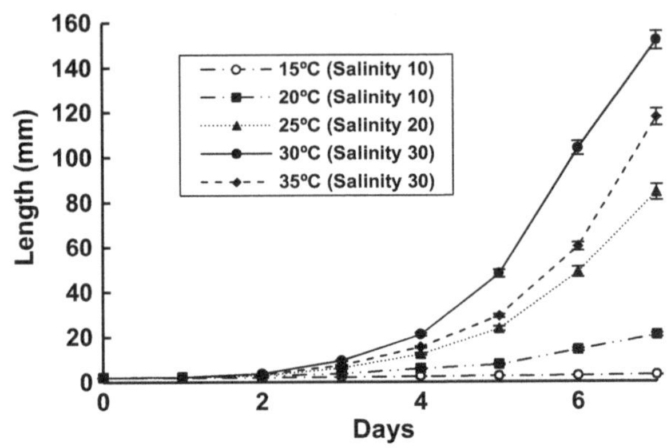

Fig. 8 Change in lengths of *Chaetomorpha* sp. cultured under optimum salinities in different water temperatures. *Error bars* show standard errors for 20 individuals

Table 6 Salinities, water temperatures, and growth rates ranges of the major green tidal seaweeds

Taxon	Salinity (psu)	Water temp. (°C)	Growth rate (% day^{-1})	Condition	References
Ulvales					
Ulva sp.	5–45		5–23	Laboratory culture	de Paula Silva et al. (2008)
U. clathrata (as *Enteromorphaclathrata*)	26–33	26–29	6–12	Outdoor culture	Copertino et al. (2009)
U. fasciata	34–41	25–33	~0.3	Chamber in pond draining fishpond	Mmochi et al. (2002)
U. intestinalis	0.6–27.7			Estuary	McAvoy and Klug (2005)
U. pertusa	6.6–66.2	20	10–23	Laboratory culture	Kakinuma et al. (2004)
U. prolifera (as *E. prolifera*)	0.3–33.1	7.5–21.6		Natural	Ohno and Takahashi (1988)
U. reticulata	23–40	27.3–29.6	1.2 (2.7)	Cage in channel	Msuya and Neori (2002)
Cladophorales					
Present species (*Chaetomorpha* sp.)	0.2–80	15–35	2–60	Laboratory culture	Present study
C. indica	5–45		5–30	Laboratory culture	de Paula Silva et al. (2008)
C. linum	20–45	25–32	<7.5	Laboratory culture	Xu and Lin (2008)
C. valida	17–29		5–25	Laboratory culture	Deng et al. (2012)
Cladophora coelothrix	15–45		12–20	Laboratory culture	de Paula Silva et al. (2008)
Rhizoclonium sp.	10–40	18–22 (in natural site)	4–19	Laboratory culture	Matsuyama-Serisawa et al. (2004)
R. riparium	0.1–34	20	2–10	Laboratory culture	Imai et al. (1997)

It is well-known that water temperature affects seaweed growth (Xu and Lin 2008). The upper lethal temperature limits of common green tidal seaweeds growing in intertidal zones, estuaries, and lagoons considered to be 33 °C and under (Table 6). In contrast, *Chaetomorpha* sp. survived well (40–100 %) and grew (50 % day^{-1}) at 35 °C, but almost all thalli died at >40 °C in the laboratory experiments. Our results, including the ROC analysis, showed that the upper lethal limit of *Chaetomorpha* sp. was 35–40 °C. Overall, the tolerance to high water temperature of this species enables it to grow in the relatively higher water temperatures commonly found in stagnant tropical waters.

Plants comprising the green tide can inordinately grow under favorable environmental conditions (Keesing et al. 2011). The maximum SGRs of plants belonging to genera related to the green tide are 23 % day^{-1} for *Ulva* sp., 23 % day^{-1} for *U. pertusa*, <20 % day^{-1} for *U. clathrata*, 12 % day^{-1} for *C. indica*, 19 % day^{-1} for *Rhizoclonium* sp., and 20 % day^{-1} for *Cladophora coelothirix* (Table 6). In contrast, the maximum SGR of the present species was approximately, 60 % day^{-1} under optimum conditions (salinity 20–30 and 30 °C), which was consistent with distribution frequencies of this alga in the field survey (Fig. 4). The maximum growth rate of *Chaetomorpha* sp. was more than two times higher than that of the aforementioned species, which enables this alga to rapidly occupy vacant niches in stagnant water.

Desiccation negatively affects seaweed growth and photosynthesis (Yamochi 2013). Even during the culture duration (7 days), terminal filament lengths were shorter in the growth pattern experiment, in which we performed daily length measurements compared with those in the SGR experiment, which was static cultured. We consider that the low growth in the growth pattern experiment was affected by desiccation.

Conclusion

Our fieldwork results indicate that *Chaetomorpha* sp. was distributed in stagnant environments widely ranging in salinity and water temperature. Laboratory experiments indicate that this species has characteristics of high euryhalinity, eurythermy, and a high growth rate. The euryhaline and eurythermal nature of this seaweed enables it to settle in fluctuating stagnant water environments from 0.1 (fresh water) to 80 (brine) at temperatures <35 °C, and the high growth nature enables it to rapidly occupy vacant niches in stagnant water.

Acknowledgments This study was performed as part of the international research project "Development of Aquaculture Technologies for Suitable and Equitable Production of Aquatic Products in Tropical Coastal Areas" conducted by the Faculty of Agricultural Technology, KMITL and Fisheries Division, JIRCAS. The field surveys were partially financed by a research grant from the Toyota Foundation (D09-R-0648). The authors would also like to sincerely thank Professor Emeritus Dr. Masao Ohno, Kochi University, Japan for giving useful comments on this manuscript. We thank the National Research Council of Thailand (NRCT), particularly, Ms. Pannee Panyawattanaporn, Chief, Section of Foreign Researcher Management, and Ms. Rapeeporn Reakspuit for providing useful advice to conduct this study in Thailand. The authors also greatly appreciate Dean Sakchai Choochote, and the Head of Fisheries group, Dr. Paveena Taveekijakarn, Faculty of Agricultural Technology, KMITL for their helpful advice. We thank Dr. Masayoshi Saito, Program Director, JIRCAS, and Dr. Hiroyuki Sudo, Project Leader, JIRCAS, for their constructive comments.

Conflict of interest The authors declare that they have no competing interests.

Authors' contributions The work presented here was carried out in collaboration between the authors. IT conceived of the study, designed methods, performed surveys and experiment and drafted the manuscript. TM participated in the sequence experiment, DA coordinated and performed field surveys, JS, CM, SK, HS, PP, IY carried out field surveys and growth experiments, MG participated in experimental coordination. YM participated in experimental coordination and helped draft manuscript. KH conceived of the study, designed methods, and drafted the manuscript. All authors read and approved the final manuscript.

References

Bolton JJ, Robertson-Andersson DV, Shuuluka D, Kandjengo L (2009) Growing *Ulva* (Chlorophyta) in intergrated systems as a commercial crop for abalone feed in South Africa: a SWOT analysis. J Appl Phycol 21:575–583. doi:10.1007/s10811-008-9385-6

Caffrey JM, Monahan C (1999) Filamentous algal control using barley straw. Hydrobiologia 415:315–318. doi:10.1023/A:1003884211027

Copertino MS, Tormena T, Seeliger U (2009) Biofiltering efficiency, uptake and assimilation rates of *Ulva clathrata* (Roth) J. Agardh (Chlorophyceae) cultivated in shrimp aquaculture waste water. J Appl Phycol 21:31–45. doi:10.1007/s10811-008-9357-x

Coppejans E, Prathep A, Leliaert F, Lewmanomont K, De Clerck O (2010) Seaweeds of Mu Ko Tha Lae Tai (SE Thailand): methodologies and field guide to the dominant species. Biodiversity Research and Training Program (BRT), Bangkok

Cruz-Suárez LE, León A, Peña-Rodríguez A, Rodríguez-Peña G, Moll B, Ricque-Marie D (2010) Shrimp/*Ulva* co-culture: a sustainable alternative to diminish the need for artificial feed and improve shrimp quality. Aquaculture 301:64–68. doi:10.1016/j.aquaculture.2010.01.021

de Paula Silva PH, McBride S, de Nys R, Paul NA (2008) Integrating filamentous 'green tide' algae into tropical pond-based aquaculture. Aquaculture 284:74–80. doi:10.1061/j.aquaculture.200807.035

Deng Y, Tang X, Huang B, Ding L (2012) Effect of temperature and irradiance on the growth and reproduction of the green macroalga, *Chaetomorpha valida* (Cladophoraceae, Chlorophyta). J Appl Phycol 24:927–933. doi:10.1007/s10811-011-9713-0

Guiry MD, Guiry GM (2014) AlgaeBase. World-wide electronic publication, National University of Ireland, Galway. http://www.algaebase.org. Accessed 8 Oct 2014

Hamano K, Tsutsui I, Songphatkaew J, Aue-umneoy D, Srisapoome P (2012) A newly developing co-culture system using discarded seaweed to enhance the production of indigenous shrimp species in Southeast Asia. In: Tanaka K, Morioka S, Watanabe S (eds) Sustainable stock management and development of aquaculture technology suitable for Southeast Asia, JIRCAS working report no. 75. Japan International Research Center for Agricultural Sciences, Tsukuba, pp 149–158

Hiraoka M, Ohno M, Kawaguchi S, Yoshida G (2004) Crossing test among floating *Ulva* thalli forming 'green tide' in Japan. Hydrobiologia 512:239–245. doi:10.1023/B:HYDR.0000020332.12641.a2

Imai M, Katayama N, Yamaguchi Y (1997) Effects of salinity on growth, photosynthesis and respiration in a freshwater alga *Rhizoclonium riparium* (Chlorophyceae, Cladophorales). Phycol Res 45:233–237. doi:10.1111/j.1440-1835.1997.tb00082.x

Kakinuma M, Kuno Y, Amano H (2004) Salinity stress responses of a sterile mutant of *Ulva pertusa* (Ulvales, Chlorophyta). Fish Sci 70:1177–1179. doi:10.1111/j.1444-2906.2004.00921.x

Keesing JK, Liu D, Fears P, Garcia R (2011) Inter- and intra-annual patterns of *Ulva prolifera* green tides in the Yellow Sea during 2007–2009, their origin and relationship to the expansion of coastal seaweed aquaculture in China. Mar Pollut Bull 62:1169–1182. doi:10.1016/j.marpolbul.2011.03.040

Lehvo A, Bäck S (2001) Survey of macroalgal mats in the Gulf of Finland, Baltic Sea. Aquat Conserv 11:11–18

Leliaert F, Boedeker C (2007) Cladophorales. In: Brodie J, Maggs CA, John DM (eds) The green seaweeds of Britain and Ireland. The British Phycological Society, London, pp 131–183

Lewin J (1966) Silicon metabolism in diatoms. V. Germanium dioxide, a specific inhibitor of diatom growth. Phycologia 6:1–12. doi:10.2216/i0031-8884-6-1-1.1

Lewmanomont K, Ogawa H (1995) Common seaweed and seagrasses of Thailand. Kasetsart University, Bangkok

Lewmanomont K, Wongrat L, Supanwanid C (1995) Algae in Thailand. Office of Environmental Policy and Planning, Thailand

Liu D, Keesing JK, Xing Q, Shi P (2009) World's largest macroalgal bloom caused by expansion of seaweed aquaculture in China. Mar Pollut Bull 8:888–895. doi:10.1016/j.marpolbul.2009.01.013

Liu D, Keesing JK, He P, Wang Z, Shi Y, Wang Y (2013) The world's largest macroalgal bloom in the Yellow Sea, China: formation and implications. Estuar Coast Shelf Sci 129:2–10. doi:10.1016/j.ecdd.2013.05.021

Maceiras R, Rodríguez M, Cancela A, Urréjola S, Sánchez A (2011) Macroalgae: raw material for biodiesel production. Appl Energy 88:3318–3323. doi:10.1016/j.apenergy.2010.11.027

Macler BA (1988) Salinity effects on photosynthesis, carbon allocation, and nitrogen assimilation in the red alga, *Gelidium coulteri*. Plant Physiol 88:690–694. doi:10.1104/pp.88.3.690

Matsuyama-Serisawa K, Serisawa Y, Tanaka J (2004) Growth, maturation and photosynthesis of the brackish water alga *Rhizoclonium* sp. (Cladophoraceae, Chlorophyta) in relation to salinity. Phycol Res 52:204–210. doi:10.1111/j.1440-183.2004.00341.x

McAvoy KM, Klug JL (2005) Positive and negative effects of riverine input on the estuarine green alga *Ulva intestinalis* (syn. *Enteromorpha intestinalis*) (Linnaeus). Hydrobiologia 545:1–9. doi:10.1007/s10750-005-1923-5

Mmochi AJ, Dubi AM, Mamboya FA, Mwandya AW (2002) Effects of fish culture on water quality of an integrated mariculture pond system. West Indian Ocean J Mar Sci 1:53–63

Morand P, Merceron M (2005) Macroalgal population and sustainability. J Coast Res 21:1009–1020. doi:10.2112/04-700A.1

Msuya FE, Neori A (2002) *Ulva reticulata* and *Gracilaria crassa*: macroalgae that can biofilter effluent from tidal fishponds in Tanzania. West Indian Ocean J Mar Sci 1:117–126

Ohno M (1993) Cultivation of the green algae, *Monostroma* and *Enteromorpha* "Aonori". In: Ohno M, Critchley AT (eds) Seaweed cultivation and marine ranching. Japan International Cooperation Agency (JICA), Kanagawa

Ohno M, Takahashi I (1988) The horizontal and vertical distribution of the food alga *Entromorpha prolifera* in Shimanto River, southern Japan, vol 10. Report of Usa Marine Biological Institute, Kochi University, pp 45–54

Provasoli L (1966) Media and prospects for the cultivation of marine algae. In: Watanabe A, Hattori A (eds) Culture and collections of algae. Japanese Society of Plant Physiologists, Tokyo

Tanaka J (2001) Taxonomy and morphology of macroalgae in brackish waters (23) Cladophoracean algae 4. Salinity tolerance of *Rhizoclonium* spp. Aquabiology 23, Seibutsu Kenkyusha, Tokyo, pp 594–598

Tsutsui I, Huỳnh QN, Nguyễn HD, Arai S, Yoshida T (2005) The common marine plants of Southern Vietnam. Japan Seaweed Association, Kochi

Tsutsui I, Kanjanaworakul P, Srisapoome P, Aue-umneoy D, Hamano K (2010) Growth of giant tiger prawn, *Penaeus monodon* Fabricius, under co-culture with a discarded filamentous seaweed, *Chaetomorpha ligustica* (Kützing) Kützing, at an aquarium-scale. Aquac Int 18:545–553. doi:10.1007/s10499-009-9274-2

Tsutsui I, Hamano K, Aue-umneoy D, Songphatkaew J, Srisapoome P, Ruangsomboon S, Klomkling S, Ganmanee M, Taveekijakarn P, Maeno Y (2012) Common underwater plants in coastal areas of Thailand. JIRCAS International Agriculture Series 21. Japan International Research Center for Agricultural Sciences, Tsukuba

Xu Y, Lin J (2008) Effect of temperature, salinity, and light intensity on the growth of the green macroalga, *Chaetomorpha linum*. J World Aquac Soc 39:847–851. doi:10.1111/j.1749-7345.2008.00223.x

Yamochi S (2013) Effect of desiccation and salinity on the outbreak of a green tide of *Ulva pertusa* in a created salt marsh along the coast of Osaka Bay, Japan. Estuar Coast Shelf Sci 116:21–28. doi:10.1016/j.ecss.2012.07.002

21

The effect of white sturgeon (*Acipenser transmontanus*) ovarian fat deposition on caviar yield and nutritional quality: introducing image processing method for sturgeon ovary fat determination

Mahmoudreza Ovissipour · Hamzah M. Al-Qadiri · Xiaonan Lu · Yaxi Hu · Carolyn F. Ross · Joel P. Van Eenennaam · Serge I. Doroshov · Barbara Rasco

Abstract Image analysis can predict the fat content of sturgeon ovaries that had been categorized as having a low, medium, and high fat content based upon the caviar yield expressed as a percent of the total ovary weight, and were correlated with the chemical measurement of total fat ($R^2 = 0.83$). The fatty acid composition of eggs was not influenced by ovary fat content. Palmitic acid (16:00) was the most abundant saturated fatty acid and oleic acid (18:1n-9) the most predominant monounsaturated fatty acids in sturgeon eggs regardless of the ovary fat content. No significant differences ($P > 0.05$) were observed in docosahexaenoic acid (DHA) and eicosapentaenoic acid (EPA) in eggs from fish with different fat ovaries. Fourier transform infrared spectroscopy (FT-IR) coupled with principal component analysis indicated no significant difference in chemical compositions in sturgeon eggs separated from ovaries of different fat contents confirming the fatty acid composition results.

Keywords White sturgeon · Ovary fat · Egg quality · Image analysis · FT-IR spectroscopy

Introduction

Caviar is salt-cured and preserved eggs of one of the many sturgeon species that have been separated from the supporting connective ovarian tissue (Bledsoe et al. 2003). Caviar is a high value seafood product, and demand for caviar on domestic markets and export exceeds 2000 tons annually and has been increasing (Bledsoe et al. 2003; Ovissipour and Rasco 2012). Commercial aquaculture of sturgeons has been developed for both meat and caviar production over the past 30 years with white sturgeon (*Acipenser transmontanus*),

M. Ovissipour · C. F. Ross · B. Rasco (✉)
School of Food Science, Washington State University, P.O. Box 646376, Pullman, WA 99164-6376, USA
e-mail: Rasco@wsu.edu

M. Ovissipour
e-mail: mrovissi@yahoo.com

H. M. Al-Qadiri
Department of Nutrition and Food Technology, Faculty of Agriculture, The University of Jordan, Amman 11942, Jordan

X. Lu · Y. Hu
Food, Nutrition and Health Program, Faculty of Land and Food Systems, The University of British Columbia, Vancouver, BC V6T 1Z4, Canada

J. P. Van Eenennaam · S. I. Doroshov
Department of Animal Science, University of California, Davis, CA 95616, USA

Adriatic sturgeon (*A. naccarii*), Siberian sturgeon (*A. baerii*), Russian sturgeon (*A. gueldenstaedti*), and beluga sturgeon (*Huso huso*) being among the most prominent (Ovissipour and Rasco 2011; Lu and Rasco 2013).

White sturgeon (*A. transmontanus*) has been commercially farmed for meat and caviar since the late 1980s in California, Idaho, British Columbia (Doroshov et al. 1997; Hung et al. 1997). White sturgeon is well adapted to culture conditions, has a fast growth rate, and high quality meat and roe. Under intensive culture conditions and rearing temperatures of 18–22 °C, female sturgeons first reach sexual maturity at live weights of 35–40 kg (Doroshov et al. 1997; Lu et al. 2011).

Commercial production of sturgeons can decrease pressure on wild populations, by increasing the opportunities for artificial propagation to enhance natural stocks, and producing high value meat, and caviar (Ovissipour and Rasco 2011, Ovissipour and Rasco 2012). White sturgeon males are sold for meat at age 3, 4, female fish are processed for caviar at age of 7–9 years, and the carcasses are sold for meat consumption. The prices of cultured white sturgeon caviars in USA range from $50 to $120 (per 30 g retail) depending upon the quality and market conditions.

The nutritional composition of cultured fish depends upon a number of endogenous and exogenous factors. Endogenous factors (mostly fish size) affect protein and ash, while exogenous factors (e.g., dietary lipid content and energy) dramatically affect tissue fat content (Shearer 1994). High dietary lipid results in fat deposition in the visceral cavity and tissues (Hung et al. 1997; Péres and Oliva-Teles 1999; Nanton et al. 2001). Commercial feed provides the essential nutrients required for gonadal development in female fish and the viability and quality of eggs (Santiago et al. 1983; Gunasekera et al. 1997). For white sturgeons, commercial feeds are limited and the use of salmonid feeds remains common (Hung et al. 1997). For instance, commercial salmonid diets with different energy levels (23.6–26.2 MJ/kg diet) were used for white sturgeon juveniles over an 8-week period, resulting in high lipid content in the fish (Hung et al. 1997). Ebrahimi et al. (2004) conducted a comprehensive study on the effect of different dietary lipid levels on beluga (*Huso huso*) fingerlings and found significantly higher fat in fish fed a high lipid diet. Lipids play an important role in fish nutrition not only as a source of energy, but also to provide essential fatty acids for the development of different tissues including gonads and eggs, and hence can influence the caviar nutritional quality (Czesny and Dabrowski 1998; Izquierdo et al. 2001). Crude lipid contents in sturgeon eggs vary from 12.5 to 27 % wet weight and are high in the n-3 fatty acids, such as eicosapentaenoic (EPA) and docosahexaenoic (DHA) acids (Bledsoe et al. 2003; Gessner et al. 2002). One of the main key roles of seafood for human is the benefits of fatty acids and their beneficial effects on cardiovascular health and neurological development which could be influenced by endogenous and exogenous factors in sturgeon (Vaccaro et al. 2005; Ghomi et al. 2013).

An important quality factor is the level of lipid in the ovarian tissue. The physiological response of the sturgeons to deposit lipid in ovarian tissue is associated with dietary lipid intake, and excessive ovarian fat can reduce caviar yield, and increase processing time because of the longer screening time required to separate connective tissue from the eggs, and additional rinsing required removing adipose tissue debris.

Image processing is a useful and fast method, which can be used for sorting and grading automatization in the meat industry (Shiranita et al. 2000). Moreover, image processing for the automatization of visual inspection has gained attraction in the meat industry and many researchers have applied it in meat grading and reported it as a useful and effective tool (Shiranita et al. 2000; Maeda et al. 2014).

One of the main challenges in the sturgeon caviar processing is the fat deposition in ovary, which can cause several problems in grading, removing the eggs from the fat tissues, and quality of the final product. Grading is conducted in the field based on the visual appearance of the ovary by skill workers depend on their experience. However, automatization in sturgeon caviar facility can provide accurate and fast grading.

Therefore, the objectives of this study were to determine the effects of fat deposition in white sturgeon ovaries on caviar yield, chemical composition and fatty acid profile. An image processing technique was used to determine the ovary fat content. In addition, Fourier transform infrared spectroscopy (FT-IR) was used to determine the differences in chemical composition in eggs recovered from fish with different ovarian fat content.

Materials and methods

Fish and sample collection

Female white sturgeons harvested for caviar were in the age range of 7–9 years from two northern California aquaculture sites (Wilton and Galt, CA, USA). Sixty ovary samples were collected throughout the processing season, on the following dates: February 2012 ($n = 5$), March 2012 ($n = 12$), April 2012 ($n = 15$), April 2012 ($n = 8$), and May 2012 ($n = 20$). All individual fish were weighed ($\pm$0.05 kg) and fork length measured ($\pm$0.5 cm). The ovaries were removed and weighed ($\pm$0.05 kg) and a 200 g subsample of the ovary was collected and frozen on dry ice for image and chemical analyses. After ovary processing, the caviar yield (in the tin) was recorded ($\pm$0.1 g). Fattiness of the ovary was categorized based upon yield of caviar from the whole ovary weight. Caviar yields of less than 50 % of the ovary weight were categorized as a high fat ovary, while yields of over 65 % were low fat ovaries. Yields that fell in between were categorized as medium fat ovaries. Also, this method was used to determine the yield of caviar. Ovary tissues were received on dry ice within 24 h by School of Food Science at Washington State University, Pullman, WA, and were immediately frozen at −30 °C until tested.

Image analysis of ovarian tissue

Fat content in ovarian tissue was determined using image analysis (ImageJ software, National Institute of Health, USA, Version 1.47a). To determine the fat ratio in an ovary sample, photographs of individual ovaries were collected using a Canon 60D camera (Canon USA., Inc. Melville, NY, USA.) from 6 dimensions of the ovary samples and then images converted to the 8-bit pictures. For each picture, a 2 cm scale was used. Providing the scale during the photography is the key part of the image processing. Because according to the software instruction, specific scale should be defined for the software. These images were then converted to threshold pictures using ImageJ software (ImageJ User Guide, 2012). Threshold option converts all pictures into black and white based on the original color differences. To find the best fit model for threshold different, limits were applied to find the optimum threshold values which were 85–120. After conversion, the black color represented low fat tissues while white color denoted the visible fat tissues (Fig. 1). In this step, the specific

Fig. 1 Digital images of low, medium and high fat ovary samples, **a**: before processing by ImageJ, **b**: after processing by imageJ (threshold pictures)

scale (which was 2 cm in current study) was defined for the software and, based on the defined scale, the area of white and black parts was automatically calculated by software. Then the white area which was representing the fat content was measured and the ratio of the white area to total area was presented as fat content.

Fat content and fatty acid analysis by gas chromatography

To determine the fat content of fish ovaries, the lipid in the ovaries was extracted according to the method by Bligh and Dyer (1959) with a slight modification. Briefly, 10 g of sturgeon ovary was mixed with 10 mL distilled water, 20 mL chloroform and 40 mL methanol, homogenized then sonicated for 2 min in iced water for better separation using an ultrasound bath (100 W, 40 kHz, Bransonic®, Process Equipment and Supply, Inc. Cleveland, OH). Then the samples were kept at 4 °C for overnight in glass tubes. Samples were centrifuged for 10 min at 3000×*g* at 4 °C. The chloroform part was separated and transferred to a glass tube and kept at 40 °C to evaporate the solvent. At the end, to completely remove the solvent, nitrogen gas was injected to samples and then the fat content was calculated based on the weight of the sample, and the fat which was obtained.

To determine the fatty acid composition in eggs, eggs were first separated from ovary tissues and then the egg lipid was extracted as described above. Fatty acid composition was determined as described by Watts and Browse (2002) with minor modification. Briefly, 50 μL of egg lipid extract was mixed with 450 μL hexane. Then, 10 μL of the mixture was added to a small screw-cap glass tube. Fatty acids were derivatized to methyl esters by adding 2.5 % H_2SO_4 in MeOH followed by heating at 80 °C for 30 min. After cooling to room temperature (i.e., 22 °C), 500 μL hexane and 1.5 mL H_2O were added and the mixture shaken vigorously. Following low speed centrifugation (1000×*g* for 1 min), two phases were clearly separated. Then, 70 μL of top layer was loaded into a vial and injected into gas chromatography (GC). Methyl esters were analyzed by GC using an Agilent 6890 series gas chromatograph equipped with 30 m × 0.32 mm Omegawax™ 320 capillary column (Supelco™, Sigma Aldrich, St. Louis, MO, USA), helium as the carrier gas at 1.4 mL/min, and a flame ionization detector. To quantify the fatty acids, internal standard (C17:0-Methyl-Ester, 10 μg for 200 mg of lipids) was added to each sample, while an external standard (Natural Menhaden oil; Matreya LLC, Pleasant Gap, PA, USA) was run as an external standard. The GC was programmed for an initial temperature of 120 °C for 1 min followed by an increase of 10 °C/min to 190 °C followed by an increase of 2 °C/min to 200 °C.

FT-IR spectroscopy and spectra collection

FT-IR spectral features of eggs separated from each individual sturgeon ovary were collected using Thermo Nicolet Avatar 360 FT-IR spectrometer (Thermo Electron Inc., San Jose, CA, USA). A sturgeon egg was put in direct contact with the attenuated total reflectance (ATR) ZnSe crystal cell (Lu et al. 2011). Eighteen spectra were collected at 22 °C for each sample at different locations on the surface of the egg.

Chemometrics and statistical analysis

To determine the influence of ovary fat on the quality of eggs, one way analysis of variance (ANOVA) (Duncan multiple test) using SPSS software, release 16.0 (SPSS Inc., Chicago, IL, USA.) at a probability level of $P < 0.05$ was used.

FT-IR spectral analyses were conducted using OMNIC (Thermo Electron Inc.) and Matlab (The MathWorks, Inc. Natick, MA, USA.). Data pre-processing algorithms, including baseline correction and normalization, were performed. Binning (2 cm^{-1}) and smoothing (Gaussian function over 10 cm^{-1}) were conducted to reduce spectral noises. After spectral processing, principal component analysis (PCA) was developed using the wavenumbers of 4000–550 cm^{-1} according to Lu et al. (2010).

Results and discussion

Chemical and image analysis of sturgeon ovary

Fat content determined by chemical and image analysis are shown in Table 1. The highest (28.7 ± 11.50 %) and lowest fat (19.85 ± 12.71 %) contents determined by image analysis were from fish that were categorized as high fat and low fat, respectively. Ovaries had highly variable amounts of adipose fat and fat interspersed between eggs in the ovarian tissue. The first step of image analysis was to examine the photographs of sturgeon ovaries in a visual determination as to whether an ovary has high, medium or low fat. Photographs from different sides of the ovaries were taken and processed by image processing software, and the results are the mean of different sides of a sample.

Chemical analysis showed significant differences ($P < 0.05$) between high fat ovary and low fat ovary. However, no significant differences were observed ($P > 0.05$) between high fat and medium fat samples, and low fat and medium fat samples.

The correlation between fat content determined by chemical analysis and image analysis is shown in Fig. 2. There was a high correlation coefficient between chemical measurements and image analysis ($R^2 = 0.83$) even though the image was from the ovary surface only and did not reflect the differences in lipid content from the surface to the interior of the ovarian mass or from the dorsal to the abdominal surface. Therefore, image analysis may be a useful, fast and inexpensive tool for sturgeon aquaculture sectors to predict the ovary fat content and sort the fish accordingly. Image analysis is one of the suitable tools that can be used to determine the quality of different food products. Many researchers have used image analysis techniques to predict the fat content of beef meat (Kuchida et al. 2000; Shiranita et al. 2000; Chambaz et al. 2003; Nakahashi et al. 2008; Maeda et al. 2014). According to the current study, image analysis could be introduced for grading and sorting of sturgeon ovary at the farms. However, further studies are required for quality grading of sturgeon ovary images from different sections of ovaries.

One of the most important influences of the ovary fat content is the effect on egg yield (%BW and OW). The results of ovary weight, egg weight and egg yield (%BW and OW) are presented in Table 2. Eggs yield (%BW and OW) results showed significant differences ($P < 0.05$) between low fat ovary and medium and high fat ovaries. The egg yield (%OW) was 75.4 % in low fat ovary compared to 38.7 % in high fat ovary.

Table 1 Fat content in sturgeon ovaries determined by chemical and image analysis

Sample	Chemical analysis (%)	Image analysis (%)	Number of samples
Low fat	28.24 ± 10.57^{a}	19.85 ± 12.71^{a}	22
Medium fat	32.63 ± 8.71^{ab}	25.03 ± 12.45^{ab}	18
High fat	37.52 ± 9.1^{b}	28.7 ± 11.50^{b}	20

Values in column with different superscripts are significantly different at $\alpha = 0.05$

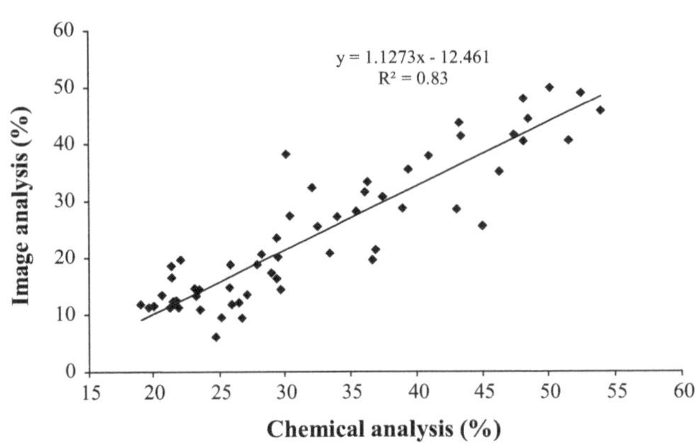

Fig. 2 Correlation between chemical and image analysis for fat content in sturgeon ovaries

Table 2 Sturgeon ovary and egg quality parameters

	Body weight (kg)	Ovary weight (kg)	Egg weight (g)	Egg yield (%BW)	Egg yield (%OW)	GSI
Low fat	35.11 ± 7.6^{a}	4.3 ± 1.2^{a}	3251 ± 900^{a}	9.22 ± 0.8^{a}	75.33 ± 2^{a}	12.22 ± 0.88^{a}
Medium fat	40.14 ± 11.8^{a}	5.5 ± 2^{a}	3136 ± 800^{a}	8.20 ± 0.4^{b}	60.34 ± 1.6^{b}	13.59 ± 0.6^{a}
High fat	47 ± 9^{a}	7 ± 2.4^{a}	2677 ± 738^{a}	5.3 ± 0.4^{b}	38.74 ± 4^{c}	14.44 ± 2^{a}

Values in column with different superscripts are significantly different at $\alpha = 0.05$

Generally, sturgeons are cultured for both meat and caviar production and an excessively fat content in the ovary would result in reduced feed conversion increasing production costs (Ovissipour and Rasco 2011). One of the major issues in female sturgeon is lipid deposition in the ovary which can decrease egg yield. In addition, this fatty tissue negatively influences caviar processing by slowing down the processing because it requires more rinsing and cleaning of eggs prior to salting and curing, compared to a lean ovary. Taken together, our current study indicated that a higher egg yield can be achieved with lower fat in the ovary.

Fatty acid composition of sturgeon eggs

Fatty acid composition of eggs separated from fish that were categorized as having low, medium and high fat ovaries is shown in Table 3. In all of the groups, palmitic acid (16:00) was the most abundant saturated fatty acid (SAFA) regardless of the ovary lipid content. Palmitic acid tends to be the predominant saturated fatty acid in cultivated sturgeons (Chen et al. 1995; Ovissipour and Rasco 2011). Oleic acid (18:1n-9) was the most predominant monounsaturated fatty acid (MUFA) as reported by others for the study of eggs and tissues from sturgeon and other fish (Ashton et al. 1993; Gallagher et al. 1998; Czesny et al. 2000; Vaccaro et al. 2005; Caprino et al. 2008; Mol and Turan 2008; Ovissipour and Rasco 2011). No significant differences ($P < 0.05$) were observed in docosahexaenoic acid (DHA) and eicosapentaenoic acid (EPA) among the ovaries in fish. The percentages of EPA and DHA were similar regardless of ovary fat levels, indicating that the optimal threshold levels had been reached. Similar results were found by other researchers for domesticated walleye (*Stizostedion vitreum*) (Czesny and Dabrowski 1998), Atlantic halibut (*Hippoglossus hippoglossus*) (Mazorra et al. 2008) and white sturgeon (*A. transmontanus*) (Caprino et al. 2008).

The ratio of n-3 to n-6 PUFA ranged from 2.7 to 3.65, which are within the typical range for freshwater fish eggs (0.5–3.8) (Caprino et al. 2008). No significant difference was observed ($P < 0.05$) in n-3, n-6 and n-3 to n-6 between different ovary groups for the fish in the current study.

FT-IR spectral features of sturgeon eggs

FT-IR spectroscopy is a type of vibrational spectroscopy and has been widely applied to determine the biochemical compositions of biological and agricultural systems, such as proteins, lipids, nucleic acids and carbohydrates moieties (Al-Holy et al. 2006; Burgula et al. 2007). Recently, FT-IR spectroscopy was applied in aquaculture to investigate the chemical components in white sturgeon female plasma, leading to a rapid and less-invasive determination of fish maturity Lu et al. (2010, 2013).

FT-IR spectral features of separated eggs from fish with low, medium and high levels of ovarian fat were determined and there was no obvious difference in egg composition (Fig. 3). The peaks around 3010 (Olefinic = CH stretching vibration: lipids, cholesterol esters), 2922 (CH_2 asymmetric stretch of methylene groups: mainly lipids), 2852 (CH_2 symmetric stretch of methylene groups: mainly lipids), 1743 (C = O of ester functional groups: triglycerides, cholesterol esters), and 1458 (CH_2 bending: mainly lipids, with little contribution from proteins) are assigned as stretching, vibration or bending of molecular bonds in lipid molecules showing that the surface of eggs is rich in lipid and there is no significant difference among eggs taken from ovaries with high or low fat content. FT-IR spectral feature of white sturgeon eggs is very highly conserved and does not appear to be affected by maturity stage of the fish (Lu et al. 2011). An unsupervised PCA model was developed to differentiate sturgeon eggs separated from ovaries with different fat content (i.e., low, medium and high fats), but no clear segregation could be obtained (Fig. 4) indicating that the

Table 3 Fatty acid composition of eggs recovered from low, medium, and high fat ovaries

Fatty acid (%wt)	Low fat ovary ($N = 22$) Mean ± SD	Medium fat ovary ($N = 18$) Mean ± SD	High fat ovary ($N = 20$) Mean ± SD
C14:0	1.74 ± 0.37	1.79 ± 0.08	2.43 ± 0.46
C16:0	18.39 ± 0.34	18.01 ± 0.52	18.02 ± 0.43
C16:1n7	5.16 ± 0.42	5.24 ± 0.35	5.75 ± 0.33
C16:2n4	0.17 ± 0.06	0.14 ± 0.03	0.27 ± 0.05
C16:3n4	7.28 ± 0.55	7.06 ± 0.57	6.09 ± 0.23
C16:4n1	0.17 ± 0.06	0.14 ± 0.03	0.27 ± 0.05
C18:0	3.19 ± 0.09	3.05 ± 0.04	2.54 ± 0.15
C18:1n9	32 ± 4.53	34.96 ± 1.54	35.10 ± 2.34
C18:1n7	0.17 ± 0.06	0.14 ± 0.03	0.27 ± 0.05
C18:2n6	6.49 ± 1.25	7.34 ± 0.63	6.75 ± 1.13
C18:2n4	0.17 ± 0.06	0.14 ± 0.03	0.27 ± 0.05
C18:3n4	0.36 ± 0.11	0.29 ± 0.04	0.36 ± 0.08
C18:3n3	1.02 ± 0.43	0.95 ± 0.35	1.14 ± 0.61
C18:4n3	0.38 ± 0.09	0.32 ± 0.06	0.49 ± 0.06
C20:1n9	1.37 ± 0.27	1.52 ± 0.38	1.79 ± 0.41
C20:4n6 (AA)	0.17 ± 0.06	0.14 ± 0.03	0.27 ± 0.05
C20:4n3	2.41 ± 0.18	2.26 ± 0.12	1.81 ± 0.12
C20:5n3 (EPA)	0.39 ± 0.17	0.34 ± 0.03	0.46 ± 0.11
C21:5n3	5.28 ± 1.34	4.50 ± 0.28	5.00 ± 0.75
C22:5n3 (DPA)	1.89 ± 0.63	1.64 ± 0.05	1.83 ± 0.34
C22:6n3 (DHA)	11.82 ± 2.00	9.95 ± 0.13	9.04 ± 1.46
∑SAFA	23.30 ± 0.83	22.84 ± 0.58	22.99 ± 0.55
∑MUFA	38.67 ± 4.29	41.88 ± 1.36	42.91 ± 2.12
∑n-3	23.18 ± 3.94	19.99 ± 0.9	19.79 ± 2.7
∑n-6	6.7 ± 1.20	7.48 ± 0.65	7.02 ± 1.08
n-3/n-6	3.64 ± 1.20	2.70 ± 0.34	2.90 ± 0.85

Low fat sample a b
Medium fat sample a b
High fat sample a b

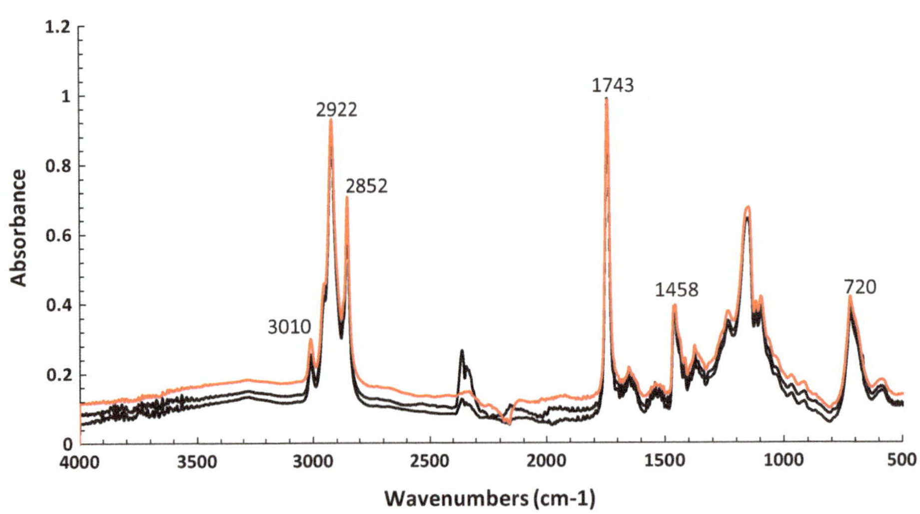

Fig. 3 FT-IR spectrum of white sturgeon eggs (*red*: low fat; *black*: medium fat; *blue*: high fat) ($n = 60$)

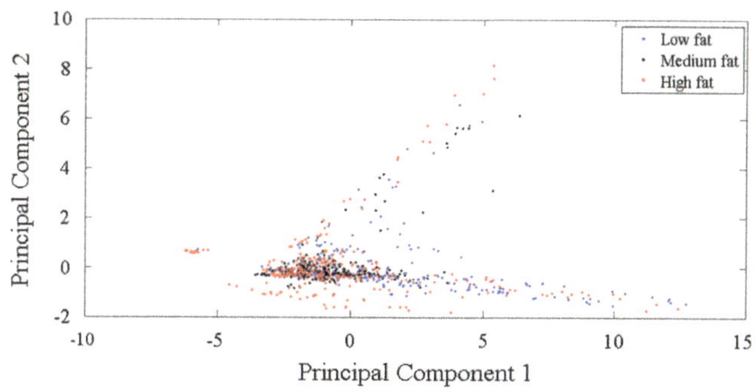

Fig. 4 Principal component analysis (PCA) of sturgeon eggs separated from ovarian samples with different levels of fat. Principal Component (PC) 1 explained 69.9 % of the variance, while PC2 explained 11.61 % of the variance ($n = 60$)

composition of sturgeon eggs was not affected by the fat content in the ovarian tissues. Raman spectroscopy has recently been applied to determine the quality of different caviars from Caspian Sea (Mohamadi Monavar et al. 2013). They indicated that Raman spectroscopy coupled with partial least squares regression models could be used to characterize fatty acid features of different species of sturgeon.

Conclusion

We used image analysis as a rapid method to determine the fat content of different ovaries. Image analysis method was correlated with the conventional Bligh and Dyer chemical extraction method for total fat ($R^2 = 0.83$). Fatty acid composition was not significantly different in eggs from low, medium and high fat ovaries. In addition, by applying FT-IR, no significant differences among fish with different ovary fat contents in protein or lipid content were observed. The current study showed that the fat content in ovary can influence the egg yield, and ovary weight, but had little influence on the macronutrient properties of the caviar.

Acknowledgments This study was funded by Western Region Aquaculture Center, U.S. Department of Agriculture, under grant number 2007-38500-18593. We would like to express our sincere thanks to Professor John A. Browse and Dr. James G. Wallis in the Institute of Biological Chemistry at WSU and for sturgeon farmers across the Western region for their technical assistance and support of this research.

Compliance with ethical standards

Conflict of interest The authors declare no competing financial interest.

Funding This study was funded by Western Region Aquaculture Center, U.S. Department of Agriculture, under grant number 2007-38500-18593.

Authors' contributions MO conducted the experiment and developed the image analysis protocol for sturgeon ovary fat determination. HA provided technical supports for FT-IR spectra collecting. XL provided technical supports for FT-IR spectra statistical analysis. YH provided technical supports for FT-IR spectra statistical analysis. CR Scientifically supported the project. JVE and SD from UC Davis conducted sturgeon aquaculture trials. BR supported the whole experiment scientifically and technically.

References

Al-Holy MA, Lin M, Cavinato AG, Rasco B (2006) The use of Fourier transform infrared spectroscopy to differentiate *Escherichia coli* O157:H7 from other bacteria inoculated into apple juice. Food Microbiol 23(2):162–168

Ashton HJ, Farkvan DO, March BE (1993) Fatty acid composition of lipids in the eggs and alevins from wild and cultured chinook salmon (*Oncorhynchus tshawytscha*). Can J Fish Aquatic Sci 50(3):648–655

Bledsoe GE, Bledsoe CD, Rasco B (2003) Caviars and fish roe products. Crit Rev Food Sci Nutr 43(3):317–356

Bligh EG, Dyer WJA (1959) A rapid method for total lipid extraction and purification. Can J Biochem. Physiol 37(8):911–917

Burgula Y, Khali D, Kim S, Krishnan SS, Cousin MA, Gore JP, Reuhs BL, Mauer LJ (2007) Review of mid-infrared Fourier transform infrared spectroscopy applications for bacterial detection. J Rapid Methods Auto Microbiol 15(2):146–175

Caprino F, Moretti VM, Bellagamba F, Turchini GM, Busetto ML, Giani I, Paleari MA, Pazzaglia M (2008) Fatty acid composition and volatile compounds of caviar from farmed white sturgeon (*Acipenser transmontanus*). Anal Chim Acta 617(1–2):139–147

Chambaz A, Scheeder MRL, Kreuzer M, Dufey PA (2003) Meat quality of Angus, Simmentral, Charolais and Limousin steers compared at the same intramuscular fat content. Meat Sci 63(4):491–500

Chen IC, Chapman FA, Wei CI, Portier KM, O'Keefe SFJ (1995) Differentiation of cultured and wild sturgeon (*Acipenser oxyrinchus desotoi*) based on fatty acid composition. J Food Sci 60(3):631–635

Czesny S, Dabrowski K (1998) The effect of egg fatty acid concentrations on embryo viability in wild and domesticated walleye (*Stizostedion vitreum*). Aquat Living Res 11(6):371–378

Czesny S, Dabrowski K, Christensen JE, Van Eenennaam JP, Doroshov SI (2000) Discrimination of wild and domestic origin of sturgeon ova based on lipids and fatty acid analysis. Aquaculture 189(1–2):145–153

Doroshov SI, Moberg GP, Van Eenennaam JP (1997) Observations on the reproductive cycles of cultured white sturgeon, *Acipenser transmontanus*. Environ Biol Fishes 48:265–278

Ebrahimi E, Pourreza J, Panamariov SV, Kamali A, Hosaini A (2004) Effect of different levels of protein and fat on growth characters and chemical composition of fingerling beluga (Huso huso L.). J Sci Technol Agric Nat Res 2:229–241 **(In Persian, Abstract in English)**

Gallagher ML, Paramore L, Alves D, Rulifson RA (1998) Comparison of phospholipid and fatty acid composition of wild and cultured striped bass eggs. J Fish Biol 52(6):1218–1228

Gessner J, Wirth M, Kirschbaum F, Kruger A, Patriche N (2002) Caviar composition in wild and cultured sturgeons–impact of food sources on fatty acid composition and contaminant load. J Appl Ichthyol 18(4–6):665–672

Ghomi MR, Nikoo M, Sohrabnejad M (2013) Effect of alive weight on body composition and fatty acid content of farmed beluga sturgeon (*Huso huso*). Intel Aqua Res 5:6

Gunasekera RM, Shim KF, Lam TJ (1997) Influence of dietary protein content on the distribution of amino acids in oocytes, serum and muscle of Nile tilapia, *Oreochromis niloticus* (L.). Aquaculture 152(1–4):205–221

Hung SSO, Storebakken T, Cui Y, Tian L, Einen O (1997) High-energy diets for white sturgeon, Acipenser transmontanus Richardson. Aquac Nutr 3(4):281–286

Izquierdo MS, Fernández-Palacios H, Tacon AGJ (2001) Effect of broodstock nutrition on reproductive performance of fish. Aquaculture 197(1–4):25–42

Kuchida A, Kono S, Konishi K, Van Vleck LD, Suzuki M, Miyoshi S (2000) Prediction of crude fat content of longissimus muscle of beef using the ratio of fat area calculated from computer image analysis: Comparison of regression equations for prediction using different input devices at different stations. J Anim Sci 78(4):799–803

Lu X, Rasco B (2013) Sturgeon (*Acipenser transmontanus*) sexual maturation and caviar quality. Rev Aquac 6(2):89–99

Lu X, Webb MAH, Talbott MJ, Van Eenennaam JP, Palumbo A, Linares-Casenave J, Doroshov SI, Struffenegger P, Rasco B (2010) Distinguishing ovarian maturity of farmed white sturgeon (*Acipenser transmontanus*) by Fourier transform infrared spectroscopy: a potential tool for caviar production management. J Agric Food Chem 58(7):4056–4064

Lu X, Webb MAH, Talbott MJ, Van Eenennaam JP, Doroshov SI, Rasco B (2011) A study of biochemical parameters associated with ovarian atresia and quality of caviar in farmed white sturgeon (*Acipenser transmontanus*) by Fourier Transform Infrared (FT-IR) Spectroscopy. Aquaculture 315(3–4):298–305

Lu X, Talbot MJ, Van Eenennaam JP, Webb MAH, Doroshov SI, Ovissipour M, Rasco B (2013) Determining ovarian maturity in farmed sturgeon (*Acipenser transmontanus*) for caviar production using Fourier transform infrared spectroscopy (FT-IR). J Aquac Res Dev 5:1. doi:10.4172/2155-9546.1000202

Maeda S, Grose J, Kato K, Kuchida K (2014) Comparing AUS-MEAT marbling scores using image analysis traits to estimate genetic parameters for marbling of Japanese Black cattle in Australia. Animal Prod Sci 54(5):557–563

Mazorra C, Bruce M, Bell JG, Davie A, Alorrend E, Jordan N, Rees J, Papanikos N, Porter M, Bromage N (2003) Dietary lipid enhancement of broodstock reproductive performance and egg and larval quality in Atlantic halibut (*Hippoglossus hippoglossus*). Aquaculture 227(1–4):21–33

Mohamadi Monavar H, Afseth N, Lozano J, Alimardani R, Omid M, Wold J (2013) Determining quality of caviar from Caspian Sea based on Raman spectroscopy and using artificial neural networks. Talanta 111:98–104

Mol S, Turan S (2008) Comparison of proximate, fatty acid and amino acid compositions of various types of fish roes. Int J Food Prop 11(3):669–677

Nakahashi Y, Maruyama S, Seki S, Hidaka S, Kuchida K (2008) Relationships between monounsaturated fatty acids of marbling flecks and image analysis traits in longissimus muscle for Japanese Black steers. J Anim Sci 86(12):3551–3556

Nanton DA, Lall SP, McNiven MA (2001) Effects of dietary lipid level on liver and muscle lipid deposition in juvenile haddock, Melanogrammus aeglefinus L. Aquac. Res 32(s1):225–234

Ovissipour M, Rasco B (2011) Fatty acid and amino acid profiles of domestic and wild beluga (*Huso huso*) roe and impact on fertilization ratio. J Aquac Res Develop. doi:10.4172/2155-9546.1000113

Ovissipour M, Rasco B (2012) Sturgeon: Conservation of Caspian Sea Stocks. J Aquac Res Develop. doi:10.4172/2155-9546.1000e104

Péres H, Oliva-Teles A (1999) Effect of dietary lipid level on growth performance and feed utilization by European seabass juveniles *Dicentrarchus labrax*. Aquac. 179(1–4):325–334

Santiago CB, Aldaba MB, Laron MA (1983) Effect of varying dietary crude protein levels on spawning frequency and growth of *Sarotherodon niloticus* breeders. Fish Res J Phil 8(2):9–18

Shearer KD (1994) Factors affecting the proximate composition of cultured fishes with emphasis on salmonids. Aquac. 119(1):63–88

Shiranita K, Hayashi K, Otsubo A, Miyajima T, Takiyama R (2000) Grading meat quality by image processing. Pattern Reco 33(1):97–104

Vaccaro AM, Buffa G, Messina M, Santulli A, Mazzola A (2005) Fatty acid composition of a cultured sturgeon hybrid (*Acipenser naccarii* × *A. baerii*). Food Chem 93(4):627–631

Watts JL, Browse J (2002) Genetic dissection of polyunsaturated fatty acid synthesis in *Caenorhabditis elegans*. Proc Natl Acad Sci 99(9):5854–5859

22

Universal primers for exon-priming intron-crossing (EPIC) PCR on ribosomal protein genes in marine animals

Seinen Chow · Takashi Yanagimoto · Yoji Nakamura

Abstract Relatively conserved exon sequences among distant aquatic animal taxa (Copepoda, Echinodermata, and Teleostei) were determined for 12 ribosomal protein genes, and 19 primer pairs were designed for an exon-primed intron-crossing polymerase chain reaction strategy. The universal utility of these primers was evaluated in distant animal species (sea urchin, squid, crab, shark, and tuna). Fragment amplification was confirmed for at least three species for all primer pairs, and single fragment amplification was observed for at least one species in 16 primer pairs. Primer sets presented here may serve as an initial step for isolating single-copy nuclear DNA sequences in a wide variety of marine animals.

Keywords Universal primers · Ribosomal protein genes · Intron · Single-copy nuclear loci · Marine animals

Introduction

Introns may be a good source for finding DNA polymorphisms in the eukaryotic genome as they rapidly accumulate mutations without greatly affecting gene function. Exon-primed intron-crossing (EPIC) polymerase chain reaction (PCR) has been developed and applied to find genetic polymorphisms in a wide variety of animals (Lessa 1992; Palumbi and Baker 1994; Corte-Real et al. 1994; Daguin et al. 2001; Chow and Takeyama 2000; Chow et al. 2007). Primers designed to target conserved areas may have a wide utility across a broad range of animal taxa (Chow and Hazama 1998; Hassan et al. 2002; Chow and Nakadate 2004; Pinho et al. 2010; Jennings and Etter 2011). The universality of the primers for EPIC PCR depends on the conservation of exonic sequences, and the efficiency of PCR amplification may be primarily affected by gene copy number, including pseudogenes. Genes encoding ribosomal proteins (RPs) may be highly suitable for designing universal primers, as RPs are the most highly conserved genes among eukaryotes and are typically encoded by a single gene (Wool 1979; Wool et al. 1995). Indeed, the coding sequences of RP genes are reported to be highly conserved among humans, fruit flies, and nematodes (Yoshihama et al. 2002). In this study, we designed 19 primer pairs in exons of 12 RP genes and demonstrated the universal utility of these primer pairs using samples of distantly related aquatic animal species.

S. Chow (✉) · T. Yanagimoto · Y. Nakamura
National Research Institute of Fisheries Science, 2-12-4 Fukuura, Kanazawa, Yokohama, Kanagawa 236-8648, Japan
e-mail: chow@affrc.go.jp

T. Yanagimoto
e-mail: yanagimo@affrc.go.jp

Y. Nakamura
e-mail: yojnakam@affrc.go.jp

Materials and methods

Eighty different ribosomal protein genes were mapped to human chromosomes (Uechi et al. 2001), and cDNA data for the homologous genes of the copepod (*Lepeophtheirus salmonis*, *Acartia pacifica*, or *Calanus helgolandicus*), sea urchin (*Strongylocentrotus purpuratus* or *Paracentrotus lividus*), and fish (*Danio rerio* or *Ictalurus punctatus*) were obtained from the GenBank database. RP genes that were not found in all three of these reference animal taxa were not analyzed. BLAST searches (Altschul et al. 1990) using these reference cDNA sequences against the draft genome sequences of the Pacific bluefin tuna (*Thunnus orientalis*; accession

Table 1 Sequences of primers for amplifying 19 introns of 12 ribosomal protein genes

Pair	Gene	Primer	Orientation	Primer sequence (5–3′)	Distance[a]
A	S2	RPS2exCF	Forward	CCHGTNCCCAAGAAGCTGCT	70
	S2	RPS2exDR	Reverse	GGTYTCCTTCCASAGRTCAG	43
B	S3	RPS3ex1F	Forward	AAATCTCVAAGAAGAGRAAG	2
	S3	RPS3ex2R	Reverse	ACRCCGGAGTANCCATCCTC	62
C	S3	RPS3ex2F	Forward	GAGGATGGNTACTCCGGYGT	53
	S3	RPS3ex3R	Reverse	CKCTTCTGVACMACAGCGGT	48
D	S3	RPS3ex3F	Forward	ACCGCTGTKGTBCAGAAGMG	30
	S3	RPS3ex4R	Reverse	GTCGCAACCTTCTCAGCRTA	24
E	S3	RPS3ex4F	Forward	TAYGCTGAGAAGGTTGCGAC	74
	S3	RPS3ex5R	Reverse	AASCKCAGRACACCATARCA	6
F	S14	RPS14exAF	Forward	GCCTCMTTYAACGAYACCTT	29
	S14	RPS14exBR	Reverse	CKGTCRGCCTTYACCTTCAT	30
G	S17	RPS17exBF	Forward	AGWGGMATCTCCATCAARCT	49
	S17	RPS17exCR	Reverse	GGRTCRACCTCGATGATSTC	18
H	S23	RPS23exAF	Forward	TCACACGCCAARGGCATYGT	15
	S23	RPS23exBR	Reverse	ATRGCAGAGTTRGGCTGCTT	14
I	S23	RPS23exBF	Forward	AAGAACGGCAARAAGATCAC	42
	S23	RPS23exCR	Reverse	TGACCYTTACGWCCRAATCC	26
J	S27	RPS27exAF	Forward	GCTAARATYCARGAYAAGGA	4
	S27	RPS27exBR	Reverse	GTGCGKCCRTCYTCCAGCTG	43
K	L3	RPL3exAF	Forward	GGCTACAAGGCYGGYATGAC	39
	L3	RPL3exBR	Reverse	CAGTTYTTGTAGAAKCGACG	149
L	L3	RPL3exBF	Forward	GCGTCGMTTCTACAARAACT	2
	L3	RPL3exCR	Reverse	GTGAARGCCTTCTTCTYGGA	9
M	L3	RPL3exDF	Forward	CAAGGGWCACGGRTDCAAGG	2
	L3	RPL3exER	Reverse	GTCTTRCGGGGMAGCTTCTT	25
N	L8	RPL8exBF	Forward	AGTTCATCTACTGCGGCAAG	37
	L8	RPL8exCR	Reverse	ATGAYGGTDCCYTCAGGCAT	5
O	L9	RPL9exCF	Forward	RAACTTCYTGGGRGAGAAGT	110
	L9	RPL9exDR	Reverse	ACDGTGGTGGCYTGCTGGAT	10
P	L11	RPL11exAF	Forward	CARACGCCHGTCTTCTCCAA	5
	L11	RPL11exBR	Reverse	AAGGATCKYACAGTGTAGCG	4
Q	L11	RPL11exBF	Forward	CGCTACACTGTRMGATCCTT	87
	L11	RPL11exCR	Reverse	CCCARRTCGATGTGYTCCTG	65
R	L17	RPL17exAF	Forward	GAGWCMGCTCAGGCCATCAA	132
	L17	RPL17exBR	Reverse	TCGGCRCTCTTCTTRGGCCA	8
S	L24	RPL24exBF	Forward	AGMCGCAARCACAAGAAGGG	13
	L24	RPL24exCR	Reverse	ACYTCAGGCTTCTGRTTCCT	85

[a] Number of nucleotides between 3′ end of the primers and start or end of the target introns

nos. BADN01000001-BADN01133062 in the DNA Data Bank of Japan) (Nakamura et al. 2013) detected partial sequences of 31 RP genes among 35 bluefin tuna scaffolds. For each of the 31 genes detected, a multiple sequence alignment was constructed by ClustalX version 1.83 (Thompson et al. 1997), and conserved regions suitable for designing universal primers were surveyed. Crude DNA samples of the long-spined sea urchin (*Diadema savignyi*), neon flying squid (*Ommastrephes bartramii*), golden crab (*Chaceon granulatus*),

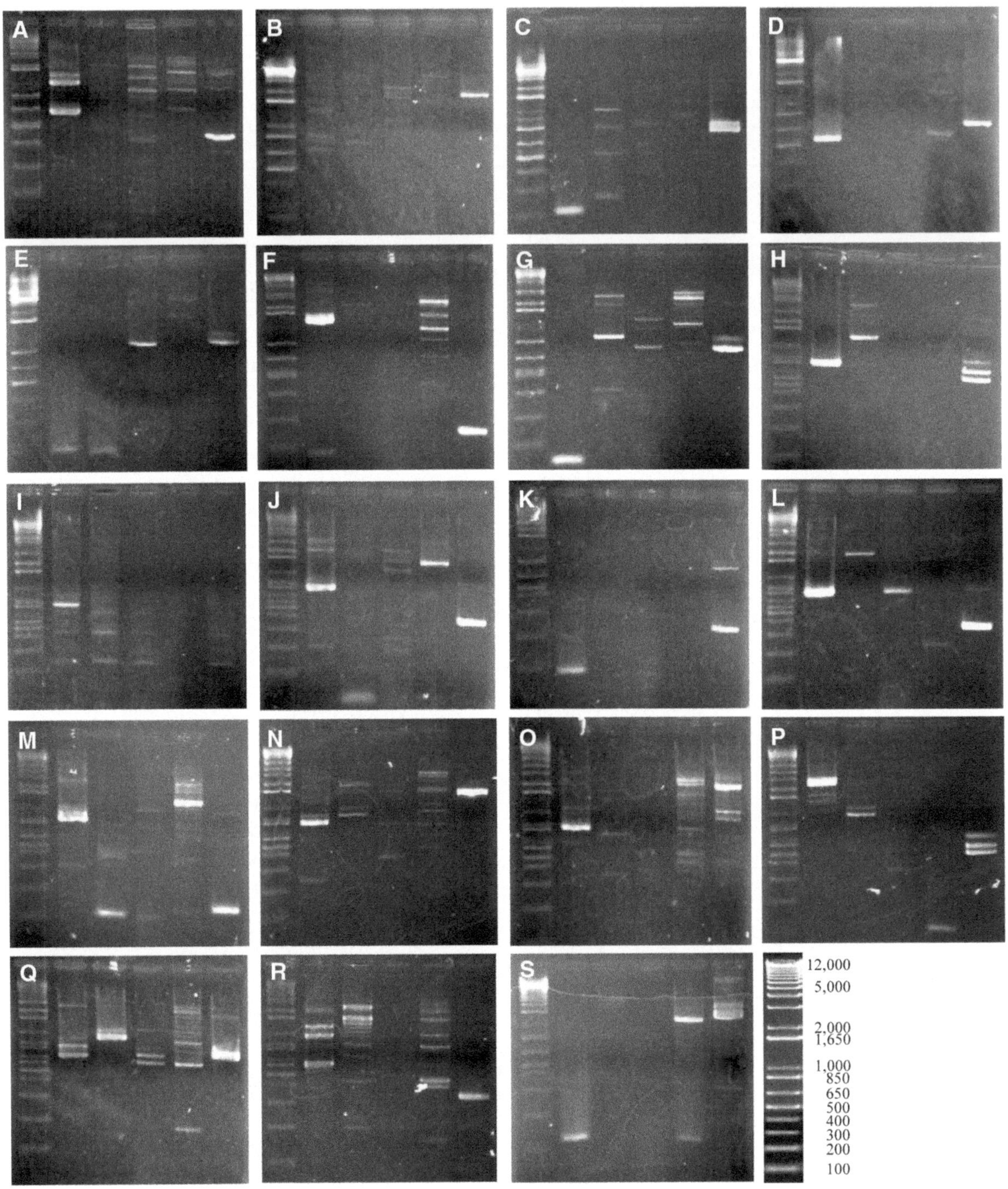

Fig. 1 Agarose gel electrophoresis images of the PCR products obtained by 19 primer pairs using a 55 °C annealing temperature. **a–s** Correspond to the primer pairs shown in Table 1. The *left-most lane* (*1st lane*) of each gel image is the size marker (1 Kb Plus DNA Ladder, Invitrogen), and the *2nd–6th lanes* correspond to long-spined sea urchin (*Diadema savignyi*), neon flying squid (*Ommastrephes bartramii*), golden crab (*Chaceon granulatus*), mako shark (*Isurus oxyrinchus*), and skipjack tuna (*Katsuwonus pelamis*), respectively

mako shark (*Isurus oxyrinchus*), and skipjack tuna (*Katsuwonus pelamis*) archived in our laboratory were used for evaluating the universal utility of the primers. PCR amplifications were performed in a 12 μL final volume containing 1 μL of template DNA (20–50 ng/μL), 1.2 μL 10× buffer, 1 mM of each deoxynucleotide phosphate (dNTP), 0.4 μM of each primer, and 0.5 units of EX Taq polymerase (Takara, Japan). The same reaction conditions were applied for all primer pairs in which the reaction mixtures were preheated to 94 °C

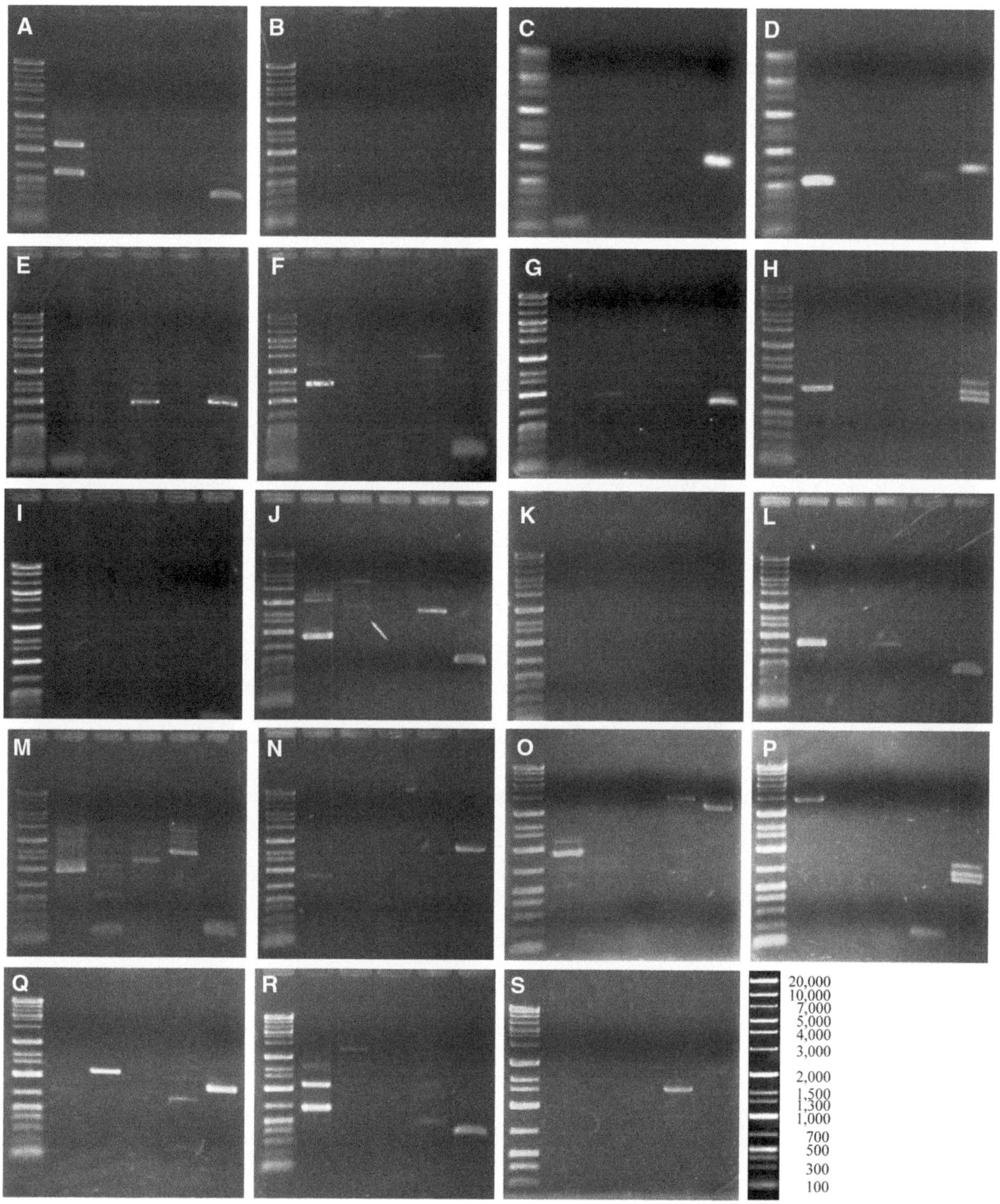

Fig. 2 Agarose gel electrophoresis images of the PCR products obtained by 19 primer pairs at a 58 °C annealing temperature. The *left-most lane* (*1st lane*) of each gel image is the size marker (Gene Ladder Wide 1, Wako), and the species used are the same as those in Fig. 1

for 4 min, followed by 35 amplification cycles (94 °C for 30 s, 55 or 58 °C for 30 s, and 72 °C for 50 s), with a final extension at 72 °C for 7 min. PCR products were electrophoresed on a 1.5 % agarose gel (UltraPure Agarose, Invitrogen, Carlsbad CA), stained with ethidium bromide and photographed.

Results and discussion

Relatively conserved exon sequences among distant reference animal taxa were determined for 12 genes in which 19 primer pairs (A–S) were designed for EPIC PCR (Table 1). To design the primers, we adopted a primer length of 20 bases, a moderate GC content (approximately 50 %), and usually less than three degenerated bases, and because the conserved regions among the variety of taxa were not extensive, there were limited options for selecting primer positions.

At a lower annealing temperature (55 °C), amplification of single to multiple fragments ranging from approximately 100–4000 bp (fragments smaller than 100 bp were not counted) was observed in at least three taxa for each primer pair (Fig. 1). The number of amplified fragments decreased at the higher annealing temperature (58 °C), to the extent that no amplified fragments were obtained for three primer pairs (Fig. 2b, i, and k). Amplification failure may be attributable to primer mismatch, large numbers of processed pseudogenes (Duvov and Perry 1984), and/or non-specific annealing, which may be responsible for multiple fragment amplification. At the higher annealing temperature, the number of primer pairs which amplified a single fragment was 13 pairs in tuna, followed by 8 in sea urchin, 6 in shark, 3 in squid, and only 2 in crab. Higher scores observed in tuna and sea urchin are reasonable, since RP gene data of teleost and sea urchin were used to design primers. The lowest result in crab may indicate distant relationship between copepod and decapod.

Two to three amplicons (13 in total) from five primer pairs (D, E, G, L, and Q) showing a single fragment amplification at the higher annealing temperature (Fig. 2) were selected and subjected to direct nucleotide sequencing using PCR primers. Partial or complete nucleotide sequences were successfully determined for eight (one each from primer pairs D, E, and G; two from L; and three from Q) out of 13 amplicons. Conserved exon sequences and expected exon–intron motifs for the respective RP genes were confirmed in six amplicons from D, E, G, and Q, whereas these were not determined in two amplicons from L due to the short distance between the primer annealing site and an intron (see Table 1). All sequences are available in DDBJ and GenBank under the accession numbers LC055480–LC055485, LC055565, LC055566.

Although intensive investigation into the optimal PCR conditions was not performed, amplification failure and multiple fragment amplification may be improved by controlling the annealing temperature or modifying the primer sequences for taxa of interest.

Thus, the present study offers new nuclear primers potentially applicable to a wide variety of marine animal taxa, which may be an initial step toward isolating single-copy nuclear DNA sequences in a wide variety of marine animals. Subsequent polymorphism detection may also provide a more informative database for genetic species identification and population study.

Acknowledgments This work was partially supported by a Grant-in-Aid for Scientific Research on Priority Areas (C) (No. 25450292) from the Ministry of Education, Science, Sports, and Culture, Japan.

References

Altschul SF, Gish W, Miller W, Myers EW, Lipman DJ (1990) Basic local alignment search tool. J Mol Biol 215:403–410

Chow S, Hazama K (1998) Universal PCR primers for S7 ribosomal protein gene introns in fish. Mol Ecol 7:1255–1256

Chow S, Nakadate M (2004) PCR primers for fish *G6PD* gene intron and characterization of intron length variation in the albacore *Thunnuns alalunga*. Mol Ecol Notes 4:391–393

Chow S, Takeyama H (2000) Nuclear and mitochondrial DNA analyses reveal four genetically separated breeding units of the swordfish (*Xiphias gladius*). J Fish Biol 56:1087–1098

Chow S, Clarke S, Nakadate M, Okazaki M (2007) Boundary between the north and south Atlantic populations of the swordfish (*Xiphias gladius*) inferred by a single nucleotide polymorphism at calmodulin gene intron. Mar Biol 152:87–93

Corte-Real HBSM, Dixon DR, Holland PWH (1994) Intron-targeted PCR: a new approach to survey neutral DNA polymorphism in bivalve populations. Mar Biol 120:407–413

Daguin C, Bonhomme F, Borsa P (2001) The zone of sympatry and hybridization of *Mytilus edulis* and *M. galloprovincialis*, as described by intron length polymorphism at locus *mac-1*. Heredity 86:342–354

Duvov KP, Perry RP (1984) The gene family encoding the mouse ribosomal protein L32 contains a uniquely expressed intron-containing gene and an unmutated processed gene. Cell 37:457–468

Hassan M, Lemaire C, Fauvelot C, Bonhomme F (2002) Seventeen new exon-primed intron crossing polymerase chain reaction amplifiable introns in fish. Mol Ecol Notes 2:334–340

Jennings BM, Etter RJ (2011) Exon-primed, intron-crossing (EPIC) loci for five nuclear genes in deep-sea protobranch bivalves: primer design, PCR protocols and locus utility. Mol Ecol Res 11:1102–1112

Lessa EP (1992) Rapid survey of DNA sequence variation in natural populations. Mol Biol Evol 9:323–330

Nakamura Y, Mori K, Saitoh K, Oshima K, Mekuchi M, Sugaya T, Shigenobu Y, Ojima N, Muta S, Fujiwara A, Yasuike M, Oohara I, Hirakawa H, Chowdhury VS, Kobayashi T, Nakajima K, Sano M, Wada T, Tashiro K, Ikeo K, Hattori M, Kuhara S, Gojobori T, Inoue K (2013) Evolutionary changes of multiple visual pigment genes in the complete genome of Pacific bluefin tuna. PNAS 110:11061–11066

Palumbi SR, Baker CS (1994) Contrasting population structure from nuclear intron sequences and mtDNA of humpback whales. Mol Biol Evol 11:426–435

Pinho C, Rocha S, Carvalho BM, Lopes S, Mourão S, Vallinoto M, Brunes TO, Haddad CFB, Gonçalves H, Sequeira F, Ferrand N (2010) New primers for the amplification and sequencing of nuclear loci in a taxonomically wide set of reptiles and amphibians. Cons Genet Res 2:181–185

Thompson JD, Gibson TJ, Plewniak F, Jeanmougin F, Higgins DG (1997) The CLUSTALX windows interface: flexible strategies for multiple sequence alignment aided by quality analysis tools. Nucl Acid Res 25:4876–4882

Uechi T, Tanaka T, Kenmochi N (2001) A complete map of the human ribosomal protein genes: assignment of 80 genes to the cytogenetic map and implications for human disorders. Genomics 72:223–230

Wool IG (1979) The structure and function of eukaryotic ribosomes. Ann Rev Biochem 48:719–754

Wool IG, Chan Y-L, Gluck A (1995) Structure and evolution of mammalian ribosomal proteins. Biochem Cell Biol 73:933–947

Yoshihama M, Uechi T, Asakawa S, Kawasaki K, Kato S, Higa S, Maeda N, Minoshima S, Tanaka T, Shimizu N, Kenmochi N (2002) The human ribosoma protein genes: sequencing and comparative analysis of 73 genes. Genome Res 12:379–390

23

Co-culture with *Chaetomorpha* sp. enhanced growth performance and reduced feed conversion ratio of the giant tiger prawn, *Penaeus monodon*

Isao Tsutsui · Jaruwan Songphatkaew · Chonlada Meeanan · Dusit Aue-umneoy · Halethichanok Sukchai · Piyarat Pinphoo · Sirimas Klomkling · Monthon Ganmanee · Hiroyuki Sudo · Kaoru Hamano

Abstract The increase in global demand for fishery products has led to a fivefold increase in aquaculture production since 1990. Commercial feed is the highest production cost in intensive aquaculture, and residual commercial feed leads to eutrophication; hence it is important to find a low-cost alternative that has less environmental impact. We investigate the use of the filamentous green algae, *Chaetomorpha* sp., as a raw feed for giant tiger prawns. The giant tiger prawn, *Penaeus monodon* was grown in monoculture, and in co-culture with *Chaetomorpha* sp. to investigate the potential benefits of co-culturing. Five 20-day-old giant tiger prawn juveniles were released in 70-L monoculture and co-culture tanks, and the specific growth rate (SGR) and feed conversion ratio (FCR) were measured after 10 weeks. The final mean body weight of co-cultured prawns was approximately 50 % heavier than that of monocultured prawns. The SGR in co-culture tanks was 4.79 ± 0.08 % day^{-1}, which was higher than that in monoculture tanks (4.14 ± 0.27 % day^{-1}). The FCR was 38.9 % lower in co-culture than in monoculture tanks. The protein content of *Chaetomorpha* sp. obtained from proximate analysis was almost the same or a little lower than other filamentous green seaweeds; however, the *Chaetomorpha* sp. has higher fiber and gross energy. These results show that *Chaetomorpha* sp. has potential for reducing feed costs in prawn intensive aquaculture through co-culturing.

Keywords Cladophoraceae · Feed conversion ratio · Growth rate · Intensive aquaculture · Penaeidae · Proximate analysis

---

I. Tsutsui · H. Sudo
Fisheries Division, Japan International Research Center for Agricultural Sciences (JIRCAS), Tsukuba, Ibaraki, Japan

I. Tsutsui (✉) · J. Songphatkaew · D. Aue-umneoy · P. Pinphoo · M. Ganmanee
Department of Animal Production and Fisheries, Faculty of Agricultural Technology, King Mongkut's Institute of Technology Ladkrabang (KMITL), Bangkok, Thailand
e-mail: cong@affrc.go.jp

C. Meeanan · H. Sukchai · S. Klomkling
Shrimp Co-culture Research Laboratory (SCORL), KMITL, Bangkok, Thailand

K. Hamano
Research Center for Marine Invertebrates, National Research Institute of Fisheries and Environment of Inland Sea, Onomichi, Hiroshima, Japan

Introduction

Aquaculture has risen in importance worldwide over the past two decades because of increasing global demand for fisheries products. World aquaculture production has rapidly increased from approximately 13 million tons in 1990 to approximately 66 million tons in 2012 (FAO 2014). In 2012, the total world aquaculture production reached almost 42 % of total fishery production (FAO 2014).

Reduction of the amount of commercial feed used in intensive aquaculture is useful both from an environmental aspect and for cost management. Residual commercial feed causes eutrophication in intensive aquaculture; therefore, reduction of commercial feed helps prevent eutrophication inside and outside of aquaculture environments. From another point of view, commercial feed represents about 43–68 % of the total culture cost, which makes it the highest production cost in intensive aquaculture (Chandra et al. 2010; Hung and Quy 2013). However, it is first necessary to find a no or low-cost alternative to enable reduction of commercial feed usage.

Trials using several species of filamentous green algae (*Ulva*, *Chaetomorpha*, and *Cladophora*) as raw or dried feed for aquacultured crustaceans and fish have recently been successfully conducted. Integrated aquaculture of brackish water prawns or freshwater herbivorous fish from tropical zones with species from the genus *Ulva* has resulted in increased growth of aquaculture species (Limsuwan et al. 2008; Cruz-Suárez et al. 2010; Gamboa-Delgado et al. 2011; Siddik and Anh 2015). Although other filamentous green seaweeds including *Chaetomorpha* and *Cladophora* may be as easily applied in intensive aquaculture systems as *Ulva* species because of their euryhalinity, so far only few studies have investigated their application as raw feed for aquaculture species (Tsutsui et al. 2010).

An unidentified species of *Chaetomorpha* grows abundantly in the central Thailand. This alga has extremely wide euryhalinity and eurythermy (Tsutsui et al. 2015), making it of potentially higher benefit to intensive shrimp aquaculture ponds than *Ulva* species; however, the use of *Chaetomorpha* sp. in aquaculture has not yet been tested. This study aims to investigate the effect of *Chaetomorpha* sp. as raw feed on the growth of giant tiger prawns, and assess *Chaetomorpha* sp. as a potential alternative for reducing feeding cost in intensive shrimp culture. This study will also detail the proximate analysis of *Chaetomorpha*, which has not yet been reported.

Materials and methods

All research activities in this study were permitted by the National Research Council of Thailand (NRCT) (Project ID 2011/005). Care and handling of the experimental animals were conducted in accordance with institute regulations and Japanese law (Kurosawa 2008). The experiment was performed in a semi-outdoor experimental facility with a transparent roof and natural temperature at the Shrimp Co-culture Research Laboratory (SCORL), King Mongkut's Institute of Technology Ladkrabang (KMITL), Thailand.

Experimental shrimp and seaweed

A batch of 15-day-old giant tiger prawn post-larvae (PL_{15}) was purchased from a commercial nursery in Chon Buri Province, Thailand. After transportation to SCORL, the PL_{15} were acclimated in a 2000-L fiberglass tank with seawater of salinity 20 PSU and gentle aeration until becoming 20-day-old juveniles (J_{20}). Commercial grow-out pellets with >40 % crude protein (Vital prawn pellets, Higashimaru Co. Ltd., Kagoshima, Japan) were provided three times a day.

The *Chaetomorpha* sp. used in this study have been allocated accession numbers AB759541 and AB819998 in the DNA Data Bank of Japan (DDBJ) (Tsutsui et al. 2015). It was originally collected at a salt swamp in Samut Sakhon Province, Thailand, and cultured at SCORL in 70-L plastic tanks at a seawater salinity of 20 PSU.

Experimental design

Three replications each were made of *Chaetomorpha* sp. in monoculture (control) and in co-culture of giant tiger prawns and *Chaetomorpha* sp. (experimental treatment). The 70-L plastic control and treatment tanks

(40 cm long × 57 cm wide × 30 cm high) were randomly arranged, and connected to each other by polyvinyl chloride pipes. The water was circulated through all tanks with a flow rate of about 30 L h^{-1} to maintain the same water conditions, and was passed through an external filter to improve water quality. Water was adjusted to an approximate salinity of 20–25 PSU. Water temperature was recorded but not adjusted, and varied from 26.7 to 31.5 °C during the experiment.

Five J_{20} individuals (body weight 0.39 ± 0.03 g) were released in each tank, creating a density of about 20 individuals per m^2. Commercial pelleted feed (vital prawn) was fed three times daily in each tank. Feeding quantity was recorded every time for each tank; quantity was adjusted according to the appetite of the juveniles, which was shown as feed remaining on the bottom of the tank. Excreta and remaining shrimp feed were removed from the tanks every morning. *Chaetomorpha* sp. was maintained to cover about 30–50 % of the water surface or bottom of the co-culture tanks. Seaweed was freely grazed by shrimp in co-culture tanks. Any seaweed, including diatoms, growing in the monoculture tanks (controls) was inhibited and scraped off from the walls with a brush every week.

Measurement of growth and feed conversion ratio (FCR)

The individual body weight (wet weight) of each juvenile in each tank after 10 weeks was measured and recorded using a digital balance (PL602-S, Mettler-Toledo, Schwerzenbach, Switzerland). Specific growth rate (SGR) and FCR were calculated by the following formulae:

$$\text{SGR}\ (\%\ \text{d}^{-1}) = (\ln W_t - \ln W_i) \times D^{-1} \times 100, \quad (1)$$

where W_t is terminal wet weight, W_i is initial wet weight, and D is culture duration

$$\text{FCR} = F/(W_t - W_i), \quad (2)$$

where F is feed given, W_t is terminal wet weight, and W_i is initial wet weight

Analysis of proximate composition and energy

Gross energy and proximate composition of *Chaetomorpha* sp. cultured at SCORL were analyzed by standard methods based on AOAC (2012) by Central Laboratory (Thailand) Co. Ltd., Bangkok, Thailand. Protein content was measured by Kjeldahl method on AOAC official method 981.10 using a factor of 6.25. Ash was measured after incineration at 600 °C for 2 h based on AOAC official method 942.05. Fat was extracted with ether using a Soxhlet apparatus and weighed following AOAC official method 920.39. Crude fiber was measured by the loss on ignition of the residue after digestion with 0.313 N sulfuric acid and 0.255 N sodium hydroxide based on AOAC official method 978.10. Carbohydrates were calculated by subtracting the protein, lipid and ash levels from 100 %. Gross energy was initially calculated as calories (protein, 4 kcal g^{-1}; fat, 9 kcal g^{-1}; carbohydrate, 4 kcal g^{-1}), and then converted to joules.

Statistics analysis

Data were shown as mean ± standard deviation (SD). Differences in SGR and FCR between control and treatment tanks were tested via the Tukey–Kramer honestly significant difference (HSD) test using JMP 10 software (SAS Institute, Cary, NC, USA); $P < 0.05$ was considered to denote significant difference.

Results

Growth

Mean body weight of shrimp from co-culture treatment tanks at the end of 10 weeks was 11.20 ± 0.652 g, which was significantly higher and approximately 50 % heavier than shrimp from monoculture tanks

Table 1 Growth performance, SGR, and FCR of giant tiger prawn juveniles in monoculture and in co-culture with *Chaetomorpha* sp.

	Monoculture	Co-culture
Mean initial weight (g) of individual shrimp	0.39 ± 0.026[a]	0.39 ± 0.034[a]
Mean terminal weight (g) of individual shrimp	7.15 ± 1.278[a]	11.20 ± 0.652[b]
Final weight difference compared with control (%)	NA	156.6
SGR (% day^{-1})	4.14 ± 0.27[a]	4.79 ± 0.08[b]
FCR	2.39 ± 0.28[a]	1.46 ± 0.62[b]
FCR reduction rate compared with control (%)	NA	38.9

Values are shown as mean value ± standard deviation from triplicate data

Different superscript labels within the same row indicate significant difference between means (Tukey–Kramer HSD test, $P < 0.05$)

NA not applicable, *SGR* specific growth rate, *FCR* feed conversion ratio

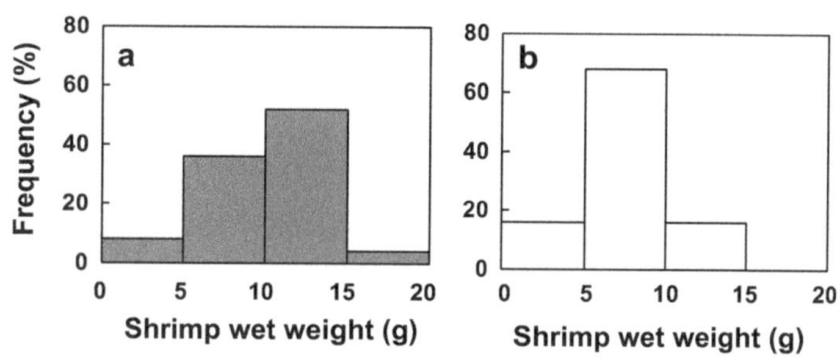

Fig. 1 Size variation of giant tiger prawns at 10 weeks in (**a**) co-culture with *Chaetomorpha* sp. and in (**b**) monoculture

(7.15 ± 1.278 g) (Table 1). The size frequency of shrimp in the co-culture treatment tanks showed that the ratio of larger sizes was higher than that in the monoculture tanks at the end of the experimental period (Fig. 1). SGR in the co-culture treatment tanks was 4.79 ± 0.08 % day^{-1}, which was also significantly higher than that in control tanks (4.14 ± 0.27 % day^{-1}) (Table 1).

FCR

FCR was significantly lower in co-culture (1.46 ± 0.62) than in monoculture tanks (2.39 ± 0.28) (Table 1). FCR reduction rate in co-culture tanks was 38.9 % compared with control tanks (Table 1).

Proximate composition

The proximate composition of wind-dried *Chaetomorpha* sp. per 100 g dry weight (DW) was: crude protein 20.4 ± 0.16 g, fat 0.8 ± 0.03 g, ash 14.0 ± 0.05 g, carbohydrate 64.8 g (including crude fiber 21.8 ± 0.06 g) (Table 2). Gross energy in 100 g DW of *Chaetomorpha* sp. was 1455.0 ± 0.35 kJ (Table 2).

Discussion

Research into improving the growth and/or feed efficiency of various aquaculture species using filamentous green seaweeds as raw or dried feed has been increasing since around 2010 (Tsutsui et al. 2007; Limsuwan et al. 2008; Ergün et al. 2009; Cruz-Suárez et al. 2010; Tsutsui et al. 2010; Gamboa-Delgado et al. 2011; Siddik and Anh 2015; Siddik et al. 2014). For species related to the seaweed investigated in the present study, *Ulva clathrata* improved growth of whiteleg shrimp by 107–164 % (Cruz-Suárez et al. 2010; Gamboa-Delgado et al. 2011), *U. intestinalis* improved growth of whiteleg shrimp and giant tiger prawns (without

Table 2 Gross energy and proximate composition of *Chaetomorpha* sp. and related species (per 100 g DW)

Species	Present species (*Chaetomorpha* sp.)	*Ulva clathrata*	*Ulva prolifera*	*Ulva* sp.
Energy (kJ)	1455.0 ± 0.35	1045.6[a]	1381.3[a]	1312.1[a]
Protein (g)	20.4 ± 0.16	20.7 ± 3.1	20.0	26.6
Fat (g)	0.8 ± 0.03	1.5 ± 0.3	0.3	0.7
Ash (g)	14.0 ± 0.05	38.4 ± 5.5	17.9	22.5
Carbohydrate[b] (g)	64.8	39.4	61.8	50.8
Crude fiber[c] (g)	21.8 ± 0.06	5.6 ± 0.5		
Reference	Present study	Cruz-Suárez et al. (2010)	MEXT (2010)	MEXT (2010)

Values are shown as mean value ± standard deviation from triplicate data. Proximate composition was re-calculated minus moisture content from original data

DW dry weight, *MEXT* Ministry of Education, Culture, Sports, Science and Technology, Japan

[a] For comparison, these values were re-calculated from the original data by applying the same gross energy coefficient (protein, 4 kcal g^{-1}; fat, 9 kcal g^{-1}; carbohydrate, 4 kcal g^{-1})

[b] Carbohydrate = 100 − (protein + fat + ash); therefore, no SD applied

[c] Crude fiber is treated as a component of carbohydrate

specific data) (Limsuwan et al. 2008), and *Chaetomorpha ligustica* increased SGR of giant tiger prawns by up to 157 % at the early juvenile stage (Tsutsui et al. 2010). Although direct comparisons with the studies cited above cannot be made because of differences in co-cultured species, our results suggest that the *Chaetomorpha* sp. is able to improve shrimp growth by almost the same amount as that of the other filamentous green seaweeds under co-culture conditions.

Previous studies have found that the filamentous green seaweed *Ulva* spp. can reduce FCR for shrimp and herbivorous fish (Cruz-Suárez et al. 2010; Siddik and Anh 2015). For example, *U. clathrata* reduced the FCR for whiteleg shrimp by approximately 47.5–62.1 % (Cruz-Suárez et al. 2010), and *U. intestinalis* reduced the FCR for spotted scat, red tilapia, and giant gourami by 26–58 % (Siddik and Anh 2015). Our study found that *Chaetomorpha* sp. as a raw feed reduced the FCR for giant tiger prawns by approximately 40 % under co-culture conditions, indicating that *Chaetomorpha* sp. can reduce artificial feed expenses as efficiently as *Ulva* species. This is the first report of FCR reduction for a farmed species by the genus *Chaetomorpha* under co-culture conditions, as previous reports have only described improvement in shrimp growth (Tsutsui et al. 2010).

In Table 2, the extremely high content of ash in *U. clathrata* (38.4 %) is believed to be salt contamination on the surface of the thalli. Considering the difference in ash content, proximate composition analysis showed that the protein content of *Chaetomorpha* sp. was the same or a little less than the closely related *Ulva* species. Conversely, *Chaetomorpha* sp. had higher energy and fiber content, but lower ash level (Table 2). Artificial commercial feed containing about 30–60 % protein is generally used in intensive shrimp aquaculture (Shigeno et al. 1972; Cruz-Suárez et al. 2010; Tsutsui et al. 2010; Hung and Quy 2013). The protein content of *Chaetomorpha* sp. and several species of *Ulva* are much lower than that of artificial feeds. However, despite its lower protein content, the enhancement of shrimp growth and reduced FCR showed that the *Chaetomorpha* sp. contributed as effectively as commercial feed to shrimp growth.

It has previously been reported that mysids and *Acetes* shrimp living in mangrove estuaries have cellulose-degrading capabilities (Niiyama et al. 2012). Although it has never been reported whether Penaeidae species have similar cellulose-degrading abilities, there is a high probability of this, as Penaeidae belong to the same taxonomic class (Malacostraca) as mysids and *Acetes* shrimp, and, like mysids and *Acetes* shrimp, are omnivorous. If the giant tiger prawn does have cellulose-degrading abilities, the *Chaetomorpha* sp. seaweed is more likely to contribute to the growth of this shrimp, as this alga has a higher fiber content than *Ulva* species. In addition, *Chaetomorpha* sp. as a supplemental raw feed may be more advantageous to growth enhancement and FCR reduction of giant tiger prawns owing to its higher gross energy content compared with *Ulva* species. However, the possible contribution of cellulose to growth enhancement and FCR reduction in giant tiger prawn requires future, detailed dietary studies.

Almost all filamentous green seaweeds currently used for growth enhancement of aquaculture species as raw or dried dietary feed are from the genus *Ulva*; species used for this application include *U. clathrata* for whiteleg shrimp, (Cruz-Suárez et al. 2010; Gamboa-Delgado et al. 2011), *U. intestinalis* for whiteleg shrimp (Limsuwan et al. 2008), *U. intestinalis* for spotted scat, red tilapia, and giant gourami (Siddik and Anh 2015), *U. rigida* for Nile tilapia (Ergün et al. 2009), and *Ulva* sp. for Nile tilapia (Siddik et al. 2014). The frequent use of *Ulva* spp. may be due to their euryhalinity and eurythermy (Msuya and Neori 2002; Kakinuma et al. 2004; McAvoy and Klug 2005; Tsutsui et al. 2012), which facilitates their application to variable aquaculture environments with a wide range of salinities, and to their widespread availability. The *Chaetomorpha* sp. investigated in this study has a wider salinity tolerance range than *Ulva* species; *Chaetomorpha* sp. can grow in salinities of 0.2–80 PSU and has a high growth rate maximum of <60 % day^{-1} (Tsutsui et al. 2015). Therefore, *Chaetomorpha* sp. may be better and more suitable than *Ulva* species in intensive shrimp aquaculture ponds.

Conclusion

The *Chaetomorpha* sp. that grows abundantly in the central Thailand enhanced the growth of giant tiger prawns and reduced the FCR when used as a raw supplemental feed under co-culture conditions. The protein content of this species is almost the same or a little less than other filamentous green seaweeds, such as *Ulva* species; however, it has the advantage of higher fiber and gross energy levels. Thus, *Chaetomorpha* sp. has the potential to reduce commercial feed costs in the shrimp aquaculture industry through co-culturing. This species of *Chaetomorpha* may be more suitable and better for shrimp co-culture systems worldwide than *Ulva* species, which are currently used as raw or dried supplemental feeds.

Author's contribution The work presented here was carried out in collaboration among the authors. IT and KH conceived the study. The methods were designed by IT, JS, and KH. The experiment was performed by IT, JS, CM, DA, HS, PP, and SK. The manuscript was drafted by IT, KH, and HS. JS carried out data analysis. MG and HS participated in experimental coordination. All authors read and approved the final manuscript.

Acknowledgments The authors thank the National Research Council of Thailand (NRCT) for providing useful advice on conducting this study in Thailand. We also greatly appreciate the helpful advice of Dean Sakchai Choochote, and the Head of Fisheries group, Dr. Paveena Taveekijakarn, Faculty of Agricultural Technology, King Mongkut's Institute of Technology Ladkrabang (KMITL). We thank Dr. Masayoshi Saito, Program Director, Japan International Research Center for Agricultural Sciences (JIRCAS) and Dr. Yukio Maeno, ex-Director, Fisheries Division, JIRCAS for their constructive comments. Our appreciation also goes to Mr. Prachara Manoromchatchawan and Ms. Patumwadee Reuangsawang, Faculty of Agricultural Technology, KMITL for their devoted support of the experiment.

Conflict of interest The authors declare that they have no competing interests.

References

AOAC (2012) Official methods of analysis, 19th edn. AOAC International, Maryland

Chandra KJ, Sarker D, Khaleque MA, Das DR (2010) Economic analysis of floodplain aquaculture at Daudkandi upazilla in Comilla. J Bangladesh Agril Univ 8:323–332

Cruz-Suárez LE, León A, Peña-Rodríguez A, Rodrgíuez-Peña G, Moll B, Ricque-Marie D (2010) Shrimp/*Ulva* co-culture: a sustainable alternative to diminish the need for artificial feed and improve shrimp quality. Aquaculture 301:64–68

Ergün S, Soyutük M, Güroy B, Güroy D, Merrifield D (2009) Influence of *Ulva* meal on growth, feed utilization, and body composition of juvenile Nile tilapia (*Oreochromis niloticus*) at two levels of dietary lipid. Aquacult Int 17:355–361

Food and Agriculture Organization of the United Nations (FAO) (2014) FAO year book 2012 Fishery and Aquaculture Statistics, Rome

Gamboa-Delgado J, Peña-Rodríguez A, Ricque-Marie D, Cruz-Suárez LE (2011) Assessment of nutrient allocation and metabolic turnover rate in pacific white shrimp *Litopenaeus vannamei* co-fed live macroalgae *Ulva clathrata* and inert feed: dual stable isotope analysis. J Shellfish Res 30:969–978

Hung LT, Quy OM (2013) On farm feeding and feed management in whiteleg shrimp (*Litopenaeus vannamei*) farming in Viet Nam. In: Hasan MR, New MB (eds) On-farming feeding and feed management in aquaculture. FAO Fisheries and aquaculture technical paper No. 583. Rome, FAO, pp 337–357

Kakinuma M, Kuno Y, Amano H (2004) Salinity stress responses of a sterile mutant of *Ulva pertusa* (Ulvales, Chlorophyta). Fish Sci 70:1177–1179

Kurosawa TM (2008) Japanese regulation of laboratory animal care with 3Rs. In: Proceedings of 6th world congress on alternatives and animal use in the life sciences. Japanese Society for Alternatives to Animal Experiments 14, pp 317–321

Limsuwan C, Kawsuralikhit C, Choocherd N, Hongrat P, Banlerdate D (2008) Co-culture shrimp and *Enteromorpha* sp. National Research Council of Thailand, Bangkok (in Thai)

McAvoy KM, Klug JL (2005) Positive and negative effects of riverine input on the estuarine green alga *Ulva intestinalis* (syn. *Enteromorpha intestinalis*) (Linnaeus). Hydrobiologia 545:1–9

Ministry of Education, Culture, Sports, Science and Technology (MEXT), JAPAN (2010) Standard tables of food composition in Japan (2010) MEXT, Tokyo

Msuya FE, Neori A (2002) *Ulva reticulata* and *Gracilaria crassa*: macroalgae that can biofilter effluent from tidal fishponds in Tanzania. Western Indian Ocean J Mar Sci 1:117–126

Niiyama T, Hanamura Y, Tanaka K, Toyohara H (2012) Occurrence of cellulose activities in mangrove estuarine Mysids and *Acetes* shrimp. JIRCAS Work Rep 75:35–39

Shigeno K, Kumada K, Deshimaru O, Aramaki T, Kuroki K, Kitaue K (1972) Studies on the artificial diets of prawn-I. Relationship between the feed efficiency and crude protein in the diets. Bull Jpn Soc Sci Fish 38:101–106

Siddik MAB, Anh NTN (2015) Preliminary assessment of the Gut weed *Ulva intestinalis* as food for herbivorous fish. Int Aquat Res. 7:41–46

Siddik MAB, Nahar A, Rahman NM, Anh NTN, Nevejan N, Bossier P (2014) Gut weed, *Enteromorpha* sp. as a partial replacement for commercial feed in Nile tilapia culture. World J Fish Mar Sci 6:267–274

Tsutsui I, Aue-umneoy D, Srisapoome P, Hamano K (2007) Possible implications of the co-cultivation of black tiger shrimp and Cladophoraceae species on a Southeast Asian shrimp farm. Abstract of 21th International Seaweed Symposium, p 77, International Seaweed Symposium, Kobe

Tsutsui I, Kanjanaworakul P, Srisapoome P, Aue-umneoy D, Hamano K (2010) Growth of giant tiger prawn, *Penaeus monodon* Fabricius, under co-culture with a discarded filamentous seaweed, *Chaetomorpha ligustica* (Kützing) Kützing, at an aquarium-scale. Aquacult Int 18:545–553

Tsutsui I, Hamano K, Aue-umneoy D, Songphatkaew J, Srisapoome P, Ruangsomboon S, Klomkling S, Ganmanee M, Taveekijakarn P, Maeno Y (2012) Common underwater plants in coastal areas of Thailand. JIRCAS International Agriculture Series 21, JIRCAS, Tsukuba

Tsutsui I, Miyoshi T, Aue-umneoy D, Songphatkaew J, Meeanan C, Klomkling S, Sukchai H, Pinphoo P, Yamaguchi I, Ganmanee M, Maeno Y, Hamano K (2015) High tolerance of *Chaetomorpha* sp. to salinity and water temperature enables survival and growth in stagnant waters of central Thailand. Int Aquat Res 7:47–62

24

Isolation of a novel lectin from the dorsal spines of the devil stinger, *Inimicus japonicus*

Hideyuki Nakagawa · Kuniko Nagasaka · Hitomi Sakai · Kozue Edo · Mitsuko Shinohara · Kiyoshi Ohura

Abstract A novel lectin was purified from the dorsal spines of the devil stinger, *Inimicus japonicus* using a combination of affinity chromatography techniques. A single band was detected on a native PAGE gel with a relative molecular mass of 97 kDa. The N-terminal partial amino acid of the intact 75 kDa subunit of the 97 kDa lectin was found to be DHEDS. The agglutination of rabbit erythrocytes by the 97 kDa lectin was inhibited most effectively by methyl α-D-mannoside. The 97 kDa lectin stimulated mitogenesis in murine splenocytes. This is the first study to examine the dorsal lectin of *I. japonicus* and one of the very few studies on venom lectins from venomous scorpaeniform fish. These results suggest that the devil stinger, *I. japonicus*, may be a novel resource for biologically active substances.

Keywords Devil stinger · *Inimicus japonicus* · Dorsal spine · Lectin · Agglutination · Mitogenic activity · Murine splenocytes

---

H. Nakagawa (✉) · H. Sakai
Division of Environmental Symbiosis, Graduate School of Integrated Arts and Sciences, Tokushima University, Tokushima 770-8502, Japan
e-mail: sea-hide@tokushima-u.ac.jp

H. Sakai
e-mail: sakai.hitomi@tokushima-u.ac.jp

K. Nagasaka
Fukae Nagasaka Clinic, Etajima 737-2214, Japan
e-mail: fkengski@ace.ocn.ne.jp

K. Edo
Junior College Division, Department of Science for Human Health, Shikoku University, Tokushima 771-1192, Japan
e-mail: kozue-edo@shikoku-u.ac.jp

M. Shinohara · K. Ohura
Department of Pharmacology, Osaka Dental University, Hirakata 573-1121, Japan
e-mail: shino.m@gol.com

K. Ohura
e-mail: ohura@cc.osaka-dent.ac.jp

Background

A large number of venomous and poisonous animals exist in aquatic environments worldwide. More than 200 of the approximately 22,000 species of fish in the ocean are considered to be venomous (Halstead 1988; Russell 1996). Most of these venomous fish are non-migratory, slow moving, and mainly live in shallow waters in protected habitats (Maretic 1988). Venomous scorpaeniform fish include the lionfish and scorpionfish from the family Scorpaenidae, devil stinger and stonefish from the family Synanceiidae, and waspfish from the family Tetrarogidae (Kiriake et al. 2013). These fish possess 11–17 dorsal, 2 pelvic, and 3 anal spines, with the venom secretory complex being located within the anterolateral grooves of these spines (Russell 1965; Halstead 1988; Haddad et al. 2003; Smith and Wheeler 2006; Andrich et al. 2010).

The devil stinger *Inimicus japonicus*, which belongs to the family Synanceiidae, has 17 dorsal, 1 pelvic, and 2 anal spines, which contain venom glands that are covered by an integumentary sheath (Tange 1954). *I. japonicus*, a valuable demersal scorpaenid fish, is widely distributed along the coastal areas of eastern Asia at a depth of between 10 and 200 m (Wang et al. 2013). The body of *Inimicus* is covered in warts or skin lumps, with many skin tubercle glands similar to the stonefish. Envenomation occurs when people carelessly handle or step on these fish, and are stung by the dorsal spines. Envenomation appears immediately as intense, sharp, and persisting local pain, and swelling around the sting (Auerbach 1991; Yamamoto et al. 2010). Symptoms depend on the amount of venom injected. Systemic effects including dizziness, fever, and delirium have been reported (Auerbach 1991). However, only a limited number of studies have investigated the toxicity of *I. japonicus*. Therefore, we herein examined the dorsal venom of the devil stinger, *I. japonicus* using column chromatography and, for the first time, separated a novel lectin that induced mitogenic activity.

Methods

Isolation of a dorsal lectin

Inimicus japonicus (18 specimens, average size of 20 cm) were collected by local fishermen from the coast of Hiroshima Prefecture and Tokushima Prefecture, Japan in May 2003 (Fig. 1a, b). The collected fish were transported alive or frozen to our laboratory. The dorsal spines (a total of 17) of *I. japonicus* were cut from their base, and the dorsal venom protein was extracted with 0.15 M NaCl as reported previously (Nagasaka et al. 2009). Briefly, in the first step of purification, the venom protein was applied to a Phenyl Sepharose CL-4B (GE Healthcare, Uppsala, Sweden) affinity chromatographic column (2 ml) equilibrated with 16 mM Tris-HCl buffer containing 2 M NaCl (pH 7.4). The sample was rinsed and washed with the same buffer containing 0.01 M NaCl at a flow rate of 20 ml/h (Fig. 1a). The 2-ml elution fractions were collected and analyzed for absorption at 280 nm and agglutinating activity. Each of the unbound and bound fractions was pooled and

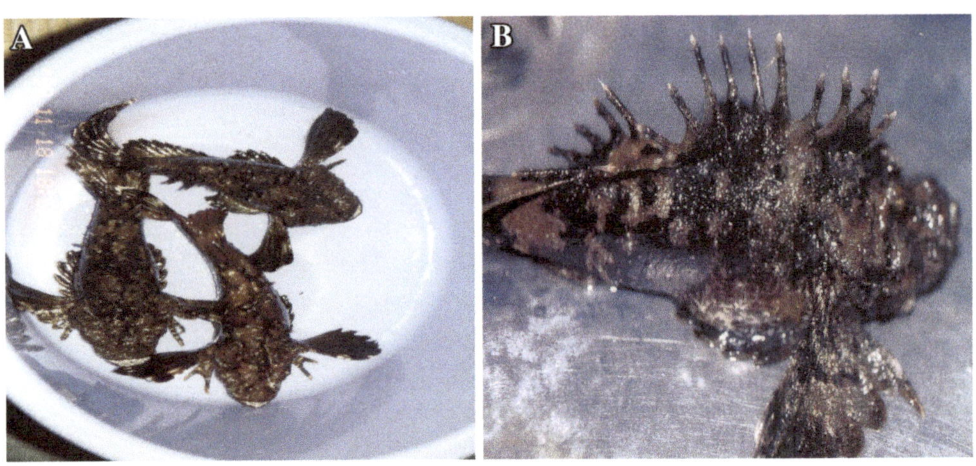

Fig. 1 The devil stinger *Inimicus japonicus* as seen from above (**a**) and a specimen with erect dorsal spines (**b**)

analyzed for electrophoresis. The final step of purification was achieved using a Concanavalin A-Sepharose 4B (Sigma-Aldrich, Missouri, USA) column (2 ml) equilibrated with 20 mM Tris-HCl buffer containing 0.4 M NaCl (pH 7.4). The unbound fraction (the PS-I fraction) was rinsed and washed with the same buffer, and eluted with the same buffer containing 100 mM methyl-α-D-mannoside in the buffer at a flow rate of 20 ml/h (Fig. 1b). Elution fractions (2 ml) were collected and analyzed for absorption at 280 nm and agglutinating activity. Each of the unbound (PS-I-ConA-I fraction) and bound (PS-I-ConA-II fraction) fractions was pooled and analyzed for electrophoresis. PS-I-ConA-II was then used as the purified lectin. Protein content was measured according to the method of Bradford (1976) using bovine albumin as a standard.

Electrophoresis

Polyacrylamide gel electrophoresis (PAGE) was performed as described by Davis (1964) using a 4–20 % gradient gel. Sodium dodecyl sulfate (SDS)-PAGE was carried out by the method of Laemmli (1970) using a 10–20 % gradient gel. Protein samples were heated in the presence of 2-mercaptoethanol for 4 min at 98 °C. The gels were stained with Coomassie brilliant blue.

Glycoprotein staining

The glycoprotein sugar moieties of the sample protein were detected in the SDS-PAGE gel using the GelCode Glycoprotein kit (Pierce Biotech., Inc., IL, USA). This kit detects sugars that occur in glycoproteins, including galactose, mannose, glucose, *N*-acetylglucose, *N*-acetylgalactosamine, sialic acid, fucose, and xylose.

Assay of agglutinating activity

Agglutinating activity was assayed using rabbit erythrocytes on microtiter plates. Twenty-five microliters of a 2 % (v/v) suspension of erythrocytes in 6.4 mM phosphate-buffered saline (PBS) was added to 50 μl of a serial twofold dilution of the sample. The plates were incubated at room temperature for 1 h. The results obtained were expressed by the minimum concentration of the sample (μg/ml) required for positive agglutination. The inhibition of agglutination was expressed as the minimum concentration of each sugar required to inhibit the agglutinating activity of the sample.

Mitogenic activity

Mitogenic activity in murine splenocytes was determined using a cell culture assay with a dye, the tetrazolium salt 3(4,5-dimethylthiazol-2-yl)-2,5-diphenyltetrazolium bromide (MTT) (Nakagawa et al. 1997). Splenocytes were collected from female ddY mice and suspended in RPMI-1640 medium supplemented with penicillin and streptomycin (100 μg/ml and 100 U/ml). Splenocytes (5×10^6 cells/ml) with or without concanavalin A (1 μg/ml) and samples were plated on flat-bottom microplates and incubated at 37 °C in a humidified atmosphere containing 5 % CO_2 for 68 h. Ten microliters of the MTT tetrazolium salt solution (5 mg/ml) was then introduced into each well, and formazan was extracted from the cells with 10 % sodium sulfate (SDS) after 4 h. The optical density of each well was measured spectrophotometrically with a microplate reader (Thermo Fisher Scientific, MultiskanGo, Yokohama, Japan) at 570 nm.

N-terminal amino acid sequencing

Approximately, 3 μg of the sample protein was subjected to SDS-PGE, followed by electroblotting on a polyvinylidene difluoride membrane. The membrane was then stained with Ponceau S and destained. The protein band was excised and subjected to automated Edman degradation using the Shimadzu Model PPSQ-30 protein sequencer (Shimadzu Corp., Kyoto, Japan).

Statistical analysis

Data are expressed as the mean or mean ± standard deviation (SD). Statistical analyses were performed using the SPSS version 16.0 software package (SPSS, Chicago, Inc., IL, USA). Statistical analyses of the results obtained were performed using Dunnett's multiple comparison test when various experimental groups were compared to the control groups, and the Student's t test was used for paired groups. $P < 0.05$ was considered significant.

Results and discussion

Dorsal venom was previously reported to induce weak agglutination in rabbit erythrocytes and stimulated mitogenesis in murine splenocytes (Nakagawa et al. 2011). The venom possessed a glycoprotein that was stained by SDS-PAGE using the GelCode Glycoprotein kit, indicating that it contained one band corresponding to a protein with an apparent mass of 75 kDa (data not shown). Therefore, we herein attempted to isolate a lectin from the dorsal spines of *I. japonicus* using a combination of affinity chromatography techniques (Fig. 2a, b). Dorsal venom was applied to a Phenyl Sepharose CL-4B column (2 ml) equilibrated with 16 mM Tris-HCl buffer containing 2 M NaCl (pH 7.4). The sample was rinsed with the same buffer and then eluted with the same buffer containing 0.01 M NaCl at a flow rate of 20 ml/h. Figure 2a shows the elution pattern with two protein peaks. The first peak (the PS-I fraction) induced agglutinating activity at a dose of 6.25 μg/ml, while the second peak (the PS-II fraction) induced this activity at a higher dose of 200 μg/ml (data not shown). SDS-PAGE analysis of the PS-I fraction showed two main bands that corresponded to proteins with apparent masses of 30 and 97 kDa (Fig. 2a). The PS-I fraction was applied to a Concanavalin A-Sepharose 4B column (2 ml) equilibrated with 20 mM Tris-HCl buffer containing 0.4 M NaCl (pH 7.4) for purification. The column was rinsed thoroughly with the same buffer and then eluted with 100 mM methyl-α-D-mannoside in the buffer at a flow rate of 20 ml/h.

As shown in Fig. 2b, a native PAGE analysis of the unbound fraction (the PS-I-ConA-I fraction) identified two bands corresponding to proteins with apparent masses of 30 and 97 kDa. On the other hand, the bound fraction (the PS-I-ConA-II fraction) showed a single discrete band corresponding to a protein with an apparent mass of 97 kDa. The PS-I-Con A-II fraction was a glycoprotein fraction that contained mannose residues. This glycoprotein fraction was designated as a 97 kDa lectin. The recovery of the 97 kDa lectin in terms of protein content accounted for 0.35 % of the dorsal venom. SDS-PAGE showed that the 97 kDa lectin was mainly composed of 75 and 30 kDa subunits (Fig. 2b). The 75 kDa subunit of the 97 kDa lectin was subjected to a partial amino acid sequence analysis. The N-terminal amino acid of the 75 kDa subunit was aspartic acid. The partial amino acid sequence was determined up to 5 residues, and found to be DHEDS. In the case of the waspfish *Hypodytes rubripinnis*, Karatoxin, a novel cytotoxic protein (110 kDa), was successfully purified from the dorsal spines of this fish (Nagasaka et al. 2009; Shinohara et al. 2010). The N-terminal partial amino acid sequence of the 76 kDa subunit of Karatoxin was shown to be DQHDDxPxxAPDPG. As the partial amino acid sequence of the 75 kDa subunit of the 97 kDa lectin was only analyzed up to 5 residues, it is currently difficult to determine whether it shares homology with the 76 kDa subunit of Karatoxin. Therefore, we are now attempting to improve the purifying method as well as the recovery of the 97 kDa lectin from the dorsal venom. Kiriake et al. (2013) recently identified the toxin of *I. japonicus* as a 160 kDa heterodimer composed of 80 kDa α- and β-subunits by cDNA cloning. This toxin was very similar to those of the lionfish *P. lunulata* and waspfish *H. rubripinnis*. The venoms of most venomous fish have been proposed to have similar toxic properties and molecules with similar structures (Saunders 1960; Russell 1965; Church and Hodgson 2002). Therefore, more detailed studies on the structure of the 97 kDa lectin are needed to elucidate its sequence homology to Karatoxin and piscine venoms such as those from the stonefish.

The agglutinating activity of the 97 kDa lectin was inhibited most effectively by methyl α-D-mannoside and, to a lesser extent, by D-mannose, and D-Glucose, suggesting that the hydrogen groups at C-1, C-3, and C-4 of the pyranose ring structure influenced sugar binding to the lectin (Table 1). Figure 3 shows the results of mitogen responses in murine splenocytes in the presence of the dorsal venom and PS-I-ConA fractions (the PS-I-ConA-I fraction and PS-I-ConA-II fraction). As shown in Fig. 3a, the dorsal venom induced mitogenic activity in murine splenocytes at doses ranging from 25 to 50 μg/ml. However, a higher dose (100 μg/ml) of

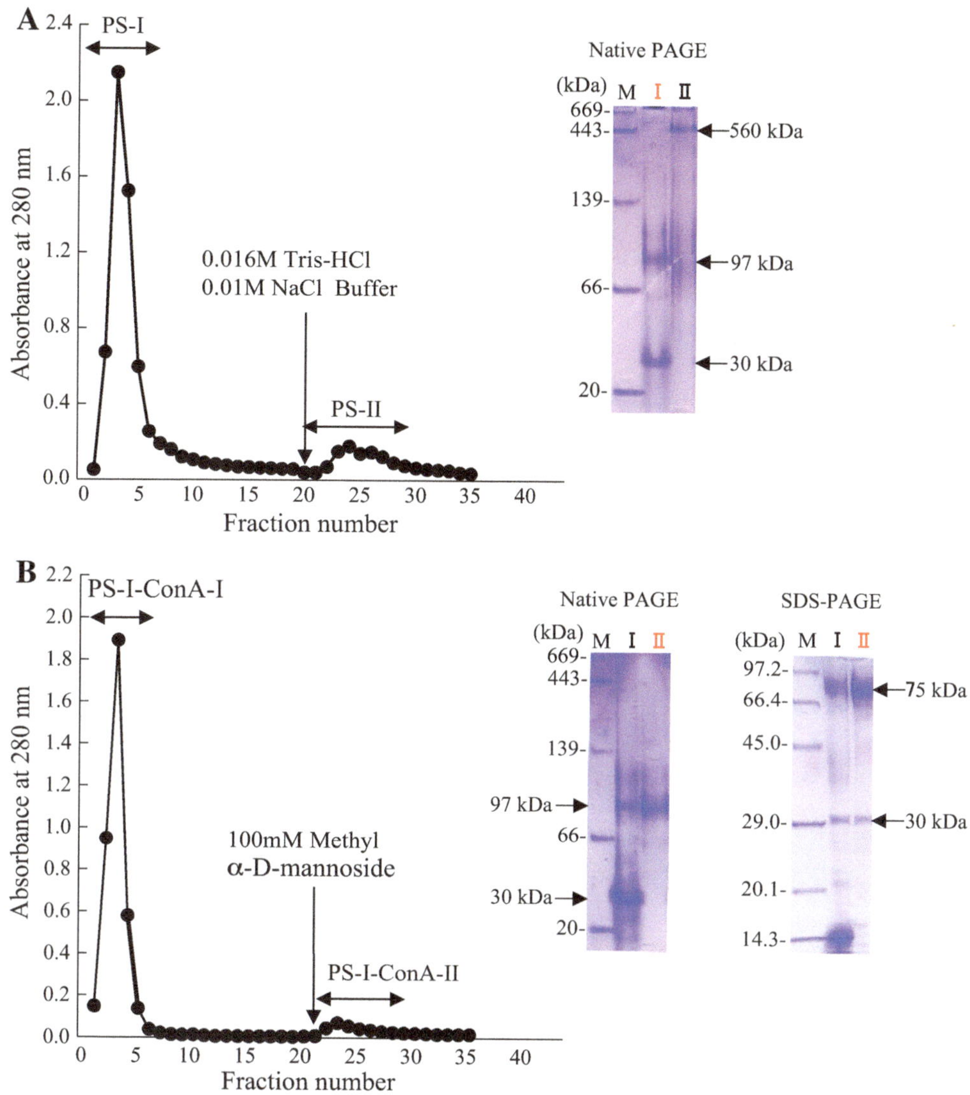

Fig. 2 Isolation of a novel lectin from dorsal spines of *Inimicus japonicus*. The isolation procedure was described in detail in "Methods". **a** The first purification step used Phenyl Sepharose CL-4B. **b** The second purification step used Concanavalin A-Sepharose 4B. Inset panels show native PAGE and SDS-PAGE of affinity chromatographic fractions. *M* mol wt markers

Table 1 Sugar inhibition of agglutinating activity of 97 kDa lectin

Sugar	Minimum effective concentration (mM)
D-Mannose	3.1
Methyl a-D-mannoside	0.78
D-Glucose	12.5
N-Acetylglucosamine	12.5
Others[a]	NI 100

NI non-inhibitory

[a] D-Galactose, D-Fucose, L-Glucose, D-Galactosamine, Lactose

the dorsal venom significantly decreased mitogenic activity, suggesting the presence of another biologically active component in addition to lectin component(s) in the venom. Most piscine venoms have been shown to exhibit potent cytolytic activity (Shier 1988; Church and Hodgson 2002). We also observed cytotoxic activity in human leukemia cells (K562) by the dorsal venom (unpublished data). The PS-I-ConA-I fraction and

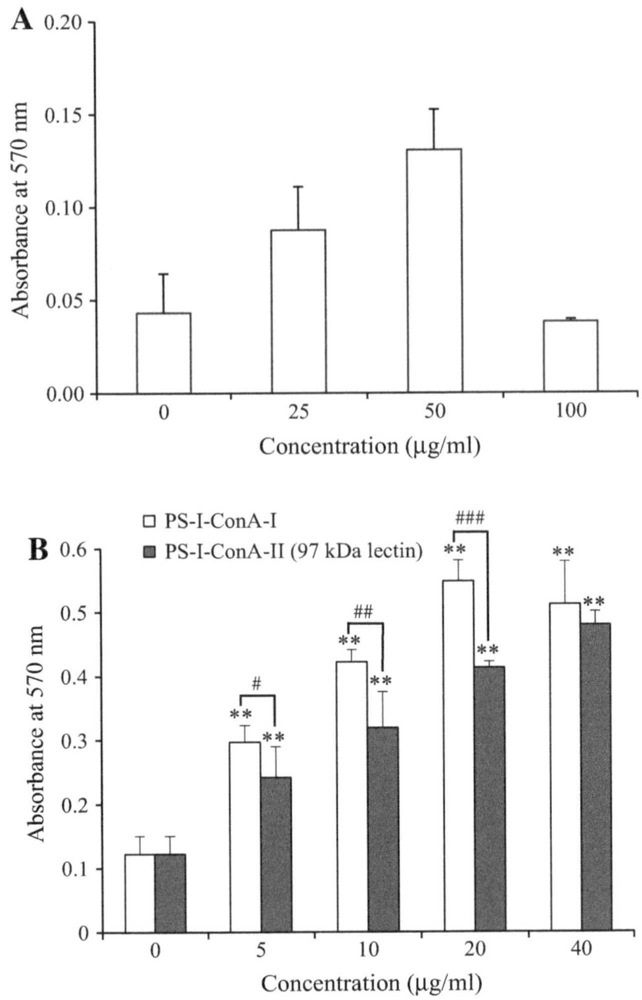

Fig. 3 Mitogen responses of the dorsal venom (**a**) and PS-I-ConA fractions (**b**) in murine splenocytes. Splenocytes (5×10^6 cells/ml) were incubated with the dorsal venom, PS-I-ConA-I and PS-I-ConA-II (97 kDa lectin), for 68 h and the incubation was continued with MTT for 4 h in a CO_2 humidified atmosphere. Data show the mean ± SD of two experiments performed in triplicate (**a**) or the mean ± SD of 3–4 experiments performed in triplicate (**b**). $^{**}P < 0.01$, significantly different from the negative control. $^{\#}P < 0.05$, $^{\#\#}P < 0.01$, $^{\#\#\#}P < 0.001$, significant difference between two groups according to the Student's unpaired *t* test

97 kDa lectin effectively induced mitogenesis in a dose-dependent manner in murine splenocytes (Fig. 3b). The mitogenic activity induced by the PS-I-ConA-I fraction was slightly higher than that induced by the 97 kDa lectin. This suggests that there is a quantitative and/or qualitative difference in the lectin component between the two fractions. Further biological and structural studies on the PS-I-ConA-I fraction may lead to the identification of more novel lectin(s).

Lectins are a large group of proteins that reversibly bind specific carbohydrates and possess at least one non-catalytic domain (Drickamer 1988). Although many lectins have been isolated from various plants, bacteria, and animals, including invertebrates (Drickamer and Taylor 1993; Hatakeyama et al. 1994; Nakagawa et al. 2003; Sharon and Lis 2004; Malagoli et al. 2010), very few have been detected in marine vertebrates, particularly venomous fish (Satoh et al. 2002; Nagasaka et al. 2009). In the present study, we successfully isolated and partially characterized the 97 kDa lectin from *I. japonicus* as a novel piscine lectin. The 97 kDa lectin from *I. japonicus* venom exhibited mitogenic activity in murine splenocytes, suggesting that it may contribute to the local and systemic effects observed on envenomation such as swelling, severe pain, and fever. We more recently observed chemotactic activity in guinea-pig neutrophils (unpublished data). Thus, the 97 kDa lectin may affect inflammatory and immunomodulatory processes. The agglutination induced by the 97 kDa lectin was inhibited by D-mannose and methyl α-D-mannoside. These results suggest that the 97 kDa lectin may induce mitogenic activity by binding the mannose-containing carbohydrate chains present on the surface of murine splenocytes. Further studies are needed to clarify the mechanisms underlying the biological activities of the 97 kDa lectin. Moreover, an investigation of the structural features of this lectin is needed to elucidate the physiological significance of venom from *I. japonicus*. The present results suggest that the 97 kDa lectin from *I. japonicus* venom is a source of biologically active substances, which may have applications as research tools.

Conclusions

A novel lectin was isolated from the dorsal spines of *I. japonicus* using a combination of affinity chromatography techniques. Native PAGE showed that the dorsal lectin was a glycoprotein with a molecular mass of 97 kDa.

The agglutinating activity of the 97 kDa lectin was effectively inhibited by methyl α-D-mannoside. The 97 kDa lectin may exhibit mitogenic activity by binding to a specific carbohydrate chain, such as mannose moieties on the cell surface. Our results suggest that the venom of the devil stinger, *I. japonicus* is a potent resource for piscine lectins.

Author contributions HN prepared the manuscript, and supervised the experiments. KN participated in the experiment design and manuscript writing. HS conducted the purification of dorsal lectin. KE conducted the mitogenic activity and agglutination activity assay. MS participated in the interpretation of the results. KO participated in the study design. All authors read and approved the final manuscript.

Acknowledgments The authors are indebted to Mr. Hidenori Yuasa for collecting the *I. japonicus* specimens, and Mr. Takefumi Hosotani, Mr. Masashi Miyata, and Ms. Atsumi Harada for their assistance purifying the dorsal venom lectin from the fish.

Conflict of interest The authors declare that they have no competing interests.

References

Andrich F, Garnielli JBT, Cassoli JS, Lautner RQ, Santos RAS, Pimenta AMC, de Lima ME, Figueiredo SG (2010) A potent vasoactive cytolysin isolated from *Scorpaena plumieri* scorpionfish venom. Toxicon 56:487–496

Auerbach PS (1991) Marine envenomations. N Engl J Med 325:486–493

Bradford MM (1976) A rapid and sensitive methods for quantitation of microgram quantities of protein utilizing the principle of protein-dye binding. Anal Biochem 72:248–254

Church JE, Hodgson WC (2002) The pharmacological activity of fish venoms. Toxicon 40:1083–1093

Davis BJ (1964) Disc electrophoresis II. Method and application to human serum proteins. Ann NY Acad Sci 121:404–427

Drickamer K (1988) Two distinct classes of carbohydrate-recognition domains in animal lectins. J Biol Chem 263:9557–9560

Drickamer K, Taylor ME (1993) Biology of animal lectins. Annu Rev Cell Biol 9:237–264

Haddad V Jr, Marins IA, Makyama HM (2003) Injuries caused by scopionfishes (*Scorpaena plumieri* Bloch, 1789 and *Scopaena brasiliensis* Cuvier, 1829) in the Southwestern Atlantic Ocean (Brazilian coast): epidemiologic, clinic and therapeutic aspects of 23 stings in humans. Toxicon 42:79–83

Halstead BW (1988) Poisonous and venomous marine animals of the world, 2nd edn. Darwin Press, Princeton, pp 1168. (Plates pp 288)

Hatakeyama T, Kohzaki H, Nagatomo H, Yamazaki N (1994) Purification and characterization of four Ca2+-dependent lectins from the marine invertebrate, *Cucumaria echinata*. J Biochem 116:209–214

Kiriake A, Suzuki Y, Nagashima Y, Shiomi K (2013) Proteinaceous toxins from three species of scorpaeniform fish (lionfish *Pterois lunulata*, devil stinger *Inimicus japonicus* and waspfish *Hypodytes rubripinnis*): close similarity in properties and primary structures to stonefish toxins. Toxicon 70:184–193

Laemmli UK (1970) Cleavage of structural protein during the assembly of the head of the bacteriophage T4. Nature 227:680–685

Malagoli D, Sacchi S, Ottaviani E (2010) Lectins and cytokines in celomatic invertebrates: two tales with the same end. Inv Surv J 7:1–10

Maretic Z (1988) Fish Venom. In: Tu AT (ed) Handbook of natural toxins: marine toxins and venoms, vol 3. Marcel Dekker, New York, pp 445–477

Nagasaka K, Nakagawa H, Satoh F, Hosotani T, Yokoigawa K, Sakai H, Sakuraba H, Ohshima T, Shinohara M, Ohura K (2009) A novel cytotoxic protein, Karatoxin, from the dorsal spines of the redfin velvetfish, *Hypodytes rubripinnis*. Toxin Rev 28:260–265

Nakagawa H, Yamaguchi C, Hayashi H (1997) Biologically active substances from sea urchins. J Nat Toxins 6:193–202

Nakagawa H, Tanigawa T, Tomita K, Tomihara Y, Araki Y, Tachikawa E (2003) Recent studies on the pathological effects of purified sea urchin toxins. J Toxicol Toxin Rev 22:633–649

Nakagawa H, Edo K, Sakai H, Watanabe Y, Shinohara M, Ohura K (2011) Mining on novel lectins from some venomous fish and sea urchins. In: Program abstracts of the 9th IST Pacific meeting on animal, plant and microbial toxins. Vladivostok, Suppl: P10

Russell FE (1965) Marine toxins and venomous and poisonous marine animals. In: Russell FE (ed) Advances in marine biology, vol 2. Academic Press, London, pp 137–151

Russell FE (1996) Toxic effects of animal toxins. In: Klassen CD (ed) Casarett and doull's toxicology—the basic science of poisons. McGraw-Hill, Sydney, pp 801–839

Satoh F, Nakagawa H, Yamada H, Nagasaka K, Nagasaka T, Araki Y, Tomihata Y, Nozaki M, Sakuraba H, Ohshima T, Hatakeyama T, Aoyagi H (2002) Fishing for bioactive substances from scorpionfish and some sea urchins. J Nat Toxins 11:297–304

Saunders PR (1960) Pharmacological and chemical studies of the venom of stonefish (Genus *Synacceija*) and other scorpionfishes. Ann NY Acad Sci 90:789–804

Sharon N, Lis H (2004) History of lectins: from hemagglutinins to biological recognition molecules. Glycobiology 14:53–62

Shier WT (1988) Cytotoxic Effect of Marine Toxins and Venoms. In: Tu AT (ed) Handbook of natural toxins: marine toxins and venoms, vol 3. Marcel Dekker, New York, pp 477–491

Shinohara M, Nagasaka K, Nakagawa H, Edo K, Sakai H, Kato K, Iwaki F, Ohura K, Sakuraba H (2010) A novel chemoattractant lectin, Karatoxin, from the dorsal spines of the small scorpionfish *Hypodytes rubripinnis*. J Pharmacol Sci 113:414–417

Smith WL, Wheeler WC (2006) Venom evolution widespread in fishes: a phylogenetic road map for the bioprospecting of piscine venoms. J Hered 97:206–217

Tange Y (1954) On the morphology of the poison apparatus in Japanese fish with remarks on its toxicity. VII. The poison apparatus in *Inimicus japonicus* (Cuvier et Valenciennes). Yokohama Med Bull 5:234–242

Wang Y, Li L, Cui G, Lu W (2013) Ontogenesis from embryo to juvenile and salinity tolerance of Japanese devil stinger *Inimicus japonicus* during early life stage. Springer Plus 2:289. http://www.sprigerplus.com/content/2/1/289

Yamamoto R, Suzuki M, Hori S, Aikawa N (2010) Stonefish "Okoze" envenomation during food preparation. Keio J Med 59:19–22

25

Acute toxicity and risk assessment of three commercial detergents using the polychaete *Capitella* sp. C from Chetumal Bay, Quintana Roo, Mexico

R. G. Uc-Peraza · V. H. Delgado-Blas

Abstract The acute toxicity of three formulations of commercial detergent (ROMA®, FOCA® and BLANCA NIEVES®) was evaluated using the polychaete *Capitella* sp. C in static bioassays over a 48-h exposure period. The probit method was used to determine the median lethal concentration (LC_{50}) of each formulation as a whole as well as the LC_{50} of the active ingredient, linear alkylbenzene sulfonate (LAS), using 95 % confidence intervals. The formulations and LAS showed LC_{50} values of between 70.79 and 147.91 ppm and 15.48 and 22.38 ppm, respectively, at 48 h. FOCA® was the most toxic detergent, followed by BLANCA NIEVES® and finally ROMA®. The variation in the toxicity of the three detergents could have been caused both by differences in the relative concentrations of the anionic surfactant LAS contained in each formulation and the presence of other ingredients (enzymes, sodium silicate, sodium tripolyphosphate, bleachers and perfumes) which can also increase formulation toxicity. Correlation analysis revealed that percent mortality of *Capitella* sp. C increased with increase in the concentrations of the detergent over the 48-h exposure period. The risk quotient was greater than one for all three evaluated detergents, indicating that there is a high risk that they adversely affect the aquatic biota, particularly sediment-dwelling organisms such as the test species.

Keywords Polychaeta · *Capitella* · Bioassays · Acute toxicity · Detergents

Introduction

The contamination of aquatic ecosystems is increasing at an alarming rate as a consequence of the discharge of untreated sewage of urban and industrial origin into coastal zones, rivers, streams and lakes. Among the contaminants found in these sewage effluents are organic pollutants such as detergents, which can cause toxicity problems for the aquatic biota found in the receiving water bodies (Uc-Peraza and Delgado-Blas 2012). Furthermore, the phosphorus component of detergents contributes to eutrophication which results in an increase in the cell numbers of some algal species leading to decreased species diversity (Tkachenko and Kutsyn 2002; Markina and Aizdaicher 2007).

Detergents and surfactants are among the extensively used chemicals at home as well as in industry (Azizullah et al. 2011). Currently, the world production of detergent amounts to 10 millions of tons per years (IMRG 2012), and the worldwide total annual production of surfactants is estimated to exceed 15 million tons

To the memory of Jonatán Delgado Martínez, poet, composer, musician and rock drummer.

R. G. Uc-Peraza (✉) · V. H. Delgado-Blas
División de Ciencias e Ingeniería, Universidad de Quintana Roo, 77019 Chetumal, Quintana Roo, Mexico
e-mail: russ_net42@hotmail.com; russell.peraza@gmail.com

(Van Bogaert et al. 2007; Henkel et al. 2012, 2014). As a result, large quantities of detergents and their components enter the environment (Warne and Schifko 1999; Liwarska-Bizukojc et al. 2005). McAvoy et al. (1993) estimated that 5 % of all linear alkylbenzene sulfonate (LAS) produced in the United States reach the aquatic environment. In Mexico, 463,965 tons of detergents are manufactured annually (Sobrino-Figueroa 2013) most of which eventually find their way into freshwater and marine ecosystems (Cserháti et al. 2002).

The toxic effect of a detergent depends on its mode of action, the toxicity of the active ingredient (surfactant) in relation to its chemical and physical structure and the response of the test organism (Ofojekwu et al. 1999). LAS is the most common anionic surfactant used in the formulation of domestic and industrial detergents, and is considered an important potential pollutant (Ainsworth 1992). However, despite the fact that several studies have been undertaken to determine the toxicity of LAS in aquatic organisms, little is known about the toxic effects of detergent as a whole (Markina and Aizdaicher 2007).

The detergents we use in everyday life are complex mixtures of various compounds. A typical detergent consists of surfactants, builders, anti-redeposition agents, zeolite, alkaline agents, corrosion inhibitors, processing aids, colorants, fragrances, oxygen bleach, suds control agents, opacifiers, bleaching agents, enzymes and other minor constituents (Bajpai and Tyagi 2007). These components can interact either antagonistically, additively or synergistically to increase or diminish the toxicity of the detergent in the aquatic environment (Warne and Schifko 1999). Thus, evaluating the individual effects of surfactants or other ingredients does not reflect either the actual or net influence of a detergent on the aquatic environment, it is thus necessary to estimate the biological effects of detergent formulations as a whole (Markina and Aizdaicher 2007; Azizullah et al. 2011).

Polychaetes are generally the most abundant taxon in benthic communities and are commonly utilized as indicator species of environmental conditions in marine ecosystems (Dean 2008). *Capitella capitata* (Fabricius 1780), a species of Capitellidae Grube, 1862, is an opportunistic polychaete that has often been used as an indicator of organic pollution in marine sediments (Reish 1959; Bellan 1967; Pearson and Rosenberg 1978). It has also been extensively employed as a test organism in toxicological bioassays (Reish and LeMay 1991). The *C. capitata* species complex includes at least 50 cryptic siblings that differ mainly in their enzyme and general protein patterns, ecophysiological characteristics and reproduction modes (Méndez et al. 2000). The feeding characteristics of the adults and juveniles mean that they play an important role in the recycling and elimination of toxic substances associated with sediments. Ecotoxicological studies performed with metals (Reish 1988; Reish and LeMay 1991; Méndez and Green-Ruiz 2005, 2006) and other pollutants (Méndez 2005, Méndez et al. 2008, 2013) have registered diverse adverse effects on this species complex.

The introduction of toxic chemical substances, such as detergents, to the aquatic environment could thus negatively impact on the population dynamics of this species in their natural ecosystems. However, in Mexico, ecotoxicological studies of commercial detergent formulations are limited. The aim of this study was to determine the mean lethal concentration (LC_{50}) of three commercial detergents and their ecological risk in sediments, using the polychaete *Capitella* sp. C as the test species. We hope that the information gained will add to the available knowledge about toxicity levels of these detergents in aquatic environments in Mexico. Such knowledge is crucial for the development and implementation of policies associated with the protection of aquatic life in Chetumal Bay, Quintana Roo, Mexico.

Materials and methods

Worms

Adult *Capitella* sp. C individuals were collected from Chetumal Bay, Quintana Roo, Mexico (18°30′47.89″N, 88°16′30.87″W) in a zone free from wastewater discharges. The worms were collected with a PVC corer (0.018 m^2). Deep cores 20 cm deep were taken and then sieved using a 0.5-mm open mesh. The polychaetes were transported to the laboratory in vials containing water from the sampling site. In addition, sediment from the site was collected and transported in sealed polyethylene bags, with the purpose of using it during the adaptation period of the test organisms as well as in the bioassays. In the laboratory, undamaged organisms, 20–25 mm long were selected (polychaetes that were fragmented, pale in color or with reduced motility were discarded). Specimens were identified as *Capitella* sp. C with the aid of a stereoscopic microscope and an optical microscope using the keys given by Garza-García (2009). Individual *Capitella* sp. C were then placed in 3 L, constantly aired aquariums, together with water and sediment from the sampling site and maintained at room temperature (25 ± 1 °C) in a natural light–dark cycle for 2 days prior to the bioassays to allow them to adapt to the conditions (APHA et al. 1992). No food was provided either during acclimation or over the ecotoxicological test period.

Chemicals

Three commercial detergents labeled as biodegradable were evaluated, all of which are among the most commonly used domestic detergents in Mexico: ROMA®, FOCA® and BLANCA NIEVES®. In all cases, the active ingredient was the anionic surfactant LAS. Their usage and other ingredients that make up the detergents are listed in Table 1. It is important to note that neither the LAS concentrations nor those of the other components are stated on the product container: this datum is confidential and inaccessible to the consumer. Nevertheless, the proportions of some of the main constituents of detergents are generally: surfactant 15 %, poly-phosphate + silicate 30 %, sodium perborate 20 %, fluorescent pigment 0.1 %, sodium sulfate 20 % and enzymes 0.5 % (Sobrino-Figueroa 2013).

A stock solution was prepared for each detergent at 0.3 % (3 g) in 1 L of distilled water. This was then used to obtain serial dilutions (uniformly in a 0.5 logarithmic scale) as follows: 15.62, 31.25, 62.5, 125, 250 ppm (test concentrations). The LAS concentration was then calculated for each dilution of each detergent using the analysis method established by the Official Mexican Guideline NMX-AA-039-SCFI-2001 (Secretaria de Económica 2001).

Sediment

The sediment used for the ecotoxicological tests was previously analyzed (very fine sand with 0.81 % organic matter) and sieved with a 0.5-mm mesh to extract any organisms possibly present in the sample. The sediment samples were then refrigerated at 4 °C for 48 h before testing to avoid potential oxidation–reduction reactions and to eliminate any other undesired organisms.

Toxicity tests

Bioassays were of the static type without renewal of the test solution and for each assay a control (without detergent, 0 ppm) and five concentrations were used, each with three replicates. Tank dimensions were 10 × 15 × 22 cm. A total of 100 g of previously treated sediment were placed in each tank (replicate). Saline water was then added and 10 *Capitella* sp. C individuals were randomly selected from a total of 180 organisms for each bioassay chamber and placed in each tank. After waiting 45 min to allow the organisms to bury into the sediment (5 mm thick), the toxic solution was added to give a total volume of 2 L per replicate.

For each bioassay chamber, mortality readings were performed at 1, 2, 4, 8, 18, 24, 36 and 48 h of exposure. Dead organisms were immediately removed after examination. Individuals were considered dead when they fulfilled the following criteria: pale color; swollen and showing no movement on the surface of the sediment (APHA et al. 1992; Uc-Peraza and Delgado-Blas 2012); unresponsive to physical stimulation. Tests were rejected if the survival rate of the control group was 90 % or less. The following physico-chemical parameters were measured before and after the bioassays: temperature, pH, salinity and dissolved oxygen.

Table 1 Summary of the use and ingredients of the LAS-containing commercial detergents tested

Detergents	Detergent use	Ingredients
ROMA®	Multi-propose	Phosphates, sodium silicate, anti-redeposition agents, bleachers and perfume
FOCA®	Laundry	Proteolytic enzyme, phosphates, sodium silicate, anti-redeposition agents, bleachers and perfume
BLANCA NIEVES®	Multi-propose	Sodium silicate, carbonate, sulfate, sodium tripolyphosphate, anti-redeposition agents and perfume

Statistical analysis

The LC_{50} was calculated using the probit method (APHA et al. 1992). The results were then graphed as regression curves with probit units against the logarithm base 10 of detergent concentration using the software SigmaPlot 12.0, with 95 % confidence intervals. The values obtained from the acute toxicity tests were then subjected to a descriptive analysis to assess the general behavior of the data. A one-way analysis of variance (ANOVA, alpha = 0.05) was then performed to analyze differences in percent mortality of *Capitella* sp. C at 48 h among the concentrations of the detergents tested and among replicates. Pearsońs product–moment correlation analyses were then done to assess the relationship between detergent concentration and percent mortality.

Characterization of ecological risk

According to USEPA guidelines, ecological risk assessment (ERA) is defined as a process that evaluates the likelihood of adverse ecological effects on ecosystems exposed to one or more stressors (US EPA 1998; Gao et al. 2013). Risk assessment is thus, essentially, a simple comparison of predicted environmental concentration (PEC) with predicted no-effect concentration (PNEC), normally expressed as the "risk quotient" (RQ = PEC/PNEC) (Jensen et al. 2001), which is then used to determine ecological risk (Cristale et al. 2013; Gao et al. 2013). The principle is that when the risk quotient is greater than or equal to one there is a stronger likelihood of an impact. In contrast, a risk quotient of less than one indicates that the likelihood of an effect is low, and thus of less concern (Lam and Gray 2001). In this study, we examined the ERA by taking the PEC to be equivalent to the LAS value in sediments of 5.3 ppm (Sánchez-Peinado 2007). The PNEC was then calculated using the following equation: $PNEC = LC_{50}/AF$, where LC_{50} is the value obtained from the acute toxicity tests and AF is the assessment factor (100). For data interpretation, an RQ value ≥ 1 indicates a potentially high risk, and an RQ value <1 a nonsignificant (low) risk.

Results

The average values of the physico-chemical parameters of the water used in the bioassays showed little variation (Table 2). However, salinity was the only parameter that remained constant at 10 ‰.

The LC_{50} values at 48 h for the three detergents tested ranged from 70.79 to 147.91 ppm (Figs. 1, 2, 3), giving the following acute toxicity sequence: FOCA® > BLANCA NIEVES® > ROMA®. Table 3 shows the LC_{50} for the active ingredient (LAS) in each detergent, with their corresponding 95 % confidence intervals. These ranged from 15.48 and 22.38 ppm. It can be appreciated that FOCA® produced the lowest LC_{50} values, both for the formulation as a whole and the active ingredient.

The percentage mortality of *Capitella* sp. C exposed to different concentrations of the three formulations of commercial detergents at 48 h is presented in Table 4. It can be seen that no mortality was registered throughout the bioassays in any of the controls used (0 ppm). However, 3–27 % mortality was recorded for the most dilute detergent concentrations: 15.62 and 31.25 ppm, and between 20 and 57 % mortality was registered for 62.5 and 125 ppm. The solutions with the highest detergent concentration (250 ppm) produced the highest mortality percentages for all three formulations tested as follows: ROMA (70 %), BLANCA NIEVES® (87 %) and FOCA®(93 %). The ANOVA detected significant differences ($p < 0.05$) in percent mortality among the concentrations of each detergent. However, there were no significant differences ($p > 0.05$) between replicates or formulations. The correlation coefficient between detergent concentration and percent

Table 2 Means and standard errors for the physico-chemical parameters measured in the bioassays

Detergents	Temperature (°C)	Dissolved oxygen (mg/L)	pH	Salinity (‰)
ROMA®	24.97 ± 0.04	4.81 ± 0.05	8.15 ± 0.04	10
FOCA®	25.00 ± 0.07	3.98 ± 0.11	7.54 ± 0.06	10
BLANCA NIEVES®	25.02 ± 0.02	3.56 ± 0.09	7.88 ± 0.09	10

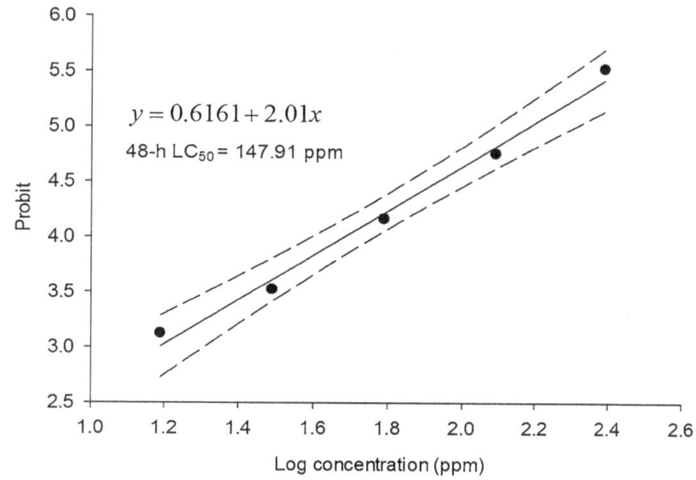

Fig. 1 Probit for the percent mortality of *Capitella sp.* C exposed to varying concentrations of ROMA® for 48 h under laboratory conditions

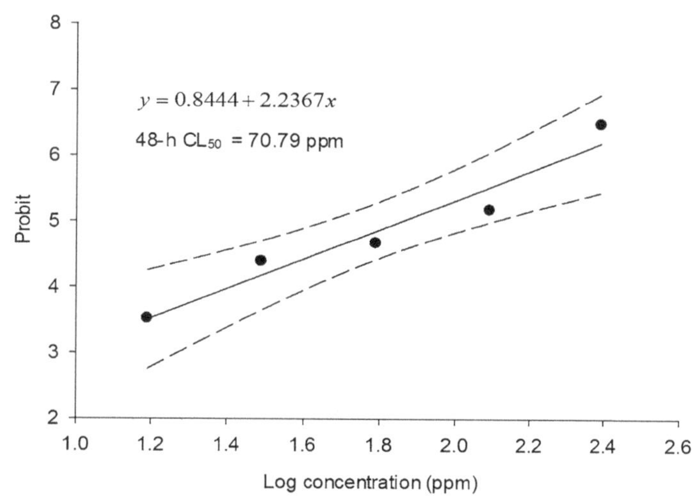

Fig. 2 Probit for the percent mortality of *Capitella sp.* C exposed to varying concentrations of FOCA® for 48 h under laboratory conditions

mortality was statistically significant ($p < 0.05$) in all cases. This means that percent mortality of *Capitella* sp. C increased with increase in the concentrations of the detergents tested over the assay period.

The RQ values calculated for the three commercial detergents were greater than one in all cases (Table 5). The highest RQ value was 7.48 (FOCA®) and the lowest 3.58 (ROMA®) with an average of 5.37. There is thus a high risk that these detergents will cause damage to aquatic ecosystems, particularly sediments.

Discussion

The three commercial detergents tested contain different toxic chemical compounds that may adversely affect aquatic ecosystems. It is known that detergents can be toxic to aquatic invertebrates, fish and plants (Warne 1995). Ecotoxicological studies with fish (Okoli-Anunobi et al. 2002; Omotoso and Fagbenro 2005; Sobrino-Figueroa 2013), algae (Aizdaicher and Markina 2006; Markina and Aizdaicher 2007; Azizullah et al. 2012), crustaceans and bacteria (Pedrazzani et al. 2012) have shown that these organisms are affected morphologically, physiologically and biochemically when exposed to detergents. It has been documented that the use of household washing detergents and softeners can be toxic for aquatic organisms in concentrations from 0.07 up

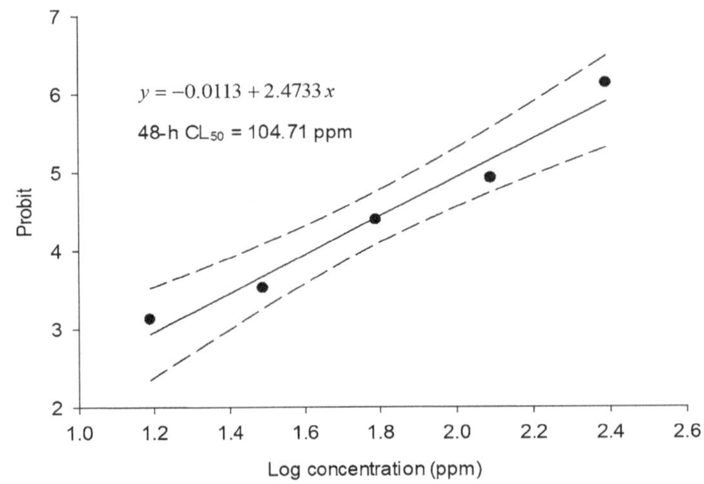

Fig. 3 Probit for the percent mortality of *Capitella sp.* C exposed to varying concentrations of BLANCA NIEVES® for 48 h under laboratory conditions

Table 3 LAS equivalent LC_{50} values at 48 h of *Capitella* sp. C. exposed to varying concentrations of ROMA®, FOCA® and BLANCA NIEVES® for 48 h under laboratory conditions

Detergents	LC_{50} (ppm)	Confidence intervals (95 %)
ROMA®	22.38	±13.42
FOCA®	15.48	±10.32
BLANCA NIEVES®	19.05	±12.55

Table 4 Percent mortality of *Capitella* sp. C. exposed to varying concentrations of ROMA®, FOCA® and BLANCA NIEVES® for 48 h under laboratory conditions

Concentration (ppm)	Mortality (%)		
	ROMA®	FOCA®	BLANCA NIEVES®
250	70	93	87
125	40	57	47
62.5	20	37	27
31.25	7	27	7
15.62	3	7	3
Control	0	0	0

Table 5 Ecological risk to the sediment of the detergents tested

Detergents	PNEC	RQ	Risk level
ROMA®	1.47	3.58	High
FOCA®	0.70	7.48	High
BLANCA NIEVES®	1.04	5.06	High

to 35.4 ppm (Ankley and Burkhard 1992; Pettersson et al. 2000). According to the literature, aquatic organisms are negatively affected by the presence of active ingredients in detergents, such as anionic and non-ionic surfactants, at concentrations ranging from 0.0025 to 300 ppm, and 0.3 to 200 ppm, respectively (Liwarska-Bizukojc et al. 2005). This may be more or less severe depending on the organism tested. In this study, the three detergents evaluated gave LC_{50} values of 70.79–147.91 ppm (formulation as a whole) and

15.48–22.38 ppm (active ingredient only) against *Capitella* sp. C after 48 h exposure, and the percent mortalities registered for the different concentrations of the detergents ranged from 3 to 93 %.

FOCA®, with LC_{50} values of 70.79 (whole formulation) and 15.48 ppm (active ingredient), was the most toxic detergent tested and also produced the highest percent mortality (93 %). These results are similar to those reported by Uc-Peraza and Delgado-Blas (2012) who examined the acute toxicity of these same commercial brands of detergents against other polychaete species. It is known that the toxicity of the detergents is due to the action the surfactants (Markina and Aizdaicher 2007) together with the other ingredients contained in the formulation (Sobrino-Figueroa 2013). It has been demonstrated that surfactants present in detergents can negatively affect living cells in different ways by damaging cell membranes (Mikolajczyk and Diehn 1978; Chawla et al. 1987), attaching to proteins, and affecting cell physiological and biochemical processes (Markina and Aizdaicher 2007; Azizullah et al. 2012). At low concentrations, detergents produce changes in cell membrane permeability, thus altering the function of the respiratory system in aquatic species (Sobrino-Figueroa 2013). In particular, anionic surfactants can solubilize and denature proteins and alter enzyme activity by binding or disrupting enzyme structure (Argese et al. 1994). In polychaetes, the anionic surfactant LAS can cause morphological damages (Conti 1987). Regarding other ingredients, it is known that enzymes (protease and lipase), sodium silicate, perfumes and bleachers (sodium perborate) can contribute to an increase in detergent toxicity for a wide range of aquatic organisms (Warne 1995; Warne and Schifko 1999; Markina and Aizdaicher 2007; Pedrazzani et al. 2012; Sobrino-Figueroa 2013).

BLANCA NIEVES® was less toxic to *Capitella sp.* C than FOCA®, possibly because this formulation does not contain enzymes and bleachers. However, it does contain other ingredients such as sodium silicate and perfumes, which may have contributed to the (lower) toxicity of the detergent. Furthermore, it cannot be ruled out that the other ingredients in the formulation, such as carbonates, sulfates, sodium and anti-redeposition agents may have contributed to its toxicity. Finally, ROMA® was found to be the least toxic of the three detergents tested.

When evaluating detergent toxicity, studies have often focused on surfactants as these are considered to be among the most toxic compounds contained in these materials, and also represent a high percentage, 15–40 %, of the total number of ingredients in the formulations (Scheibel 2004). Warne and Schifko (1999) found that surfactants contributed between 10.4 and 98.8 % of the toxicity of detergents they analyzed with a mean contribution of 40.7 %. Thus, the low toxicity of ROMA® might be due to the low concentrations of surfactant it contains, as well as the absence of other toxic ingredients (enzymes and sodium tripolyphosphate). In conclusion, the variations in toxicity between the detergents tested could be caused both by the concentration of the anionic surfactant LAS in the formulation and the presence of other ingredients (enzymes, sodium silicate, sodium tripolyphosphate, bleachers and perfumes).

The LAS LC_{50} values reported by this study with *Capitella sp.* C are similar to those registered by Uc-Peraza and Delgado-Blas (2012). These authors reported LAS toxicity (LC_{50}) values of between 12.88 and 14.12 ppm against *Laeonereis culveri* at 48 h. Similar values were also reported by Conti (1987) who determined the toxicity of two anionic surfactants: sodium dodecyl sulfate (SDS) and LAS in another polychaete species, *Arenicola marina* (SDS: LC_{50} 15.2 ppm; LAS: LC_{50} 12.5 ppm, at 48 h). This shows that *Capitella sp.* C is less sensitive to anionic surfactants than *A. marina* and *L. culveri* in acute toxicity tests.

Comparing our results with those for other groups of aquatic organisms, we can observe that *Capitella* sp. C is more resistant to detergents than crustaceans and fish. For example, Warne (1995) examined the toxicity of 24 detergents to the cladoceran *Ceriodaphnia* cf. *dubia* and found that toxicity levels (measured as EC_{50} at 48 h, immobilization) varied between 1.6 and 70.3 ppm. In another study, Pettersson et al. (2000) evaluated the toxicity of 26 commercial detergents using *Daphnia magna* as the test organism, reporting values between 4 and 85 ppm (EC_{50} at 48 h) for 25 of them. In fish, Okoli-Anunobi et al. (2002) calculated the acute toxicity of a commercial detergent to *Oreochromis niloticus* finding an LC_{50} of 9.77 ppm at 96 h. Omotoso and Fagbenro (2005) determined the toxicity of three commercial detergents with this same species registering LC_{50} values of between 12.04 and 41.88 ppm at 96 h. However, if we look at studies performed with mollusk species as test organisms, it can be observed that *Capitella* sp. C is more sensitive. For example, Iannacone and Alvariño (2002) undertook an ecotoxicological evaluation of three commercial detergents against different species of mollusks and reported LC_{50} values of 201.07 ppm (*Melanoides tuberculata*), 82.93 ppm (*Physa venustula*) and 71.41 ppm (*Heleobia cumingii*) at 48 h. The differences in sensitivity between polychaetes and mollusks may be explained by the way in which the detergent acts on the organism. In polychaetes

(*Arenicola marina*), anionic detergents (LAS) cause severe damage mainly at the level of the epidermis and in the region of the gills (Conti 1987). This is because, in the majority of cases, the soft bodies of the polychaetes are exposed to the environment and thus come into direct contact with pollutants. In contrast, the hard resistant shells of mollusks afford some protection from contaminants making them more tolerant. It is important to mention that the comparisons made here are based on a limited number of experiments, and the response of different species within taxa may vary with their biochemistry and physiology, the age of the test organisms, the detergent evaluated and the conditions under which the bioassays are conducted. More toxicity tests performed under similar conditions with a wide variety of aquatic species are thus needed.

Our results of the correlation analysis show that *Capitella* sp. C percent mortality increased significantly ($p < 0.05$) with increase in the concentrations of the detergents during the 48-h exposure period, coinciding with the results of Uc-Peraza and Delgado-Blas (2012). This observation also agrees with Okwuosa and Omoregie (1995) and Ogundele et al. (2005) who determined the acute toxicity of anionic surfactants in fish. In another study, Aizdaicher and Markina (2006) evaluated the toxicity of two detergents and one anionic surfactant to an algal species (*Plagioselmis prolonga*) and also observed that the effects on the test organisms increased with increasing concentrations of these substances.

The risk quotient value of the three LAS-containing detergents indicates that they are high-risk substances likely to cause damage to aquatic biota, particularly sediment-dwelling organisms such as *Capitella* sp. C. This is because the motility and feeding characteristics of these organisms make them more vulnerable to the impact of detergents. It has been demonstrated that 20–30 % of surfactants entering the sea environment easily accumulate in sediments and can affect bottom-dwelling biota (Marín et al. 1991; Marin et al. 1994; Markina and Aizdaicher 2007). The effects are particularly acute in those zones where there are outfalls of raw sewage or sewage effluent, as is the case at several sites within the urban shoreline zone at Chetumal, Quintana Roo, where clandestine discharges of soapy water have been detected. Surfactants and other ingredients contained in detergents are common components of the domestic and municipal residual output which eventually reaches the natural environment causing diverse toxic effects to the aquatic organisms present (Ankley and Burkhard 1992; Pettersson et al. 2000, Azizullah et al. 2011). Warne and Schifko (1999) mention that surfactants and sodium silicates are the main contributors to detergent toxicity, and were components of the three detergents evaluated in this study. Surfactants are used as cleaning agents while sodium silicates act as water softeners (Table 1). Specifically, the toxic effects of the anionic surfactant LAS have been examined in several experiments with different groups of aquatic organisms providing evidence of its toxicity (Marín et al. 1991; Bao-Quey and Dar-Yi 1994; Jorgensen and Christoffersen 2000; Rosety et al. 2001; Hampel et al. 2001; Christoffersen et al. 2003; Stefanoni and Abessa 2008; Coelho and Rocha 2010). However, it has been determined that the ecological risk of LAS to aquatic organisms is low (Fendinger et al. 1994; Van de Plassche et al. 1999; Versteeg et al. 1999; Temara et al. 2002). Despite this, both this study and that undertaken by Uc-Peraza and Delgado-Blas (2012) show that when the effects of the surfactant and the other ingredients contained in detergents are considered as a whole, there is a high risk that the aquatic biota will be adversely affected. Commercial detergents are complex mixtures of different compounds which can interact either antagonistically, additively or synergistically, thus modifying their toxicity to the aquatic environment (Warne and Schifko 1999). The observed effects can thus be considered as the net impact of the detergent as a whole (Azizullah et al. 2011). Studies undertaken using different aquatic organisms (Aizdaicher and Markina 2006; Sobrino-Figueroa 2013) have demonstrated that detergent formulations taken as a whole may be more toxic than the active ingredients they contain, with significant ecological consequences for shore ecosystems both at high (10 ppm) and low (0.1–1 ppm) concentrations (Aizdaicher and Markina 2006). Nevertheless, given that the results obtained in this study with *Capitella* sp. C cannot be generalized with other taxa, we recommend that ecotoxicological bioassays at sub-lethal doses to test for chronic toxicity are performed, both with polychaetes and other groups of aquatic organisms representative of the aquatic biota in Chetumal Bay, Quintana Roo, Mexico.

Conclusions

Three formulations of commercial detergents tested showed LC_{50} values (48 h exposure) for *Capitella* sp. C between 70.79–147.91 ppm and 15.48–22.38 ppm corresponding to the formulation and the active ingredient, respectively. The percentage of mortality among the concentrations of the three detergents was 3–93 %. FOCA® was the most toxic detergent of all. The variation in the toxicity of the three detergents could have been caused both by differences in the relative concentrations of LAS contained in each formulation and the presence of other ingredients (enzymes, sodium silicate, sodium tripolyphosphate, bleachers and perfumes) which can also increase formulation toxicity. Finally, the ecological risk assessment for the three detergents indicates that there is a high risk to the aquatic biota, especially sediment-dwelling organisms such as the test species.

Acknowledgments This study was done with the support of the División de Ciencias e Ingeniería of the Universidad of Quintana Roo, México. We would like to thank two anonymous reviewers whose comments and observations have substantially improved this manuscript.

References

Ainsworth SJ (1992) Soaps and detergents. Chem Eng News 70(3):27–37

Aizdaicher NA, Markina ZV (2006) Toxicity effects of detergents on the alga *Plagioselmis prolonga* (Cryptophyta). Russ J Mar Biol 32:45–49

Ankley GT, Burkhard P (1992) Identification of surfactants as toxicants in a primary effluent. Environ Toxicol Chem 11(9):1235–1248

APHA, Awwa, WPCF (American Public Health Association, American Water Works Association and Water Pollution Control Federation) (1992) Métodos normalizados para el análisis de aguas potables y residuals. Ediciones Díaz de Santos, Madrid

Argese E, Marcomini A, Miana P, Bettiol C, Perin G (1994) Submitochondrial particle response to linear alkylbenzene sulfonates, nonylphenol polyethoxylates and their biodegradation derivatives. Environ Toxicol Chem 13(5):737–742

Azizullah A, Richter P, Häder DP (2011) Toxicity assessment of a common laundry detergent using the freshwater flagellate *Euglena gracilis*. Chemosphere 84(10):1392–1400

Azizullah A, Richter P, Jamil M, Häder DP (2012) Chronic toxicity of a laundry detergent to the freshwater flagellate *Euglena gracilis*. Ecotoxicology 21(7):1957–1964

Bajpai D, Tyagi VK (2007) Laundry detergents: an overview. J Oleo Sci 56(7):327–340

Bao-Quey H, Dar-Yi W (1994) Effects of linear alkylbenzene sulfonate (LAS) on the respiratory functions of Tigerperch (*Terapon jarbua*). Zool Stud 33(3):205–210

Bellan G (1967) Pollution et peuplement benthiques sur substrat meuble dans la region de Marseille. 1 Partie. Le secteur de Cortiu. Rev Int Océanogr Méd 6–7:53–87

Chawla G, Viswanathan PN, Devi S (1987) Biochemical studies on the toxicity of linear alkylbenzene sulphonate to *Scenedesmus quadricauda* in culture. Environ Exp Bot 27:311–319

Christoffersen K, Hansen BW, Johansson LS, Krog E (2003) Influence of LAS on marine calanoid copepod population dynamics and potential reproduction. Aquat Toxicol 63(4):405–416

Coelho KS, Rocha O (2010) Assessment of the potential toxicity of a linear alkylbenzene sulfonate (LAS) to freshwater animal life by means of cladoceran bioassays. Ecotoxicology 19(4):812–818

Conti E (1987) Acute toxicity of three detergents and two insecticides in the lugworm, *Arenicola marina* (L.): a histological and a scanning electron microscopic study. Aquat Toxicol 10:325–334

Cristale J, Katsoyiannis A, Sweetman AJ, Jones KC, Lacorte S (2013) Occurrence and risk assessment of organophosphorus and brominated flame retardants in the River Aire (UK). Environ Pollut 179:194–200

Cserháti T, Forgacs E, Oros G (2002) Biological activity and environmental impact of anionic surfactants. Environ Inter 28(5):337–348

Dean HK (2008) The use of polychaetes (Annelida) as indicator species of marine pollution: a review. Rev Biol Trop 56(4):11–38

Fendinger NJ, Versteeg DJ, Weeg E, Dyer S, Rapaport RA (1994) Environmental behavior and fate of anionic surfactants. In: Baker L (ed) Environmental chemistry of lakes and reservoirs. American Chemical Society, Washington, pp 527–557

Gao P, Li Z, Gibson M, Gao H (2013) Ecological risk assessment of nonylphenol in coastal waters of China based on species sensitivity distribution model. Chemosphere 104:113–119

Garza-García ME (2009) Capitellidae Grube, 1862. In: de León-González JA, Bastida-Zabala JR, Carrera-Parra LF, García-Garza ME, Peña-Rivera A, Salazar-Vallejo SI, Solís-Weiss V (eds) Poliquetos (Annelida: Polychaeta) de México y América Tropical. Monterrey, pp 101–114

Hampel M, Moreno-Garrido I, Sobrino C, Lubian LM, Blasco J (2001) Acute toxicity of LAS in marine microalgae: esterase activity and inhibition growth as endpoints of toxicity. Ecotoxicol Environ Saf 48(3):287–292

Henkel M, Müller MM, Kügler JH, Lovaglio RB, Contiero J, Syldatk C, Huasmann R (2012) Rhamnolipids as biosurfactants from renewable resources: concepts for next-generation rhamnolipid production. Process Biochem 47(8):1207–1219

Henkel M, Sydatk C, Hausmann R (2014) The prospects for the production of Rhamnolipids on renewable resources. In: Kosaric N, Sukan FV (eds) Biosurfactants: production and utilization-processes technologies, and economics, 1st edn. CRC Press, Florida, pp 83–99

Iannacone J, Alvariño L (2002) Efecto del detergente doméstico alquil aril sulfonato de sodio lineal (LAS) sobre la mortalidad de tres caracoles dulceacuícolas en el Perú. Ecol Aplic 1:81–87

Info Mine Research Group (2012) Synthetic detergents and cleaning products in the CIS and Baltic Countries: production, market and forecast. Info Mine Market Research Group. http://www.infomine.ru/files/catalog/166/file_166_eng.pdf. Accessed 24 August 2015

Jensen J, Løkke H, Holmstrup M, Krogh PH, Elsgaard L (2001) Effects and risk assessment of linear alkylbenzene sulfonates in agricultural soil. 5. Probabilistic risk assessment of linear alkylbenzene sulfonates in sludge-amended soil. Environ Toxicol Chem 20(8):1690–1697

Jorgensen E, Christoffersen K (2000) Short-term effects of linear alkylbenzene sulfonate on freshwater plankton studied under field conditions. Environ Toxicol Chem 19(4):904–911

Lam PK, Gray JS (2001) Predicting effects of toxic chemicals in the marine environment. Mar Pollut Bull 42(3):169–173

Liwarska-Bizukojc E, Miksch K, Malachowska-Jutsz A, Kalka J (2005) Acute toxicity and genotoxicity of five selected anionic and nonionic surfactants. Chemosphere 58(9):1249–1253

Marin MG, Pivotti L, Campesan G, Turchetto M, Tallandini L (1994) Effects and fate of sediment-sorbed linear alkylbenzene sulphonate (LAS) on the bivalve mollusc *Mytilus galloprovincialis* Lmk. Wat Res 28:85–90

Marín MG, Bressan M, Brunetti R (1991) Effects of linear alkylbenzene sulphonate (LAS) on two marine benthic organisms. Aquat Toxicol 19(3):241–248

Markina ZV, Aizdaicher NA (2007) Influence of laundry detergents on the abundance dynamics and physiological state of the benthic microalga *Attheya ussurensis* (Bacillariophyta) in laboratory culture. Russ J Mar Biol 33(6):391–398

McAvoy DC, Eckhoff WS, Rapaport RA (1993) Fate of linear alkylbenzene sulfonate in the environment. Environ Toxicol Chem 12(6):977–987

Méndez N (2005) Effects of teflubenzuron on larvae and juveniles of the polychaete Capitella sp. B from Barcelona, Spain. Water Air Soil Pollut 160:259–269

Méndez N, Green-Ruiz C (2005) Preliminary observations of cadmium and copper effects on juveniles of the polychaete *Capitella* sp. Y (Annelida: Polychaeta) from Estero del Yugo, Mazatlán, México. Rev Chil Hist Nat 78:701–710

Méndez N, Green-Ruiz C (2006) Cadmium and copper effects on larval development and mortality of the polychaete *Capitella* sp. Y from Estero del Yugo, Mazatlán, México. Water Air Soil Pollut 171:291–299

Méndez N, Linke-Gamenick INEZ, Forbes VE (2000) Variability in reproductive mode and larval development within the *Capitella capitata* species complex. Invert Reprod Develop 38(2):131–142

Méndez N, Anguas-Cabrera DN, García-de la Parra LM (2008) Effects of methamidophos on sediment processing and body mass of *Capitella sp.* Y from Estero del Yugo, Mazatlán, México. J Exp Mar Biol Ecol 361:92–97

Méndez N, Lacorte S, Barata C (2013) Effects of the pharmaceutical fluoxetine in spiked-sediments on feeding activity and growth of the polychaete *Capitella teleta*. Mar Environ Res 89:76–82

Mikolajczyk E, Diehn B (1978) Morphological alteration in Euglena gracilis induced by treatment with CTAB (Cetyltrimethylammonium bromide) and Triton X-100: correlations with effects on photophobic behavior response. J Protozool 25:461–470

Ofojekwu PC, Okwuosa VN, Omoregie E (1999) Effect of alkylbenzene sulphonate (ABS) detergent on the weight gain of the toothed carp, *Aphyosemion gairdneri* (L.). J Environ Sci 3(2):234–238

Ogundele O, Ihuahi JA, Omojowo FS, Bitrus P (2005) Toxicity of linear alkylbenzene sulphonate (LAS) detergent to *Clarias gariepinus* fingerlings. http://aquaticcommons.org/4036/1/273.pdf. Accessed 24 January 2015

Okoli-Anunobi CAIN, Ufodike EBC, Chude LA (2002) Lethal effects of the detergent, Elephant Blue® on the Nile Tilapia, *Oreochromis niloticus* (L.). J Aquat Sci 17(2):95–97

Okwuosa VN, Omoregie EO (1995) Acute toxicity of alkylbenzene sulphonate (ABS) detergent to the toothed carp, *Aphyosemion gairdneri* (L.). Aquacult Res 26(10):755–758

Omotoso FO, Fagbenro OA (2005) A comparative study on the toxicity of three commercial detergents on the survival of the Nile Tilapia, *Oreochromis niloticus*. J Agric Res Develop 4(2):139–147

Pearson TH, Rosenberg R (1978) Macrobenthic succession in relation to organic enrichment and pollution of the marine environment. Oceanogr Mar Biol Ann Rev 16:229–311

Pedrazzani R, Ceretti E, Zerbini I, Casale R, Gozio E, Bertanza G, Gelatti U, Donato F, Feretti D (2012) Biodegradability, toxicity and mutagenicity of detergents: integrated experimental evaluations. Ecotoxicol Environ Saf 84:274–281

Pettersson A, Adamsson M, Dave G (2000) Toxicity and detoxification of Swedish detergents and softener products. Chemosphere 41(10):1611–1620

Reish DJ (1959) An ecological study of pollution in Los Angeles-Long Beach Harbors, California. Occas Papers Allan Hancock Found Publ 22:1–117

Reish DJ (1988) The use of toxicity testing in marine environmental research. In: Soule DF, Kleppel GS (eds) Marine organisms as indicators. Springer, New York, pp 231–245

Reish DJ, LeMay JA (1991) Toxicity and bioconcentration of metals and organic compounds by polychaeta. Ophelia Suppl 5:653–660

Rosety M, Ordóñez FJ, Rosety-Rodríguez M, Rosety JM, Rosety I, Carrasco C, Ribelles A (2001) Acute toxicity of anionic surfactants sodium dodecyl sulphonate (SDS) and linear alkylbenzene sulphonate (LAS) on the fertilizing capability of gilthead (*Sparus aurata* L.) sperm. Histol Histopathol 16(3):839–843

Sánchez-Peinado MDM (2007) Efectos biológicos de los sulfonatos de alquilbenceno lineales (LAS) en suelo agrícola: biotransformación y estudios de biodiversidad. http://hdl.handle.net/10481/1555. Accessed 13 March 2015

Scheibel J (2004) The evolution of anionic surfactant technology to meet the requirements of the laundry detergent industry. J Surfact Deterg 7(4):319–328

Secretaria de Económica (2001) Análisis de aguas–Determinación de sustancias activas al azul de metileno (SAAM) en aguas naturales, potables, residuales y residuales tratadas—Método de Prueba. http://www.conagua.gob.mx/CONAGUA07/Noticias/NMX-AA-039-SCFI-2001.pdf. Accessed 22 November 2014

Sobrino-Figueroa AS (2013) Evaluation of oxidative stress and genetic damage cause by detergents in the zebrafish *Dario rerio* (Cyprinidae). Comp Biochem Physiol Part A Mol Integr Physiol 165(4):528–532

Stefanoni MF, Abessa DMS (2008) Lysossomal membrane stability of the brown mussel *Perna perna* (Linnaeus) (Mollusca, Bivalvia) exposed to the anionic surfactant linear alkylbenzene sulphonate (LAS). Panam J Aquat Sci 3(1):6–9

Temara A, Carr G, Webb S, Versteeg D, Feijtel T (2002) Marine risk assessment: linear alkylbenzene sulponates (LAS) in the North Sea. Mar Pollut Bull 42(8):635–642

Tkachenko FP, Kutsyn EB (2002) Detergent effect upon amino-acid composition of protein in the green alga *Cladophora vagabunda* (L.) Hoek. Gidrobiologicheskii Zh 38(3):94–98

Uc-Peraza RG, Delgado-Blas VH (2012) Determinación de la concentración letal media (CL_{50}) de cuatro detergentes domésticos biodegradables en *Laeonereis culveri* (Webster, 1879) (Polychaeta, Annelida). Rev Int Contam Ambient 28(2):137–144

US EPA (US Environmental Protection Agency) (1998) Guidelines for ecological risk assessment. US Environmental Protection Agency. EPA/630/R095/002F Washington, DC

Van Bogaert IN, Saerens K, De Muynck C, Develter D, Soetaer W, Vandamme EJ (2007) Microbial production and application of sophorolipids. Appl Microbiol Biotechnol 76:23–24

van de Plassche EJ, de Bruijn JHM, Stephenson RR, Marshall SJ, Feijtel TCJ, Belanger SE (1999) Predicted no-effect concentrations and risk characterization of four surfactants: linear alkyl benzene sulfonate, alcohol ethoxylates, alcohol ethoxylated sulfates and soap. Environ Toxicol Chem 18(11):2653–2663

Versteeg LD, Belanger SE, Carr GJ (1999) Understanding single species and model ecosystem sensitivity: a data based comparison. Environ Toxicol Chem 18(6):1329–1346

Warne MSJ (1995) Acute toxicity of laundry detergents to on Australian Cladoceran (*Ceriodaphnia dubia*). Aust J Ecotoxicol 1(2):127–135

Warne MSJ, Schifko AD (1999) Toxicity of laundry detergent components to a freshwater cladoceran and their contribution to detergent toxicity. Ecotoxicol Environ Saf 44(2):196–206

26

Atmospheric deposition coupled terrestrial export of organic carbon in Ganga River (India): linking cross-domain carbon transfer to river DOC

Jitendra Pandey · Anand V. Singh · Rachna Singh · Pooja Kaushik · Usha Pandey

Abstract The atmosphere–land–water connectivity of dissolved organic carbon (DOC) is not altogether accounted in major rivers of India despite recent researches highlighting its importance. We studied the coupled effect of atmospheric deposition (AD) and local land use on DOC buildup in Ganga River for a period of 6 years. The AD-OC input increased consistently over time and there was over 1.5- to 1.8-fold increase in 2012 relative to 2007. Microbial activity and water soluble organic carbon (WSOC) in sub-catchment and DOC in land surface runoff increased consistently over time along the gradient of AD input. The river DOC showed positive correlation ($R^2 = 0.24$–0.84; $p < 0.001$) with AD-OC and runoff DOC and WSOC showed positive correlation ($R^2 = 0.96$; $p < 0.001$) with soil microbial activity. Principal component analysis segregated study sites into four groups demarcating source relationships. Our study, that forms the first report on atmosphere–land–water transfer of organic carbon in Ganga River, suggests that future climate models should include region-specific time series data on changing state of atmosphere– land–water connectivity and associated shift in carbon balance of major rivers for more accurately predicting the climate change drivers.

Keywords Atmospheric deposition · Climate change drivers · Dissolved organic carbon · Ganga River · Land use · Surface runoff

Introduction

The human alteration of carbon cycle is a key node of perturbed Earth System functions such as climate change, global transport of pollutants, changing biogeochemical cycles, ecosystem productivity and plant species shift (Reich et al. 2006; Ramanathan and Garmichael 2008; Jurado et al. 2008; Langley and Megonigal 2010; Pandey 2011; Pandey et al. 2014a). Despite the fact that atmosphere–land–water connectivity is central to global C cycling, most of the prediction models did not explicitly consider atmosphere–land–water linkages in regional and global C budgets. Surface water especially rivers respond differently than uplands to environmental perturbations and have relevance in constructing a complete regional C budget (Cole et al. 2007; Buffam et al. 2011). Biogeochemical cycling of carbon (C) across the terrestrial–aquatic interface is

J. Pandey (✉) · A. V. Singh · R. Singh · P. Kaushik
Ganga River Ecology Research Laboratory, Environmental Science Division, Centre of Advanced Study in Botany, Banaras Hindu University, Varanasi 221005, India
e-mail: jiten_pandey@rediffmail.com

U. Pandey
Department of Botany, Faculty of Science and Technology, Mahatma Gandhi Kashi Vidyapith University, Varanasi 221002, India

dynamically linked with the water cycle. Not only is the movement of C controlled in a major way by the movement of water, but also the processes of transformations between organic and inorganic forms are strongly influenced by the movement of water.

The knowledge of regional sensitivity of landscapes, whether as C sink or source, the magnitude of C storage or flux and vulnerability to anthropogenic perturbations are all important for understanding regional C budget. Because freshwater covers a small fraction of the Earth's surface area, inland freshwater ecosystems have rarely been considered as potentially important component of regional or global scale C budget (Cole et al. 2007). Under the rising anthropogenic pressure, data on changing state of atmosphere–land–water connectivity coupled with climatic heterogeneity provide valuable information on which the possible linkages of climate change drivers can be keyed. Further, although the oxidized form of carbon (CO_2) has received major scientific attention globally, regional and global scale data on atmosphere–land–water linkages of reduced carbon, especially in the form of organic carbon, are very scarce (Buffam et al. 2011). In India, the National Carbon Project (NCP), which aims at improving, understanding and quantification of net C balance, has been implemented with three major domains namely, vegetation C pool, soil C pool and vegetation–atmospheric fluxes (Dadhwal 2012). This mega project did not, however, consider C pool of surface waters. To date, no data so far are available from India addressing explicitly the human-induced changes in atmosphere–land–water C transfer and associated shift in C balance in major rivers of the country. The newly established Ministry of Water Resources, River Development and Ganga Rejuvenation by the Government of India may warrant watershed-scale research initiatives to understand how the transfer of terrestrial C to river ecosystems is being influenced by land use change and AD inputs.

Atmospheric deposition of OC in catchment may alter the quality and quantity of soil organic matter and consequently the quality of runoff emerging from these land surfaces. The responses of catchment C pool to AD-OC, however, may vary depending on catchment properties including land use change and soil nutrient status (Aber et al. 1998; Kopácèk et al. 2013). The coupled effect of terrestrial C and AD-OC may influence the storage and flux of C in receiving water bodies. Therefore, long-term time series data integrating cross-domain processes are important for predicting specific climate change drivers and designing action plans for integrated river basin management (IRBM). The available literature suggests that the atmospheric deposition contributes significantly to land OC pool. Willey et al. (2000) reviewed the measurement of dissolved organic carbon (DOC) in rain water and estimated that of the total global AD-OC input of 400, 310 Tg C yr^{-1} corresponds to land input. Therefore, for understanding and quantification of net regional C balance, data on AD-OC and associated shift in land–water transfer to major rivers are important. However, there is a general dearth of studies addressing explicitly the human-induced changes in atmosphere–land–water transfer of organic carbon and associated shift in C balance in major rivers of India. This study presents data on atmospheric deposition of OC; effect of AD-OC and land use on land surface runoff DOC; and the coupled effect of AD-OC and land derived DOC on the concentration of DOC in Ganga River. We hypothesized that, together with atmospheric- OC, the land derived DOC in surface runoff enhances the DOC concentration in Ganga River. To test this hypothesis, we evaluated, if the pattern of microbial activity in sub-catchment, terrestrial C flushing through runoff and DOC in the river varied over time along the gradient of AD-OC.

Materials and methods

Study area

The data presented here are the results of six consecutive years (March 2007 to February 2013) of study conducted at seven sites along a 35 km long stretch of Ganga River at Varanasi (25°18′N lat. and 83°1′E long.) (Figure 1). The Ganga River together with Brahmaputra–Meghna river system is second only to Amazon with total drainage area of 1.75 million km^2. The Indo-Gangetic Plains, which constitute the most fertile agricultural belt in India, represent the highest population density in the country. During the last two decades the region witnessed unprecedented growth of urban-industrial sectors. As a result, the Ganges basin, along its 2525 km course from Gangotri in Himalaya to its confluence with Bay of Bengal, receives massive input of air-borne carbon and other pollutants. Each site presented in Fig. 1 represents three sub-sites considered separately for collecting atmospheric deposition samples (5–20 m distance from the river bank), surface runoff (at discharge outlet adjacent to river; 5–20 m distance from the river bank) and mid-stream river water samples (considered in the mid-stream centrally to the mean width of the respective site). Site selection was based on local land uses and the nature of emission sources (Table 1). Site 1 is relatively in a natural state and all other sites were invariably human disturbed. Land use at Site 2 constitutes agriculture with natural patches; Site 3 predominantly agricultural; Site 4 anthropogenic dredging; Site 5 and 6 urban and at Site 7 it is characterized by urban with mosaic of woodland. Samples were collected for six consecutive years and for data presentation the times series was divided as: Year 2007, (March 2007 to February 2008); 2008,

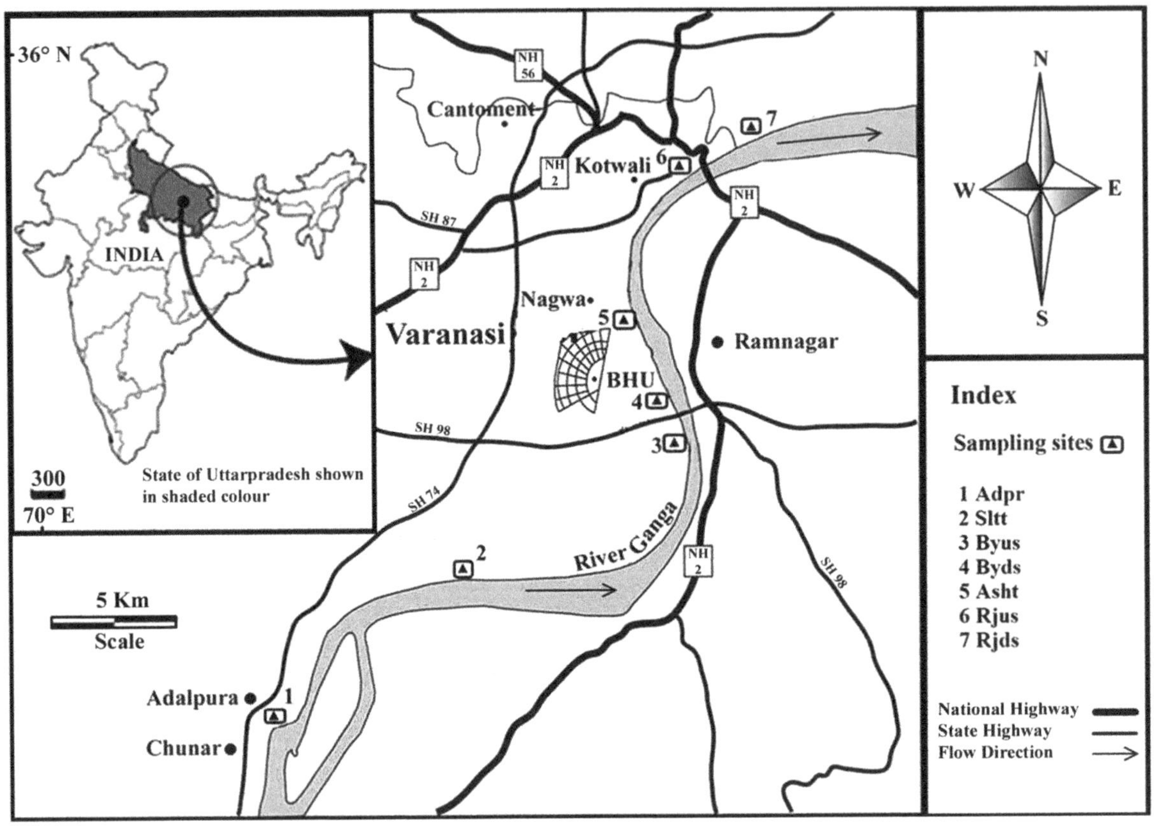

Fig. 1 Location of the study sites at Ganga River. *1* Adalpura (Adpr); *2* Shultankeshwar ghat (Sltt); *3* Bypass upstream (Byus); *4* Bypass downstream (Byds); *5* Assi ghat (Asht); *6* Rajghat upstream (Rjus); *7* Rajghat downstream (Rjds)

Table 1 Description of major emission source, river attributes and sub-catchment characteristics at seven study sites

No.	Name	Code	Mean Depth (m)	Average velocity (ms^{-1})	Emission source	Area (km^2)	Land use
1.	Adalpura	Adpr	21	0.40 ± 0.02	Long-range transport	84	Natural patches
2.	Shultankeshwar Ghat	Sltt	21	0.40 ± 0.03	Long-range transport	114	Agriculture with natural patches
3.	Bypass upstream	Byus	12	0.39 ± 0.02	Havy duty vehicle	540	Rural and agricultural
4.	Bypass downstream	Byds	12	0.39 ± 0.01	Havy duty vehicle	62	Anthropogenic dredging
5.	Assi Ghat	Asht	14	0.26 ± 0.01	Urban emission	72	Residential, urban
6.	Rajghat upstream	Rjus	20	0.36 ± 0.02	Urban; biomass burning	85	Urban
7.	Rajghat downstream	Rjds	18	0.32 ± 0.01	Urban emission	90	Urban; woodland

(March 2008 to February 2009); 2009, (March 2009 to February 2010); 2010, (March 2010 to February 2011); 2011, (March 2011 to February 2012); and 2012, (March 2012 to February 2013).

The climate is tropical monsoonal, marked by a hot and dry summer, a humid rainy season and a cold winter season. During summer, day temperature varied between 29 and 46.2 °C. During winter, night temperature at rare occasion drops below freezing. Mean annual rainfall varied between 870 and 1130 mm, relative humidity between 27 and 83 % (summer) and 58 and 99 % (rainy season). About 90 % of the total annual rainfall is received in rainy season. Wind direction shifts from predominantly westerly and south-westerly in October through April to easterly and north-westerly in remaining months. The soil is alluvial fluvisol associated with recurrent floods or long wetness, recent sedimentation and high natural fertility. Daily climate data consisting of precipitation, temperature, wind and humidity were taken from Meteorological Unit, Department of Geophysics, Banaras Hindu University.

Measurements

Atmospheric deposition

Atmospheric deposition samples were collected at monthly intervals following an identical protocol. Samples were collected using bulk samplers made up of 5L high density polyethylene bottle connected to a Teflon funnel (diameter: 15 cm) with PVC needles devised on the top to avoid bird nesting. Samplers were placed at 5–20 m distance from the river bank and maintained at a height of 2 m to avoid collection of re-suspended matters. In order to prevent large changes in the concentrations, thymol was used as a biocide in the collection buckets (Gillet and Ayers 1991). At the end of the sampling period the funnel was rinsed with double distilled water to collect deposits from the funnel walls. A 50 ml of sub-sample of atmospheric deposit collected in the bottle and a 50 ml sub-sample of the rinsing water was filtered using pre-combusted Whatman glass fiber filter (0.7 μm) and the organic carbon was analyzed following $KMnO_4$ digestion procedure (Michel 1984). The samples were screened for contamination using high soluble reactive P (>0.2 mg) as an indicator (Lohse et al. 2008).

Catchment soil

Microbial activity in catchment soil was measured in terms of substrate-induced respiration (SIR). For this purpose, soil samples (0–10 cm depth) were collected monthly from three randomly selected blocks (plot size, 20 × 20 m) from each study site. Soil sampling blocks were selected in riparian zone of the respective sites at a distance of 50–100 m from the river bank. Microbial activity was measured by amending 20 g of soil sample (moisture content adjusted to 55 %) with 120 mg of glucose at the beginning of incubation (Anderson and Domsch 1978). The CO_2 produced was trapped in 0.5 N NaOH and determined by titrating excess of NaOH with 0.5 N HCl. Water soluble organic carbon (WSOC) in aqueous extract of soil was determined following dichromate oxidation titration (Ciavatta et al. 1991).

Water chemistry

We collected mixed surface runoff samples to represent a wash-out effect characteristic of different site functions. For this purpose, sites were selected considering different land uses such as residential, agricultural, roads, anthropogenic dredging, woodland etc. and runoff samples were collected at a discharge outlet adjacent to the river (5–20 m distance from the river bank). There was only one exit to discharge the runoff from each area marked for the purpose. The strategy was to collect surface runoff to represent characteristics of this mixed area. Three such outlets, armed with artificial barriers, were installed in each sub-catchments. The flow velocity was measured using a flow meter and the runoff was determined considering flow velocity and cross area of the discharge outlet. The samples were collected manually during rain event using pre-sterilized sampling bottles. The sampling was initiated with first flush, and based on the duration of surface runoff, the sampling interval was kept 10–30 min. River water samples were collected in replicates from mid-stream of the river monthly from each site, directly below the surface (15–25 cm depth), in acid-rinsed 5L plastic containers.

The NO_3^-–N in water samples was estimated using brucine sulphanilic acid method (Voghel 1971), ammonia-N using Nessler's reagent method (Maiti 2001) and total dissolved nitrogen (TDN) following high temperature persulphurate digestion. Dissolved organic nitrogen (DON) was computed as TDN minus dissolved inorganic nitrogen (DIN), where DIN represents the sum of NO_3^-–N and NH_4^+–N (Perakis and Hedin 2002). Since most of the N was inorganic, the C: N ratio presented here represents DOC: TDN. Dissolved organic carbon (DOC), the fraction of organic carbon filtered through a 0.7 μm pore size filter, in runoff and river water samples was estimated using a $KMnO_4$ digestion procedure (Michel 1984). For this purpose, a known volume of filtered water was mixed with acidified N/80 potassium permanganate and incubated at 37 °C. Organic carbon was estimated by titrating to quantify oxygen after 4 h of incubation (APHA 1998). As one mole of oxygen (32 g) is released for each mole of C (12 g) fixed, the factor 0.375 was used to convert titration results to organic carbon. For quality assurance, the analytical accuracy of the data generated in the laboratory was validated by repeated calibration using different concentrations of standard potassium hydrogen phthalate solution. Repeated calibration showed $KMnO_4$ oxidation efficiency of 75 % and therefore, an efficiency factor of 1.33 was used to ensure analytical accuracy.

Statistical analysis

Significant effects of site and time series were tested using analysis of variance (ANOVA). Samples were collected in replicates and all the measurements were considered in ANOVA model. To justify uncertainty of the mean, arithmetic means are accompanied by ±1SE. Coefficient of variation (cv) with least significant difference ($\alpha = 0.05$) across time was computed for expressing data variability. The Mann–Kendall test and Sen's slope estimator (XLSTAT 2013) were used for detecting the trend direction and magnitude of change in seasonal atmospheric deposition data sets. Principal component analysis (PCA) was used to construct the overall spatial and temporal profile of OC for the entire study period. Correlation coefficient (R^2) and regression analyses were used for testing the extent of linearity in relationships. Confidence limits above 95 ($\alpha < 0.05$) were considered as significant. SPSS package (version 16) was used for statistical analysis.

Results

The atmospheric deposition of organic carbon (AD-OC) increased over time and there was over 1.5- to 1.8-fold increase in AD-OC in 2012 relative to 2007 (Fig. 2). At Rjus and Rjds sites, more than 80 % of atmospheric deposits contained OC above 237.5 μmol l^{-1}. Organic carbon concentration in rain water followed a trend similar to AD-OC (Fig. 2). Seasonally, deposition of OC was lowest in monsoon and highest in winter (Fig. 3). Mann–Kendall test and Sen's slope showed steep slopes with significant seasonality ($p < 0.001$; Fig. 3). Seasonally, the AD-OC input varied between 3.4 kg ha^{-1} $season^{-1}$ (Adpr, rainy season) and 11.9 kg ha^{-1} $season^{-1}$ (Rjds, summer season). Month-wise, the AD-OC was highest in December (Byds, Asht, Rjus and Rjds) and January (Adpr, Sltt and Byus) (Fig. 4) although the differences in AD-OC fluxes between these months were not significant. On spatial scale, deposition of OC was found to be the lowest at Adpr site. Spatio-temporal variations in AD-OC were statistically significant ($p < 0.001$; ANOVA).

Microbial activity (measured in terms of substrate-induced respiration, SIR) in top soil of the sub-catchment increased consistently overtime (Fig. 5). The SIR increased consistently along the pollution gradient from Adpr (3.84–5.70 μg CO_2–C g^{-1} h^{-1}) to Rjds (6.81–11.74 μg CO_2–C g^{-1} h^{-1}) and this trend was common throughtout the study period. Water soluble organic carbon (WSOC) showed temporal and spatial trends similar to that of SIR (Fig. 5). The differences in SIR and WSOC with respect to site and year were significant ($p < 0.001$; ANOVA). Concentration of DOC in runoff varied with site and increased consistently over time from 2007 to 2012 (Fig. 2). The increase in runoff DOC over time was highest at Rjds, although the overall flux of DOC was highest at Byus (115–163 kg ha^{-1} $year^{-1}$). Spatial and temporal differences in runoff DOC were significant ($p < 0.001$; ANOVA). Similar to the runoff, river DOC increased consistently over time (Fig. 2). Year-wise increase in river DOC was lowest at Adpr and highest at Rjds. On spatial scale, river DOC increased along the study gradient from Adpr (2007; 2.82–4.36 mg L^{-1}) to Rjds (2007; 3.97–6.84 mg L^{-1}). This trend was common throughout the study period. Concentration of river DOC was lowest in low flow (summer) and highest in high flow (rainy season) (Fig. 3). Unlike AD-OC, the river DOC

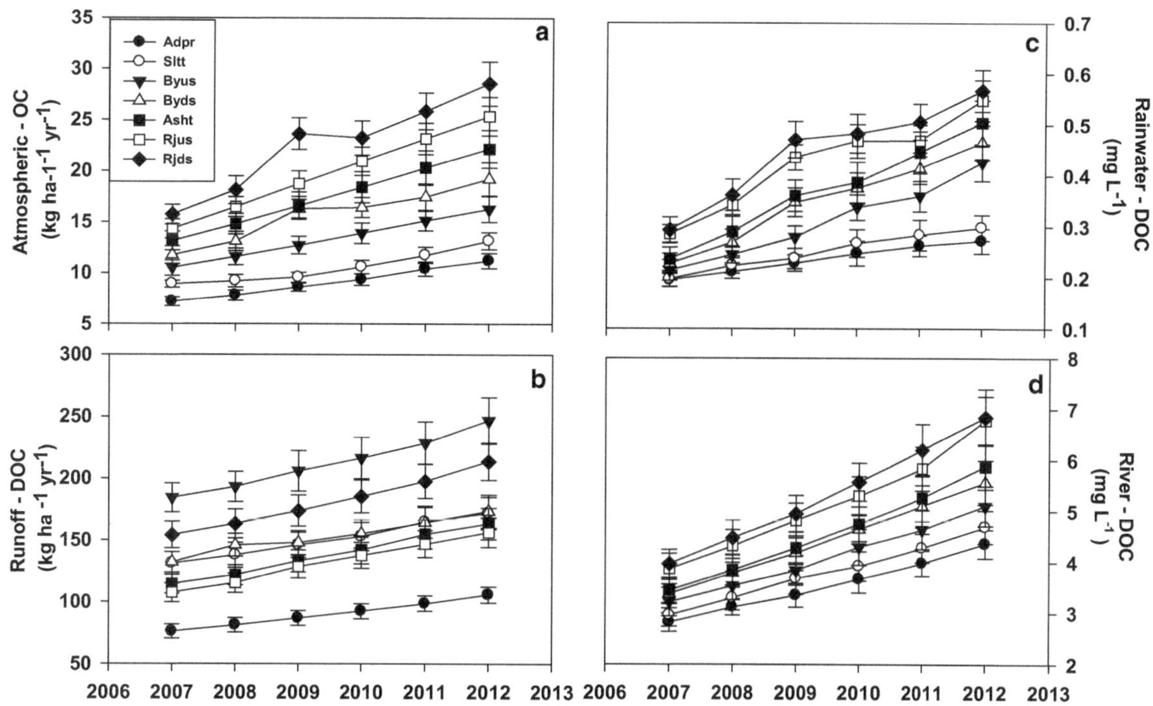

Fig. 2 Year-wise trends in **a** atmospheric deposition of organic carbon, **b** runoff DOC, **c** rainwater DOC; and **d** river DOC at seven study sites. Values are mean ± 1SE. For atmospheric deposition and river DOC $n = 36$, for runoff $n = 12$

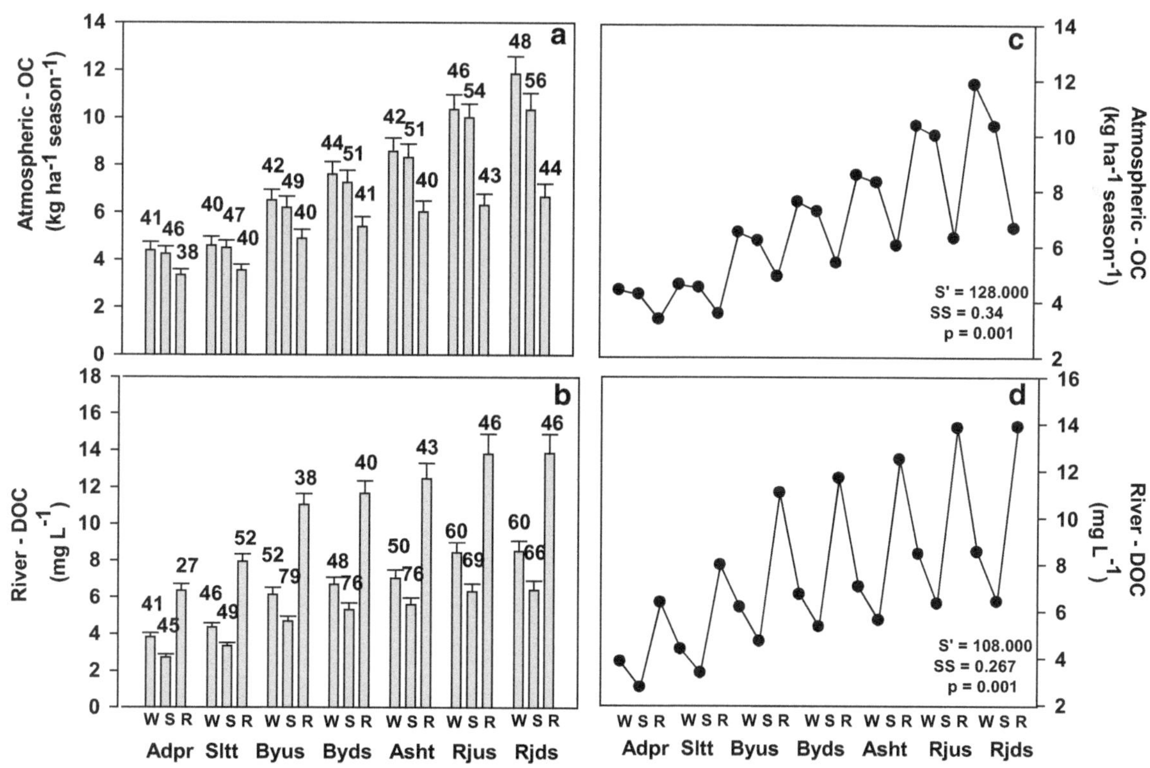

Fig. 3 Seasonal trends in **a** atmospheric deposition of organic carbon and, **b** river DOC and, Mann–Kendall test with Sen's slope statistics for **c** atmospheric OC, and **d** river DOC at seven study sites. Seasonal trend and statistics is based on final study year data and the numbers *above bars* **a**, **b** represent percent difference with initial study year records. Values are mean (n = 12) ± 1SE. *S'* Sen's estimate, *SS* Sen's slope

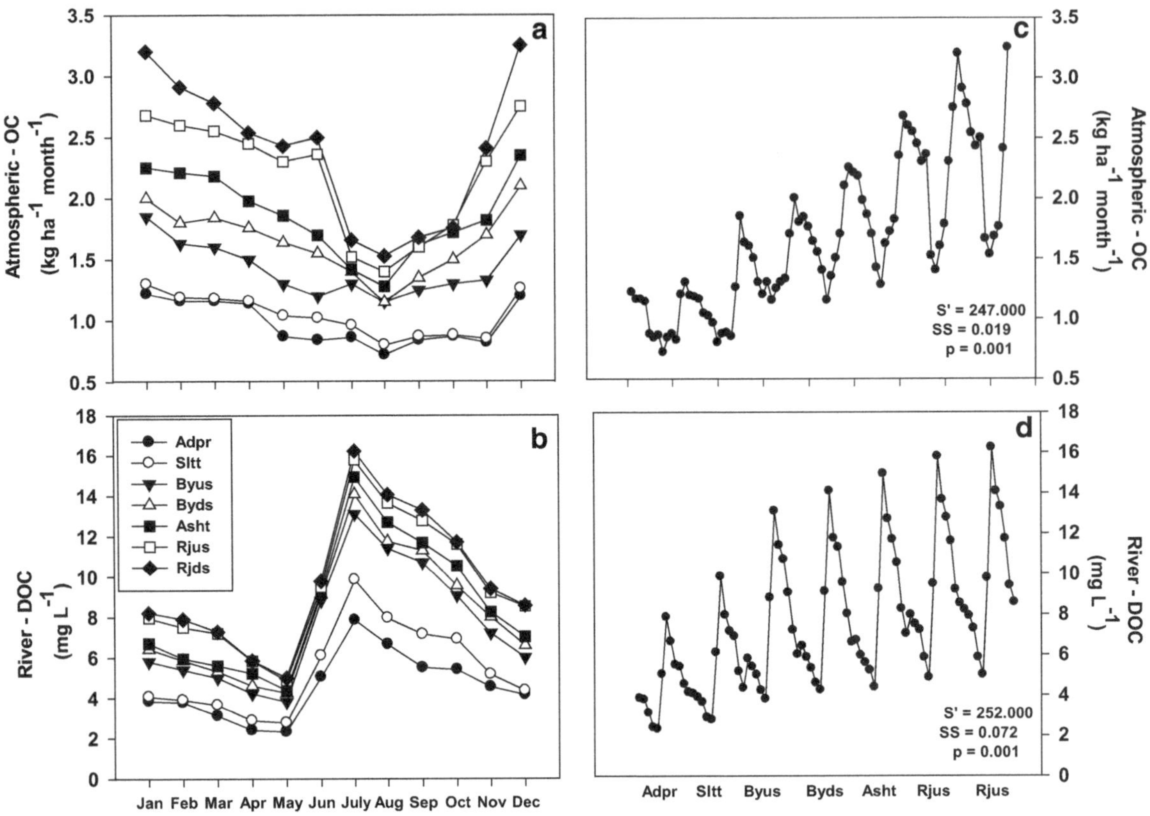

Fig. 4 Month-wise ($n = 3$) trends and time series analysis of atmospheric OC (**a**, **c**) and river DOC (**b**, **d**). Time series statistics (**c**, **d**) is based on final study year data representing seven localities and 12 months. *S'* Sen's estimate; *SS* Sen's slope

concentration was highest in July irrespective of site (Fig. 4). The runoff DOC and river DOC showed significant ($p < 0.001$) positive correlation with AD-OC (Fig. 6). Similarly, runoff DOC showed significant ($p < 0.001$) positive correlation with AD-OC and soil microbial activity (Fig. 6). Principal component analysis (PCA) showed two PCs accounting 88 % of total variance with PC1 for 71 % and PC2 for another 17 % of total variance. The PCA clustering segregated seven sampling sites into four groups of similar characteristics (Fig. 7).

Discussion

An important problem likely to be continued to constrain water resource management in India is a consistently rising trend in atmospheric deposition of pollutants. Atmospheric particulates travel long distances from emission sources and reach waterways via direct deposition on water surface and via land deposition and subsequent runoff (Pandey et al. 2013, 2014b). Role of such inputs becomes important in areas characterized by rapid land use change and unprecedented increase in atmospheric deposition (AD). One of the objectives of the present study was to assess the extent to which atmospheric fluxes of OC are influenced by urban sources and how they vary with regional climate conditions. We present data on site-wise variations in atmospheric fluxes of OC to explain the influence of urban sources; inter-annual variations to explain rising trends over time and; seasonal (intra-annual) variations and month-wise trends to explain within a year influence of climatic variables (temperature, rainfall, wind etc.). We found significant seasonality and a consistently rising trend in AD-OC overtime as well as along the study gradient. Despite site-wise variations, a steadiness of the trend over time indicates strong local control connected with local changes in human activities. From about a decade, the region is witnessing a massive rise in population coupled urban development. Urban-industrial activities intensify as one travels from Site 1 to Site 7. In addition, the city side of the river receives emission

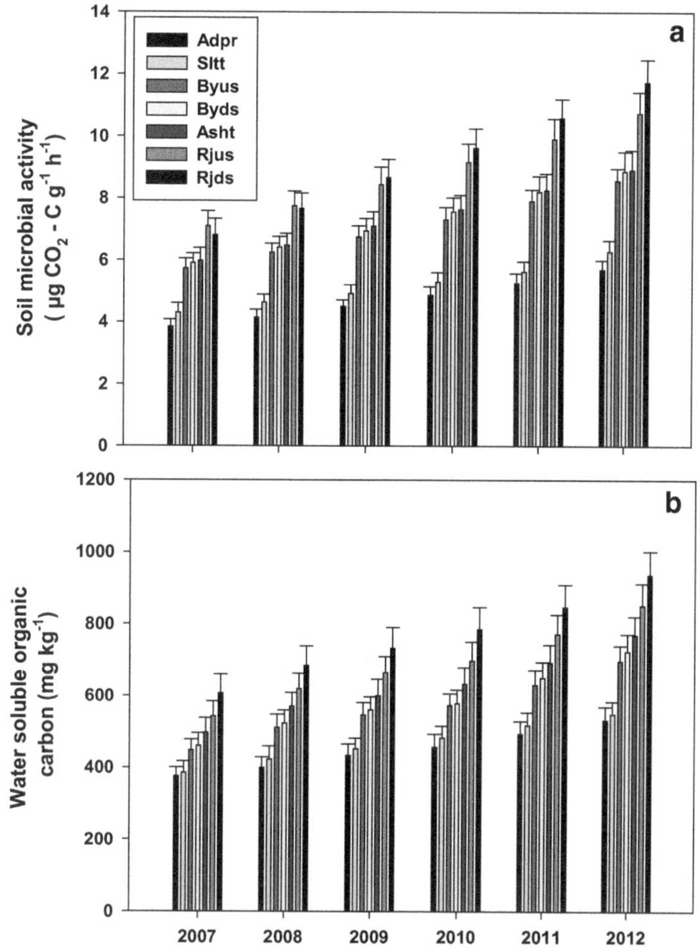

Fig. 5 Microbial activity (measured as substrate-induced respiration, SIR, **a**) and water soluble organic carbon (WSOC, **b**) in top 10 cm soil at different study sites. Values are mean ($n = 36$) ± 1SE

from burning of over 36,000 dead bodies with over 25,000 tons of dry wood for cremation annually. Thus, additional supply of carbon from biomass burning, which is continuously rising, could enhance AD-OC at downstream sites. Biomass burning has been identified as an important source of atmospheric OC in many parts of the world (Taylor 2010; Murillo et al. 2013). The AD-OC flux measured in the present study was in the range reported for north temperate Lake district (Buffam et al. 2011) and Atlantic Ocean (Jurado et al. 2004) but was higher than those reported for an arid metropolitan area of Phoenix, Central Arizona, USA (Lohse et al. 2008).

Urbanization is increasing rapidly in developing countries and is predicted to sizably enhance the atmospheric deposition of pollutant aerosols to cities and down-wind (Pandey et al. 2013). Lovett et al. (2000) suggested that urban atmosphere acts as a scrubber removing aerosols and vapors with dust and depositing them in and around the cities. For this reason, atmospheric deposition generally declines with distance from the city. The scrubbing effects of urban atmosphere may be further enhanced by the large mass of water along the city as the case of Ganga River in the present study. Further, the insoluble mineral particles act as cloud condensation nuclei and consequently enhance the overall deposition (Yin et al. 2002). In the present study, the AD-OC increased with proximity to urban core. Further, the particulate emission derived from site-specific activities (cooking, abraded streets, vehicle frequency, building construction, excavation and nearby agricultural lands), which are continuously rising due to population coupled urban development, may generate strong local control. The Mann–Kendall test and the Sen's slope estimates showed significant seasonality with increasing AD-OC over the year. The variances in the residuals were almost constant. Intra-annual trends in AD-OC indicated long-range transport during summer, offsite scavenging during monsoon and enhanced

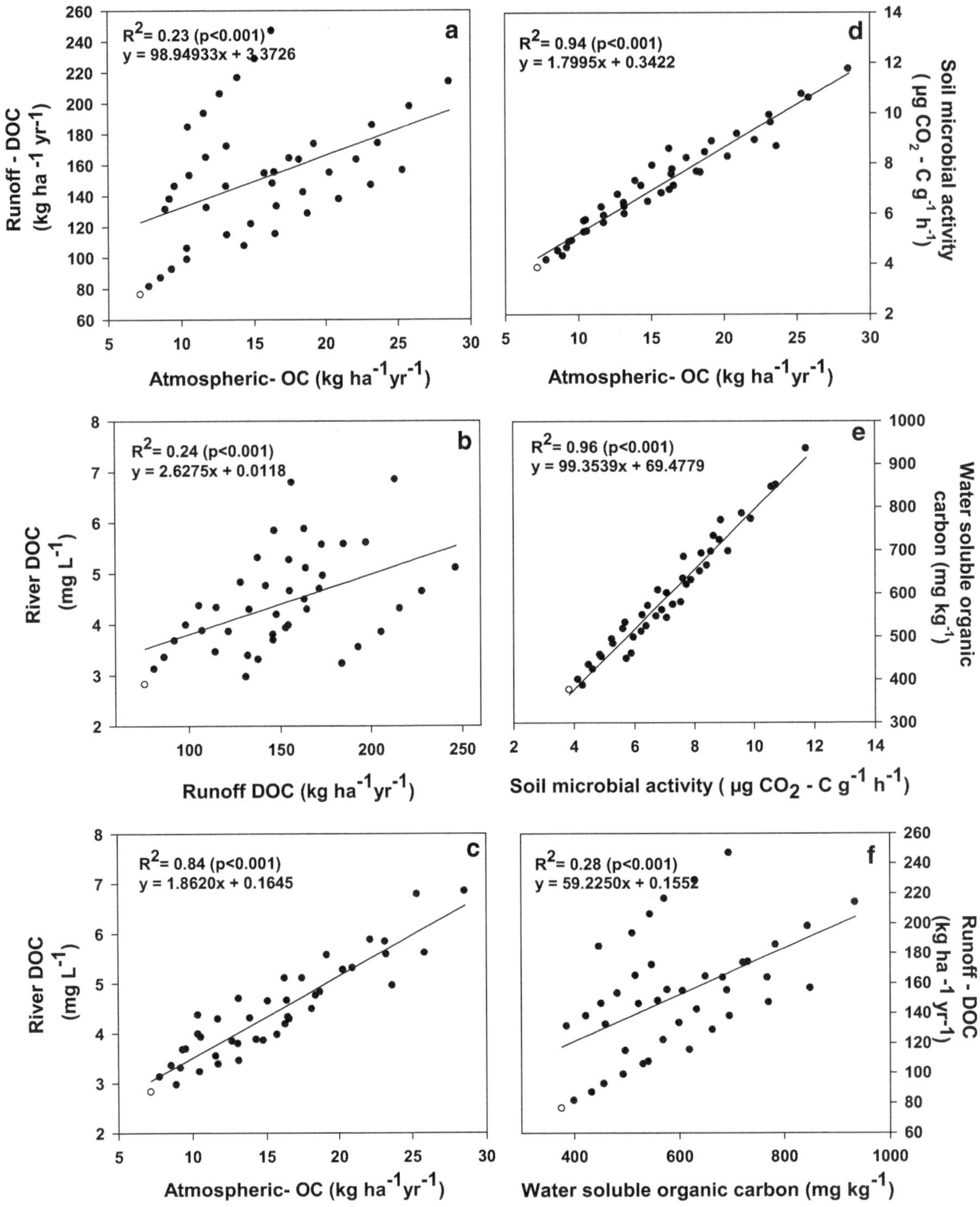

Fig. 6 Significant relationship between **a** atmospheric OC and runoff DOC, **b** runoff DOC and river DOC, **c** atmospheric OC and river DOC, **d** atmospheric OC and soil microbial activity, **e** soil microbial activity and water soluble organic carbon, and **f** water soluble organic carbon and runoff DOC

deposition rates during winter season (Lohse et al. 2008). As reported elsewhere (Lohse et al. 2008; Murillo et al. 2013), OC deposition showed the influence of urban emission and scrubbing effect and peaked in winter likely because of source effects coupled with low temperature, calm wind, low mixing heights and concurrent formation of inversion layer in winter season.

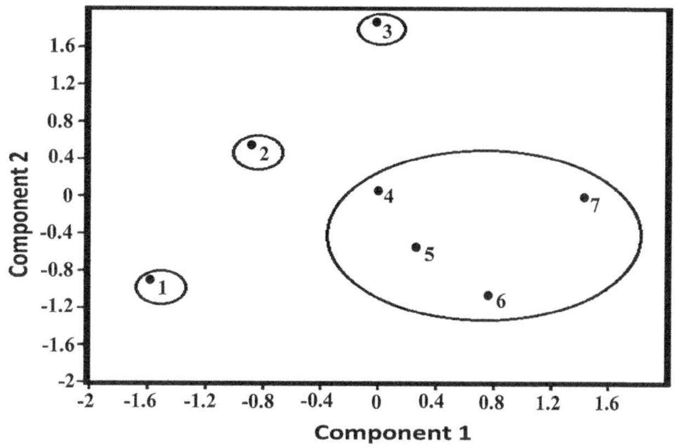

Fig. 7 Principal component analysis (PCA) showing the grouping of seven sampling sites into four clusters of similar characteristics

The observed trend in runoff DOC could be the result of a coupled effect of atmospheric deposition and local land use. Correlation analyses showed that the runoff DOC responded positively to AD-OC. Further, land use and hydrology are linked with each other in a complex way at multiple spatial and temporal scales (Matheussen et al. 2000). Hydrological release of carbon from land depends on geologic parent materials, topography, biota, land use, changes in soil chemistry and, microbial consumption or release of OC (Pandey and Pandey 2009). Site-wise differences in runoff DOC indicate such possible effects as the values were lowest for cemented terrains and highest for sites witnessing agricultural activity followed by woodland. The DOC of terrestrial origin is sensitive to watershed properties including organic matter quality and microbial activity in the catchment (Kalbitz et al. 2000; Tranvik and Jansson 2002; Kalbitz et al. 2007; Porcal et al. 2009). Previous studies (summarized by Gilvear et al. 2002) have shown strong links between hydrology and ecological quality of river ecosystems. Hydrological factors may influence DOC production and transport through a number of ways (Evans et al. 2005, 2006). In the present study, a corresponding trend in surface runoff DOC and river DOC indicates the role of hydrological factors influencing DOC production and export. Hydrological drivers coupled with strong local controls could explain a monotonic rise in DOC observed in this study. Evans et al. (2006), while evaluating the climate change hypothesis and alternative drivers of rising DOC export from organic soils, argued that if hydrology is the principal driver of increased freshwater DOC, a concurrent monotonic trend would be expected.

In a study combining in vitro manipulation with mesocosm and field experiments, Fenner and Freeman (2011) observed that dry periods stimulate bacterial growth and enzymatic degradation of organic matter in peat lands and subsequent re-wetting enhances carbon losses to receiving waters. In the region representing this study, although extreme weather events are not common, seasonal cycle of temperature and precipitation could similarly influence organic matter and microbial activity in the sub-catchment. Possibly for this reason, river DOC was highest in July indicating the first flush effect. We found enhanced microbial activity and a corresponding rise in water soluble organic matter (WSOC) in response to atmospheric AD-OC input over time along the study gradient. As observed in present study, increased microbial activity in soil enhances the production of dissolved organic carbon (Kalbitz et al. 2000, 2007). It seemed that the consistently rising runoff DOC observed in this study has an important link to microbial processing of organic matter in dry seasons followed by pulsed response and subsequent lateral flow in rainy season (Fenner and Freeman 2011; Ritson et al. 2014). Elevated inputs of OC may further enhance microbial activity and pulsed responses to rainfall (Sponseller 2007). Although long-term studies are required to confirm this issue, our data clearly indicate increasing AD-OC linked soil stimulation as a possible mechanism addressing causative relationship of rising DOC in land surface runoff. This highlights a need for additional research into AD driven changes in catchment processes considering other factors potentially operating at smaller scales especially in the Indo-Gangetic plains.

The solute chemistry of a river depends on constituents added through surface discharge as well as through atmospheric deposition, and data on these issues are important to uncover cross-domain carbon linkages and to design action plan for integrated river basin management (IRBM). Our data indicate a highly dynamic and tightly coupled linkage between AD inputs, runoff chemistry and river water chemistry. The common mechanisms used to explain increases in freshwater DOC include climate change (Freeman et al. 2001), elevated CO_2 (Freeman et al. 2004), hydrologic factors (Evans et al. 2005), changing atmospheric deposition chemistry (Monteith et al. 2007; Pandey and Pandey 2013), enhanced pelagic production (Kirchman et al. 1991) and declining sulfate deposition (Evans et al. 2006). In the present study, variation in river DOC could be the result of a combined influence of AD inputs, hydrological point discharge, non-point discharge and sewage inputs. We found a positive relationship ($R^2 = 0.24$; $p < 0.001$) between runoff DOC and river DOC indicating that hydrological flushing is an important vector of river DOC. Seasonal variability in river DOC could be linked to catchment flushing as river DOC increased significantly in monsoon high flows. In dry seasons, when surface runoff remains at minimum, changes in mid-stream DOC may be attributed to AD-OC, pelagic production and sewage input. Earlier studies have shown positive relationships between discharge and DOC concentration in river systems (Westhorpe and Mitrovic 2012). Large rain event often contributes a large proportion of total flow and accounts for a significant proportion of total DOC exported from a catchment (Buffam et al. 2001). In this study, DOC concentration during low discharge periods were fairly low and increased substantially during high discharge events ranging from 11.7 to 16.2 mg L^{-1}. Significant positive relationship ($R^2 = 0.59$, $p < 0.001$) was found between DOC concentrations and river discharge indicating that during period of high discharge allochthonous sources dominate. Discharge appeared highest in August while river DOC peaked in July indicating first flush effect. The overall results of annual trend, however, suggest the role of a mixture of terrestrial, atmospheric and aquatic sources (allochthonous and phytoplankton derived C pool together with AD-OC and anthropogenic point sources discharge). Correlative evidence showed that the DOC in runoff ($R^2 = 0.23$; $p < 0.001$) and in river ($R^2 = 0.84$; $p < 0.001$) was positively related to AD-OC. The relationships of runoff DOC with AD-OC, river DOC and WSOC (Fig. 6a, b, f) appeared weak and showed quite linear dependence of several groups reflecting site-specific inter-annual variations. Large variances associated with the main line of best fit representing entire data for inter-annual and between-site variations are due probably to site characteristics such as vegetation, soil, land use etc. and associated variations, in addition to AD-OC, resulting in linear dependence of several groups and low value of R^2. These relatively independent groups explaining 58–76 % variances have relevance identifying site-specific effect of runoff. Principal component analysis (PCA) of entire data set identified four distinct groups of sampling locations. Group 1 showed negative relationship with both the PCs reflecting the role of natural controls at Site 1 (Adpr). The deviation of Site 2 and 3 (upstream to urban core) from rest of the sites indicated a combined influence of AD and runoff inputs. The fourth group clustered rest of the sites indicating the influence of AD, sewage and runoff inputs at downstream sites.

For evaluating the effect of changing state of atmosphere–land–water transfer, we tested whether river DOC was positively related to AD-OC and runoff DOC and whether the microbial activity and WSOC were positively related to AD-OC. The river DOC showed characteristic synchrony with AD-OC and runoff DOC and the latter showed strong synchrony with microbial activity and WSOC. We used C:N ratio as an indicator of terrestrial C input. Emeis et al. (2000) argued a C:N ratio >10 indicates high proportion of terrestrial organic matter. Meyers (2003) referred C:N > 20 to terrestrial origin. In the present study, despite a declining trend over time, the C:N ratios were invariably high (>22) indicating terrestrial influence (Elser et al. 2000). It seems that if the rising trend of atmospheric inputs is continuing, as expected for countries like India, it would enhance the runoff C transport and, in turn, the DOC in surface waters. Data on these issues are important in understanding the ongoing changes in the biogeochemical cycle of carbon and options for managing carbon cycle for mitigating future climate change. Our study showed that the hydrological transport of carbon as influenced by land use and AD-OC is an important vector of DOC to Ganga River in Varanasi region. Since the Ganges is an oceanic river, atmospheric and terrestrial carbon subsidies bear a particular significance.

Conclusions

This study showed that the Varanasi region is receiving increasingly high input of organic carbon (OC) through atmospheric deposition (AD). Changing land use pattern coupled with rising input of AD-OC may alter the atmosphere–land–water transfer causing a shift in carbon balance in the river. Regional carbon fluxes and pools, therefore, would be misrepresented without appropriately considering the changing state of trans-ecosystem linkages. Thus, the models that incorporate carbon data based only on soil C pool, vegetation C pool and vegetation–atmospheric–C fluxes may over- or underestimate the future carbon budget. Our study, that forms the first report on atmosphere–land–water–C linkages in Ganga River and identifies region-specific landscape elements which are more vulnerable to AD inputs and runoff C transfer, suggests the need for large-scale inter-regional time series data on carbon balance as influenced by land use change and increasing AD-OC inputs. Long-term data on these issues would help predicting future state of climate change drivers and have relevance from climate change perspective and for designing action plan for integrated river basin management.

Acknowledgements We thank Coordinator, Centre of Advanced Study in Botany, Banaras Hindu University for laboratory facilities. We are grateful to reviewers for their critical and helpful comments. The work was supported by the University Grant Commission, New Delhi (Grant No. 32-383/2006 SR and as fellowship to PK), National Academy of Sciences, India (as fellowship to AVS) and Banaras Hindu University (as fellowship to RS).

References

Aber J, McDowell W, Nadelhoffer K, Magill A, Berntson G, Kamakea M, McNulty S, Currie W, Rustad L, Fernandez I (1998) Nitrogen saturation in temperate forest ecosystems. Hypothesis revisited. BioScience 48:921–934

American Public Health Association (1998) Standard methods for the examination of water and wastewater. APHA, Washington, DC

Anderson JPE, Domsch KH (1978) A physiologically active method for the quantitative measurement of microbial biomass in soils. Soil Biol Biochem 10:215–221

Buffam I, Galloway JN, Blum LK, McGlathery KJ (2001) A stormflow/baseflow comparison of dissolved organic matter concentrations and bioavailability in an Appalachian stream. Biogeochemistry 53:269–306

Buffam I, Turner MG, Desai AR, Hanson PC, Rusak JA, Lottig NR, Stanley EH, Carpenter SR (2011) Integrating aquatic and terrestrial components to construct a complete carbon budget for a north temperate lake district. Glob Change Biol 17:1193–1211

Ciavatta C, Govil M, Antisari VL, Sequi P (1991) Determination of organic carbon in aqueous extract of soil and fertilizers. Commun Soil Sci Plant Anal 22:795–807

Cole J, Prairie Y, Caraco N, McDowell W, Tranvik L, Striegl R, Duarte C, Kortelainen P, Downing J, Middelburg J, Melack J (2007) Plumbing the global carbon cycle: integrating inland waters into the terrestrial carbon budget. Ecosystems 10:171–184

Dadhwal VK (2012) Assessment of Indian carbon cycle components using earth observation systems and ground inventory. Int Arch Photogram Remote Sens Spat Inform Sci 39:249–254

Elser JJ, Fagan WF, Denno RF, Dobberfuhl DR, Folarin A, Huberty A, Interlandi S, Kilham SS, McCauley E, Schulz KL, Siemann EH, Sterner RW (2000) Nutritional constraints in terrestrial and freshwater food webs. Nature 408:578–580

Emeis KC, Struck U, Leipe T, Pollehne F, Kunzendorf H, Christiansen C (2000) Changes in the C, N, P burial rates in some sediments over the last 150 years: relevance to P regeneration rates and the phosphorus cycle. Marine Geol 167:43–59

Evans CD, Monteith DT, Cooper DM (2005) Long-term increases in surface water dissolved organic carbon observations, possible causes and environmental impacts. Environ Pollut 137:55–71

Evans CD, Chapman PJ, Clark JM, Monteith DT, Cresser MS (2006) Alternative explanations for rising dissolved organic carbon export from organic soils. Glob Change Biol 12(11):2044–2053

Fenner N, Freeman C (2011) Drought-induced carbon in peatlands. Nat Geosci 4:895–900

Freeman C, Evans CD, Monteith DT, Reynolds B, Fenner N (2001) Export of organic carbon from peat soils. Nature 412:785

Freeman C, Fenner N, Ostle NJ, Kang H, Dowrick DJ, Reynolds B, Lock MA, Sleep D, Hughes S, Hudson J (2004) Export of dissolved organic carbon from peat lands under elevated carbon dioxide levels. Nature 430:195–198

Gillet RW, Ayers GP (1991) The use of thymol as a biocide in rain water samples. Atmos Environ 25A:2677–2681

Gilvear DJ, Heal KV, Stephen A (2002) Hydrology and ecological quality of Scottish river ecosystems. Sci Total Environ 294:131–159

Jurado E, Jaward FM, Lohmann R, Jones KC, Simó R, Dachs J (2004) Atmospheric dry deposition of persistent organic pollutants to the Atlantic and inferences for the global oceans. Environ Sci Technol 38:5505–5513

Jurado E, Dachs J, Duarte CM, Simó R (2008) Atmospheric deposition of organic and black carbon to the global oceans. Atmos Environ 42:7931–7939

Kalbitz K, St Solinger, Park J-H, Michalzik B, Matzner E (2000) Controls on the dynamics of dissolved organic matter in soils: a review. Soil Sci 165:277–304

Kalbitz K, Meyer A, Yang R, Gerstberger P (2007) Response of dissolved organic matter in the forest floor to long-term manipulation of litter and throughfall inputs. Biogeochemistry 86:301–318

Kirchman DL, Suzuki Y, Garside C, Ducklow HW (1991) High turnover rates of dissolved organic carbon during a spring phytoplankton bloom. Nature 352:612–614

Kopácěk J, Cosby BJ, Evans CD, Hruška J, Moldan F, Oulehle F, Šantrůčková H, Tahovská K, Wright RF (2013) Nitrogen, organic carbon and sulphur cycling in terrestrial ecosystems: linking nitrogen saturation to carbon limitation in soil microbial processes. Biogeochemistry 115:33–51

Langley JA, Megonigal JP (2010) Ecosystem response to elevated CO_2 levels limited by nitrogen- induced plant species shift. Nature 466:96–99

Lohse KA, Hope D, Sponseller R, Allen JO, Grimm NB (2008) Atmospheric deposition of carbon and nutrients across an arid metropolitan area. Sci Total Environ 402:95–105

Lovett GM, Traynor MM, Pouyat RV, Carreiro MM, Zhu WX, Baxter JW (2000) Atmospheric deposition to oak forests along an urban-rural gradient. Environ Sci Technol 34:4294–4300

Maiti SK (2001) Handbook of methods in environmental studies. In: Water and wastewater, vol 1. ABD Publisher, Jaipur

Matheussen B, Kirschbaum RL, Goodman IA, O'Donnell GM, Lettenmaier DP (2000) Effects of land cover change on stream flow in interior Columbia River Basin (USA and Canada). Hydrol Proc 14:867–885

Meyers PA (2003) Applications of organic geochemistry to paleolimnological reconstruction: a summary of examples from the Laurentian Great Lakes. Org Geochem 34:261–289

Michel P (1984) Ecological methods for field and laboratory investigation. Tata McGraw–Hill Publ Comp, New Delhi

Monteith DT, Stoddard JL, Evans CD, deWit HA, Forsius M, Høgasen T, Wilander A, Skjelkvale BL, Jeffries DS, Vuorenmaa J, Keller B, Kopacek J, Vesely J (2007) Dissolved organic carbon trends resulting from changes in atmospheric deposition chemistry. Nature 450:537–540

Murillo JH, Marin JFR, Roman SR, Guerrero VHB, Arias DS, Ramos AC, Gonzalez BC, Baumgardner DG (2013) Temporal and spatial variations in organic and elemental carbon concentrations in PM10/PM2.5 in the metropolitan area of Costa Rica, Central America. Atmos Pollut Res 4:53–63

Pandey J (2011) The influence of atmospheric deposition of pollutant elements on cross-domain causal relationships at three tropical freshwater lakes of India. Lakes Reser Res Manag 16:111–119

Pandey J, Pandey U (2009) Microbial processes at land water interface and cross-domain causal relationships as influenced by atmospheric deposition of pollutants in three freshwater lakes in India. Lakes Reser Res Manag 14:71–84

Pandey U, Pandey J (2013) Impact of DOC trends resulting from changing climatic extremes and atmospheric deposition chemistry on periphyton community of a freshwater tropical lake of India. Biogeochemistry 112:537–553

Pandey J, Singh AV, Singh A, Singh R (2013) Impacts of changing atmospheric deposition chemistry on nitrogen and phosphorus loading to Ganga River (India). Bull Environ Contam Toxicol 91:184–190

Pandey J, Pandey U, Singh AV (2014a) The skewed N:P stoichiometry resulting from changing atmospheric deposition chemistry drives the pattern of ecological nutrient limitation in Ganges River. Cur Sci 107(06):956–958

Pandey J, Pandey U, Singh AV (2014b) Impact of changing atmospheric deposition chemistry on carbon and nutrient loading to Ganga River: integrating land-atmosphere-water components to uncover cross-domain carbon linkages. Biogeochemistry 119:179–198

Perakis SS, Hedin LO (2002) Nitrogen loss from unpolluted South American forests mainly via dissolved organic compound. Nature 145:416–419

Porcal P, Koprivnjak JF, Molot LA, Dillon PJ (2009) Humic substances-part 7: the biogeochemistry of dissolved organic carbon and its interactions with climate change. Environ Sci Pollut Res 16:714–726

Ramanathan V, Garmichael G (2008) Global and regional climate change due to black carbon. Nature Geosci 1:221–227

Reich PB, Hobbie SE, Lee T, Ellsworth DS, West JB, Tilman D, Knops JNH, Naeem J, Trost J (2006) Nitrogen limitation constrains sustainability of ecosystem response to CO_2. Nature 440:922–925

Ritson JP, Graham NJD, Templeton MR, Clark JM, Gough R, Freeman C (2014) The impact of climate change on the treatability of dissolved organic matter (DOM) in upland water supplies: a UK perspective. Sci Total Environ 473–474:714–730

Sponseller RA (2007) Precipitation pulses and soil CO_2 flux in a Sonoran Desert Ecosystem. Glob Change Biol 13:426–436

Taylor D (2010) Biomass burning, humans and climate change in southeast Asia. Biodiv Cons 19:1025–1042

Tranvik LJ, Jansson M (2002) Climate change: terrestrial export of organic carbon. Nature 415:861–862

Voghel AI (1971) A text book of quantitative inorganic analysis, 4th edn. The english language book society, Longman

Westhorpe DP, Mitrovic SM (2012) Dissolved organic carbon mobilisation in relation to variable discharges and environmental flows in a highly regulated lowland river. Mar Freshwater Res 63:1218–1230

Willey JD, Kieber RJ, Eyman MS, Avery GB Jr (2000) Rainwater dissolved organic carbon: concentrations and global flux. Glob Biogeochem Cycle 14:139–148

Yin Y, Wurzler S, Levin Z, Reisin TG (2002) Interactions of mineral dust particles and clouds: effects on precipitation and cloud optical properties. J Geophys Res 107:AAC 19-1–AAC 19-14

27

Bloom of the two dinoflagellates *Ceratium furca* and *Diplopsalis lenticula* in a mangrove estuary of Thailand

Tatsuya Yurimoto · Dusit Aue-umneoy · Chonlada Meeanan · Isao Tsutsui

Abstract Knowledge of the planktonic species that appear in mangrove estuaries in tropical areas is still poor. We believe that case report data on the appearance of species is important for clarifying the ecosystem of the mangrove estuary. We report in this paper on the occurrence of blooms and discuss the factors that influence blooming, based on an observed bloom consisting of two dinoflagellates in the mangrove estuary of Samut Songkhram, Thailand, in December 2012. From fluorescence microscopic observation of the armors of the dinoflagellates after calcofluor white staining, we identified these dinoflagellates as *Ceratium furca* and *Diplopsalis lenticula*. In addition, from the annual data on air temperature and precipitation for that year, it was found that the bloom occurred in early dry season. Additionally, diatoms that inhibit the growth of dinoflagellates were observed at low densities. These results suggested that the two dinoflagellates, *C. furca* and *D. lenticula*, grew during disappearance of diatoms.

Keywords Dinoflagellate · *Ceratium furca* · *Diplopsalis lenticula* · Bloom · Mangrove estuary

Introduction

The dinoflagellates include many harmful and/or toxic species that cause damage to fisheries, and there are a large number of case reports about the distributions and the bloom they cause (Tangen 1977; Carder and Steward 1985; Azanza and Max Taylor 2001; Townsend et al. 2001; Su-Myat et al. 2012a, b). Some species have established laboratory culture strains and their growth characteristics have been investigated (Heaney and Eppley 1981; Parkhill and Cembella 1999; Hwang and Lu 2000; Navarro et al. 2006; Yamatogi et al. 2009). In general, it is considered that competition between organisms (predators, prey and niche) and environmental factors (water temperature, salinity, light, wind, ocean currents, nutrients, etc.) determine the growth of dinoflagellates (Bomber et al. 1988; Anderson 1994; Siu et al. 1997; Straile 1997; Paerl 1997; Smayda 2002; Yurimoto et al. 2006, 2007; Yamatogi et al. 2009). For example, in the case of the species that causes paralytic shellfish poisoning, *Gymnodinium catenatum*, it is involved in wind direction and estuary circulation in the

T. Yurimoto (✉) · I. Tsutsui
Japan International Research Center for Agricultural Sciences, Ohwashi, Tsukuba, Ibaraki 305-8686, Japan
e-mail: yurimoto@outlook.com

D. Aue-umneoy · C. Meeanan
King Mongkut's Institute of Technology Ladkrabang, Chalongkrung Rd. Ladkrabang, Bangkok 10520, Thailand

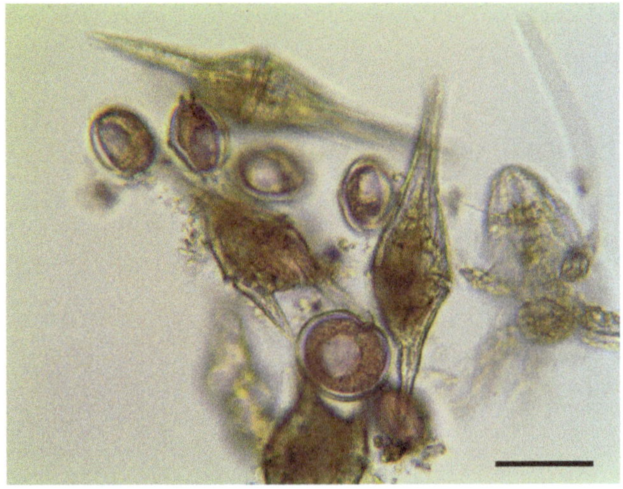

Fig. 1 Microphotograph of an accumulated plankton sample concentrated using a 20 μm mesh sieve, collected from St. 1 of Samut Songkhram in December 2012. *Scale bar* = 50 μm

inner part of the bay, to form a red tide when the growth rate is slow in winter (Abo and Miyamura 2005). In addition, it is easy to form a red tide because dinoflagellates and raphidophytes such as *Cochlodinium* spp., *Chattonella* spp. etc., have the ability to swim vertically, and so they can take advantage of a stratified environment to access nutrients, as is likely to occur in a thermocline or halocline in an estuary area during the rainy or summer seasons (Kimura et al. 1999; Park et al. 2001).

Mangrove forests in the tropics are formed around the intertidal zone of estuaries or the inner part of the bay (Lugo and Snedaker 1974). Fine mud is supplied from terrestrial areas through rivers, and it forms mud flats around the river mouth (Furukawa and Wolanski 1996). The coastal area experiences large environmental change due to the mixing of seawater and water flowing in from rivers, and turbidity is high in this area (Cyrus and Blaber 1992). Extreme environmental changes occur in the rainy season in the tropics due to the large climate change from the dry season (Goodbody 1961; Devassy and Goes 1988, 1989). For this reason, the appearance of phytoplankton around the estuaries of mangrove forests shows wide variation, because diatoms are likely to be dominated by the silicate supply from terrestrial areas in the rainy season, and dinoflagellates are likely to appear during low supply of silicate in the dry season (Biswas et al. 2010). However, little is known about the plankton species that appear in the tropical mangrove estuary, so data based on the accumulation of case reports on the occurrence of the various species is important to clarify the ecosystem of the mangrove estuary. In this paper, we report on the occurrence of blooms and discuss the factors that influence them, based on our observations of the blooms of two species of small- and large-celled dinoflagellates (Fig. 1) in the mangrove estuary of Samut Songkhram province in December 2012, with the aim of clarifying the dominance of dinoflagellates in a mangrove estuary as a changing environment from wet to dry seasons, and to find new research issues about plankton ecology in there.

Materials and methods

Survey area

Samut Songkhram province located in the inner west coast of the Gulf of Thailand (Fig. 2a), is a popular area for solar salterns in Thailand (Murakami 1983). In general, salterns require long sunshine hours and low rainfall. The province is known as a low rainfall region in Thailand because the monsoon from the Indian Ocean is blocked by mountains on the border of Thailand and Myanmar (Murakami 1983). Mangrove forests are present along wide area on the coast and huge mud flats and wide shallow coast are formed (Fig. 2c). For this reason, this area is popular for bivalve aquaculture. Aquaculture of the blood cockle is located in the tidal flat area close to the mangrove forests, and green mussels or oysters are farmed by hanging or pole cultures in the offshore area.

The research site is located approximately in the middle of two estuaries, formed by the Mae Klong River in the north and the Bang Tabun River in the south, both of them susceptible to fresh water (Fig. 2b). Three sampling stations depending on each different situation were set in the estuary (Fig. 2c): Station 1 was in an

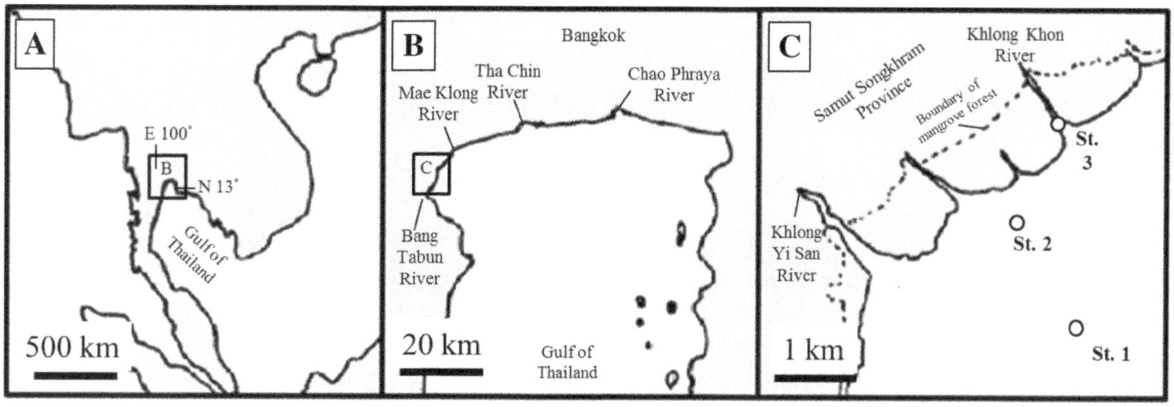

Fig. 2 Three sampling stations in the coastal waters of Samut Songkhram, Thailand, December 2012. **a** Location of the inner part of the Gulf of Thailand. **b** Location of the coastal area of Samut Songkhram. **c** Location of the three sampling stations in the coastal waters of Samut Songkhram

offshore area of the blood cockle aquaculture ground and near the mussels and oyster farming area (depth about 2 m), station 2 was in blood cockle aquaculture farm area near tidal flat (depth about 1 m) and station 3 was in a waterway in a mangrove forest (depth about 1 m).

Sample collection

Surface water for the plankton sample was taken with a 500 ml bottle from a fishing boat at three stations in coastal waters of Samut Songkhram province, Thailand, in December 2012 (Fig. 2). These samples were brought back to the laboratory, and each sample was concentrated in 2 ml of seawater using a sieve of 20 μm mesh width. These samples were fixed with formalin solution and stored at room temperature until used for observation by light or fluorescence microscopes.

Plankton observation

Two species of the dinoflagellates that was present in particularly high densities (Fig. 1) were observed after fluorescent staining of the armor forms on the cell surface with calcofluor white solution (Fritz and Triemer 1985). Cell densities were recorded using a Sedgewick-rafter counting chamber (1 ml) under a light microscope, and other plankton present was recorded at the genus level.

Weather information

For weather information in the survey area in January–December 2012, we referred to the annual climate data in Bangna Agromet near Bangkok City and close to Samut Songkhram (Japan Meteorological Agency http://www.data.jma.go.jp/gmd/cpd/monitor/index.html).

Results

Cell observation

Photo images from the light and fluorescence microscopes of the large-celled species of dinoflagellate are shown in Fig. 3a, b. A light microscope observation of this cell was compared with articles by Yamaji (1966) and Fukuyo et al. (1990). The upper body of the observed cell was long and narrow, and formed an apical horn. It became gradually thicker towards the base, where it formed an epitheca. In addition, the lower body was extending almost parallel to the rear, and two dorsal horns were observed in the

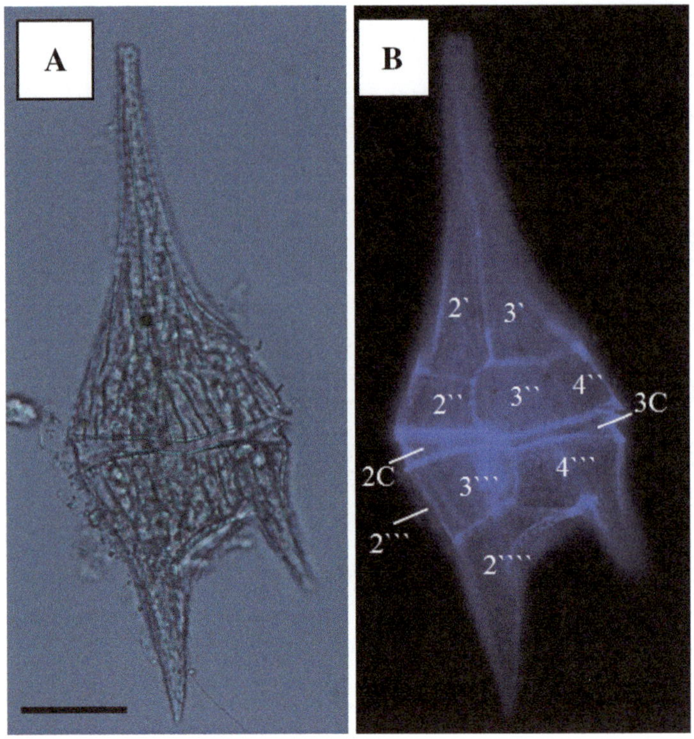

Fig. 3 Microphotographs of the large dinoflagellate identified as *Ceratium furca*. **a** The cell image from a light microscope. **b** The cell image from a fluorescence microscope. Kofoidian system of tabulation (Gómez et al. 2010): ` apical, `` precingular, ``` postcingular, ```` antapical and *c* cingular. *Scale bar* = 25 μm

hypotheca. The left dorsal horn was a little thicker and about 2-fold longer than the right horn. A fluorescence microscope image of the cell is shown in Fig. 3b. The armor forms were stained with the fluorescence solution. It was highly consistent with the illustration by Yamaji (1966), and this large cell was identified as *Ceratium furca*.

Photo images of the small-celled dinoflagellate with light and fluorescence microscopes are shown in Fig. 4a, b. Comparison of the morphology revealed by light microscope observation of the small cell was made with the descriptions of Lebour (1922), Yamaji (1966) and Gribble (2006). The observed cell was a flat disc shape (Fig. 4a), the side was oval, and the transverse groove was annular (Fig. 4b). In addition, the cytoplasm was a pink color.

The fluorescence microscope images are shown in Fig. 4c, d. The shape of each armor was compared with the illustrations and electron microscope photographs published by Lebour (1922) and Gribble (2006), respectively. The upper shell, consisting of three plates, met in the center and the second and third plates were almost the same form. Only the center plate was small. The small plate formed an isosceles triangle, such that the two upper sides were long and the bottom side was short (Fig. 4c). The hypotheca formed a large single plate (Fig. 4d). Based on these observations, this small-celled dinoflagellate was identified as *Diplopsalis lenticula*.

Number of cells

Cell numbers of the two species observed during the bloom are shown in Fig. 5. The results show significant differences in cell numbers between the stations. Although both *C. furca* and *D. lenticula* at St. 1 were present in high densities (almost the same, at about 2,000 cells/L) in the three stations, St. 2 *D. lenticula* was measured at over 3,000 cells/L and *C. furca* at only 200 cells. This result is indicating that *D. lenticula* comprises over 90 % in both the species. In addition, although *D. lenticula* was present at only 50 cells/L at St. 3, *C. furca* was measured at 680 cells/L. This result is indicating that *C. furca* comprises over 90 % in both the species.

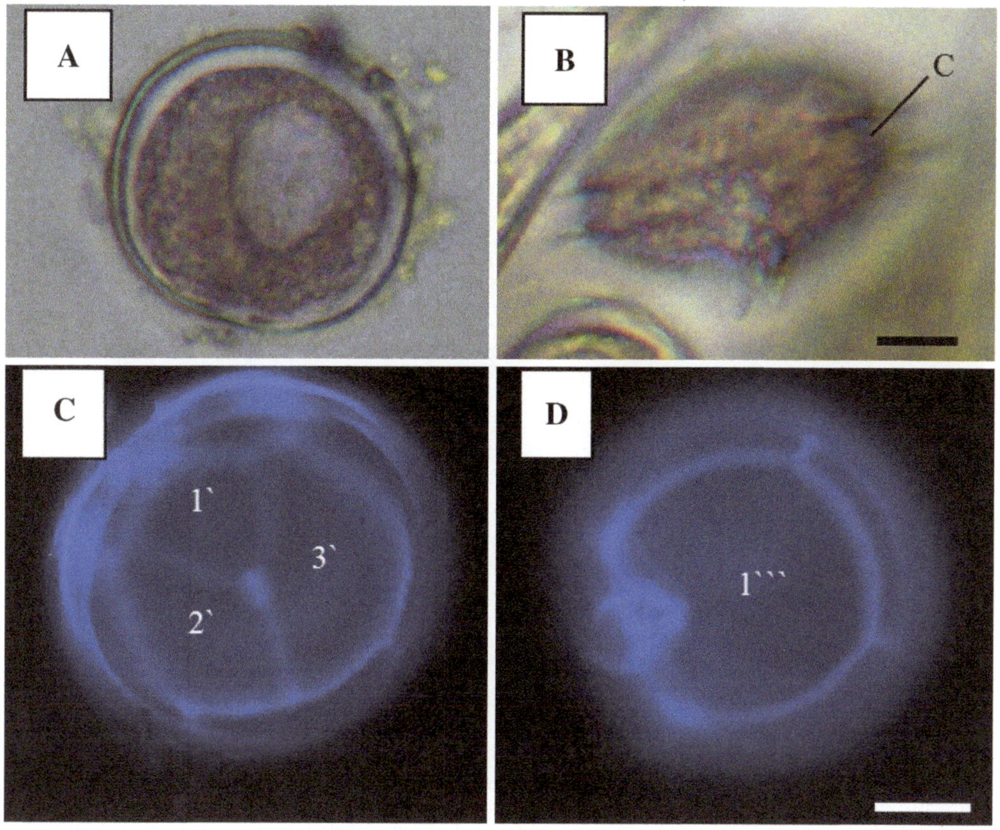

Fig. 4 Microphotographs of the small dinoflagellate identified as *Diplopsalis lenticula.* **a**, **b** The cell images from a light microscope. **c**, **d** The cell images from a fluorescence microscope. Kofoidian system of tabulation (Gómez et al. 2010): ` apical, ``` postcingular and *c* cingular. *Scale bar* = 25 μm

Fig. 5 Cell densities of the two dinoflagellates *Ceratium furca* (*filled square*) and *Diplopsalis lenticula* (*open square*) at each sampling station of Samut Songkhram in December 2012

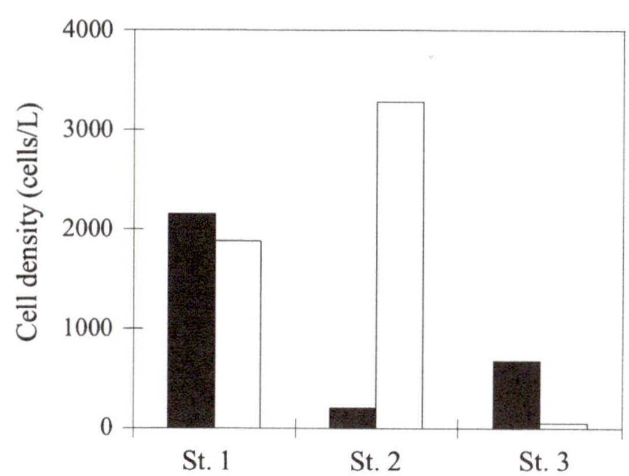

Other plankton

Other plankton observed were almost exactly the same species at St. 1 and 2 and at low density (Fig. 6). Diatoms identified were *Chaetoceros* spp. (300 and 600 cells/L, Fig. 6a), *Pseudo-nitzschia* sp. (180 and 270 cells/L, Fig. 6b), *Rhizosolenia* sp.(10 and 20 cells/L, Fig. 6c), and *Leptocylindrus* sp.(30 and 50 cells/L, Fig. 6d), along with a zooplankton, *Copepoda* (160 and 500 ind./L, Fig. 6e). St. 3, located in the mangrove forests, was a different situation from the other two stations. No diatoms were observed, and *Copepoda* (30 ind./L, Fig. 6e) and detritus-like particles (50 μm < size particle; 60 pieces, Fig. 6f) was observed.

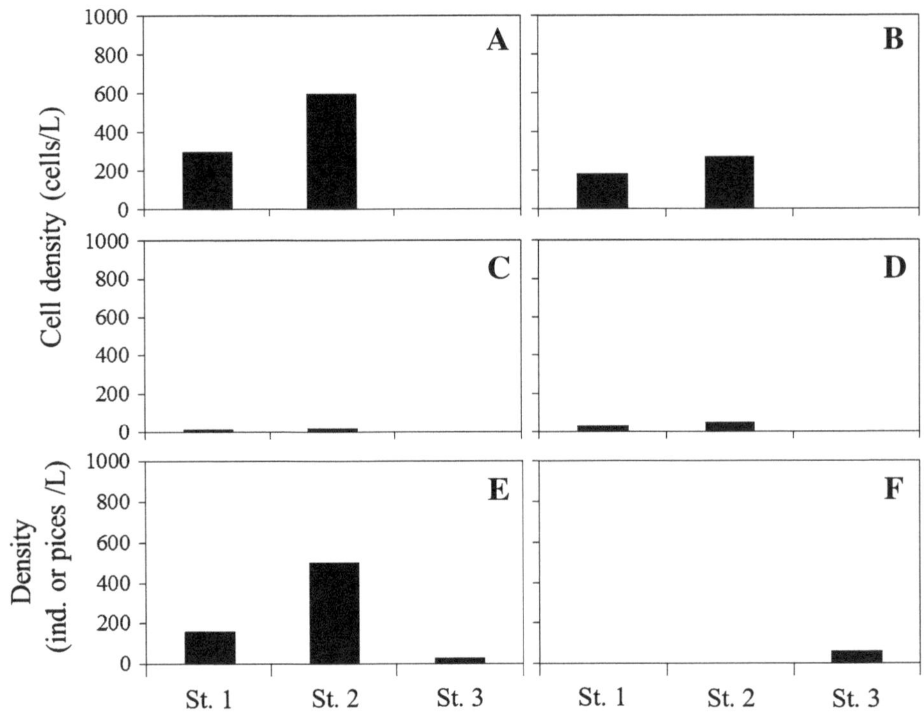

Fig. 6 Densities of phytoplankton, zooplankton and detritus-like particles in the samples from the three stations of Samut Songkhram in December 2012. **a** Cell densities of *Chaetoceros* spp. in the stations. **b** Cell densities of *Pseudo-nitzschia* sp. in the stations. **c** Cell densities of *Rhizosolenia* sp. in the stations. **d** Cell densities of *Leptocylindrus* sp. in the stations. **e** Densities of *Copepoda* in the stations. **f** Detritus-like particles (50 μm < size particle) in the stations

Weather information

The monthly average air temperature in 2012 showed the highest value of 31.1 °C in April and the lowest value of 28.0 °C in January, and in December it was 28.9 °C (Fig. 7). The monthly rainfall maximum was 455 mm in September and the minimum was 0 mm in December. The volume reduced from 226 to 128 mm during October and November.

Discussion

The weather information indicate that November was a transition period from rainy season to dry season, and that December, when sampling was carried out, can be classified as early dry season because the data show cool temperatures and low rainfall. Normally, because silicate is supplied from the rivers during the rainy season, diatoms predominate and dinoflagellates are in the minority (Wang et al. 2006). However, when the volume of silicate from the rivers is reduced in the dry season, the number of diatom cells is reduced (Wang et al. 2006). Dinoflagellates, which can swim and are able to take in nutrients efficiently, can form a bloom by growing in the niche vacated by the diatoms. We observed a bloom of *C. furca* and *D. lenticula* in Samut Songkhram during the early dry season, and the growth of these two species suggested that it was the result of the coastal environment changing due to seasonal variation. Although we could not collect weather information on the wind direction around Samut Songkhram during the blooming event, we discovered that the wind direction from the northeast dominates in this region in December, based on literature describing the monsoon in Thailand (Ogino 1967). This means the northeast monsoon highly develops as a dominant, because the tropical front move to the south, lower than the equator, and the

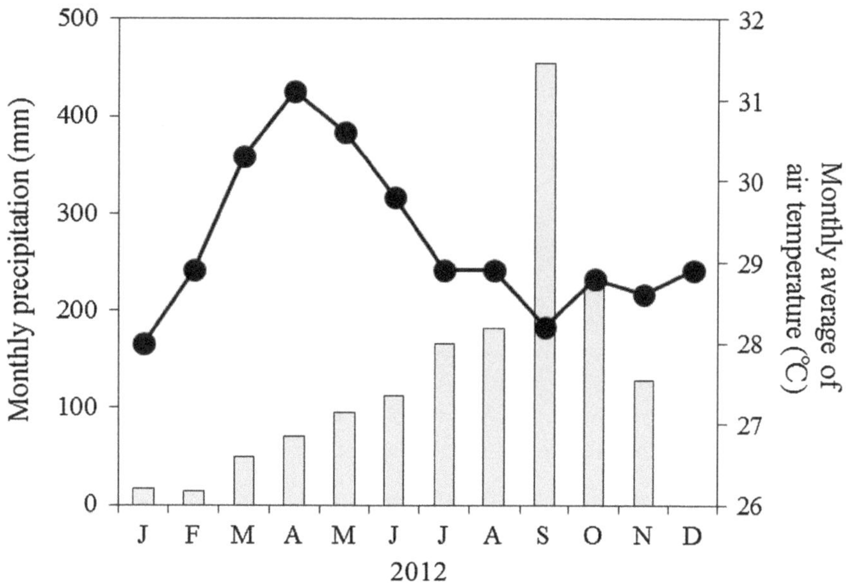

Fig. 7 Annual changes in monthly average air temperature (*line graph*) and monthly precipitation (*bar graph*) around Bangkok City (Bangna Agromet) in 2012. Data obtained from the Japan Meteorological Agency: http://www.data.jma.go.jp/gmd/cpd/monitor/index.html

northeast trade winds blow near the equator. Additionally, it is due to the pressure pattern changing that high pressure in the continental interior develops in low latitudes region of the Northern Hemisphere (Ogino 1967). Given the topography of the survey area and information on wind direction, because the waters are located on the west coast of the inner part of the Gulf of Thailand, it is highly possible that the formation of high cell densities of the two dinoflagellates species was due to drifting to the west coast caused by monsoon winds blowing towards the southwest from the northeast.

C. furca is distributed in oceans around the world, from tropical to frigid zones (Fukuyo et al. 1990). The genus *Ceratium* has been classified into surface distribution species (surface species) and deep layer distribution species (shade species) in a water column. *C. furca* is known as a surface species and often forms a red tide in the inner part of the bay (Fukuyo et al. 1990). This species is distributed across a wide temperature zone, confirmed as 6.6–29.4 °C in its natural habitat (Graham and Bronikowsky 1944; Fukuyo et al. 1990). The optimal growth conditions of this species have been confirmed by culture experiments as having a wide range: the optimal water temperature ranges from 18 to 28 °C and the salinity ranges from 17 to 34 PSU. For this reason, this species is considered to be always capable of growing in the environment of Samut Songkhram, and it seems likely that the growth observed was due to the vacated niche resulting from the disappearance of the diatoms in the conditions following the transition from rainy season to dry season. In addition, cell density of *C. fruca* that was observed in this study was about the 2,000 cells/L. In recent years, Karthik et al. (2014) had reported the extremely high density of the 20,000 cells/ml over *C. fruca* in the Andaman Sea. Our survey result was very low density compared with the case of Andaman Sea and the cells in Samut Songkhram was not affecting density in life and death on fish.

D. lenticula has also been found across a wide salinity distribution (Gopinathan 1972). For this reason, the low cell densities found at St. 3 were considered to be due to the physical dilution effect of river water rather than growth inhibition due to low salinity. *C. furca* is a mixotrophic dinoflagellate, which can obtain nutrition from both photosynthesis and predation of organisms, whereas *D. lenticula* is a heterotrophic dinoflagellate, which can feed only by predation. For this reason, *D. lenticula* requires organisms as food for growth. However, other than the two dinoflagellates observed, only some species of diatoms and zooplankton were observed at low densities. It was suspected predation of *C. furca* by *D. lenticula* for the reason, but we were not able to find the *D. lenticula* phagocyted *C. furca* in these samples. In addition, although it was suspected to involve nano and picoplankton in the growth of *D. lenticula*, we were not able to capture the occurrence of nano and picoplankton because the collected plankton sample was concentrated with a 20 µm mesh sieve in the sample preparation. For this reason, to make a clear understanding of the mutual relationship between *C. furca* and *D. lenticula*, we need to focus on these points in further study. In this paper, it is only captured one red tide event and each event is issue for a study. It is necessary to consider in a future study, other parameters to determine their influence in the occurrence of these events.

Conclusion

A bloom event comprising two species of dinoflagellate was observed in a mangrove estuary of the Samut Songkhram Province in Thailand in December 2012. These dinoflagellates were identified as *Ceratium furca* and *Diplopsalis lenticula*, and the bloom occurred during the early dry season. Additionally, diatoms that inhibit the growth of dinoflagellates were observed at low densities. These results suggested the two dinoflagellates, *C. furca* and *D. lenticula*, grew during disappearance of diatoms.

Author contribution T. Yurimoto carried out sampling and the plankton observation, and wrote the draft manuscript. D. Aue-umneoy, C. Meeanan and I. Tsutsui organized and supported the field survey, and discussed the results and the written manuscript. All authors read and approved the final version of the manuscript.

Acknowledgments This work was partially supported by JST/JICA, science and technology research partnership for suitable development (SATREPS).

Conflict of interest All authors declared that they have no conflict of interest.

References

Abo K, Miyamura K (2005) Characteristics of water movements during winter in Inokushi Bay and Its effects on the growth of the toxic dinoflagellate *Gymnodinium catenatum*. Bull Coast Oceanogr 42:161–169

Anderson DM (1994) Red tides. Sci Am 271:62–68

Azanza RV, Max Taylor FJR (2001) Are *Pyrodinium* blooms in the Southeast Asian region recurring and spreading? A view at the end of the millennium. AMBIO J Hum Environ 30:356–364

Biswas H, Dey M, Ganguly D, De TK, Ghosh S, Jana TK (2010) Comparative analysis of phytoplankton composition and abundance over a two-decade period at the land–ocean boundary of a tropical mangrove ecosystem. Estuar Coast 33:384–394

Bomber JW, Guillard RR, Nelson WG (1988) Roles of temperature, salinity, and light in seasonality, growth, and toxicity of ciguatera-causing *Gambierdiscus toxicus* Adachi et Fukuyo (Dinophyceae). J Exp Mar Biol Ecol 115:53–65

Carder KL, Steward RG (1985) A remote-sensing reflectance model of a red-tide dinoflagellate off west Florida. Limnol Oceanogr 30:286–298

Cyrus DP, Blaber SJM (1992) Turbidity and salinity in a tropical northern Australian estuary and their influence on fish distribution. Estuar Coast Shelf Sci 35:545–563

Devassy VP, Goes JI (1988) Phytoplankton community structure and succession in a tropical estuarine complex (central west coast of India). Estuar Coast Shelf Sci 27:671–685

Devassy VP, Goes JI (1989) Seasonal patterns of phytoplankton biomass and productivity in a tropical estuarine complex (west coast of India). Proc Plant Sci 99:485–501

Fritz L, Triemer RE (1985) A rapid simple technique utilizing calcofluor white M2R for the visualization of dinoflagellate thecal plates. J Phycol 21:662–664

Fukuyo Y, Takano H, Chihara M, Matsuoka K (eds) (1990) Red tide organisms in Japan—an illustrated taxonomic guide. Uchida Rokakuho (publisher), Tokyo, p 407

Furukawa K, Wolanski E (1996) Sedimentation in mangrove forests. Mangrove Salt Marsh 1:3–10

Gómez F, Moreira D, López-García P (2010) *Neoceratium* gen. nov., a new genus for all marine species currently assigned to *Ceratium* (Dinophyceae). Protist 161:35–54

Goodbody I (1961) Mass mortality of a marine fauna following tropical rains. Ecology 42:150–155

Gopinathan CP (1972) Seasonal abundance of phytoplankton in the Cochin backwater. J Marine Biol Assoc India 14:568–577

Graham HW, Bronikowsky N (1944) 6. *Ceratum fruca* (Ehrenberg) Dejardin. The genus *Ceratium* in the Pacific and north Atlantic Oceans. Scient. Res. Cruise VII Carnegie 1928-1929. Biol 5:18–19

Gribble KE (2006) The ecology, life history, and phylogeny of the marine thecate heterotrophic dinoflagellates *Protoperidinium* and *Diplopsalidaceae* (Dinophyceae), Doctoral dissertation, Massachusetts Institute of Technology and Woods Hole Oceanographic Institution

Heaney SI, Eppley RW (1981) Light, temperature and nitrogen as interacting factors affecting diel vertical migrations of dinoflagellates in culture. J Plankton Res 3:331–344

Hwang DF, Lu YH (2000) Influence of environmental and nutritional factors on growth, toxicity, and toxin profile of dinoflagellate *Alexandrium minutum*. Toxicon 38:1491–1503

Karthik R, Padmavati G, Jayabarathi R (2014) Occurrence of dinoflagellate bloom of *Ceratium furca* in the coastal waters of south Andaman. Int J Curr Res 6:4906–4910

Kimura T, Watanabe M, Kohata K, Sudo R (1999) Phosphate metabolism during diel vertical migration in the raphidophycean alga, *Chattonella antiqua*. J Appl Phycol 11:301–311

Lebour MV (1922) Plymouth peridinians: I. *Diplopsalis lenticula* and its relatives. J Marine Biol Assoc UK New Ser 12:795–812

Lugo AE, Snedaker SC (1974) The ecology of mangroves. Annu Rev Ecol Syst 5:39–64

Murakami M (1983) Salt manufacture in Thailand. Bull Soc Sea Water Sci Jpn 37:171–182

Navarro JM, Muñoz MG, Contreras AM (2006) Temperature as a factor regulating growth and toxin content in the dinoflagellate *Alexandrium catenella*. Harmful Algae 5:762–769

Ogino K (1967) A climatological classification of Thailand, with special reference to humidity. Acad J Southeast Asian Stud 5:500–531

Paerl HW (1997) Coastal eutrophication and harmful algal blooms: importance of atmospheric deposition and groundwater as "new" nitrogen and other nutrient sources. Limnol Oceanogr 42:1154–1165

Park JG, Jeong MK, Lee JA, Cho KJ, Kwon OS (2001) Diurnal vertical migration of a harmful dinoflagellate, *Cochlodinium polykrikoides* (Dinophyceae), during a red tide in coastal waters of Namhae Island, Korea. Phycologia 40:292–297

Parkhill JP, Cembella AD (1999) Effects of salinity, light and inorganic nitrogen on growth and toxigenicity of the marine dinoflagellate *Alexandrium tamarense* from northeastern Canada. J Plankton Res 21:939–955

Siu GK, Young ML, Chan DKO (1997) Environmental and nutritional factors which regulate population dynamics and toxin production in the dinoflagellate *Alexandrium catenella*. In: Asia-Pacific Conference on Science and Management of Coastal Environment. Springer, Netherlands, pp 117–140

Smayda TJ (2002) Adaptive ecology, growth strategies and the global bloom expansion of dinoflagellates. J Oceanogr 58:281–294

Straile D (1997) Gross growth efficiencies of protozoan and metazoan zooplankton and their dependence on food concentration, predator-prey weight ratio, and taxonomic group. Limnol Oceanogr 42:1375–1385

Su-Myat, Maung-Saw-Htoo-Thaw, Matsuoka K, Khin-Ko-Lay, Koike K (2012a) Phytoplankton surveys off the southern Myanmar coast of the Andaman Sea: an emphasis on dinoflagellates including potentially harmful species. Fish Sci 78:1091–1106

Su-Myat, Yurimoto T, Hinode K, Takata Y, Mohd Nor Azman A, Alias M, Maeno Y, Koadama M, Koike K, Matsuoka K (2012b) Finding of toxic *Gymnodinium catenatum* Graham and *Alexandrium tamiyavanichii* Balech (Dinophyceae) from coastal waters of Selangor, Peninsular Malaysia. Malays Fish J 11:32–41

Tangen K (1977) Blooms of *Gyrodinium aureolum* (Dinophygeae) in North European waters, accompanied by mortality in marine organisms. Sarsia 63:123–133

Townsend DW, Pettigrew NR, Thomas AC (2001) Offshore blooms of the red tide dinoflagellate, *Alexandrium* sp., in the Gulf of Maine. Cont Shelf Res 21:347–369

Wang Z, Qi Y, Chen J, Xu N, Yang Y (2006) Phytoplankton abundance, community structure and nutrients in cultural areas of Daya Bay, South China Sea. J Mar Syst 62:85–94

Yamaji I (1966) Illustration of the marine plankton of Japan. Hoikusha publishing, Osaka, p 538

Yamatogi T, Kitahara S, Ura K, Yurimoto T (2009) Occurrence and growth characteristics of the toxic and noxious dinoflagellate *Alexandrium catenella* (Whedon and Kofoid) Balech in coastal waters of Nagasaki Prefecture, Japan. Bull Plankton Soc Jpn 56:111–119

Yurimoto T, Maeno Y, Kimoto K (2006) Distribution of several *Dinophysis* species off Western Kyushu in winter. Aquac Sci Suisanzoshoku 54:455–464

Yurimoto T, Maeno Y, Kimoto K, Sato S, Kodama M (2007) Occurrence of two toxic dinoflagellate species, *Alexandrium catenella* and *Gymnodinium catenatum*, off Western Kyushu, Japan, in winter. Proc Int Conf Molluscan Shellfish Saf ICMSS 6:210–219

28

Seasonal dynamics in the trophic status of water, floral and faunal density along some selected coastal areas of the Red Sea, Tabuk, Saudi Arabia

Abid Ali Ansari · Sulaiman Al Ghanim · Subrata Trivedi · Hasibur Rehman · Zahid Khorshid Abbas · Shalini Saggu

Abstract Eutrophication is posing a threat to the coastal marine ecosystems. The consequence of anthropogenic induced eutrophication of waters has resulted in severe deterioration of surface waters. In this research a preliminary data was collected on eutrophication by environmental monitoring on some selected coastal areas of the Red Sea near Tabuk, Saudi Arabia. Sampling sites were selected from Haql, Sharmaa and Duba (five for each station) and water samples collected in four different seasons (Oct 2013, Jan 2014, April 2014 and July 2014) to study seasonal dynamics of water quality parameters (especially in reference to the trophic status of water). Only one coastal site from each station was found with some eutrophic characteristics. Water samples collected from these eutrophic sites were more alkaline, turbid, with higher nutrient and low dissolved oxygen contents as compared to other sampling sites. Dynamics in relative densities of coastal flora and fauna were also found to be related with coastal water quality. Human settlements, fish markets, boating clubs and tourisms near these coastal areas of the Red Sea were observed as the primary causes of nutrient loading; run-offs during rains carrying nutrients also aggravate the problem. The results of this study may help in estimating the intensity of the problem and as a forecast for time frame to take an action for the conservation of affected coastal marine ecosystems and restoration of degraded coastal areas of the Red Sea.

Keywords Eutrophication · Coastal ecosystems · Red Sea · Bio-indicators

Introduction

Coastal eutrophication is one of the major threats to the marine ecosystems. The European Union (EU) Water Framework Directive (WFD) is given to prevent the deterioration, protect aquatic systems and to promote the sustainable use of water (Andersen et al. 2006). Hypoxia is one of the common effects of eutrophication in coastal marine ecosystems and is becoming an increasingly prevalent problem worldwide. The causes of hypoxia are associated with excess nutrient inputs from both point and non-point sources. Changes in climate, particularly temperature, may also affect the susceptibility of coastal marine ecosystems to hypoxia (Feuchtmayr et al. 2009). Hypoxia is a particularly severe disturbance because it causes death of biota and catastrophic changes in the ecosystem (Conley et al. 2009). The discovery of trophic development of waters and their terminology dates back to the early 20th century. The availability of water for human consumption will be one of the great issues of 21st century. It is predicted that eutrophication is increase as threats to water resources and ecosystems (Johanson et al. 2001; Liu et al. 2011; Ansari et al. 2014; Naeem et al. 2014).

A. A. Ansari (✉) · S. A. Ghanim · S. Trivedi · H. Rehman · Z. K. Abbas · S. Saggu
Department of Biology, Faculty of Science, University of Tabuk, Tabuk, Saudi Arabia
e-mail: aansari@ut.edu.sa; aaansari40@gmail.com

The consequences of anthropogenic induced eutrophication of waters however, were not anticipated until they became evident in the 1940s and 1950s. Severe deterioration of waters at that time raised remarkable public concern and triggered expanding scientific interest (Vollenweider 1970, Officer and Ryther 1980, Vollenweider and Kerekes 1980, Nixon 1995, Dokulil and Teubner 2011, Jeppesen et al. 2005). The biologists are studying it extensively in order to prevent eutrophication of aquatic bodies around the world (Yang et al. 2008). Eutrophication of coastal waters is a serious environmental problem with high costs for society globally. This is a development which demands immediate environmental action along many coastal sites (Lindahl and Kollberg 2009; Smith 2007). The consequence of anthropogenic induced eutrophication of waters has resulted in the contamination of surface water and related management issues which in turn has raised the brows of environmentalist across the world (Ansari et al. 2011a; Ansari and Khan 2011; Ansari and Gill 2014; Ansari and Khan 2014).The work on eutrophication was started by Vollenweider in 1968 and Rohlich in 1969. It is now evident that the nutrients are primarily responsible for limiting productivity in lakes, rivers and coastal areas (Likens 1972). Eutrophic ecosystems depend on many factors, including retention time, temperature, nutrients loading, hydraulic regimes, plant harvesting, and light intensity. The species compositions of phytoplankton are regulated by pH, light, temperature, turbulence, salinity and nutrients contents of waters (El-Shafai et al. 2007, Ansari and Khan 2009b; Lu et al. 2010; Devlin and Witham 1986).Initially, P and C have drawn the attention of limnologists and ecologists as primary causes of eutrophication. However, now it is proved that N contributes in increasing eutrophication especially in estuaries where it is a limiting factor under normal conditions. Therefore, it is evident that N is a primary cause of eutrophication in many coastal ecosystems (Paerl et al. 2003). Phosphorus has been considered the principal limiting nutrient for primary production of phytoplankton in aquatic bodies (Philips 2002; Sharpley et al. 2003). Various studies showed that the anthropogenic nutrients input are directly related with intensity of coastal eutrophication and degradation of coastal areas (Polomski et al. 2009). Nonpoint sources of nutrients are more responsible than the point sources for eutrophication in many regions (Carpenter 2005). Eutrophication generate significant alterations in structure and functioning of a wetland ecosystem (Davis et al. 2003) and alter the physicochemical and biological characteristics of soil and water (Corstanje and Reddy 2006).

A regular environmental monitoring and assessment of eutrophication will provide a research data on level of degradation of marine ecosystems in coastal areas of the red sea. The work will be an ecological vision on this global problem of coastal eutrophication and its causes, consequences and controls, in special reference to coastal areas of Red Sea in the boundary of Kingdom of Saudi Arabia.

Materials and methods

Three coastal stations of the Red Sea viz. Haql (29° 17′9.9″ N 34° 56′ 18.9″ E), Sharmaa (28° 1′ 27.9″ N 35° 16′ 9.9″ E) and Duba (27° 20′ 57.3″ N 35° 41′ 46.2″ E) were selected for environmental monitoring and assessment of coastal eutrophication (Fig. 1). After every 500 meters five sampling sites from each station (Haql, Sharmaa and Duba) were selected starting from the residential area of the city near to the sea coast. Water samples were collected in four different seasons (Oct 2013, Jan 2014, April 2014 and July 2014) to study the seasonal variations in water quality parameters. Physico-chemical parameters (pH, Turbidity, DO, Magnesium, Calcium, SiO_3, Phosphate and Nitrate) of collected waters samples were measured following APHA (2005). Data was analyzed statistically for the significance of research following Gomez and Gomez (1984) and computer software (SPSS for statistical analysis). Qualitative survey for the density of flora and fauna along the selected coastal areas of the Red Sea was also carried out in four different seasons of Tabuk. To determine the distribution of some plant and animal species along the coastal areas studied in this research, density was calculated. Five different sites were considered as replicates for each station.

Density in terms of occurrence and water surface occupied was expressed in four qualitative density classes. Negative (−) used to indicate non-occurrence and occurrence of the species. Single + indicates low density with sporadic occurrence, two ++ signs indicate for a medium density, three +++ shows high density of species with the dominance by a single species at particular coastal area of the Red Sea.

Fig. 1 Coastal stations of the Red Sea selected for environmental monitoring and assessment of coastal eutrophication

Results

Table 1 showed physico-chemical characteristics of water samples collected from five different sites along Red Sea coast at Haql, Tabuk, Saudi Arabia in October 2013 and January, April, July 2014. All the water samples were found with alkaline pH, but maximum 8.5 was recorded at site S1 in the month of July 2014. Water temperature was at its peak (22 °C) at site S4 in the month of July 2014. Water of S1 site was found more turbid than the samples from other coastal sites at Haql. Dissolved oxygen contents in water samples were significantly low at site S1 whereas optimum at rest of the sampling sites. Highest amount of oxygen in water was found at S5 in the month of April 2014. Magnesium, Calcium and Silica contents were significantly high in water samples collected from site S1. The nutrients contents viz. phosphates and nitrates in water samples were recorded significantly high at site S1. The statistical analysis of water quality parameters revealed that the symptoms of coastal eutrophication were more prominent in waters at site S1 only, whereas the water samples showed improvement in quality with the distance from site S1 (Table 1).

Table 2 showed physico-chemical characteristics of water samples collected from five different sites along Red Sea coast at Sharmaa, Tabuk, Saudi Arabia in October 2013 and January, April, July 2014. Most alkaline pH 8.8 was recorded at site S1 in the month of October-2013. Water temperature was of its maximum (20 °C) at site S2 and S3 in the month of July 2014. The water temperature was as low as 7 °C in January-2014 at site S5. Water at S1 site was found more turbid than the samples from other coastal sites at Sharmaa. Dissolved oxygen contents in water samples were significantly low at site S1 whereas higher oxygen level in water observed at rest of the sampling sites. Magnesium, calcium and silica nitrates and phosphates contents in water were significantly higher at site S1 as compared to the other sampling site showing a decreasing trend with the distance from site S1 (Table 2).

Table 3 showed physico-chemical characteristics of water samples collected from five different sites along Red Sea coast at Duba, Tabuk, Saudi Arabia in October 2013 and January, April, July 2014. The water was most alkaline with a pH 8.7 at site S1 recorded in the month of January-2013. Highest water temperature of 21 °C was recorded at sites S3 and S4 in the month of July 2014. The water temperature goes 8 °C in January-2014 at site S2. Water turbidity was high at site S1 as compared to the coastal sites at Duba. There was low content of dissolved oxygen in water samples from site S1 whereas significantly higher amount of oxygen was available in the water at rest of the sampling sites. Magnesium, calcium and silica nitrates and phosphates contents in water were significantly higher at site S1 in the month of January 2014, as compared to the other sampling site showing a decreasing trend with the distance from site S1 (Table 3).

Table 1 Physico-chemical characteristics of water samples collected from five different sites along Red Sea coast at Haql, Tabuk, Saudi Arabia in October 2013 and January, April, July 2014

Parameters	S1				S2				S3			
	Oct-13	Jan-14	April-14	July-14	Oct-13	Jan-14	April-14	July-14	Oct-13	Jan-14	April–14	July -14
pH	8.3 ± 0.4	8.2 ± 0.2	8.4 ± 0.4	8.5 ± 0.3	8.1 ± 0.2	8.1 ± 0.2	8.2 ± 0.4	7.4 ± 0.2	7.3 ± 0.2	7.9 ± 0.4	7.7 ± 0.3	7.8 ± 0.2
Temp °C	18	8	16	20	19	9	17	21	20	8	18	21
Turbidity (NTU)	13 ± 3	14 ± 4	14 ± 5	13 ± 3	12 ± 5	12 ± 4	13 ± 2	12 ± 3	11 ± 2	11 ± 2	11 ± 3	11 ± 4
DO (mg/l)	3.21 ± 1.1	3.05 ± 1.2	2.95 ± 0.92	3.46 ± 0.92	3.54 ± 0.81	3.38 ± 0.97	2.82 ± 0.75	4.45 ± 1.88	3.75 ± 0.87	4.88 ± 1.25	4.65 ± 1.1	3.58 ± 0.90
Mg (mg/l)	122 ± 5.6	240 ± 9.8	242 ± 6.8	232 ± 9.2	222 ± 7.5	178 ± 9.4	165 ± 9.1	109 ± 3.9	195 ± 9.9	184 ± 8.2	137 ± 8.6	98 ± 5.8
Ca (mg/l)	88.4 ± 9.2	74.5 ± 7.2	154.9 ± 11.8	76.7 ± 9.1	68.8 ± 8.8	62.7 ± 5.4	57.6 ± 4.1	52.24 ± 6.7	58.9 ± 13.2	56.7 ± 10.4	49.4 ± 8.2	48.24 ± 4.5
SiO_3 (μg/l)	5.8 ± 1.4	6.7 ± 1.5	7.2 ± 1.0	7.1 ± 1.2	5.6 ± 1.4	3.98 ± 1.2	5.5 ± 0.9	4.98 ± 1.3	3.4 ± 1.2	5.01 ± 1.2	3.5 ± 1.0	5.14 ± 1.8
P (μg/l)	30.4 ± 4.30	19.38 ± 1.42	32.54 ± 3.25	23.5 ± 1.22	14.52 ± 2.26	22.52 ± 3.21	17.22 ± 2.21	11.54 ± 0.61	10.18 ± 0.31	12.43 ± 1.21	8.38 ± 0.75	14.35 ± 2.95
N (μg/l)	68.6 ± 5.63	95.8 ± 5.39	55.8 ± 8.45	47.79 ± 3.39	29.85 ± 4.48	44.26 ± 1.1	22.35 ± 4.63	27.66 ± 3.39	10.86 ± 4.45	11.23 ± 1.2	17.34 ± 2.63	13.17 ± 1.1

Parameters	S4				S5				CD at 5 %
	Oct-13	Jan-14	April-14	July-14	Oct-13	Jan-14	April-14	July-14	
pH	8.3 ± 0.3	7.4 ± 0.1	7.8 ± 0.3	7.5 ± 0.2	7.2 ± 0.2	7.5 ± 0.3	7.4 ± 0.3	7.5 ± 0.2	0.2
Temp °C	18	10	17	22	19	10	17	20	2
Turbidity (NTU)	12 ± 2	13 ± 3	12 ± 2	8 ± 2	10 ± 3	8 ± 2	10 ± 3	10 ± 3	2.0
DO (mg/l)	4.42 ± 1.0	5.58 ± 1.61	3.92 ± 0.96	5.54 ± 0.85	5.13 ± 1.3	5.86 ± 1.80	6.29 ± 1.97	5.43 ± 1.12	1.5
Mg (mg/l)	82 ± 4.8	49 ± 5.9	82 ± 6.6	69 ± 5.8	79 ± 7.8	54 ± 5.1	22 ± 3.3	73.6 ± 8.2	8.5
Ca (mg/l)	33.8 ± 7.3	35.5 ± 6.4	48.64 ± 6.8	38.24 ± 8.9	25.5 ± 6.2	30.9 ± 12.7	29.5 ± 5.2	25.9 ± 11.5	12.7
SiO_3 (μg/l)	4.5 ± 0.9	2.98 ± 1.0	4.2 ± 1.1	1.98 ± 0.8	3.7 ± 1.1	1.0 ± 0.6	2.7 ± 1.4	3.9 ± 1.1	0.8
P (μg/l)	13.28 ± 1.21	5.32 ± 0.54	2.53 ± 0.45	1.52 ± 0.41	4.24 ± 1.28	2.24 ± 0.15	2.55 ± 0.26	1.42 ± 0.38	0.35
N (μg/l)	12.84 ± 1.2	4.85 ± 0.19	3.7 ± 1.03	4.6 ± 1.45	8.86 ± 1.0	3.42 ± 0.85	4.5 ± 1.1	2.42 ± 0.08	1.8

Data Mean ± SD, *CD* Critical Difference at 5 % probability, *DO* Dissolved Oxygen, *Mg* Magnesium, *Ca* Calcium, *P* Phosphates, *N* Nitrates

Table 2 Physico-chemical characteristics of water samples collected from five different sites along Red Sea coast at Sharmaa, Tabuk, Saudi Arabia in October 2013 and January, April, July 2014

Parameters	S1				S2				S3			
	Oct-13	Jan-14	April-14	July-14	Oct-13	Jan-14	April-14	July-14	Oct-13	Jan-14	April–14	July-14
pH	8.8 ± 0.4	8.7 ± 0.3	8.1 ± 0.2	8.5 ± 0.3	8.6 ± 0.4	8.2 ± 0.2	8.4 ± 0.4	7.8 ± 0.2	8.2 ± 0.3	8.2 ± 0.4	7.9 ± 0.2	8.4 ± 0.2
Temp °C	17	8	15	18	16	9	16	20	18	10	18	20
Turbidity (NTU)	16 ± 3	17 ± 2	15 ± 3	16 ± 3	15 ± 4	13 ± 4	13 ± 2	14 ± 4	13 ± 2	12 ± 2	14 ± 3	13 ± 4
DO (mg/l)	2.98 ± 0.61	2.95 ± 0.62	2.91 ± 0.71	3.2 ± 0.76	3.4 ± 0.91	3.2 ± 0.87	3.58 ± 0.19	3.4 ± 0.92	3.1 ± 0.80	3.2 ± 0.86	4.1 ± 0.76	5.0 ± 0.60
Mg (mg/l)	272 ± 9.5	196 ± 6.3	265 ± 12.8	284 ± 10.8	280 ± 9.8	250 ± 9.9	222 ± 8.5	132 ± 7.6	125 ± 5.8	110 ± 6.6	154 ± 6.5	195 ± 9.1
Ca (mg/l)	77.7 ± 9.14	76.9 ± 11.7	80.9 ± 12.6	65.9 ± 11.8	69.7 ± 12.1	56.64 ± 8.2	49.24 ± 7.6	60.24 ± 6.9	59.4 ± 10.2	68.8 ± 9.2	67.6 ± 9.12	76.8 ± 11.5
SiO_3 (μg/l)	7.2 ± 1.9	8.5 ± 2.4	8.9 ± 1.5	7.2 ± 1.5	7.9 ± 2.3	6.5 ± 1.1	6.5 ± 1.2	5.5 ± 1.4	5.2 ± 1.2	6.5 ± 1.6	4.8 ± 1.4	4.5 ± 1.5
P (μg/l)	35.5 ± 4.24	40.4 ± 7.34	41.2 ± 4.2	28.3 ± 5.43	18.1 ± 2.42	24.5 ± 4.21	17.7 ± 2.25	15.3 ± 2.51	15.2 ± 2.32	12.2 ± 1.54	10.2 ± 1.28	19.4 ± 2.33
N (μg/l)	55.9 ± 3.39	57.5 ± 3.24	98.8 ± 6.39	68.7 ± 4.45	79.2 ± 5.62	34.83 ± 4.48	31.2 ± 3.1	15.13 ± 2.2	25.2 ± 2.63	17.7 ± 2.11	17.8 ± 2.09	27.7 ± 3.63

Parameters	S4				S5				CD at 5 %
	Oct-13	Jan-14	April-14	July-14	Oct-13	Jan-14	April-14	July-14	
pH	8.4 ± 0.4	8.5 ± 0.5	7.5 ± 0.1	7.7 ± 0.1	7.5 ± 0.3	7.8 ± 0.4	7.9 ± 0.3	7.2 ± 0.1	0.3
Temp °C	17	8	15	17	17	7	15	19	3
Turbidity (NTU)	14 ± 2	12 ± 4	12 ± 2	14 ± 4	13 ± 3	12 ± 3	11 ± 3	11 ± 4	3.0
DO (mg/l)	5.2 ± 1.1	4.6 ± 0.51	4.4 ± 0.83	4.25 ± 0.45	4.8 ± 0.78	4.1 ± 0.28	4.5 ± 0.18	3.8 ± 0.90	1.2
Mg (mg/l)	110 ± 8.8	95 ± 5.4	95 ± 6.6	78 ± 4.8	66 ± 5.9	89 ± 6.5	31 ± 2.9	58 ± 5.4	9.1
Ca (mg/l)	45.24 ± 4.5	38.1 ± 4.2	37.5 ± 5.2	58.4 ± 9.5	40.5 ± 6.4	38.4 ± 9.7	33.5 ± 6.2	30.9 ± 11.5	13.3
SiO_3 (μg/l)	6.3 ± 0.9	5.5 ± 1.2	2.5 ± 0.8	4.5 ± 0.7	4.2 ± 1.1	3.3 ± 0.8	3.1 ± 0.6	1.8 ± 0.9	1.4
P (μg/l)	15.3 ± 2.47	8.4 ± 1.55	10.8 ± 3.36	4.61 ± 0.15	5.7 ± 1.48	3.1 ± 0.33	1.42 ± 0.38	3.1 ± 0.35	0.64
N (μg/l)	11.82 ± 2.05	6.2 ± 1.45	5.5 ± 0.93	13.4 ± 2.45	6.5 ± 1.59	6.7 ± 3.2	2.8 ± 0.09	4.2 ± 0.41	2.24

Data Mean ± SD, *CD* Critical Difference at 5 % probability, *DO* Dissolved Oxygen, *Mg* Magnesium, *Ca* Calcium, *P* Phosphates, *N* Nitrates

Table 3 Physico-chemical characteristics of water samples collected from five different sites along Red Sea coast at Duba, Tabuk, Saudi Arabia in October 2013 and January, April, July 2014

Parameters	S1				S2				S3			
	Oct-13	Jan-14	April-14	July-14	Oct-13	Jan-14	April–14	July-14	Oct-13	Jan-14	April-14	July-14
pH	8.2 ± 0.2	8.7 ± 0.5	8.0 ± 0.1	7.6 ± 0.1	7.8 ± 0.3	7.8 ± 0.2	8.8 ± 0.4	7.3 ± 0.2	7.5 ± 0.2	7.4 ± 0.3	7.3 ± 0.2	7.8 ± 0.3
Temp °C	18	9	16	19	17	8	15	20	18	10	18	21
Turbidity (NTU)	13 ± 3	14 ± 2	13 ± 4	11 ± 4	12 ± 4	12 ± 5	12 ± 4	12 ± 3	09 ± 3	11 ± 2	11 ± 2	10 ± 2
DO (mg/l)	3.43 ± 0.5	3.13 ± 1.1	3.54 ± 1.1	3.49 ± 1.2	3.84 ± 0.6	4.32 ± 1.2	3.92 ± 1.5	4.15 ± 1.1	3.85 ± 0.9	4.68 ± 1.8	4.22 ± 0.9	4.55 ± 0.9
Mg (mg/l)	223 ± 9.1	249 ± 4.5	219 ± 4.9	102 ± 2.9	129 ± 4.8	132 ± 5.5	184 ± 6.9	163 ± 9.6	172 ± 7.4	143 ± 9.5	148 ± 7.8	149 ± 5.4
Ca (mg/l)	74.4 ± 9.6	76.7 ± 8.1	64.5 ± 9.4	52.4 ± 9.7	68.8 ± 8.8	68.8 ± 10.1	45.4 ± 6.7	54.9 ± 11.2	67.4 ± 9.2	38.8 ± 7.5	58.2 ± 10.5	38.5 ± 9.8
SiO_3 (μg/l)	3.8 ± 1.2	8.7 ± 1.7	7.8 ± 1.4	5.9 ± 1.2	5.8 ± 1.1	6.8 ± 1.8	6.5 ± 1.9	4.6 ± 1.5	4.5 ± 1.3	3.6 ± 1.2	4.8 ± 1.4	4.7 ± 1.5
P (μg/l)	17.5 ± 1.3	32.54 ± 4.28	22.2 ± 0.65	25.2 ± 0.46	21.4 ± 0.64	17.00 ± 2.0	12.4 ± 0.57	15.3 ± 1.12	12.3 ± 0.73	12.5 ± 0.29	12.54 ± 1.28	5.31 ± 0.54
N (μg/l)	92.41 ± 5.93	117.79 ± 8.3	79.28 ± 3.25	37.88 ± 6.28	27.68 ± 4.4	21.15 ± 3.51	22.34 ± 3.2	15.42 ± 2.8	29.75 ± 4.45	22.34 ± 4.74	11.32 ± 1.11	19.95 ± 4.51

Parameters	S4				S5				CD at 5 %
	Oct-13	Jan-14	April-14	July-14	Oct-13	Jan-14	April-14	July-14	
pH	8.1 ± 0.4	8.3 ± 0.4	8.2 ± 0.2	8.1 ± 0.1	7.9 ± 0.2	8.2 ± 0.3	7.9 ± 0.2	7.7 ± 0.1	0.2
Temp °C	18	10	16	21	19	11	17	19	4
Turbidity (NTU)	10 ± 4	13 ± 5	10 ± 3	11 ± 3	09 ± 3	08 ± 3	11 ± 4	09 ± 2	3.0
DO (mg/l)	4.58 ± 1.2	3.54 ± 0.7	5.36 ± 1.6	4.24 ± 0.9	5.36 ± 1.9	6.23 ± 1.0	5.64 ± 1.3	5.72 ± 1.4	1.4
Mg (mg/l)	89 ± 5.3	72 ± 4.8	92 ± 3.6	42 ± 5.3	54 ± 5.9	25 ± 3.8	69 ± 6.9	33 ± 10.1	8.9
Ca (mg/l)	30.4 ± 5.2	48.6 ± 9.8	46.4 ± 9.8	32.5 ± 4.5	42.7 ± 7.7	35.4 ± 10.7	29.5 ± 6.8	35.8 ± 6.4	12.9
SiO_3 (μg/l)	3.3 ± 1.2	5.2 ± 1.2	3.6 ± 1.1	3.7 ± 0.9	1.67 ± 0.7	2.8 ± 0.9	2.5 ± 0.8	2.5 ± 0.9	1.4
P (μg/l)	4.31 ± 0.33	7.38 ± 0.52	9.32 ± 0.54	9.21 ± 0.49	1.22 ± 0.17	1.22 ± 0.38	2.38 ± 0.41	1.32 ± 0.25	0.31
N (μg/l)	10.84 ± 0.9	5.82 ± 1.1	3.73 ± 0.95	3.86 ± 1.9	2.14 ± 0.9	2.35 ± 0.9	4.85 ± 1.01	4.35 ± 1.63	1.8

Data Mean ± SD, *CD* Critical Difference at 5 % probability, *DO* Dissolved Oxygen, *Mg* Magnesium, *Ca* Calcium, *P* Phosphates, *N* Nitrates

Table 4 Relative density of some flora and fauna along the selected coastal sites of the Red Sea

Sampling stations	Haql				Sharmaa				Duba			
Months	Oct-13	Jan-14	Apr-14	July-14	Oct-13	Jan-14	Apr-14	July-14	Oct-13	Jan-14	Apr-14	July-14
Plant samples												
Gastroclonium ovatum	+	+	+	++	+	−	−	−	+	−	+	−
Padina pavonica	++	+	++	+++	++	+	−	+	++	++	+	−
Hildenbrandia rubra	+	++	+	+	−	−	+	++	+	−	−	+
Nemalion helminthoides	+	+	+	+	−	−	++	+	−	+	+	−
Polyides rotundus	+	+	+	++	+	−	−	−	+	−	+	+
Cladophora rupestris	++	++	+	+++	−	−	+	−	−	−	−	+
Ulva lactuca	+++	++	+++	++	++	++	+++	++	+++	++	+	++
Enteromorpha flexuoca	++	++	+	+++	+	+	+	+	+	++	+++	+++
Animal samples												
Lottia limulata	+++	++	+++	++	++	+	++	+	+	++	++	+
Nerita fulgurans	++	+	++	+++	+	++	+	++	++	+	+	+
Puncturella longifissa	+	+	+++	++	−	−	−	+	−	+	−	−
Istiblennius dussumieri	+	+	+	++	−	+	−	−	++	++	+++	++
Gerres sps.	+	−	+	−	−	−	−	−	+	+	+	+

Table 4 showed the density of some flora and fauna along the selected coastal sites of the Red Sea. Among the diffrent plant species sudied along the selected coastal sites *Ulva lactuca, Padina pavonica* and *Enteromorpha flexuoca* showed higher densitities in most of the seasons at Haql station as compared to the other two stations at Sharmaa and Duba. *Ulva lactuca* showed its highest density at five selected sites of Haql station in October-2013 and April-2014. The densities of *Lottia limulata* and *Boleophthalmus sps.* were higher among the different animal spcies studied at coastal sites. *Lottia* showed its maximum density at Haql station in the month of October-2013, whereas *Boleophthalmus* at Duba station in April 2014 (Table 4).

Discussion

Many coastal waters of developed nations have experienced widespread and rapid eutrophication (the increase in supply of organic matter) during the last half of the 20th century. This has resulted in increased phytoplankton production, decreased water clarity, often-severe depletion of dissolved oxygen in bottom waters, loss of seagrasses and, in some cases, declines or changes in the quality of fisheries production (Andersen and Conley 2009; Sayyed 2008; Boesch and Brinsfield 2000). Recent national assessments documented that the nitrogen-driven coastal eutrophication is widespread and increasing in the United States. An aquatic system takes thousands of years to become eutrophic which is a natural process. However, a high rate of input of nutrients due to anthropogenic activities significantly enhance the condition in a very short period of time (Ansari and Khan 2002, 2006a, 2007). This is called cultural eutrophication. High chemical or physical stress also enhance the struggle for survival in eutrophic systems. As a result the diversity of organisms declines in eutrophic than in oligotrophic systems (Ansari 2005; Ansari and Khan 2006b, 2009a).

Continuous evaporation from the surface of Red Sea results into an increase in concentration of magnesium, calcium, silica, nitrates and phosphates and Khomayis (2002). Inorganic nitrogen in the form of nitrates and nitrites can be used as pollution indicator to correlate with phytoplankton composition. Inorganic nitrogen in an aquatic ecosystem stimulates the development, maintenance and proliferation of primary producers and results into eutrophication (Camargo and Alonso 2006, Fang et al. 2007)

This significant coastal pollution problem includes impacts including increased areas and severity of hypoxic and anoxic waters; alteration of food webs; degradation and loss of sea grass beds, kelp beds and coral reefs; loss of biodiversity; and increased incidences and duration of harmful algal blooms (Scavia and Bricker 2006). Increase in phytoplankton species, water temperature and frequent decomposition of dead planktons and other organic matter results in depletion of oxygen level (Calliaria et al. 2005). Light reduction in the water column and enhanced organic matter (OM) load into the sediments are two main consequences of eutrophication in marine coastal areas (Olive et al. 2009). In eutrophicated coastal waters seagrass decline was found associated with light intensity (Burkholder et al. 2007). Various treatments of light showed a significant effect on photosynthesis and acclimation in aquatic macrophytes (Jensen et al. 2006; Karlsson et al. 2009).

Temperature is one of important environmental factor for directly related with the functioning of an aquatic ecosystem (Ansari and Khan 2008, 2009b; Lau and Lane 2002; Shen 2002). Temperature is recognized as an important environmental factor in aquatic ecosystems as it regulates all the biological activities like metabolism, growth, development and reproduction. Seasonal variations in temperature strongly influence these biological activities. Long-term, changes in temperature of the surrounding environment directly affect the evolution, physiology and behavior of aquatic organisms (Begon et al. 1990, Regier et al. 1990).

In addition to more favorable temperatures during a particular season, there is some phonological specificity in developing plant and animal biomass (Picard et al. 2005). The temperature values recorded during spring appeared to be suitable for seaweeds growth. These data are in accordance of those of Kebede and Ahlgren (1996) reported 30 °C as the optimum temperature growth and development of aquatic plants. pH controls absortion of nutrients and bichemical reactions taking place in living organisms. It is observed that more alkaline pH enhances the growth of phytoplankton (Ansari and Khan 2008, 2009b; Ansari et al. 2011b, 2011c).

Density represents the numerical strength of a species in the community. The number of individuals of the species in a unit area is its density. Density gives an idea of degree of competition (Sharma 2012). The high nutrient concentrations enhance the excessive growth of phytoplankton and macrophytes in aquatic ecosystem (Rovira and Pardo 2006, Ansari and Khan 2013). Eutrophication causes undesirable changes in species composition of an aquatic ecosystem (Romermann et al. 2008) and has been identified as the major threat to the survival of aquatic ecosystems and biodiversity (Murphy et al. 2003, Murphy 2002). Abdul-Aziz et al. (2003) reported blooming of phytoplankton in May and August influenced by temperature and nutrients in coastal waters of Arabian Gulf, Saudi Arabia. Species composition of the benthic phytoplankton in marine fouling area of Sharm Obhour (North Obhour), Red Sea was reported by Al-Harbi and Khomayis (2005). Khomayis (2002) studied the annual cycle of nutrient and chlorophyll-a in the coastal water of Jeddah, Red Sea and revealed the role of inorganic nutrients as limiting factors for the growth of phytoplankton.

Most of the countries around the globe have now realized the serious threat posed by eutrophication. The water bodies of countries like India, China, Bangladesh, Pakistan, Indonesia, Switzerland, Poland, Austria, Denmark, Croatia, Ireland, France, Greece, Estonia, Russia, Italy, Turkey, Japan great lakes states of the USA and Canada are under the direct threat of eutrophication (Ansari 2005). In the mid-20th century, eutrophication had been recognized as one of the causative factors of pollution in European and North American lakes and reservoirs. Since then, it has become more widespread and surveys showed that 54 % of lakes in Asia, 53 % in Europe, 48 % in North America, 41 % in South America and 28 % in Africa are in eutrophic state (Colin et al. 2007). In Spain 80 % of the lakes, 70 % of the reservoirs and 60 % river sites were eutrophic in the 1990s with hypertrophy increasing downstream (Alvarez-Cobelas et al. 2001; Khan and Ansari 2005; Khan et al. 2014). Three large coastal marine ecosystems (Chesapeake Bay, Northern Gulf of Mexico, and Danish Straits) showed repeated hypoxic events have led to an increase in susceptibility of further hypoxia and accelerated eutrophication (Conley et al. 2009). Large-scale changes resulting from coastal eutrophication have been documented for continental shelf waters in the Gulf of Mexico, Mediterranean, Black and North Seas, relatively confined seas such as the Baltic and Seto Inland Sea, large bays such as the Chesapeake Bay and Long Island Sound and numerous smaller estuaries and lagoons (Boesch and Brinsfield 2000).

The Red Sea is a narrow inland sea separating the Arabian Peninsula, western Asia, from northeastern Africa. It extends northwest from the strait of Bab el Mandeb to Suez, Egypt, for a distance of 1900 km (1200 mile). The northern extremity is divided by the Sinai Peninsula into the gulfs of Suez and Aqaba. The Suez Canal connects the Red Sea with the Mediterranean Sea and Bab El-Mandeb connects it with the Gulf of Aden, an arm of the Arabian Sea. The seasonal variations of the algal flora in the different water sources in Saudi Arabia in the Red Sea had been studied by only few workers (Touliabah et al. 2010). Kingdom of Saudi Arabia is sharing its boundary with seven countries and three water bodies. In the west Gulf of Aqaba and Red Sea form a coastal border of almost 1800 kilometres, which have a huge human settlement in terms of many towns and cities (viz. Al Lith, Al Qunfudhah, Al Wajh, Duba, Haql, Jeddah, Jazan, Rabigh, Thuwal, Yanbu). This large human population settled on the coasts significantly contributing nutrients to the coastal areas in the form of different wastes and causing coastal eutrophication. A survey on scientific literature shows a very little work has been done on the eutrophication of coastal areas in the boundary of Saudi Arabia. So, we have selected the present topic and taken a task for the preliminary research on the coastal eutrophication and its causes, consequences and controls. One time study cannot provide a forecast to imply effective control measures, much more data from the monitoring of coastal marine ecosystems for a longer duration is required to reach any conclusion. However, the results of this research may give a initial sign of the problem that can help in preventing a huge capital loss in order to control coastal eutrophication as compare to the condition when the problem is completely established. The study may be helpful in planning the research projects to combat the coastal eutrophication to restore and conserve the marine ecosystems and degraded coastal areasof the Red Sea in Saudi Arabia.

The consequences of coastal eutrophication caused by excessive anthropogenic inputs of nutrients are: Increase in biomass of phytoplankton, zooplanktons and macrophytes, domination of algal bloom forming species which is toxic to consumers of eutrophic ecosystem, increase in biomass of consumer species due to increase of biomass of benthic and epiphytic algae, anoxic conditions in the aquatic ecosystem results in frequent fish kills, decline in species diversity of aquatic ecosystem, decrease in harvestable fish biomass, decrease in water transparency. Problems related to the quality of water and its aesthetic value. In addition to increasing scientific knowledge, answers to these questions related to eutrophication can influence strategies for ecosystem restoration or rehabilitation (Hecky 2009, Smits 2005).

Acknowledgments The authors would like to acknowledge financial support for this work, from the Deanship of Scientific Research (DSR), University of Tabuk, Tabuk, Saudi Arabia, under the No. S-0021-1436.

References

Abdul-Aziz PK, Al-Tisan IA, Daili MA, Green TN, Dalvi AGI, Javeed MA (2003) Chlorophyll and plankton of the RSA coastal waters of Saudi Arabia bordering a desalination plant. Desalination 154:291–302

Al-Harbi SM, Khomayis HS (2005) Seasonal variations of chlorophyll a and micro-nutrientsin Obhur Creek, Red Sea. J King Abdul-Aziz Univ Mar Sci 16:91–104

Alvarez-Cobelas M, Cirujano S, Sanchez-Carrillo S (2001) Hydrological and botanical man-made changes in the Spanish wetland of Las Tablas de Daimiel. Biol Conserv 97:89–98

Andersen JH, Conley DJ (2009) Eutrophication in coastal ecosystems, towards better understanding and management strategies. The Springer, Netherlands **264 pp**

Andersen JH, Schluter L, Rtebjerg G (2006) Coastal eutrophication: recent developments in definitions and implications for monitoring strategies. J Plankton Res 28(7):621–628

Ansari AA (2005) Studies on the role of selected household detergents in the eutrophication of freshwater ecosystem. PhD thesis, Aligarh Muslim University, Aligarh

Ansari AA, Gill SS (2014) Eutrophication: causes, consequences and control. Springer, The Netherlands, p 264

Ansari AA, Khan FA (2002) Nutritional status and quality of water in a waste water pond of Aligarh showing blooms of *Spirodela polyrrhiza* L. (Shleid). J Ecophysiol Occupl Health 2:185–189

Ansari AA, Khan FA (2006a) Studies on the role of some selected nutrient sources in the eutrophication of freshwater ecosystem. Nat Environ Poll Technol 5:47–52

Ansari AA, Khan FA (2006b) Growth responses of Spirodela polyrrhiza treated with a common detergent at varying temperature and pH conditions. Nat Environ Poll Technol 5:399–404

Ansari AA, Khan FA (2007) Eutrophication studies on some freshwater ponds of Aligarh. Ind J Appl Pure Biol 22:21–26

Ansari AA, Khan FA (2008) Remediation of eutrophied water using Lemna minor in controlled environment. Afr J Aqua Sci 33:275–278

Ansari AA, Khan FA (2009a) Eutrophication studies on Jeffery canal of Aligarh International Conference on Emerging Technologies in Environmental Science and Engineering. International Conference, Aligarh Muslim University, Aligarh, India 845-849

Ansari AA, Khan FA (2009b) Remediation of eutrophied water using *Spirodela polyrrhiza* (L.) Shleid in controlled environment. Pan-Am J Aqua Sci 4:52–54

Ansari AA, Khan FA (2011) Nutrients phytoremediation of eutrophic waters using *Eichhornia crassipes* in a controlled environment. Int J Environ Sci 2:241–246

Ansari AA, Khan FA (2013) Response of *Eichhornia crassipes* to eutrophic waters receiving nutrients from various sources. Int J Environ Sci 4(1):39–45

Ansari AA, Khan FA (2014) Household detergents causing eutrophication in freshwater ecosystems. In: Ansari AA, Gill SS (eds) Eutrophication: causes, consequences and control, vol 2. Springer, The Netherlands, pp 139–164

Ansari AA, Gill SS, Khan FA (2011a) Eutrophication: Threat to aquatic ecosystems. In: Ansari AA, Gill SS, Lanza GR, Rast W (eds) Eutrophication: Causes, Consequences and Control. Springer, The Netherlands, pp 143–170

Ansari AA, Gill SS, Lanza GR, Rast W (2011b) Eutrophication: Causes, Consequences and Control. Springer, The Netherlands **394 pp**

Ansari AA, Khan FA, Gill SS (2011c) Aquatic plant diversity in eutrophic ecosystems. In: Ansari AA, Gill SS, Lanza GR, Rast W (eds) Eutrophication: Causes, Consequences and Control. Springer, The Netherlands, pp 247–263

Ansari AA, Gill SS, Khan FA, Naeem M (2014) Phytoremediation systems for the recovery of nutrients from eutrophic waters. In: Ansari AA, Gill SS (eds) Eutrophication: Causes, Consequences and Control, vol 2. Springer, The Netherlands, pp 239–248

APHA (2005) Standard methods for the examination of water and wastewater, 21st edn. APHA, AWWA and WEF, New York, p 1368

Begon M, Harper JL, Townsend CR (1990) Ecology: individuals, populations, and communities, 2nd edn. Blackwell Scientific Publications, Oxford

Boesch D, Brinsfield R (2000) Coastal Eutrophication and Agriculture: Contributions and Solutions. Biological Resource Management Connecting Science and Policy, 93–115

Burkholder JM, Tomasko DA, Touchette BW (2007) Sea grasses and Eutrophication. J Exp Mar Biol Ecol 350:46–72

Calliaria D, Gomez M, Gomez N (2005) Biomass and composition of phytoplankton. Cont Shelf Res 25:197–210

Camargo JA, Alonso A (2006) Ecological and toxicological effects of inorganic nitrogen pollution in aquatic ecosystems: a global assessment. Environ Int 32(6):831–849

Carpenter SR (2005) Eutrophication of aquatic ecosystems: bistability and soil phosphorus. PNAS 102:10002–10005

Colin RB, Robert E, Wreede D (2007) Do changes in seaweed biodiversity influence associated invertebrate epifauna? J Exp Mar Biol Ecol 344:206–214

Conley DJ, Carstensen J, Vaquer-Sunyer R, Duarte CM (2009) Ecosystem thresholds with hypoxia. Hydrobiologia 629(1):21–29

Corstanje R, Reddy KR (2006) Microbial indicators of nutrient enrichment: a mesocosm study. Soil Sci Soc Am J 70:1652–1661

Davis SE, Corronado-Molina C, Childers DL, Day JW (2003) Temporally dependent C, N, and P dynamics associated with the decay of *Rhizophora mangle* L., leaf litter in oligotrophic mangrove wetlands of the Southern Everglades. Aquat Bot 75:199–215

Devlin RM, Witham FH (1986) Plant physiology, 4th edn. CBS Publishers, New Delhi

Dokulil MT, Teubner K (2011) Eutrophication and climate change: Present situation and future scenarios. In: Ansari AA, Gill SS, Lanza GR, Rast W (eds) Eutrophication: causes consequences and control. Springer, New York, pp 1–16

El-Shafai SA, El-Gohary FA, Nasr FA, Van der Steen NP, Gijzen HJ (2007) Nutrient recovery from domestic wastewater using a UASB-duckweed ponds system. Biores Technol 98:798–807

Fang YY, Babourina O, Rengel Z, Yang XE, Pu PM (2007) Ammonium and nitrate uptake by the floating plant *Landoltia punctata*. Ann Bot 99:365–370

Feuchtmayr H, Moran R, Hatton K (2009) Global warming and eutrophication: effects on water chemistry and autotrophic communities in experimental hypertrophic shallow lake mesocosms. J Appl Ecol 46:713–723

Gomez KA, Gomez AA (1984) Statistical procedures for agricultural research, 2nd edn. Wiley and Sons, New York, pp 680

Hecky RE (2009) Eutrophication: more nitrogen data needed. Science 324:721–772

Jensen JP, Pedersen AR, Jeppesen E, Søndergaard M (2006) An empirical model describing the seasonal dynamics of phosphorus in 16 shallow eutrophic lakes after external loading reduction. Limnol Oceanogr 51:791–800

Jeppesen E, Søndergaard M, Jensen JP, Lauridsen TL (2005) Shallow Lakes: Effects of Nutrient Loading and How to Remedy Eutrophication. I: Encyclopedia of Life Support Systems (EOLSS): Developed under the Auspices of the UNESCO. Oxford, UK

Johanson N, Revenga C, Echeverria J (2001) Managing water for people and nature. Science 292:1071–1072

Karlsson J, Bystrom P, Ask J (2009) Light limitation of nutrient-poor lake ecosystems. Nature 460:506–509

Kebede E, Ahlgren G (1996) Optimum growth conditions and light utilization efficiency of *Spirulina platensis* (=*Arthrospira fusiformis*) (Cyanophyta) from Lake Chitu. Ethiopia Hydrobiol 332(2):99–102

Khan FA, Ansari AA (2005) Eutrophication: an ecological vision. Bot Rev 71:449–482

Khan FA, Naushin F, Rehman F, Masoodi A, Irfan M, Hashmi F, Ansari AA (2014) Eutrophication: Global scenario and local threat to dynamics of aquatic ecosystems. In: Ansari AA, Gill SS (eds) Eutrophication: causes, consequences and control, vol 2. Springer, The Netherlands, pp 17–28

Khomayis HS (2002) The annual cycle of nutrient salts and chlorophyll a in the coastal waters of Jeddah, Red Sea. J King Abdul-Aziz Univ Mar Sci 13:131–145

Lau SSS, Lane SN (2002) Biological and chemical factors influencing shallow lake eutrophication: a long-term study. Sci Tot Environ 3:167–181

Likens GE (1972) Nutrients and eutrophication: the limiting nutrient controversy. Limnol Oceanogr 1:321–328

Lindahl O, Kollberg S (2009) Can the EU agri-environmental aid program be extended into the coastal zone to combat eutrophication? Hydrobiologia 629(1):59–64

Liu Q, Hu C, Long W, Wang H, Zhou S (2011) International symposium on water resource and environmental protection (ISWREP), 3: 2239–2242

Lu Q, He ZL, Graetz DA, Stoffella PJ, Yang X (2010) Phytoremediation to remove nutrients and improve eutrophic stormwaters using water lettuce (*Pistia stratiotes* L.). Environ Sci Pollut Res 17:84–96

Murphy KJ (2002) Plant communities and plant diversity in softwater lakes of northern Europe. Aquat Bot 73:287–324

Murphy KJ, Dickinson G, Thomaz SM (2003) Aquatic plant communities and predictors of diversity in a subtrophical river flood plain: the upper Rio Parana Brazil. Aquat Bot 77:257–276

Naeem M, Idrees M, Khan MMA, Moinuddin Ansari AA (2014) Task of mineral nutrients in eutrophication. In: Ansari AA, Gill SS (eds) Eutrophication: causes, consequences and control, vol 2. Springer, The Netherlands, pp 223–238

Nixon SW (1995) Coastal marine eutrophication: a definition, social causes, and future concerns. Ophelia 41:199–219

Officer CB, Ryther JH (1980) The possible importance of slllconIn marine eutrophication. Mar Ecol Prog Ser 3:83–91

Olive I, Garcıa-Sanchez MP, Brun FG, Vergara JJ (2009) Interactions of light and organic matter under contrasting resource simulated environments: the importance of clonal traits in the seagrass Zostera noltii. Hydrobiologia 629(1):199–208

Paerl HW, Valdes LM, Pinckney JL (2003) Phytoplankton photopigments as indicators of estuarine and coastal eutrophication. Bio Sci 53:953–964

Philips EJ (2002) Algae and eutrophication. In: Bitton G (ed) Encyclopedia of environmental microbiology. John Wiley and Sons, New York

Picard CR, Fraser LH, Steer D (2005) The interacting effects of temperature and plant community type on nutrient removal in wetland microcosms. Biores Technol 96:1039–1047

Polomski RF, Taylor MD, Bielenberg DG (2009) Nitrogen and phosphorus remediation by three floating aquatic macrophytes in greenhouse-based laboratory-scale subsurface constructed wetlands. Water Air Soil Pollut 197:223–232

Regier HA, Holmes JA, Pauly D (1990) Influence of Temperature Changes on Aquatic Ecosystems: an Interpretation of Empirical Data. Trans Am Fisher Soc 119(2):374–389

Rohlich GA (1969) Eutrophication: causes, consequences, correctives. National Academy of Sciences, Washington **307**

Romermann C, Tackenberg O, Poschlod AJKP (2008) Eutrophication and fragmentation are related to species' rate of decline but not to species rarity: results from a functional approach. Biodivers Conserv 17:591–604

Rovira JL, Pardo P (2006) Nutrient pollution of waters: eutrophication trends in European marine and coastal environments. Contrib Sci 3:181–186

Sayyed MK (2008) Marine environment (In: Arab Environment: Future Challenges) (eds.) Tolba M.K. and Saab N.W. A report of Arab Forum for Environment and Development (AFED) 75-94

Scavia D, Bricker SB (2006) Coastal eutrophication assessment in the United States. Biogeochemistry 79:187–208

Sharma PD (2012) Ecology and Environment, 12th Edition. Rastogi Publications, India, pp 708

Sharpley AN, Daniel T, Sims T, Lemunyon J, Stevens R, Parry R (2003) Agricultural Phosphorus and Eutrophication, 2nd ed, U.S. Department of Agriculture, Agricultural Research Service, ARS–149, pp 44

Shen DS (2002) Study on limiting factors of water eutrophication of the network of river in plain. J Zheji Univ Agric Life Sci 28:94–97

Smith VH (2007) Using primary productivity as an index of coastal eutrophication: the units of measurement matter. J Plankton Res 29:1–6

Smits EP (2005) Phytoremediation. Annual RevPlant Biol 56:15–39

Touliabah HE, Abu El-Kheir WS, Kuchari MG, Abdulwassi NIH (2010) Phytoplankton composition at Jeddah Coast-Red Sea, Saudi Arabia in relation to some ecological factors. J King Abdul-Aziz Univ Sci 22(1):115–131. doi:10.4197/Sci.22-1.9

Vollenweider RA (1968) The scientific basis of lake eutrophication, with particular reference to phosphorus andnitrogen as eutrophication factors. Technical report OECD, Paris, p 159

Vollenweider RA (1970) Models for calculating integral photosynthesis and some implications regarding structural properties of the community metabolism of aquatic systems. Prediction and measurements of photosynthetic productivity. Proc IBP/PP Meeting Trebon, Wageningen, pp 455–472

Vollenweider RA, Kerekes J (1980) The loading concept as basis for controlling eutrophication philosophy and preliminary results of the OECD programme on Eutrophication. Prog Water Tech 17:5–38

Yang X, Wu X, Hao H, He Z (2008) Mechanisms and assessment of water eutrophication. J Zhejiang Univ Sci B 9(3):197–209

Permissions

All chapters in this book were first published in IAR, by Springer; hereby published with permission under the Creative Commons Attribution License or equivalent. Every chapter published in this book has been scrutinized by our experts. Their significance has been extensively debated. The topics covered herein carry significant findings which will fuel the growth of the discipline. They may even be implemented as practical applications or may be referred to as a beginning point for another development.

The contributors of this book come from diverse backgrounds, making this book a truly international effort. This book will bring forth new frontiers with its revolutionizing research information and detailed analysis of the nascent developments around the world.

We would like to thank all the contributing authors for lending their expertise to make the book truly unique. They have played a crucial role in the development of this book. Without their invaluable contributions this book wouldn't have been possible. They have made vital efforts to compile up to date information on the varied aspects of this subject to make this book a valuable addition to the collection of many professionals and students.

This book was conceptualized with the vision of imparting up-to-date information and advanced data in this field. To ensure the same, a matchless editorial board was set up. Every individual on the board went through rigorous rounds of assessment to prove their worth. After which they invested a large part of their time researching and compiling the most relevant data for our readers.

The editorial board has been involved in producing this book since its inception. They have spent rigorous hours researching and exploring the diverse topics which have resulted in the successful publishing of this book. They have passed on their knowledge of decades through this book. To expedite this challenging task, the publisher supported the team at every step. A small team of assistant editors was also appointed to further simplify the editing procedure and attain best results for the readers.

Apart from the editorial board, the designing team has also invested a significant amount of their time in understanding the subject and creating the most relevant covers. They scrutinized every image to scout for the most suitable representation of the subject and create an appropriate cover for the book.

The publishing team has been an ardent support to the editorial, designing and production team. Their endless efforts to recruit the best for this project, has resulted in the accomplishment of this book. They are a veteran in the field of academics and their pool of knowledge is as vast as their experience in printing. Their expertise and guidance has proved useful at every step. Their uncompromising quality standards have made this book an exceptional effort. Their encouragement from time to time has been an inspiration for everyone.

The publisher and the editorial board hope that this book will prove to be a valuable piece of knowledge for researchers, students, practitioners and scholars across the globe.

List of Contributors

Toms C. Joseph, Roswin James and K. V. Lalitha
Microbiology, Fermentation and Biotechnology Division, Central Institute of Fisheries Technology, Cochin 682 029, Kerala, India

L. Anbu Rajan
Andaman and Nicobar Centre for Ocean Science and Technology, National Institute of Ocean Technology, Dollygunj P.O., Port Blair 744 103, Andaman and Nicobar Islands, India

P. K. Surendran
Poothuvallil, Palluruthy, Cochin, India

Abdul Wahab Mohammad and Abd Ghani Kumar
Department of Chemical and Process Engineering, Universiti Kebangsaan Malaysia UKM, Bangi, 43600 Selangor Darul Ehsan, Malaysia

Roseliza Kadir Basha
Department of Process and Food Engineering, Universiti Putra Malaysia UPM, Serdang, 43400 Selangor Darul Ehsan, Malaysia

Hadi Jamali and Nasrollah Ahmadifard
Department of Fisheries, Faculty of Natural Resources, Urmia University, P.O. Box: 46414-356, Urmia, Iran

Daruosh Abdollahi
Department of Fisheries, Gonbad Kavous University, Gonbad Kavous, Iran

Huma Vaseem and T. K. Banerjee
Department of Zoology, Banaras Hindu University (BHU), Varanasi 221005, India

Dominik Winckelmann and Franziska Bleeke
Jacobs-University Bremen, School of Engineering and Science, Campus Ring 1, 28759 Bremen, Germany
University of Applied Sciences Bremen, Am Neustadtswall 30, 28199 Bremen, Germany

Bimiya Thomas
University of Applied Sciences Bremen, Am Neustadtswall 30, 28199 Bremen, Germany

Clemens Elle
University of Applied Sciences Bremen, Am Neustadtswall 30, 28199 Bremen, Germany

Gerd Klöck
University of Applied Sciences Bremen, Am Neustadtswall 30, 28199 Bremen, Germany

Ali Karami
Laboratory of Aquatic Toxicology, Faculty of Medicine and Health Sciences, Universiti Putra Malaysia, 43400 Serdang, Selangor, Malaysia

Peyman Eghtesadi Araghi
Iranian National Institute for Oceanography and Atmospheric Science (INIOAS), Tehran, Iran

Mohd Arif Syed
Department of Biochemistry, Faculty of Biotechnology, University Putra Malaysia, 43400 Serdang, Selangor, Malaysia

Scott P. Wilson
School of Medical and Applied Sciences, CQ University, Rockhampton, Australia

Rashed Noor and Md. Sakil Munna
Department of Microbiology, Stamford University Bangladesh, 51 Siddeswari Road, Dhaka 1217, Bangladesh

Md. Faqrul Hasan and M. Majibur Rahman
Department of Microbiology, University of Dhaka, Dhaka 1000, Bangladesh

Anastasiya V. Kotova
Institute of Cytology RAS, St. Petersburg, Russia

Oleg A. Kovtun
Hydrobiological Station of Odessa National I. I. Mechnikov University, Odessa, Ukraine

Olga I. Podgornaya and Leonid S. Adonin
Institute of Cytology RAS, St. Petersburg, Russia
Department of Cytology and Histology, Faculty of Biology and Soil Sciences, St-Petersburg State University, St.Petersburg, Russia

Franziska Bleeke
Jacobs University, 28759 Bremen, Germany
Institute of Environmental Biology and Biotechnology, University of Applied Sciences Bremen, 28199 Bremen, Germany

Malgorzata Milas
Gdansk University of Technology, 80-233 Gdansk, Poland

Dominik Winckelmann and Gerd Klöck
Institute of Environmental Biology and Biotechnology, University of Applied Sciences Bremen, 28199 Bremen, Germany

Jaeyong Lee
Environmental Research Institute, Nature and Human Co. Ltd, Wonju 220-844, Korea

Jae-Ok Kim
Rural Research Institute Korea Rural Community Corporation, Ansan 426-908, Korea

Jeffrey S. Owen
Department of Environmental Science, Hankuk University of Foreign Studies, Yongin 449-791, Korea

Bomchul Kim
Department of Environmental Science, Kangwon National University, Chuncheon 200-701, Korea

Christopher Marlowe A. Caipang
School of Applied Science, Temasek Polytechnic, Singapore 529757, Singapore

Fernand F. Fagutao
Laboratory of Aquatic Animal Diseases, College of Veterinary Medicine, Gyeongsang National University, Jinju, Gyeongnam 660-701, Republic of Korea

Effrosyni Fatira and Michail Pavlidis
Department of Biology, University of Crete, 71409 Heraklion, Crete, Greece

Carlo C. Lazado
Aquaculture Genomics Research Unit, Faculty of Biosciences and Aquaculture, University of Nordland, 8049 Bodo, Norway
National Institute of Aquatic Resources, Technical University of Denmark, 9850 Hirtshals, Denmark

Christopher Marlowe A. Caipang, Han Xuan Choo, Zemin Bai, Huilin Huang and Clara M. Lay-yag
School of Applied Science, Temasek Polytechnic, Singapore 529757, Singapore

Sergey A. Piontkovski and Wardal Muna Al-Manthri
CAMS, Sultan Qaboos University, P.O. Box 34, Al-Khod 123, Muscat, Sultanate of Oman

Asyla Al-Mawali and Ahlam Al-Kharusi
Marine Fisheries Research Center, Muscat, Sultanate of Oman

Sharon Smith
RSMAS, University of Miami, 4600 Rickenbacker Causeway, Miami, FL 33149, USA

Elena Popova
Institute of Biology of the Southern Seas, 2 Nakhimov Prospect, Sevastopol, Russia

Patcharawalai Sriyasak
Faculty of Fisheries Technology and Aquatic Resources, Maejo University, Chiang Mai, Thailand

Chanagun Chitmanat, Niwooti Whangchai and Jongkon Promya
Faculty of Fisheries Technology and Aquatic Resources, Maejo University, Chiang Mai, Thailand

Louis Lebel
Unit for Social and Environmental Research, Chiang Mai University, Chiang Mai, Thailand

L. Yu. Khaliullina and G. V. Demina
Department of Botany and Plant Physiology, Institute of Fundamental Medicine and Biology, Kazan Federal University, Kremlyovskaya 18, Kazan 420008, Russia

Supatra Karnjanapratum and Soottawat Benjakul
Department of Food Technology, Faculty of Agro-Industry, Prince of Songkla University, Hat Yai, Songkhla 90112, Thailand

Muhammad Abu Bakar Siddik
Sado Marine Biological Station, Faculty of Science, Niigata University, Niigata, Japan
Department of Fisheries Biology and Genetics, Patuakhali Science and Technology University, Patuakhali, Bangladesh

Nguyen Thi Ngoc Anh
College of Aquaculture and Fisheries, Can Tho University, Can Tho, Vietnam

Samwel Mchele Limbu and Yunus Daud Mgaya
Department of Aquatic Sciences and Fisheries, University of Dar es Salaam, P. O. Box 35064, Dar es Salaam, Tanzania

Amon Paul Shoko and Mary Alphonce Kishe-Machumu
Tanzania Fisheries Research Institute, Headquarters, P. O. Box 9750, Dar es Salaam, Tanzania

Hieromin Amon Lamtane and Evantuce Dominick Shirima
Department of Animal Science and Production, Sokoine University of Agriculture, P. O. Box 3004, Morogoro, Tanzania

Huruma Fredrick Mgana
Tanzania Fisheries Research Institute, Kigoma Centre, P. O. Box 90, Kigoma, Tanzania

Pavriz Tavakoli-Kolour
Young Researchers and Elite Club, Bandar Abbas Branch, Islamic Azad University, PO Box 79159-14 1311, Bandar Abbas, Iran

Javid Kavousi
Graduate School of Engineering and Science, University of the Ryukyus, 1 Senbaru, Nishihara, Okinawa 903-0213, Japan

Hamid Rezai
Iranian National Institute for Oceanography and Atmospheric Sciences, Tehran, Iran

Isao Tsutsui and Yukio Maeno
Fisheries Division, Japan International Research Center for Agricultural Sciences (JIRCAS), Tsukuba, Ibaraki, Japan

Tatsuo Miyoshi and Kaoru Hamano
National Research Institute of Fisheries and Environment of Inland Sea, Onomichi, Hiroshima, Japan

Dusit Aue-umneoy, Jaruwan Songphatkaew, Piyarat Pinphoo and Monthon Ganmanee
Department of Animal Production and Fisheries, Faculty of Agricultural Technology, King Mongkut's Institute of Technology Ladkrabang (KMITL), Bangkok, Thailand

Chonlada Meeanan, Sirimas Klomkling and Halethichanok Sukchai
Shrimp Co-culture Research Laboratory (SCORL), King Mongkut's Institute of Technology Ladkrabang (KMITL), Bangkok, Thailand

Iori Yamaguchi
Faculty of Engineering, Fukuoka Institute of Technology (FIT), Fukuoka, Japan

Mahmoudreza Ovissipour . Carolyn F. Ross and Barbara Rasco
School of Food Science, Washington State University, P.O. Box 646376, Pullman, WA 99164-6376, USA

Hamzah M. Al-Qadiri
Department of Nutrition and Food Technology, Faculty of Agriculture, The University of Jordan, Amman 11942, Jordan

Xiaonan Lu and Yaxi Hu
Food, Nutrition and Health Program, Faculty of Land and Food Systems, The University of British Columbia, Vancouver, BC V6T 1Z4, Canada

Joel P. Van Eenennaam and Serge I. Doroshov
Department of Animal Science, University of California, Davis, CA 95616, USA

Seinen Chow, Takashi Yanagimoto and Yoji Nakamura
National Research Institute of Fisheries Science, 2-12-4 Fukuura, Kanazawa, Yokohama, Kanagawa 236-8648, Japan

Isao Tsutsui
Fisheries Division, Japan International Research Center for Agricultural Sciences (JIRCAS), Tsukuba, Ibaraki, Japan
Department of Animal Production and Fisheries, Faculty of Agricultural Technology, King Mongkut's Institute of Technology Ladkrabang (KMITL), Bangkok, Thailand

Hiroyuki Sudo
Fisheries Division, Japan International Research Center for Agricultural Sciences (JIRCAS), Tsukuba, Ibaraki, Japan

Jaruwan Songphatkaew, Dusit Aue-umneoy, Piyarat Pinphoo and Monthon Ganmanee
Department of Animal Production and Fisheries, Faculty of Agricultural Technology, King Mongkut's Institute of Technology Ladkrabang (KMITL), Bangkok, Thailand

Chonlada Meeanan, Halethichanok Sukchai and Sirimas Klomkling
Shrimp Co-culture Research Laboratory (SCORL), KMITL, Bangkok, Thailand

Kaoru Hamano
Research Center for Marine Invertebrates, National Research Institute of Fisheries and Environment of Inland Sea, Onomichi, Hiroshima, Japan

Hideyuki Nakagawa and Hitomi Sakai
Division of Environmental Symbiosis, Graduate School of Integrated Arts and Sciences, Tokushima University, Tokushima 770-8502, Japan

Kuniko Nagasaka
Fukae Nagasaka Clinic, Etajima 737-2214, Japan

Kozue Edo
Junior College Division, Department of Science for Human Health, Shikoku University, Tokushima 771-1192, Japan

Mitsuko Shinohara and Kiyoshi Ohura
Department of Pharmacology, Osaka Dental University, Hirakata 573-1121, Japan

R. G. Uc-Peraza and V. H. Delgado-Blas
Divisi6n de Ciencias e Ingenieria, Universidad de Quintana Roo, 77019 Chetumal, Quintana Roo, Mexico

Jitendra Pandey, Anand V. Singh, Rachna Singh and Pooja Kaushik
Ganga River Ecology Research Laboratory, Environmental Science Division, Centre of Advanced Study in Botany, Banaras Hindu University, Varanasi 221005, India

Usha Pandey
Department of Botany, Faculty of Science and Technology, Mahatma Gandhi Kashi Vidyapith University, Varanasi 221002, India

Tatsuya Yurimoto and Isao Tsutsui
Japan International Research Center for Agricultural Sciences, Ohwashi, Tsukuba, Ibaraki 305-8686, Japan

Dusit Aue-umneoy and Chonlada Meeanan
King Mongkut's Institute of Technology Ladkrabang, Chalongkrung Rd. Ladkrabang, Bangkok 10520, Thailand

Abid Ali Ansari, Sulaiman Al Ghanim, Subrata Trivedi, Hasibur Rehman, Zahid Khorshid Abbas and Shalini Saggu
Department of Biology, Faculty of Science, University of Tabuk, Tabuk, Saudi Arabia